IET POWER AND ENERGY SERIES 88

# Smarter Energy

**Other related titles in this series:**

Volume 73 **Wide Area Monitoring, Protection and Control Systems: The enabler for Smarter Grids** A. Vaccaro and A. Zobaa (Editors)

Volume 77 **Wide Area Monitoring of Interconnected Power Systems** Arturo Roman Messina

Volume 79 **Vehicle-to-Grid: Linking electric vehicles to the smart grid** J. Lu and J. Hossain (Editors)

# Smarter Energy

## From Smart Metering to the Smart Grid

Edited by
Hongjian Sun, Nikos Hatziargyriou,
H. Vincent Poor, Laurence Carpanini
and Miguel Angel Sánchez Fornié

The Institution of Engineering and Technology

Published by The Institution of Engineering and Technology, London, United Kingdom

The Institution of Engineering and Technology is registered as a Charity in England & Wales (no. 211014) and Scotland (no. SC038698).



First published 2016



The Institution of Engineering and Technology
Michael Faraday House
Six Hills Way, Stevenage
Herts, SG1 2AY, United Kingdom

www.theiet.org



**British Library Cataloguing in Publication Data**
A catalogue record for this product is available from the British Library

**ISBN 978-1-78561-104-9 (hardback)**
**ISBN 978-1-78561-105-6 (PDF)**

Typeset in India by MPS Limited
Printed in the UK by CPI Group (UK) Ltd, Croydon

# Contents

# Biographies of the editors

**Hongjian Sun** (SM'15) received the Ph.D. degree from The University of Edinburgh, UK, in 2010. Since 2013, he has been a lecturer in smart grids with the University of Durham, UK. He has made one contribution to the IEEE 1900.6a Standard and published four book chapters and more than 60 papers in refereed journals and conferences. His recent research interests include smart grids, wireless communications, and signal processing. He served as an Organizing Chair for a number of workshops/symposium and also on the Editorial Boards of the *Journal of Communications and Networks* and *EURASIP Journal on Wireless Communications and Networking.*

**Nikos D. Hatziargyriou** is Chairman and CEO of the Hellenic Distribution Network Operator. Since 1984 he is with the Power Division of the Electrical and Computer Engineering Department of the National Technical University of Athens, and since 1995 he is full professor in Power Systems. From 2007 to 2012, he was Deputy CEO of the Public Power Corporation (PPC) of Greece, responsible for Transmission and Distribution Networks, island DNO and the Center of Testing, Research and Prototyping. He is Fellow Member of IEEE, past Chair of the Power System Dynamic Performance Committee, Distinguished member of CIGRE and past Chair of CIGRE SC C6 "Distribution Systems and Distributed Generation". He is Chair of the EU Technology Platform on SmartGrids. He has participated in more than 60 R&D projects funded by the EC and the industry. He is the author of the book *Microgrids: Architectures and Control* and has published more than 180 journal publications and 500 conference proceedings papers.

**H. Vincent Poor**, FREng, ForMemRS, is the Michael Henry Strater University Professor at Princeton University. He has also served as the faculty of the University of Illinois and as a visiting professor at Stanford, Imperial College, and other institutions. His research interests are in wireless communications, smart grid, and related fields. Among his publications in these fields are *Smart Grid Communications and Networking* (Cambridge, 2012). He is a Fellow of the IEEE and IET and is a member of the US National Academies of Engineering and Sciences. He holds a Ph.D. in EECS from Princeton, and honorary degrees from a number of universities. In 2016, he received the John Fritz Medal.

**Laurence Carpanini** leads the development of Smart Energy Solutions for IBM in Europe. He has over 30 years' experience in the sector, providing strategic leadership, subject matter expertise, and an infectious enthusiasm for smarter energy solutions driving industry transformation. In this role, he engaged with government and regulatory bodies in establishing the smart metering market model in Great Britain, provided strategic advice to companies developing smart energy programs, and represented IBM on government and industry workgroups. He speaks regularly at conferences, is a Chartered Mechanical Engineer, and a member of the Institution of Mechanical Engineers. He holds an MBA from Warwick Business School.

**Miguel Angel Sánchez Fornié** is Director of Global Smart Grids (reporting to Global Networks CEO, IBERDROLA). He is a member of the UTC Board of Directors and President of its European division, a member of the Advisory Committee of the European platform "SMART GRIDS", and a member of the Advisory Committee of the Smart Grids Task Force (DG Energy). He is a member of the Advisory Committee of the MIT Future of the Electric Grid Study and the MIT Utility of the Future project. He is chairing the Management Board of the PRIME Alliance. Currently, he is a professor in the postgraduate course of energy in the University of Comillas.

# Preface

We are at an inflexion point in the transformation of the global energy industry, facing challenges that demand clear strategies and strong leadership. The primary challenges are well known to us all. Delivering a secure supply that supports economic and population growth; replacing aging infrastructure with affordable, reliable, flexible generation, transmission, and distribution; and meeting ever more stringent carbon reduction obligations. As a result, within the next 10 years the energy industry will look completely different. We will see, indeed we are already seeing, high concentrations of distributed, intermittent generation; increasing numbers of virtual power plants and microgrids backed up by local storage; interconnected smart appliances generating huge volumes of data through the 'Internet of Things'; widespread uptake of electric vehicles; application of everyday automated demand response; and prosumers who will need and expect new, radically different ways of engaging with the energy industry. The challenge is to develop smart energy systems, networks and regulatory models that will enable the rapid transition to this new operating paradigm.

Data is at the heart of this industry transformation and the creation of smarter energy systems. The collection of large volumes of data is a key feature of such systems, and the ability to derive intelligence from this data through the application of advanced analytics will be immensely important in responding to the challenges faced. One important source of data will come from the deployment of smart metering, which is widely accepted as the first step in the development of smart grids. Smart meters will ultimately provide insights into consumer usage and network availability, but smart meter data by itself is not enough. At medium voltage levels it contributes network data for higher levels of automation, and at low voltage levels it provides extended supervisory capability. Beyond smart meters, the proliferation of interconnected consumer devices and appliances, and rapidly increasing levels of instrumentation at all voltage levels, means that the smart grid will become a platform for collecting and handling immense volumes of data. This book sets out both practical and theoretical solutions to the problems that must be solved to make this a reality.

In the future, data will be regarded as a valuable corporate asset, representing a new basis of competitive advantage for energy companies in the transformation of industry and regulatory frameworks. Handling, analysing, and making sense of this huge variety, velocity and volume of data is complex and challenging, but will be facilitated by the advent of faster more powerful computers, smarter analytics tools and the fact that we are entering the age of cognitive computing systems. Data and analytics will therefore be critical enablers in building an optimised dynamic network, for example:

- The convergence of operating domains is driven by increasing application integration and automation. Moving from systems that rely on unidirectional flows of data and a requirement to manually assess asset health or availability, to a fully automated participatory system is a significant journey, but resolving this challenge will enable true orchestration of the network.
- Real-time optimisation of asset maintenance and reinforcement programmes will be commonplace, based on actual and predicted condition and performance; the state of interconnected assets and capacity constraints; forecast demand; current and predicted weather patterns; and availability of skilled personnel with suitable expertise.
- Automatic monitoring of the health of the grid in real time will be possible, predicting outages based on performance, load, weather conditions, and so on. Remotely sensing vulnerabilities and dynamically dispatching crews will also be possible without waiting for customers to report the fault, as happens today.
- The importance of engaging with and involving customers in the transformation journey cannot be over-emphasised. The availability of data and the way that we use it will be the key to unlocking the benefits for customers, for industry, and for society – and ultimately to underwriting the success of the smarter energy transformation.

This book contains a wealth of cutting edge papers and research results that address the challenges in building smart energy systems. It represents a powerful collaboration between leading academics and innovators in the field of smarter energy, and focuses on maximising the potential of smart grids through the innovative use of systems and data. Each chapter individually addresses a specific element of the smart energy puzzle, covering the effective treatment of smart metering data; smart grid modelling, control, and optimisation; and the importance of interoperable smart grid communications and networking, underpinned by rigorous data and cyber-security. When viewed independently each chapter will challenge existing perceptions, considering holistically the book represents a powerful statement of the future.

Laurence Carpanini
H. Vincent Poor
Nikos Hatziargyriou
Miguel Angel Sánchez Fornié
21 July 2016

*Chapter 1*

# Smart energy – smart grid research and projects overview

*Ilias Lamprinos*[1]*, Nikos Hatziargyriou*[2] *and Hongjian Sun*[3]

According to the European Technology Platform for the Electricity Networks of the Future, a smart grid is an electricity network that can intelligently integrate the actions of all users connected to it – generators, consumers and those that do both – in order to efficiently deliver sustainable, economic and secure electricity supplies. The aim of this chapter is to give an overview of the recent and ongoing research activities that are focused on the smart grid domain. The international literature on this subject is particularly rich and growing. With this in mind, the motivation and key objective for the authors when preparing this chapter was an attempt to collate the distribution grid related research and innovation activities co-funded by the European Commission, in the context of the Horizon 2020 Work Programme 2014–2015 with activities co-funded by similar initiatives in the USA and Asian and Pacific countries. This material is accompanied by information about the recent past and the near future of smart grid research and innovation in Europe. On top of this mapping of facts and trends, this chapter presents the key objectives and functional characteristics of a set of information and communications technology (ICT) tools for the distribution grid that are being implemented in the context of SmarterEMC2, a collaborative project co-funded by the European Commission.

## 1.1 The Smart Grid

### *1.1.1 Introduction*

Due to the global concerns of environmental and energy sustainability, the European Union (EU) has set an ambitious target of reducing greenhouse gas emissions by at least 80% by 2050 compared with 1990 levels. This poses immense challenges on future power systems in terms of integrating increasing numbers of intermittent

[1]Telco Software Department, Intracom SA Telecom Solutions, Athens, Greece
[2]Hellenic Electricity Distribution Network Operator S.A, Athens, Greece
[3]School of Engineering and Computing Sciences, Durham Energy Institute, Durham University, Durham, UK

renewable energy sources (RESs) and meeting ever-rising demand. Smart grid is known as a promising application of ICTs for tackling these challenges. According to the European Technology Platform for the Electricity Networks of the Future [1], a smart grid is:

> An electricity network that can intelligently integrate the actions of all users connected to it – generators, consumers and those that do both – in order to efficiently deliver sustainable, economic and secure electricity supplies. A smart grid employs innovative products and services together with intelligent monitoring, control, communication, and self-healing technologies in order to:
>
> - Better facilitate the connection and operation of generators of all sizes and technologies;
> - Allow consumers to play a part in optimising the operation of the system;
> - Provide consumers with greater information and options for choice of supply;
> - Significantly reduce the environmental impact of the whole electricity supply system;
> - Maintain or even improve the existing high levels of system reliability, quality and security of supply;
> - Maintain and improve the existing services efficiently; and
> - Foster market integration towards a European integrated market.

Essentially, different from existing grids, smart grids have to know fine-grained information about supply and demand in a real-time manner in order to integrate dispersed RESs, flexible demand and energy storage, and further make intelligent decisions on the operation of power systems. A brief comparison between existing grid and smart grid can be seen in Table 1.1.

For all smart grids, data is essential and plays an important role in the system modelling, control, and optimization. Generally speaking, data is typically a kind of

*Table 1.1 A comparison between existing grid and smart grid*

| | **Existing grid** | **Smart grid** |
|---|---|---|
| Sensing and measurement | Electromechanical<br>Few sensors | Digital<br>Sensors throughout |
| Communication and networking | One-way communications<br>Few security threats | Two-way communications<br>Massive security threats |
| Control and optimization | Centralized generations<br>Limited, centralized control | More distributed energy resources<br>Pervasive, decentralized control |
| Customers' role | Few customer choices<br>No privacy concerns | Many customer choices<br>Increasing privacy concerns |

measurements, from which information and knowledge are derived. In power systems, the following types of data can often be seen:

- data from supervisory control and data acquisition (SCADA) system, i.e. measurements of environmental parameters or power system parameters by remote equipment;
- data from phasor measurement unit (PMU), i.e. measurements of the electrical waveform; and
- data from energy consumers, i.e. measurements of energy consumption by meters.

### *1.1.2 Smart metering and data privacy*

Smart metering system is a major source of generating energy consumption data as it is capable of automatically measuring, collecting, analysing and controlling energy usage data, either on request or on a schedule. A typical smart metering system can be seen in Figure 1.1. Besides the traditional function of metering, smart metering infrastructures have the potential of supporting smart grid functionalities on the basis of their connectivity with sensors and interfacing devices to households and companies. Smart metering infrastructures are key enablers of novel smart grid services. For example, they enable domestic customers to choose tariffs and make use of intelligent appliances that help them to reduce bills and to shift demand away from peak hour, thus leading to more efficient use of energy. This smart grid functionality is often called as demand response (DR) which is at the very heart of smart grids, since it has the capability of enabling utilities to control the energy consumption at the consumers' side, thereby enabling demand to follow supply without the need of

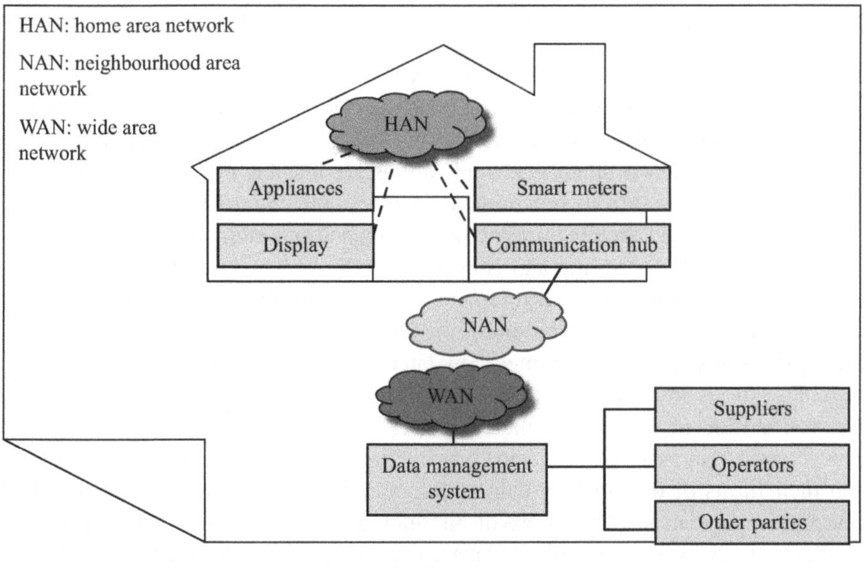


*Figure 1.1 Diagram of a smart metering system*

(or at least defer) adding new generation and transmission infrastructure [2]. The DR encompasses various electricity-pricing schemes, such as time of use (TOU) pricing, real-time pricing (RTP), and critical peak pricing (CPP), and a variety of smart control and optimization algorithms for controlling appliances in response to the time variations of generations and in particular of the RES generation.

Another important advantage of smart metering infrastructures is their capability of providing useful information about utilities for helping them to set the electricity price, predict peak demand and improve the operation and management of power grids. But, this may cause major concerns to the customers, particularly related to data privacy issues. That is, improper use of metering data could pose security threats or financial losses to the consumers. For instance, the analysis of the metering data can reveal whether a consumer is at home or not, posing certain security risks if such information falls into wrong hands. Further, by the use of data mining techniques, it is highly possible to distinguish appliance usage patterns, causing further privacy issues such as disclosing personal activities and behaviours to commercial bodies. For example, appliance manufacturers may want to understand how and why consumers used their products in certain ways in order to produce more competitive products. Without appropriate access control and authorization mechanisms, the metering data can be abused, jeopardizing smart grid operations and violating customers' privacy.

To protect the privacy of customers, it becomes essential to investigate effective approaches to address privacy concerns. For instance, in the UK, the Energy Networks Association (ENA) develops systems and processes to guide the use of smart metering data, balancing between these two aspects [3]: (1) how to make use of metering data to improve the operational efficiency of power grids and (2) how to protect consumers' privacy.

### *1.1.3 Smart grid communications, networking and security*

In addition to smart metering infrastructures, smart grids heavily rely on ICTs to handle the data from sensors, and control and management systems in order to enable DR, the integration of renewable energy and energy storage, and various automation applications. These smart grid applications are feasible only if the information exchange among the various functional units is reliable and trustful.

However, the ICT architectures of the existing power systems are limited to small-scale local system monitoring and control, typically in the form of SCADA, which cannot meet the demanding requirements for the real-time operations, intelligent management and efficient operations of the future electric power systems. On the one hand, the requirements on extremely low latency and high quality-of-service (QoS) communications are still challenging even for the advanced ICTs. On the other hand, the proliferation of sensors and metering devices has led to a massive collection of data across power systems (also known as big data), pushing smart grid ICT systems into their limits of transmitting data and computing. What is worse, the emerging power systems consist of a variety of distributed energy resources (DERs) that are spreading across vast geographical areas: connecting them together in an efficient way is extremely challenging. Hence, real-time ICT infrastructures play a vital role in future power systems, but pose a number of significant challenges. Since smart grid applications may vary in different scales, most smart grid deployments will rely on a number of different ICTs to achieve cost-effective solutions together with full coverage. Thus, identifying the best ICT infrastructure has become a significant research focus.

Due to the diverse integration of ICTs with power systems, there also exists a diversity of security threats. Generally speaking, they can be classified into passive attacks and active attacks [4]. Briefly, passive attacks aim to access the information exchanged within a specific network such as eavesdropping and passive traffic analysis, while active attacks, such as denial-of-service (DoS) attacks, could disrupt the normal functionality of a network. In some cases, a combination of passive/active attacks can be seen where passive attacks are the preparation of active attacks.

All these attacks have certain impacts on the smart grid applications in many aspects, such as data availability, data integrity, scalability, privacy, confidentiality and entity authentication. As such, many organizations focus their work on smart grid security issues including the International Electrotechnical Commission (IEC), the National Institute of Standards and Technology (NIST) and IEEE 1402. In fact, in quite some cases the traditional information technology (IT) security techniques are transplanted into the smart grids domain to meet security requirements. For instance, employing a firewall or proxy could reduce the risk of having DoS attacks. However, choosing a security measure is often difficult, since there exists a trade-off among security, cost and performance. Hence, it becomes very important to understand and model system components and their interactions in order to identify associated cyber vulnerabilities in smart grids.

### *1.1.4 Smart grid modelling, control and optimization*

With aforesaid ICT supports, smart grids should accommodate not only large centralized power generation, but also diverse DERs (e.g. solar, wind or geothermal energy) through flexible power network architectures, advanced control and optimization algorithms. To maximize the use of DERs, smart grids are aimed at optimally balancing demand and supply in real time in order to complement the intermittency of RESs. However, most of renewables are uncontrollable and distributed across a large spatial area, making the stable, reliable and flexible electricity transmission and distribution extremely challenging, particularly with high percentage of renewable penetration in large and complex power networks. As such, the development of decentralized and/or distributed control architectures and optimization algorithms become particularly important. The research objectives and topics vary when considering different perspectives, as shown in Table 1.2. The overall aim of such efforts is twofold [5]:

1. To enable optimal demand aggregation and DR to follow generations such as DERs, hence facilitating the efficient and reliable operations of future power grids.
2. To decompose centralized control and optimization problems into decentralized and/or distributed solutions, hence improving the quality of controlling dispersed DERs, flexible demands and energy storage devices.

*Table 1.2 Research objectives and topics of smart grid modelling, control and optimization*

| | Control/optimization objectives | Research topics |
|---|---|---|
| Customer perspectives | • Minimize bills<br>• Maximize satisfactory level<br>• Maximize energy efficiency | • Home energy management<br>• Appliance scheduling<br>• Smart metering |
| Operational perspectives | • Maximize profit<br>• Maximize stability | • Smart pricing<br>• Automation<br>• Demand aggregation and response |
| Economic perspectives | • Minimize cost<br>• Maximize energy security<br>• Maximize sustainability | • Supply side management<br>• Demand side management |

## 1.2 Smart grid research: mapping of ongoing activities

### *1.2.1 Europe*

The aim of this section is to map the main recent and ongoing research and innovation activities in Europe.

#### 1.2.1.1 European smart grid projects outlook, 2002–2014

The European Commission Joint Research Centre (JRC) issues periodically a report that presents an overview and insights of smart grid projects across the EU Member States. The 2014 edition [6] includes a total of 459 smart grid projects, launched from 2002 up until 2014, amounting to €3.15 billion investment. The report highlights seven domains of smart grid applications:

- ***Smart network management:*** This category includes implementations that focused on increasing the operational flexibility of the electricity grid, like substation automation, grid monitoring and control. Typically, the goal was to improve the *observability* and *controllability* of the networks.

  With regards to observability, project results confirmed that the tools developed or used to observe the network are mature and reliable. Some identified areas of improvement were related to standardization and interoperability, particularly on the communication infrastructure. At transmission level, emphasis was also placed on the development of tools for the coordinated operation of pan-European networks. In this area, some of the key themes addressed in the projects were: implementation of smart meters to collect and store, on demand and in real time, specific high quality and accurate data for each consumer and group of consumers; improving distribution grid monitoring to cope with volatile states in the grid; real-time asset monitoring; and fault identification and localization.

With regards to controllability, the report confirms that several control technologies are already highly developed and efficient. Areas for improvement include cyber security and scalability of applications from small scale to large-scale projects. In this area, some of the key themes addressed in the projects were: implementation of new capabilities for frequency control, reactive control, power flow control; controllable distribution sub-stations, controllable inverters and charging stations; development and testing of distributed generation and intelligent load controllers; smart protection selectivity (smart relays); smart auto-reconfigurable networks, easily stabilizable online tap changers; dynamic line rating; deployment of a range of leading-edge transformers across a number of low- and medium-voltage circuits, together with the use of capacitors, volt-ampere reactive (VAR) control devices and electronic boosters which when optimized together can lead to reduced losses from the power system.

- ***Integration of large-scale RES:*** Most projects in this category were concerned with the integration of RES mainly at the transmission level. Key focus areas were: tools for planning, control and operation of renewables in order to facilitate their market integration; integration of demand side management (DSM) and ancillary services by Distribution System Operators (DSOs) to support Transmission System Operator (TSO) operation; tools to forecast RES production; off-shore networks for wind power integration.
- ***Integration of DERs:*** This category includes projects that focused on new control schemes and new hardware/software solutions for integrating DERs while assuring system reliability and security. Project results showed that technical solutions for the integration of DERs become quite consolidated. Technical aspects that were considered include: implementation of voltage control and reactive power control of DERs for the provision of ancillary services; DER production forecast and active/reactive power measurement for network observability; DER protection settings for anti-islanding operation; use of storage together with distributed generation for voltage control, power flow modulation, balancing, etc.; centralized versus decentralized (e.g. agent-based) control architectures; aggregation of controllable DERs into virtual power plants (VPPs) and into microgrid configurations.
- ***Aggregation (VPP, DR):*** This group of projects focused on the implementation of aggregation mechanisms like VPPs and DR to aggregate the supply and demand flexibilities of decentralized resources taking into account various grid constraints and market signals.
- ***Smart customers and smart home:*** This group includes projects that tested smart appliances and home automation together with new tariff schemes. Such projects typically require the active participation of consumers or aimed at analysing consumer behaviour and fostering consumer involvement.
- ***Electric vehicles and vehicle2grid applications:*** Projects in this category focused on the smart integration of electric vehicles (EVs) and plug-in hybrid electric vehicles (PHEVs) in the electricity network.
- ***Smart metering:*** This category includes projects that focused on smart metering installations, which are part of a wider smart grid project with one or more additional application.

#### 1.2.1.2 European smart grid projects outlook, 2014–2015

The majority of the ongoing activities are in the form of collaborative projects that are co-funded by the European Commission, in the context of the Horizon 2020 Work Programme 2014–2015. This programme reflects the research and innovation agenda established by EU as part of its Strategic Energy Technology Plan (SET Plan) 2010–2012 [2]. The SET Plan prioritizes those technologies most relevant to the energy and climate policy objectives for 2020 [7]: wind, solar, electricity networks, carbon capture and storage (CCS), bioenergy, nuclear, fuel cells and hydrogen, and energy efficiency.

It should be noted that while preparing this chapter, the European Commission has adopted a new Strategic Energy Technology Plan (SET Plan 2015) [8] that establishes the strategic frame for the years to come. On parallel to the Horizon 2020 Work Programme 2016–2017 was announced, detailing an extensive list of research and innovation topics to be addressed by the European academic and industrial community. This work programme sets the context of the near future research activities, i.e. activities that have not yet emerged, and as such it is out of the scope of this document.

Going back to the Work Programme 2014–2015, one of the key dimensions that are addressed is that of the modernization of the European electricity grid. The aim is to fund actions that contribute to a more flexible grid, increase capacity, and introduce DR and active user involvement. Towards this direction, this work programme covers the following research dimensions:

- transmission grid and wholesale market,
- distribution grid and retail market,
- local/small-scale storage,
- large-scale energy storage, and
- next-generation technologies for energy storage.

This chapter summarizes the key aspects of the smart grid projects that were launched in 2015 as part of the Low Carbon Economy (LCE) call for proposals that is part of the Horizon 2020 Work Programme 2014–2015. The presentation of these projects follows a grouping approach that is based on the identification of the key research objective they set.

- ***Co-sharing of ICT infrastructure:*** One important research dimension that is being explored is related to the dedicated or shared use of the ICT infrastructure among the electricity and the communication sector. Various different aspects affect the decision on this matter: requirements, feasibility of the technical solutions, infrastructure deployment models (dedicated vs. shared use), cost aspects (capital expenditures CAPEX, operational expenditures OPEX), regulatory aspects, and business models. These issues are being explored in the context of the research project *Energise* [9].

  Co-sharing is a concept that is being explored also with regards to the metering infrastructure, e.g. in the research project *Flexmeter* [10]. In particular, the multi-service smart metering system concept is being explored. Such an approach, when coupled with the development of services like non-intrusive appliance load monitoring (NIALM) enables utilities to operate their core businesses in a smarter way and gives them the possibility to implement additional services for their customers. A multi-service meter system permits also to share the costs of the system itself, using common devices (e.g. gateways) to control different kind of meters and sensors.

- ***Metering data sharing and value-added services:*** Nowadays there is no unique standard format for the exchange and processing of electricity consumption data. Moreover, metering data is made available for settlement, with limited forwarding to other third parties (there are few infrastructures available that could facilitate this information exchange). Also, the data is generally available in low resolution and with considerable delays, often up to 30 days. Towards this direction Flexiciency project [11] defines a common language for data and service exchange aiming at facilitating the metering data made available with customer consent by DSOs also in real time to any interested stakeholder willing to provide services. Similarly, an open metering data interface is being implemented in the framework of the research project *Flex4Grid* [12].

With the aim of facilitating the delivery of value-added services, several projects pursue activities towards the realization of the Open Market Place, mimicking the concept of online applications market that is very popular in the mobile applications domain. For example, the *Flexiciency* project implements a Market Place that aims to serve as a pan-European platform for developing new services related to electricity retail markets, while supporting standard interfaces with the systems and platforms of DSOs, Energy Service Companies (ESCOs), aggregators, retailers and vendors. The rationale behind this approach is that interoperable solutions and data accessibility can foster market competitiveness, improve the business case and create new areas of business for new comers and existing players.

From a technical perspective, and in order to achieve this data sharing need, one option that is explored is that of cloud-based data management. This is for example the case in the *Flex4Grid* project. The cloud technology acts as glue as well as a separation between the system components (cloud collection, aggregation, storage, computing, data analytic, visualization and users' access to data and analytic results). On the other hand, it interconnects market actors since it provides same operational/market/response/demand view. Cloud technologies are promoted also by the *Empower* project [13] for grid operation needs (forecasting, grid balancing, DSM, optimization), for metering-related functions (metering data management, meters' monitoring), and for market operation. Cloud-based services are explored also in the *Flexmeter* project, e.g. DR services, DSO services, load and generation forecasting.

An overlapping research activity is related to the implementation of a low-cost smart meter that can be used in the provision of value-added services for the end users. This is the case in the *NobelGrid* project [14] where the aim is to implement a cost-effective smart metering solution that can be used in the provision of value-added services by the various different stakeholders. Such a service is for example an energy monitoring and analytics app to be offered to domestic and industrial prosumers.

Smart metering data is valuable not only for the entities that aim at offering value-added services, but also for the DSOs. For example, one of the objectives of the *UpGrid* project [15] is to integrate and process these data in the operation and management system of the DSO. Again, the DSO plays a central and pivotal role in the delivery of new services (facilitator).

- ***DR solutions:*** DR is a research and innovation thread with significant presence in European ongoing research activities. For example, in the *SmarterEMC2* project [16] DR is mainly considered from the viewpoint of the DSOs, as a means to achieve network stability. On the other hand, in the *NobelGrid* project a solution that enables other stakeholders (ESCOs, DR Aggregators and Energy Retailers) to handle the flexibility of DR-related assets in the energy markets is being implemented.
- ***Research related to the management of the distribution grid:*** Several projects deal with the implementation of tools that facilitate the DSOs in their role. For example, *NobelGrid* implements a grid management and maintenance of master framework. In *SmarterEMC2* an ICT tool that facilitates the DSOs in the management of local constraints in their grid is implemented. *UpGrid* puts the DSOs in the centre of gravity of its research and pursues activities related to the low-voltage network representation, functional specification of low-voltage dispatch, deployment of tools to support low-voltage operation of field crews, improvement in operations and maintenance of the low-voltage grid and improvement of QoS to customers. Finally, the *P2P-SmarTest* project [17] explores the peer-to-peer based control paradigm applied in the distribution networks.
- ***Research related to the role of the prosumers:*** A group of ongoing projects deals with research and innovation activities that are focused on the prosumer side and its operation either in stand-alone mode or in close collaboration with the DSO side. The *AnyPLACE* project [18] approaches this issue by developing a solution comprising of hardware and software to extend smart grid concept to the home environment, integrable with smart metering, interacting with end users, market and system operators and dealing with the control of loads (e.g. appliances) and generation (e.g. renewables). This project puts in the same context also the infrastructure that facilitates the interaction of the prosumers with other utilities (e.g. gas, water, heat).

  Quite some projects also deal with energy markets that are close to the prosumer. While the *P2P-SmarTest* project explores this aspect mainly from the technical point of view (peer-to-peer communication schemes), the *Empower* project aims at developing a complete new energy market where consumers can buy and sell "neighbourhood energy" which is produced locally by solar panels, micro wind turbines and other decentral energy production.
- ***Communication technologies:*** Peer-to-peer is a concept that is being explored in *P2P-SmarTest* project not only as a means to facilitate the integration of demand side flexibility and the optimum operation of DER and other resources within the network, but also as a technology facilitator for energy trading and exchange. Distributed Long Term Evolution (LTE)-based architecture in a hybrid scheme that also employs sensor networks is being explored.

- ***Storage:*** Several research groups deal with the enhancement of the energy management systems (EMSs) being used by DSOs and aggregators with storage and aggregation capacities. The aim is to enhance the existing management and control systems with functionality related to the management and control of these new types of assets as well as with decision support tools that take into consideration the renewable generation and storage aspects.

  For example, in the *NETfficient* project [19] the research around storage technologies focuses on second-life EV batteries and hybrid energy storage systems. From a technical viewpoint what is explored is the flow management between the low and medium voltage grid. The project aims at demonstrating a high power and fastest in response time energy storage on a medium voltage level for peak shaving, and a hybrid solution of batteries and supercapacitors on the low voltage level. Apart from the technical aspects, also the socioeconomic feasibility of deploying and exploiting storage assets at residential level is examined.

  Storage is also explored in synergy with DSM, wind and photovoltaic power. This is the case for example in the *Tilos* project [20], where challenging battery performance and its interoperability with the rest of components is explored with the aim of charting the technical barriers applying to various use cases of storage (both stand-alone and interconnected ones).

  Energy storage is explored also as a dimension of total solutions related to building energy management. This is the case in *Sensible* [21] and *Elsa* [22] projects. In these projects, storage is considered as an asset to be taken into account by a flexible and adaptable ICT-based EMS that integrates a range of diverse physical (second-life EV batteries, thermal storage) and virtual (DR) storage technologies at building, district and DSO/substation level. For the approaches that are being examined, of particular importance is the elaboration of the technical solutions for the integration of the storage with the distribution grid, as well as the investigation of the impact of storage and energy management solutions at low voltage level in mitigating intermittency and increasing grid security and stability. Various storage technologies are examined (electro-chemical, electromechanical, and thermal). Another research dimension is related to the evaluation of the current market opportunities for local-scale battery storage across Europe and globally and the development of novel business models that can ensure the diffusion and market uptake of relevant technologies.

  The economic viability of the storage-related smart grid technology is a particularly important aspect. *NiceGrid* [23] is a recently finished pilot project by Europe's largest power network operators that aimed at integrating power from rooftop solar panels into the grid. The project has shown that battery storage of renewable energy is not yet economically viable. The pilot indicated that while the technology works perfectly, it is still too expensive for wider rollout. One interesting outcome was that storage could be economically viable in remote areas where it competes with expensive diesel generators, as a back-up source of power. But in areas where gas-fired power plants exist the storage is not competitive enough.

### 1.2.2 United States of America

In the USA, major smart grid projects are funded through the US Department of Energy (DoE) and are divided into Smart Grid Investment Grants (SGIG), Smart Grid Demonstration Projects (SGDP), and Renewable and Distributed Systems Integration (RDSI) projects [24]. Particularly, the SGDPs aim at funding actions that demonstrate how a suite of existing and emerging smart grid concepts can be innovatively applied and integrated to prove technical, operational and business-model feasibility. Two types of smart grid projects are considered for the SGDP: one includes regional smart grid demonstrations to verify smart grid viability, to quantify smart grid costs and benefits, and to validate new smart grid business models at scales that can be readily replicated; the second includes energy storage technologies such as batteries, flywheels, and compressed air energy storage systems for load shifting, ramping control, frequency regulation services, distributed applications, and the grid integration of renewable resources such as wind and solar power.

Another entity that finances research related to the smart grid is the Advanced Research Projects Agency-Energy (ARPA-E), a governmental agency that aims at promoting and funding research and development of advanced energy technologies. In late 2015, ARPA-E announced a set of 12 grants meant to enable real-time coordination between distributed generation, such as rooftop and community solar assets, and bulk power generation, while proactively shaping electric load. There is also work being done to allow distributed energy assets to communicate and act in concert to solve local and system-wide grid challenges. Some of the projects funded by ARPA-E aim at supporting DER integration, while others go beyond what today's utility grid control platforms and DERs are built to handle. The key objectives of an indicative set of projects out of this group are given below [25]:

- The University of Vermont explores a new approach for DSM, called Packetized Energy Management, based on approaches used to manage data in communication networks that lack centralized control, but also need high levels of privacy.
- The University of California develops coordination algorithms and software meant to allow DERs and end loads to serve frequency regulation services as an alternative to the current situation where fast-responding grid services are performed by power plants and large-scale grid batteries.
- Arizona State University develops algorithms related to the stochastic inputs of the traditional grid power flow models. These algorithms are meant to integrate uncertainty from renewable resources, load, distributed storage and demand-response technologies, and aim at providing the system operators with real-time guidance in the coordination of DERs and DR.
- Stanford University expands its Powernet, an open-source and open architecture platform for scalable and secure coordination of consumer flexible load and DERs. Powernet is based on the principle of connecting information networks to the power network. It uses a layered architecture that enables real-time coordination of centralized resources with millions of DERs by integrating embedded sensing and computing, power electronics and networking with cloud computing. The novel

research threads are related to the networking of a residential energy gateway with existing inverters and appliances and with controlling power via smart switches that replace traditional fuses. Another research thread focuses on novel techniques of aggregation of local customer resources in order to meet local constraints and global coordination objectives.

- A consortium led by General Electric develops a novel distributed flexibility resource (DFR) technology that aggregates responsive flexible loads and DERs to provide synthetic reserve services to the grid while maintaining customer QoS. One of the key novelties of the project is that it leverages short-term and real-time weather forecasts along with other data to estimate the reserve potential of aggregate loads and DERs on a day-ahead basis. This feature is exploited in an optimization framework that enables aggregation of large numbers of flexible loads and DERs and determines the optimal day-ahead schedule to bid into the market.
- The National Renewable Energy Laboratory implements a distribution network management framework that unifies real-time voltage and frequency control at the home/DER controllers' level with network-wide energy management at the utility/aggregator level. The distributed control architecture leverages the real-time feedback control to continuously steer frequency and voltages towards optimal operating points while dynamically procuring and dispatching synthetic reserves based on current system state and forecasts of ambient and load conditions. The framework enables computationally affordable distributed implementations by decomposing network-level optimization into smaller sub-problems and applying appropriate approximations. This allows implementation across distributed low-cost microcontrollers and control is carried out via elementary operations. Responsiveness to rapidly changing conditions is facilitated by incorporating intrinsic network physics into the control formulation and by processing real-time measurements.
- The Pacific Northwest National Laboratory develops a hierarchical control framework for coordinating the flexibility of a full range of DERs, based on two separate speeds of communication. The slower of the two communicates data on economic incentives for devices to make their flexibility available to the grid, while the faster one ensures that devices actually respond appropriately, and do not interact in unexpected ways to worsen the problems they are trying to solve.
- The University of Minnesota develops techniques for centralized cloud-based and distributed peer-to-peer networks, with the aim to enable coordinated response of many local units to adjust consumption and generation of energy, satisfy physical constraints, and provide ancillary services requested by a grid operator. The project applies concepts from non-linear and robust control theory to design self-organizing power systems that effectively respond to the grid events and variability.
- A consortium led by Northwestern University develops frequency-based load control architecture to provide additional frequency response capability and allow increased renewable generation on the grid. A multi-layer control architecture makes it possible to simultaneously ensure system stability at the transmission

network level, control frequency at the local distribution network level, and maintain the QoS for individual customers at the building level, all under a single framework. At the transmission level, coordination among different areas is achieved through a centralized scheme that ensures stable frequency synchronization, while the control decisions within a single area are based on local information.

- The National Rural Electric Cooperative Association implements technology that monitors grid voltage and frequency and controls the target load in order to address excursions from grid operating targets and integrates that functionality into DR-enabled water heater controllers and smart circuit controllers.
- Eaton Corporation develops a cloud-computing solution that provides synthetic regulating reserve services to the power grid. The innovative aspect of this approach is that it separates the decision-making of synthetic regulating reserve services into two levels to significantly reduce the computational and communication complexity, thereby enabling large-scale coordinated control of a vast number of DERs and flexible loads. The system-operator level estimates and predicts reserve capacity of the distribution network and decides on the appropriate economic incentives for DERs to participate in future services. At the local level, an energy node comprised of a cluster of DERs automatically decides its own reserve services strategy, taking into account short-term net load and economic incentives. By splitting these decisions between the two levels, the solution does not require extensive communication or negotiation between the local DERs and the system operators.

### *1.2.3 Asia-Pacific*

As the smart grid's huge potential of improving power system reliability and reducing outages and costs, smart grid technologies have also attracted much attention in Asia-Pacific areas. Particularly there are significant investments in both RESs and smart meters, as shown in Table 1.3.

*Table 1.3 Some Asia-Pacific countries' investments on RESs and smart meters*

| | **RESs** | **Smart meters** |
|---|---|---|
| China | • Wind: 114.76 GW by 6/2015;<br>• Solar: 35.78 GW by 6/2015 | • 377 million by 2020 |
| India | • Wind: 24.67 GW by 10/2015;<br>• Solar: 4.23 GW by 8/2015 | • 130 million by 2020 |
| Australia | • Wind: 3.8 GW by 12/2014;<br>• Solar: 4.0 GW by 12/2014 | • 2.5 million by 2020 |
| South Korea | • Wind: 0.6 GW by 12/2014;<br>• Solar: 3.2 GW by 12/2014 | • 24 million by 2020 |

More importantly, some Asia-Pacific countries have set up a number of projects to trial smart grid technologies as follows.

#### 1.2.3.1 China

Since 2009, China has started an ambitious smart grid rollout plan for 2009–2020, with the name of "Strong and Smart Grid". Slightly different from US and EU's smart grid concepts, China puts more emphasis on the implementations of ultra-high voltage (UHV) lines [26] to address the problem of imbalanced distribution of power demand and power supply, i.e. higher demand in the east China but more resources in the west China. State Grid Corporation of China (SGCC) [27], which is the brain behind China's smart grid plan, has divided this plan into three stages as shown in Figure 1.2. This plan encompasses all aspects of smart grid, such as increasing generation and transmission capacity, installing and integrating large-scale renewable energy, and nationwide deployment of smart meters and PMUs. As of 2015, SGCC has launched over 300 pilot projects with more than €5 billion investments, installed over 2,600 PMUs for providing dynamic information to improve power operations across the country, built up over 650 smart substations of 110–750 KV, and tested smart distribution systems in 23 cities.

#### 1.2.3.2 India

India is also pursuing an aggressive smart grid plan. In November 2014, Indian Prime Minister Narendra Modi announced that $4 billion would be invested for funding smart metering programs across the country. Meanwhile, it is predicted that $21.6 billion will be invested over the period 2015–2025 for deploying smart grid related infrastructures. As of 2015, the Indian Ministry of Power has funded 14 smart grid pilot projects that will be implemented by state-owned distribution utilities for testing a range of smart grid technologies [28]. These projects cover over 250 thousands of

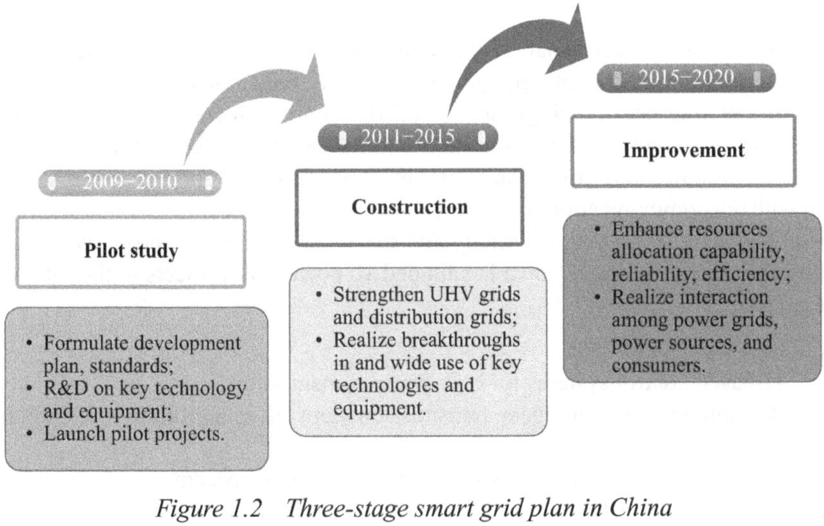


*Figure 1.2 Three-stage smart grid plan in China*

customers in 14 cities, testing a number of smart grid technologies with a primary focus on smart metering. To date, these projects have verified:

- the use of smart metering infrastructure for remote, regular and reliable reading;
- electricity consumption reading can properly support demand forecasting, peak load management and management of dynamic tariff;
- smart metering can facilitate the enabling of partial (individual/group of consumers) rather than feeder level outage remotely;
- smart metering is a good tool for enabling outage management system and integration of distributed generation and
- smart metering can improve the power quality, reliability of power supply, network control management, etc.

#### 1.2.3.3 Australia

Since 2009, the Australian Government has recognized the importance of smart grid technologies, and as a result allocated funding in the order of AUD 100 million to the Ausgrid consortium in 2010 within the *Smart Grid, Smart City Programme 2010–2013* to test a number of smart grid technologies, including Active Volt-Var Control (AVVC), fault detection, substation and feeder monitoring, Wide Area Measurement (WAM), smart metering, EVs, distributed generation and storage, and customer applications. In particular, the customer application trials have tested 22 products comprising of pricing structures, network controlled devices, feedback technologies, emerging technologies and educational products with over 7,000 residential customers. These trials were largely focused on the greater Newcastle and Sydney CBD areas. The Ausgrid consortium has rolled out 1,000 km of new fibre optic cables for connecting substations, approximately 12,000 monitoring devices throughout its electricity distribution network for reducing outages through faster fault location, and over 4,000 smart meters for collecting energy consumption data [29].

#### 1.2.3.4 South Korea

South Korea has built its smart grid plan on the robust and well-developed Internet and Electronics industry, consequently proposed to merge the concept of smart grid with that of Internet of Things (IoT). In 2011, Korea Electric Power Corporation (KEPCO), the largest electric utility in South Korea, announced that it would invest over $7 billion in smart grid technologies by 2030. As of 2015, KEPCO has signed a deal with telecommunications operator LG U+ to develop next-generation smart grid technologies by combining electricity and IoT technologies. To better marry power with IoT industries, South Korea has funded 10 power IoT projects in the following areas [30]:

1. development of EMS;
2. IT-based control systems for bulk power transmission;
3. development of intelligent transmission network monitoring and operating system;
4. development of a digital next-generation substation system;

5. development of intelligent distribution management systems;
6. development of power active telematics systems for facility monitoring;
7. development of a consumer integrated resource management system for high value-added power services;
8. development of Power Line Communication (PLC) ubiquitous technology;
9. development of power semiconductor technology for distributed generation and its application in industrial inverters and
10. development of integration EMS for the microgrid.

## 1.3 Smart grid research in Europe: what comes next?

On the basis of an integrated roadmap developed with stakeholders and Member States during the period 2013–2014 and in line with the new political priorities defined in the Energy Union, the SET Plan has been upgraded in order to accelerate the EU energy system transformation in a cost-effective way. This will be achieved by moving to smarter, more flexible, more decentralized, more integrated, more sustainable, secure and competitive ways of producing energy, transporting it and delivering it to consumers. The aim is to foster innovations that will accelerate the energy transformation and will place consumers in the centre stage.

In the upgraded Integrated SET Plan, an extensive set of priority actions have been identified which will serve to develop and integrate the innovative technologies and system solutions:

- ***Renewable technologies at the heart of the new energy system:*** The aim is to sustain technological leadership for the EU, by developing highly performing renewable technologies and their integration in the EU's energy system and to reduce the cost of key technologies.
- ***Reduced cost of key technologies:*** The aim is to coordinate research and innovation and scaled up manufacturing, and leverage regional cooperation towards achieving cost reduction of key renewable technologies.
- ***A smarter energy system, empowering the consumer:*** The aim is to create innovative technologies and services for smart homes that provide smart solutions to energy consumers.
- ***New materials and methods for buildings' energy efficiency:*** Further research and innovation is expected related to new construction materials and methods with a potential for energy savings of between 13% and 22% of the energy consumption of buildings in the next two decades. While European industry is constantly improving its energy efficiency, reducing its energy intensity by 19% between 2001 and 2011, further research and innovation is needed to increase even more the efficiency of appliances (e.g. pumps, boilers or furnaces), processes and systems (e.g. better control and energy management) exploiting advances in ICT and improved heat recovery.
- ***Carbon capture, storage and use:*** CCS reduces greenhouse gas emissions from the energy sector and carbon intensive industries. Four large-scale demonstration

projects in the EU are expected to be operational around 2020, but none has taken the final investment decision, as the business case for CCS in Europe has not yet been established.

- ***Maintain high level of safety of nuclear reactors and associated fuel cycles:*** EU leads in the adaptability of reactors to variation in demand and is constructing the world's first geological repositories for high level and heat generating long-lived radioactive waste. The closure of nuclear power stations in the EU will increase focus on decommissioning and associated radioactive waste management.
- ***Increase the resilience, security and smartness of the energy system:*** The aim is to develop and demonstrate innovative power electronics, flexible thermal generation, DR and storage, as well as efficient heating and cooling technologies (such as heat pumps and combined heat and power) to use synergies between energy vectors, new transmission technologies, new techniques for physical and cybersecurity of networks, and demand analysis including exploitation of Big Data.
- ***Boost energy storage, particularly in light of e-mobility:*** Energy storage solutions will require batteries that have higher performance (e.g. energy density), extended life, reduced costs, larger capacity and can be scaled-up to competitive manufacturing. The aim is to bring together major industrial innovators and researchers to work on these issues.
- ***Strengthen market take-up of renewable fuels:*** The aim is to foster market uptake of renewable fuels, particularly on transport modes lacking readily available sustainable fuels alternatives, such as aviation.

## 1.4 The SmarterEMC2 project

The SmarterEMC2 project aims at achieving a breakthrough in the ICT integration with power systems by enhancing various smart grid services. The main objective of the project is to develop ICT tools and propose business models that support customer side participation and RES integration, and facilitate open access in the electricity market. These tools take into account the Smart Grid Architecture Model (SGAM) [31] as well the future structure of the distribution grids as described by the relevant EU bodies and organizations.

In particular, the SmarterEMC2 project develops new tools that support the integration of consumers in the energy market. Consumers are integrated in the relevant schemes through an advanced DR management system that is able to manage consumption and effectively communicate in a multi-level hierarchically organized smart grid with all business entities (including transmission and distribution system operators, VPP owner, DR aggregator, energy hub owner, consumers, etc.). The SmarterEMC2 project also creates new services to support the integration of DG/RES through the concept of VPP that participates in the market and provides ancillary services to the network assuming a similar multi-level hierarchically organized architecture for the communication with all business entities.

### *1.4.1 Stakeholders involved in SmarterEMC2*

The list of stakeholders and their role in the context of SmarterEMC2 project is as follows:

- ***Distribution System Operator:*** Responsible for the technical operation of the distribution grid and interested on utilizing flexible loads for managing grid contingencies.
- ***Transmission System Operator:*** Responsible for the technical operation of the transmission grid, and capable to validate and activate flexibility offers and send direct control signals to DER and market agents connected to the transmission network.
- ***Virtual Power Plant Operator:*** Aggregates and manages various resources (DERs, loads, storage systems) in order to allow market participation and provide ancillary services to the grid.
- ***DR Aggregator:*** Aggregates and manages resources in order to provide ancillary services to the DSO or participate in the energy market.
- ***Energy Hub Operator:*** Aggregates and manages local microgrid customers, owners of different resources (DER, storage, loads, EVs), willing to participate in flexibility programs (DR and VPP) in an aggregated way.
- ***DER/Load Owner:*** Manages DERs, storage systems and loads that are able to provide flexibility to market actors.
- ***Smart Consumer:*** It is a consumer capable of providing flexibility to other market actors, through smart loads. DER and/or storage could be utilized as well.

The above stakeholders have a set of requirements related to ICT tools that leverage flexibility that is summarized in Table 1.4. This content comprises the input provided by the stakeholders participating in the SmarterEMC2 project.

### *1.4.2 Conceptual architecture of the SmarterEMC2 ICT ecosystem*

The conceptual architecture of the ICT tools that comprise the SmarterEMC2 ecosystem is depicted in Figure 1.3. This figure presents a high-level decomposition of these systems based on the core functionality and interactions with external systems. The components are grouped in two categories: DSO ICT Tools and DR Aggregator/VPP Operator ICT Tools, depending on which stakeholder they are addressed to.

The main components are:

- *VPP Platform*: Set of ICT tools that enables the operation of a number of components connected to the distribution grid as an aggregated single entity. The aggregation of such components allows the VPP Operator to participate as a single entity in the energy markets but also requires a set of ICT tools for the coordination of the components in order to fulfil market obligations and distribution grid requirements. Furthermore, the VPP Platform may interact with

*Table 1.4 Requirements synopsis per stakeholder*

| **Stakeholder** | **Requirements** |
| --- | --- |
| Distribution System Operator | • *Identify and solve problems in the distribution grid*: Data from field equipment can be used for evaluating the status of the grid and identifying or forecasting problems in the DG (in a time horizon of minutes to hours), using power flow algorithms.<br>• *Utilize flexibility*: DSO shall receive, validate, adjust and activate flexibility to solve problems in the distribution grid. |
| Transmission System Operator | • *Explore market-based flexibilities*: The TSO can activate feasible flexibility offers within the distribution grids, in articulation with the DSO who provides technical validation. |
| Virtual Power Plant Operator | • *Participate in energy market*: The VPP Operator aims at maximizing its profits by providing the aggregated flexibility of its DER resources portfolio to the market.<br>• *Provide ancillary services*: The VPP Operator aims at providing ancillary services to the grid through flexibility offerings. |
| Demand Response Aggregator | • *Participate in energy market*: The Demand Response Aggregator aims at maximizing its profits by providing aggregated flexibility from DR resources to the market. |
| Energy Hub Operator | • *Participate in energy market*: The Energy Hub Operator aims at maximizing its profits by providing aggregated flexibility from microgrid resources to the market.<br>• *Optimal planning and control*: the Energy Hub Operator aims at planning of supply and demand, while taking into account flexibility offerings/requests and dispatching control signals to field equipment. |
| DER Owner | • *Maximize profits:* The DER owner aims at financial benefits by providing flexibility to other market actors, through DER, storage systems and loads. |
| Smart Consumer | • *Minimize costs/maximize profits*: The Smart Consumer aims at financial benefits by providing flexibility to other market actors, through smart loads, as well as DER and storage. |

DSO tools. This interaction may take place in case the DSO provides control services to the VPP for specific distribution grid assets (e.g. the case where the DSO installs DG controllers or even DGs-like storage systems). Another case of interaction between DSO ICT tools and the VPP Platform is when the DSO takes full control of a VPP in emergency situations. This means that according to the role of the DSO, the VPP Platform might or might not interact with DSO tools.

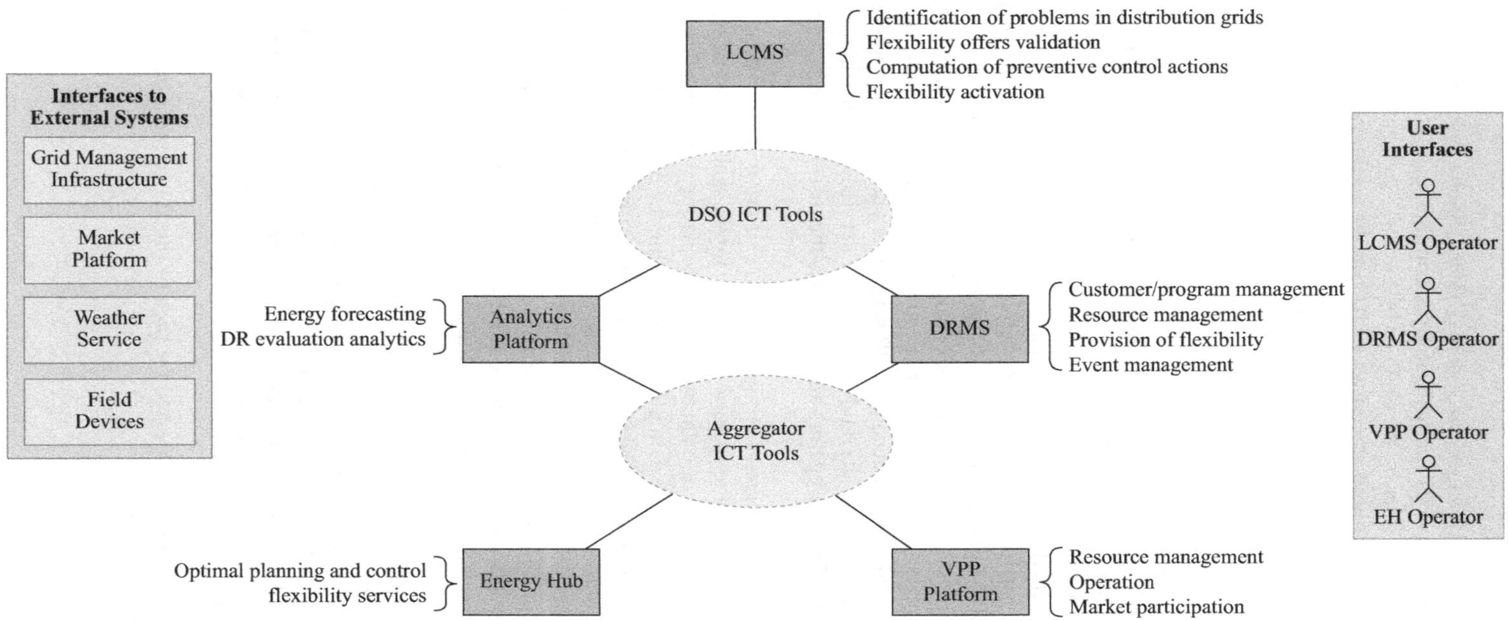


*Figure 1.3 SmarterEMC2 high-level conceptual architecture*

- *Demand Response Management System (DRMS)*: Such a system is utilized for the provision of DR services and realization of communications with DR resources. It can be operated either by the DSO or by the DR Aggregator; hence, the component has a link to both categories of ICT tools. In the first case, the DRMS manages resources for providing the necessary flexibility to DSO, for solving grid problems. In the second case, the DR Aggregator utilizes the system either for providing DR services to the DSO, or for participating in the energy market. Hence, a DRMS operated by the DR Aggregator can interoperate with DSO's DRMS and accept DR messages, while it has also the ability to offer its DR capacity to the market.
- *Energy Hub (EH)*: The system associates a certain number of prosumers that participate to the local consumption and generation of electricity with an objective of EH to optimize local energy flows and minimize the economical and the environmental costs. Due to regulatory reasons and its dimensions, EH cannot directly participate in the market; thus EH is aggregated, as single entity, under the VPP Aggregator or the DR Aggregator, to which it appears as a single entity. EH has the role of a small-scale flexibility operator. EH may define its own optimization strategy, i.e. EH estimates the optimal operating points of its prosumers taking into account VPP set points and DR Event information. It indirectly provides ancillary and grid support services, by providing flexibility to the VPP and DR Aggregators and facilitates the distributed/renewable sources integration.
- *Local Constraints Management System (LCMS)*: are a set of management and control tools to support the operation of a DSO for the specific case of solving local constraints in the distribution network in coordination with market participation. These tools make use of existing flexibility within the distribution grid to address requests from the TSO or the DSO, by controlling resources owned by the DSO or by requesting the participation of resources managed and controlled by other entities such as VPP Operators and DR Aggregators. The management and control tools are supported by an ICT infrastructure, controlled and managed by the DSO, that is designed to allow the exchange of information between the different entities and systems and capable of fulfilling the associated functional and non-functional requirements.
- *Analytics Platform (AP)*: The AP is the source of data input to the forecasting and non-real-time control of the SmarterEMC2 systems, gathering performance, market and external driver data into a single source for applications to perform analytics over it. The data managed by the AP is both historical – the AP stores performance data and the results of analytic applications used to control the SmarterEMC2 systems – and predictive – the AP acquires forecast data for, say, weather conditions.

The key functionalities supported by each of the above-mentioned components are summarized in Table 1.5.

*Table 1.5 Function synopsis per key SmarterEMC2 system*

| Component | Functionality title | Short description |
|---|---|---|
| Virtual Power Plant (VPP) Platform | Resource management | Resource management functionality involves the registration process of VPP resources and the information exchange of sensitive information in general. In addition, this functionality involves health monitoring and modelling of VPP assets as well as measurement and verification. |
| | Operation | Optimal dispatch refers to the procedure of creation, parameterization and activation of dispatch signals, according to the VPP's operation mode. It includes the calculation of flexibility, real-time and day-ahead optimization, monitoring and signalling. |
| | Market participation | The ability to calculate valid market bids by taking into account the existing constrains of the grid and the status of its resources as well as available flexibility. |
| Demand Response Management System (DRMS) | Customer/ programme management | The functionality of the system related to configuration of the DR programme definitions and parameters, operation of a variety of programs, monitoring of their performance as well as management of the enrolment of participants. |
| | Resource management | This functionality involves handling of sensitive information' exchanges (registration, availability), the operation of a monitoring system for the health status of the resources, as well as the measurement and verification process. |
| | Provision of flexibility | The ability of the DRMS to forecast its capacity to provide load/supply modifications through DR and provide them as flexibility offerings to the other parties (DSO, VPP Operator) or as bids to the energy market |
| | Event management | This functionality involves scheduling/ parameterization/dispatch of DR events, optimal handling of programme and resource selection according to selected strategies, as well as tracking participation during DR events |
| Energy Hub (EH) | Flexibility services | The capability of the EH to provide flexibility services to DRMS and VPP thus participating in the market. |
| | Optimal planning and control | The capability of calculating the optimal day-ahead schedules for microgrid resources and to adjust their operating point based on the actual energy measurement data. |

(*Continues*)

*Table 1.5 (Continued)*

| Component | Functionality title | Short description |
|---|---|---|
| Local Constraints Management System (LCMS) | Identification of problems in distribution grids | This functionality refers to the ability of the DSO to identify local problems in its distribution grid. |
| | Flexibility offers validation | The capacity of the DSO to technically validate the flexibility offers that can be feasibly activated by the TSO. |
| | Computation of preventive control actions | This functionality concerns the capacity of the DSO to define control actions to deal with local constraints issues. |
| | Flexibility activation | The ability of the TSO to activate flexibility offers technically validated by the DSO. |
| Analytics Platform (AP) | Forecasting | The ability of the system to provide energy-related forecasts to third parties. Related services include PV forecast, wind forecast and load forecast. |
| | DR evaluation analytics | The AP supports the operation of DRMS by providing metrics and calculations for the evaluation of DR programme and DR customers and the optimization of DR event creation. |

## Acknowledgements

This work was supported in part by the European Community's Horizon 2020 Framework Programme under grant agreement No. 646470, SmarterEMC2 Project. The authors would like to thank their colleagues in SmarterEMC2 project for their collaboration.

## Bibliography

[1] Available from http://www.smartgrids.eu

[2] Commission of the European Communities, COM (2009) 519, Investing in the Development of Low Carbon Technologies (SET-Plan), Brussels, 2009.

[3] DECC&OFGEM, "Smart Grid Vision and Routemap", *Smart Grid Forum*, Crown, London, 2014.

[4] Y. Mo, "Cyber-Physical Security of a Smart Grid Infrastructure", *Proceedings of the IEEE*, 100(1), pp. 195–209, Jan. 2012.

[5] E. Loukarakis, J. Bialek and C. Dent. "Investigation of Maximum Possible OPF Problem Decomposition Degree for Decentralized Energy Markets", 2015 *IEEE Power & Energy Society General Meeting*, Denver, CO, 2015, pp. 1–1.

[6] European Commission, Joint Research Centre, Institute for Energy and Transport, Smart Grid Projects Outlook 2014. Available from http://ses.jrc.ec.europa.eu/sites/ses.jrc.ec.europa.eu/files/u24/2014/report/ld-na-26609-en-n_smart_grid_projects_outlook_2014_-_online.pdf
[7] European Commission, COM (2015) 080, A Framework Strategy for a Resilient Energy Union with a Forward-Looking Climate Change Policy, Brussels, 2015.
[8] European Commission, COM (2015) 6317, Towards an Integrated Strategic Energy Technology (SET) Plan: Accelerating the European Energy System Transformation, Brussels, 2015.
[9] ENERGISE – ICT-based Energy Grid Implementation – Smart and Efficient. Available: http://project-energise.eu/. Accessed on: July 19, 2016.
[10] FLEXMETER – Flexible Smart Metering for Multiple Energy Vectors with Active Prosumers. Available: http://flexmeter.polito.it/. Accessed on: July 19, 2016.
[11] FLEXICIENCY – Energy Services Demonstrations of Demand Response, Flexibility and Energy Efficiency, Based on Metering Data. Available: http://www.flexiciency-h2020.eu/. Accessed on: July 19, 2016.
[12] Flex4Grid – Prosumer Flexibility Services for Smart Grid Management. Available: https://www.flex4grid.eu. Accessed on: July 19, 2016.
[13] EMPOWER – Local Electricity Retail Markets for Prosumer Smart Grid Power Services. Available: www.empowerh2020.eu. Accessed on: July 19, 2016.
[14] NOBELGRID – New Cost-Efficient Business Models for Flexible Smart Grids. Available: http://nobelgrid.eu. Accessed on: July 19, 2016.
[15] UPGRID – Real Proven Solutions to Enable Active Demand and Distributed Generation Flexible Integration, through a Fully Controllable Low Voltage and Medium Voltage Distribution Grid. Available: http://upgrid.eu/. Accessed on: July 19, 2016.
[16] SmarterEMC2 – Smarter Grid: Empowering SG Market Actors through Information and Communication Technologies. Available: http://www.smarteremc2.eu/. Accessed on: July 19, 2016.
[17] P2P-SmartTest – Peer to Peer Smart Energy Distribution Networks. Available: http://www.p2psmartest-h2020.eu/. Accessed on: July 19, 2016.
[18] AnyPlace – Adaptable Platform for Active Services Exchange. Available: http://www.anyplace2020.org/. Accessed on: July 19, 2016.
[19] NETFFICIENT – Energy and Economic Efficiency for Today's Smart Communities Through Integrated Multi Storage Technologies. Available: http://netfficient-project.eu/. Accessed on: July 19, 2016.
[20] TILOS – Technology Innovation for the Local Scale, Optimum Integration of Battery Energy Storage. Available: http://www.tiloshorizon.eu/. Accessed on: July 19, 2016.
[21] SENSIBLE – Storage Enabled Sustainable Energy for Buildings and Communities. Available: www.h2020-project-sensible.eu. Accessed on: July 19, 2016.

[22] ELSA – Energy Local Storage Advanced System. Available: http://elsa-h2020.eu. Accessed on: July 19, 2016.
[23] NICEGRID – Un Quartier Solaire Intelligent. Available: www.nicegrid.fr/. Accessed on: July 19, 2016.
[24] US Department of Energy (DOE) (2009). Guidebook for ARRA Smart Grid Program Metrics and Benefits. Available from www.smartgrid.gov/sites/default/files/pdfs/metrics_guidebook.pdf
[25] Network Optimized Distributed Energy Systems (NODES) Projects Description, December 2015, Available from http://arpa-e.energy.gov/sites/default/files/documents/files/NODES_Project_Descriptions.pdf
[26] "Reprint: China and the World's Greatest Smart Grid Opportunity". Available from http://www.greentechmedia.com/articles/read/enter-the-dragon-china-and-the-worlds-greatest-smart-grid-opportunity (access date: 14 Dec 2015).
[27] SGCC. Available from http://www.sgcc.com.cn/ywlm/gsgk-e/gsgk-e/gsgk-e1.shtml (access date: 14 Dec 2015).
[28] Available from http://www.indiasmartgrid.org/pilot.php
[29] Available from http://www.ausgrid.com.au/Common/Customer-Services/In-your-neighbourhood/Network-projects/Network-projects-by-area/Smart-grid-projects/Smart-grid-facts.aspx#.Vm7o0mThDjA
[30] Available from https://en.wikipedia.org/wiki/Smart_grids_in_South_Korea
[31] CEN-CENELEC-ETSI Smart Grid Coordination Group, Smart Grid Reference Architecture, November 2012. Available from http://ec.europa.eu/energy/sites/ener/files/documents/xpert_group1_reference_architecture.pdf

*Part I*

# Smart metering

# *Chapter 2*

# Privacy-preserving data aggregation in smart metering systems

*Fábio Borges*[1]

In this chapter, a survey of protocols for privacy-preserving data aggregation in smart metering systems is presented. They are necessary to protect customers' profiles from adversaries, who can predict customers' behaviour and manipulate them. The state of the art covers the entire cycle of a smart grid with real-time consolidated consumption, billing processes and verifications. However, the real-time consolidated consumption is public in the best protocol. In this work, we find an improved protocol that can hide real-time consolidated consumptions keeping them accessible only to the public utility. In addition, a complexity analysis of algorithms is presented and a simulation with millions of real-world measurements collected by thousands of smart meters is run to validate the theoretical evaluation. Furthermore, the improved protocol slightly outperforms the state of the art in the encryption algorithm, which runs in constrained smart meters. Moreover, such techniques can be applied in several other research areas.

## 2.1 Introduction

Currently, public utilities are transforming their old infrastructure in smart grids. An important step of this process is the deployment of smart meters, which are micro-computers with remote access. More than a computerised power grid, *smart grid is a network of people, computers and sensors in public infrastructures that monitors and manages the usage of commodities* as defined in Reference 1.

The majority of the smart grid projects come from power grids, but other supply networks can become smart grids [1]. Energy suppliers are pushing the development of technologies for smart grids. As a result, suppliers of other commodities can benefit from such technologies.

Smart meters are installed at customers' premises to measure customers' consumptions and send the measurements to public utilities. Besides smart grid, advanced

[1]National Laboratory for Scientific Computing (LNCC – *Laboratório Nacional de Computação Científica* in Portuguese), Petrópolis – RJ 25651-075, Brazil

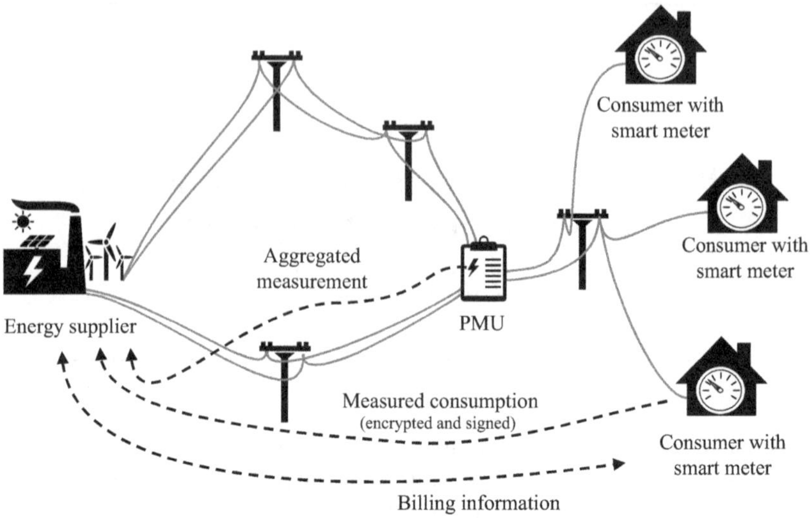


*Figure 2.1 A smart grid with supply network and information network*

metering infrastructure (AMI) and automatic meter reading (AMR) also have the capacity of collecting remote measurements [1]. Another sort of smart metering systems is called phasor measurement unit (PMU), which collects information about the amount and quality of the electricity in a distribution area. Smart metering systems generate an information network in a smart grid. Figure 2.1 depicts a smart grid with an information network in dashed lines and power lines in grey lines, where an energy supplier provides the electricity through transmission and distribution power networks and receives information from a PMU and smart meters. In contrast to a non-smart grid, the supplier can receive signed and encrypted measurements at any time by means of communication protocols to develop new services. Moreover, some protocols enable verification of measurements keeping customers' privacy.

Supplier and customers can benefit from such smart metering systems in many ways, for instance:

1. using surplus electricity by means of dynamic pricing,
2. cutting costs and eliminating peak power plants,
3. detecting energy loss and fraud [1],
4. providing fair distribution of limited resources [1],
5. detecting overload [2],
6. virtualising power lines for multiples suppliers [1],
7. providing many benefits of distributed small power sources [3].

All these benefits can be achieved without the supplier receiving the measurements $m_{i,j}$, because the necessary processing can be done over encrypted measurements $\mathfrak{M}_{i,j}$ also known as ciphertexts or cryptograms.

Although smart grids generate many benefits, security and privacy are their drawbacks. Whereas smart metering systems are connected to an information network, they can have all legacy attacks from the Internet. In addition, frequent measurements can reveal information about customers' private life. The detail of the disclosed information depends on the granularity of measurements [4]. Fine-grained measurements enable even the detection of what a customer is watching on a TV [5]. It is well known that information is power and money. Consequently, such privacy intrusion eventually creates a surveillance society and jeopardises the democracy [6].

Despite a measurement contains the consumption of all appliances in a period, techniques of non-intrusive load monitoring (NILM) can disaggregate the consumption of each appliance [7]. Such techniques are not only theoretical but also practical software implementations [8]. Similarly, smart meters are not laboratory devices but commonplace [1] even for slum in Rio de Janeiro. Smart meters in favelas can ensure fair distribution and can avoid theft of electricity.

On the one hand, smart grids bring many advantages. On the other hand, they also bring substantial disadvantages. To solve this dilemma, a satisfactory level of security can be achieved with classical cryptography and a satisfactory level of privacy can be achieved with privacy-preserving data aggregation.

Section 2.2 lists the acronyms and symbols used in this chapter. Section 2.3 introduces the necessary background to understand the security and privacy model. Section 2.4 reviews the state-of-the-art protocols to enhance privacy. In particular, a technique called asymmetric Dining Cryptographers Network (DC-Net) provides several interesting properties. Section 2.5 presents an improved solution for privacy based on this asymmetric technique. Section 2.6 shows a comparison between related works. Section 2.7 validates the theoretical results with simulations. Section 2.8 draws the conclusions.

## 2.2 Definitions

This section just lists the acronyms and symbols used in this chapter to simplify the reading.

### *2.2.1 List of acronyms*

**ADC-Net** Asymmetric DC-Net.
**AHEP** additive homomorphic encryption primitive.
**AMI** advanced metering infrastructure.
**AMR** automatic meter reading.

**DC-Net** Dining Cryptographers Network.

**NA** Not Applicable.
**NILM** non-intrusive load monitoring.

**PMU** phasor measurement unit.

**SDC-Net** Symmetric DC-Net.

**TPM** Trusted Platform Module.
**TTP** trusted third party.

### 2.2.2 List of symbols

**aggregated measurement** ($a_j$) aggregated measurement of the PMU in the round $j$.

**billed consumption** ($b_i$) consumption in Watts registered in the invoice of the meter $i$, s.t. $b_i = \sum_{j=1}^{J} m_{i,j}$.

**commitment function** ($\mathsf{Commit}$) a commitment scheme defined according to the protocol.
**committed measurement** ($\mathfrak{N}_{i,j}$) committed measurement of $m_{i,j}$, i.e., $\mathfrak{N}_{i,j} = \mathsf{Commit}(m_{i,j})$.
**consolidated committed measurement** ($\mathfrak{T}_i$) consolidated committed measurement $\mathfrak{T}_i = \prod_{j=1}^{J} \mathfrak{N}_{i,j}^{t_j}$.
**consolidated consumption** ($c_j$) decrypted consolidated consumption of the measurements in the round $j$, i.e., $c_j = \mathsf{Dec}(\mathfrak{C}_j)$.

**decryption function** ($\mathsf{Dec}$) a decryption function defined according to the protocol.
**digital signature** ($\mathfrak{S}_{i,j}$) digital signature of the meter $i$ in the round $j$.

**encrypted consolidated consumption** ($\mathfrak{C}_j$) encrypted consolidated consumption of the measurements in the round $j$, s.t. $\mathfrak{C}_j = \prod_{j=1}^{J} \mathfrak{M}_{i,j}$ in the majority of the protocols.
**encrypted measurement** ($\mathfrak{M}_{i,j}$) encrypted measurement of the meter $i$ in the round $j$, i.e., $\mathfrak{M}_{i,j} = \mathsf{Enc}(m_{i,j})$.
**encryption function** ($\mathsf{Enc}$) an encryption function defined according to the protocol.

**group** ($\mathbb{Z}_p^*$) a multiplicative group of integers $\mathbb{Z}$ modulo $p$, where $p$ is prime.

**hash function** ($\mathsf{H}$) a secure hash function s.t. it behaves as a one-way function and has collision resistance.

**integers** ($\mathbb{Z}$) the set of the integer numbers.

**least common multiple function** ($\mathsf{lcm}$) function that returns the least common multiple.

**measurement** ($m_{i,j}$) consumption measured by the meter $i$ in the round $j$.
**meter** ($i$) smart meter with identification $i$.

**number of meters** ($I$) total number of meters $i$.
**number of rounds** ($J$) total number of rounds $j$ per billed consumption $b_i$.

**open function** ($\mathsf{Open}$) a function that opens committed measurement $\mathfrak{N}_{i,j}$ and returns true iff the values are correct.

**round** ($j$) round identification.

**signature function** ($\mathsf{Sign}$) a secure function that returns a digital signature $\mathfrak{S}_{i,j}$.
**whether** ($\stackrel{?}{=}$) the values are correct whether the equation holds.

## 2.3 Background

In the Shannon security model [9], we have a sender and a receiver. In our case, the sender is a meter, which sends measurements to a receiver called supplier. The measurements can be positive for consumption and negative for generation, in case of the customer to own renewable sources. The adversaries know the cryptographic algorithms used and can access the communication channel, excluding the key exchange. Shannon security model was developed about 27 years before the first algorithm for asymmetric cryptography [10]. Even though, it is current and valid for symmetric and asymmetric cryptographic algorithms. Figure 2.2 depicts Shannon security model for a smart metering system. The adversaries have two limitations. They cannot access the keys, and neither messages unencrypted, i.e., before encryption and after decryption. A box in Figure 2.2 limits the adversaries' knowledge. Thus, we assume that meters

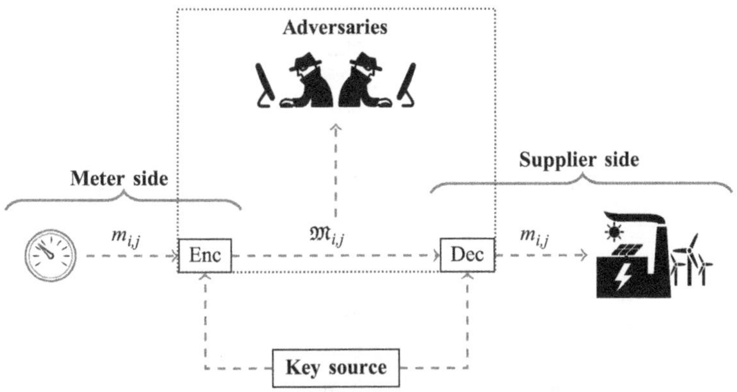


*Figure 2.2 Shannon security model*

have a Trusted Platform Module (TPM), where the cryptographic functions are computed.

In contrast to security, privacy has a different model with its own peculiarity. To protect privacy, the protocols should do more than just protecting a measurement in terms of Shannon, because adversaries might know information about the supplier side. The goal is to hide and protect the measurements from the supplier, which is the receiver. In this case, the supplier can outsource some services and be sure that no employee can leak information. Such protection might sound a contradiction, but there are many solutions for this problem. First, notice that the supplier needs information about the billed consumption and the real-time consolidated consumption in a geographic area, i.e., the consumption of a set of meters subject to a real-time constraint. This consumption is addressed as consolidated consumption $c_j$.

In some countries, the billed consumption $b_i$ comes from a measurement collected on a monthly base while in others on a yearly base. For dynamic pricing, it is necessary at least one measurement for each price. However, the supplier does not need to know them. The billed consumption $b_i$ and consolidated consumption $c_j$ are used to achieve the seven benefits aforementioned. Notice that the supplier does not need individual measurements, but only billed consumptions $b_i$ and consolidated consumptions $c_j$, which are data aggregation of measurements $m_{i,j}$ from customers' consumptions. Table 2.1 shows the measurements $m_{i,j}$ that should be hidden and the last line and last column that should be known by the supplier. The interval for the measurement of all meters is called round. In Table 2.1, the measurements $m_{i,j}$ should be hidden information when the billed consumptions $b_i$ and consolidated consumptions $c_j$ should be accessible.

*Table 2.1 Accessible and hidden information*

| | **1** | **2** | **...** | ***j*** | **...** | ***J*** | $b_i$ |
|---|---|---|---|---|---|---|---|
| 1 | $m_{1,1}$ | $m_{1,2}$ | $\cdots$ | $m_{1,j}$ | $\cdots$ | $m_{1,J}$ | $\sum_{j=1}^{J} m_{1,j}$ |
| 2 | $m_{2,1}$ | $m_{2,2}$ | $\cdots$ | $m_{2,j}$ | $\cdots$ | $m_{2,J}$ | $\sum_{j=1}^{J} m_{2,j}$ |
| $\vdots$ | $\vdots$ | $\vdots$ | $\ddots$ | $\vdots$ | $\ddots$ | $\vdots$ | $\vdots$ |
| $i$ | $m_{i,1}$ | $m_{i,2}$ | $\cdots$ | $m_{i,j}$ | $\cdots$ | $m_{i,J}$ | $\sum_{j=1}^{J} m_{i,j}$ |
| $\vdots$ | $\vdots$ | $\vdots$ | $\ddots$ | $\vdots$ | $\ddots$ | $\vdots$ | $\vdots$ |
| $I$ | $m_{I,1}$ | $m_{I,2}$ | $\cdots$ | $m_{I,j}$ | $\cdots$ | $m_{I,J}$ | $\sum_{j=1}^{J} m_{I,j}$ |
| $c_j$ | $\sum_{i=1}^{I} m_{i,1}$ | $\sum_{i=1}^{I} m_{i,2}$ | $\cdots$ | $\sum_{i=1}^{I} m_{i,j}$ | $\cdots$ | $\sum_{i=1}^{I} m_{i,J}$ | |

Keeping customers' privacy, even the supplier cannot discover the measurements $m_{i,j}$ with a NILM technique and relate them with the customers. Even if the number of meters $I$ is relatively small and the supplier can infer the measurements $m_{i,j}$, at least, the supplier should not be able to identify which measurement belongs to which meter. Similarly as mentioned, the supplier should keep customers' privacy to safeguard against leakages, scandal, complaints, etc.

Counter-intuitively, the number of meters $I$ and the number of rounds $J$ should be as large as possible to ensure the effectiveness of privacy-preserving protocols by means of data aggregation [1]. Such protocols are composed of three algorithms, namely: encryption, aggregation and decryption. Each meter $i$ encrypts its measurement $m_{i,j}$. Subsequently, the encrypted measurements $\mathfrak{M}_{i,j}$ are aggregated generating an encrypted consolidated consumption $\mathfrak{C}_j$, which is sent for the supplier to decrypt it. After the decryption, the supplier has a consolidated consumption $c_j$. Figure 2.3 depicts the privacy model for data aggregation with a box delimiting the adversaries' knowledge. However, protocols might present different assumptions, e.g., some protocols assume that the adversaries have no access to the encrypted measurements $\mathfrak{M}_{i,j}$ [11–13].

Besides data aggregation, one can use a battery to protect privacy. It works very well to mask small fluctuations in the consumption pattern. In addition, batteries help to improve the power grid's load balancing. However, batteries are expensive and they cannot keep the same privacy level that we have nowadays, because measurements are collected monthly or yearly in a non-smart grid. Batteries should keep the houses for a month and yet adversaries can detect when a customer usually has vacation. Moreover, batteries create a trade-off [14–16] between privacy and money, because batteries should recharge on the cheapest electricity prices but recharging in such moments might leak private information.

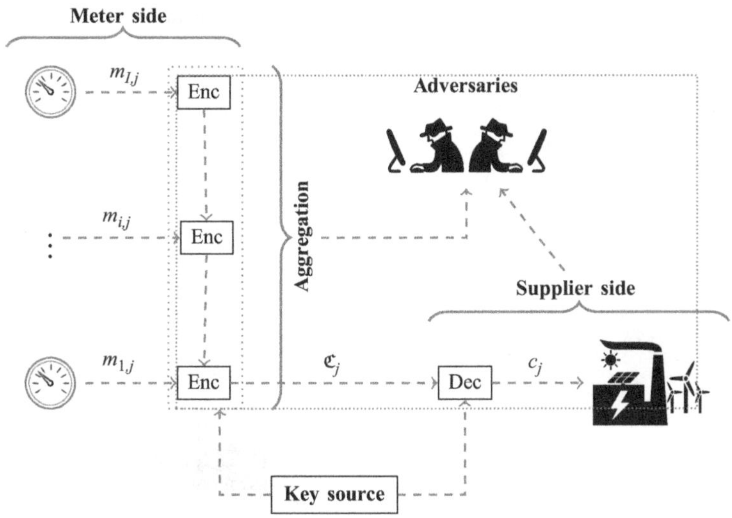


*Figure 2.3 Data aggregation privacy model*

## 2.4 State-of-the-art protocols

This section reviews the state of the art of privacy-preserving protocols, which use data aggregation, namely: homomorphic encryption, commitments, Symmetric DC-Nets (SDC-Nets) and Asymmetric DC-Nets (ADC-Nets).

### 2.4.1 Homomorphic encryption

Additive homomorphic encryption primitives (AHEPs) enable the addition of measurements $m_{i,j}$ over encrypted measurements $\mathfrak{M}_{i,j}$. Thus, an operation over encrypted measurements $\mathfrak{M}_{i,j}$ results in the addition of measurements $m_{i,j}$, i.e.,

$$\mathsf{Enc}(m_{1,j}) \cdot \mathsf{Enc}(m_{i,j}) \cdot \mathsf{Enc}(m_{I,j}) = \mathfrak{M}_{0,j} \cdot \mathfrak{M}_{i,j} \cdot \mathfrak{M}_{I,j} = \mathsf{Enc}(m_{1,j} + m_{i,j} + m_{I,j})$$

The act of adding the measurements $m_{i,j}$ by encrypted measurements $\mathfrak{M}_{i,j}$ is known as aggregation, which can be computed in several ways, for instance, without third party using in-network aggregation [17], using third party [18] and using network devices [19]. Figure 2.4 depicts an abstract aggregator, which receives encrypted measurements $\mathfrak{M}_{i,j}$ and sends encrypted consolidated consumptions $\mathfrak{C}_j$ to a supplier.

The most suitable AHEP is Paillier encryption [20], where measurements $m_{i,j}$ are in $\mathbb{Z}_n$ and $n$ is the product of two safe primes $p$ and $q$, while encrypted measurements $\mathfrak{M}_{i,j}$ are in $\mathbb{Z}_{n^2}$. Thus, the encryption function $\mathsf{Enc}(m_{i,j})$ of a measurement $m_{i,j}$ is given by

$$\begin{aligned} \mathsf{Enc} : \mathbb{Z}_n \times \mathbb{Z}_n^* &\longmapsto \mathbb{Z}_{n^2} \\ \mathsf{Enc}(m_{i,j}, r_{i,j}) &\longmapsto g^{m_{i,j}} \cdot r_{i,j}^n \mod n^2 \end{aligned} \tag{2.1}$$

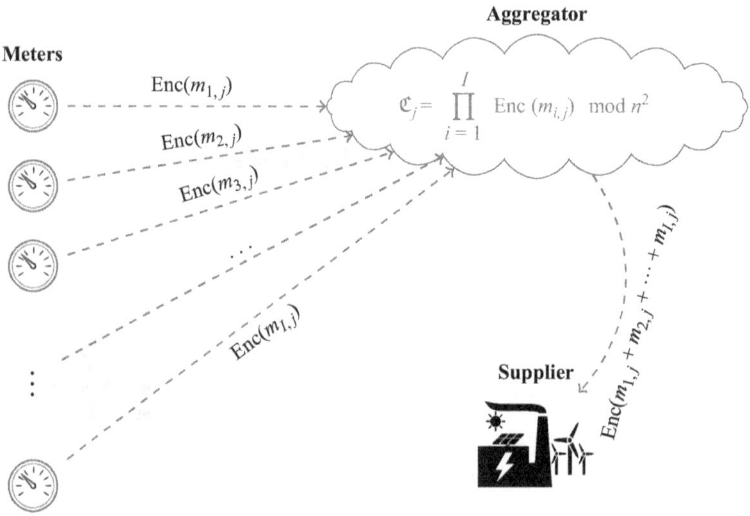


*Figure 2.4 Abstract aggregator for AHEP*

where $g$ is a random number chosen by the supplier in the setup, s.t. $n$ divides the order of $g \in \mathbb{Z}^*_{n^2}$ and $r_{i,j}$ is a random number chosen by meters $i$ in rounds $j$.

Consequently, the aggregation is given by

$$\begin{aligned}\mathfrak{C}_j &= \prod_{i=1}^{I} \mathsf{Enc}(m_{i,j}, r_{i,j}) = g^{m_{1,j}} r^n_{1,j} \cdots g^{m_{I,j}} \cdot r^n_{I,j} \mod n^2 \\ &= g^{\sum_{i=1}^{I} m_{i,j}} \prod_{i=1}^{I} r^n_{i,j} \mod n^2 = \mathsf{Enc}\left(\sum_{i=1}^{I} m_{i,j}, r_{i,j}, \prod_{i=1}^{I} r_{i,j}\right) \\ &= \mathsf{Enc}(c_j, R)\end{aligned} \tag{2.2}$$

where $R = \prod_{i=1}^{I} r_{i,j}$.

To transform an encrypted consolidated consumption $\mathfrak{C}_j$ into a consolidated consumption $c_j$, we just apply the decryption function given by

$$\begin{aligned}\mathsf{Dec} &: \mathbb{Z}_{n^2} \mapsto \mathbb{Z}_n \\ \mathsf{Dec}(\mathfrak{C}_j) &\mapsto \mathrm{L}(\mathfrak{C}^\lambda_j \mod n^2) \cdot d \mod n,\end{aligned} \tag{2.3}$$

where $\mathrm{L}(u) = (u-1)/n$, $d = \mathrm{L}(g^\lambda \mod n^2)^{-1}$ and $\lambda = \lambda(n) = \mathsf{lcm}(p-1, q-1)$ is defined by Carmichael's function.

Notice that the random number values are not relevant to the decryption function, because of the following property:

$$\forall\, w \in \mathbb{Z}^*_{n^2}, \quad \begin{cases} w^\lambda \equiv 1 \mod n \\ w^{n\lambda} \equiv 1 \mod n^2 \end{cases} \tag{2.4}$$

We are used to saying that schemes dependent on random numbers are probabilistic encryption schemes. AHEPs should be probabilistic encryption, because encryption functions of the same measurement values should return different encrypted measurements, i.e.,

$$\mathfrak{M}_{i,j} \neq \mathfrak{M}_{\alpha,\beta} \quad \text{and} \quad m_{i,j} = \gamma \quad \forall\, i, j, \alpha, \beta, \gamma$$

For the sake of simplicity, we are used to omitting the random number in the notation. Thus, we write $\mathsf{Enc}(m_{i,j})$ instead of $\mathsf{Enc}(m_{i,j}, r_{i,j})$ and we have:

$$\mathsf{Dec}\left(\prod_{i=1}^{I} \mathsf{Enc}(m_{i,j}) \mod n^2\right) = \sum_{i=1}^{I} m_{i,j} \mod n = c_j \tag{2.5}$$

We also omit to write the keys in the functions, but the private key for Paillier is $d$ and the public key is $\{n, g\}$.

## 2.4.2 Commitments

Commitment schemes provide verification of the values previously sent. In addition, some schemes are homomorphic. In contrast to homomorphic encryption, they cannot encrypt and decrypt measurements, but they can commit and verify them. A checking function of a committed measurement is also known as an open function of a commitment. Thus, commitment schemes can enable billing verification, i.e., supplier and each customer associated to a meter $i$ can verify whether the billed consumption $b_i$ is correct.

Pedersen commitment [21] is additive homomorphic and information-theoretic secure. The commit function is given by

$$\mathfrak{M}_{i,j} = \mathsf{Commit}(m_{i,j}, r_{i,j}) = g^{m_{i,j}} \cdot h^{r_{i,j}} \mod p \tag{2.6}$$

where $g$ is a generator, $h$ is generated by $g$ and $r_{i,j}$ is a random number in $\mathbb{Z}_q^*$ such that $q$ divides $p-1$ and $p$ and $q$ are large primes. The values of $g$ and $h$ are public, but the values of $m_{i,j}$ and $r_{i,j}$ are private. It was used to construct a protocol that enables dynamic pricing, where the supplier does not need to receive the measurements $m_{i,j}$ [22]. Other protocols can enable dynamic pricing with similar strategy. To clarify the strategy, the definition of billed consumption $b_i$ is changed for this protocol, namely:

$$b_i = \sum_{j=1}^{J} m_{i,j} \cdot t_j \tag{2.7}$$

where $t_j$ is the electricity tariff sent from the supplier to meters for a round $j$. In particular, this protocol assumes a private component $\text{PC}_i$ connected to a meter $i$ and has data aggregation in the lines of Table 2.1, instead of columns.

For every round $j$, a meter $i$ sends its measurement $m_{i,j}$, a random number $r_{i,j}$, its commitment $\mathfrak{M}_{i,j} = \mathsf{Commit}(m_{i,j})$ and its digital signature $\mathfrak{S}_{i,j}$ for $\mathfrak{M}_{i,j}$ to its $\text{PC}_i$, which computes $b_i$ and

$$r' = \sum_{j=1}^{J} r_{i,j} \cdot t_j \tag{2.8}$$

Afterwards, each $\text{PC}_i$ sends the billed consumption $b_i$, $r'$, committed measurements $\mathfrak{M}_{i,j}$ and digital signatures $\mathfrak{S}_{i,j}$ to the supplier, which computes:

$$\mathfrak{T}_i = \prod_{j=1}^{J} \mathfrak{M}_{i,j}^{t_j} = \prod_{j=1}^{J} \mathsf{Commit}(m_{i,j})^{t_j} \tag{2.9}$$

Figure 2.5 depicts the communication between a meter $i$, its private component and its supplier. Adversaries neither manipulate the tariff $t_i$ nor spoof the communication between a meter and its PC.

At the end of the protocol, the supplier uses the digital signatures $\mathfrak{S}_{i,j}$ to verify if the PC computed the commitments correctly. Moreover, a pair of meter and

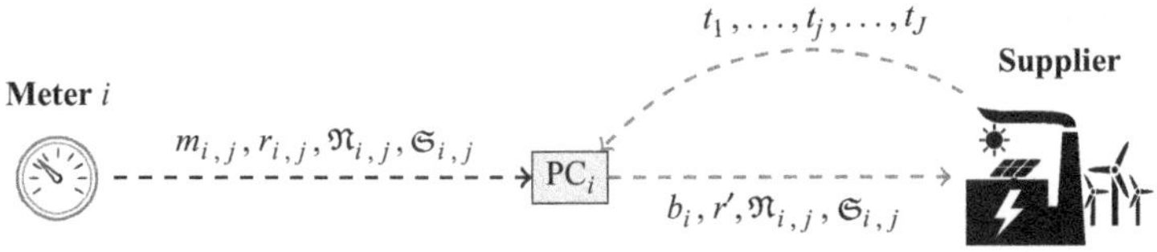


*Figure 2.5 Commitment for billing data aggregation*

supplier can verify whether the billed consumption $b_i$ is correct with the open function given by

$$\mathsf{Open}(\mathfrak{T}_i, b_i, r') \Rightarrow g^{b_i} \cdot h^{r'} \stackrel{?}{=} g^{\sum_{i=1}^{I} m_{i,j}} \cdot h^{\sum_{i=1}^{I} r_{i,j}} = \mathfrak{T}_i \tag{2.10}$$

Therefore, the function returns true iff the values are correct.

With slight changes in this protocol, the pricing can float even by meter $i$. We just need to replace $t_j$ with $t_{i,j}$ to enable positive pricing discrimination.

In summary, we just need to consider the consumption in monetary value instead of watts to enable dynamic pricing or positive pricing discrimination. For that reason, they are discussed only in this protocol. The remainder of this chapter does not present the equations for dynamic pricing nor positive pricing discrimination.

### *2.4.3 Symmetric DC-Net (SDC-Net)*

SDC-Net is a protocol for privacy-preserving data aggregation introduced in 1988 and can be unconditional secure also known as information-theoretic secure [23]. Its name came from Dining Cryptographers Network (DC-Net) and it is symmetric because the technique is based on symmetric cryptography. The protocol was introduced by a problem in which cryptographers want to know if one of them paid a restaurant bill, but they do not want to disclose who paid. To solve this problem, they send a symmetric key to each other and agree on sending either one or zero. Who paid sends one and the others send zero. Afterwards, they add the sent keys to the subtraction of the keys received. The cryptographer that paid also adds one to the result. Finally, each cryptographer reveals the result and the sum of the results is one if one of the cryptographers paid and zero otherwise.

This section presents two of many ways to use SDC-Net in smart grid scenarios [24]. Instead of adding zeros and ones, the protocols for smart grid add measurements. The first one is called Low-Overhead Protocol (LOP), which uses 32 bits of a hash function of the symmetric keys $k_{i,o}$ concatenated with the rounds $j$, where a meter $i$ sent the key to the meter $o$. Mathematically, we have:

$$\mathfrak{M}_{i,j} = \mathsf{Enc}(m_{i,j}) = m_{i,j} + \sum_{o \in \mathcal{M}-\{i\}} \mathsf{H}(k_{i,o} \,||\, j) - \mathsf{H}(k_{o,i} \,||\, j) \tag{2.11}$$

where $||$ denotes string concatenation, $\mathcal{M}$ is the set of meters and $\mathsf{H}$ is the hash function.

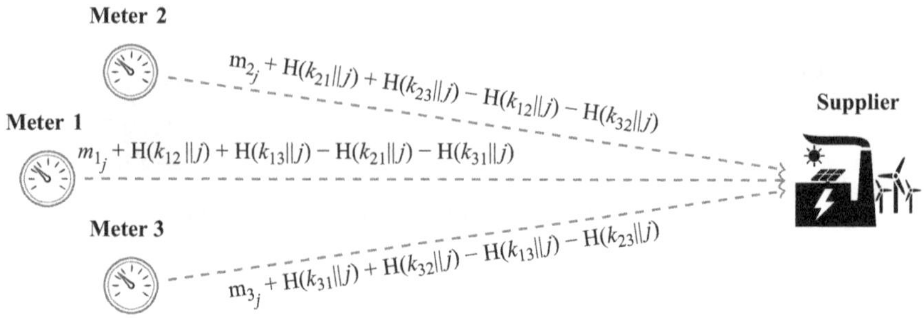


*Figure 2.6 LOP: an SDC-Net*

The decryption happens along with the aggregation, i.e.,

$$c_j = \sum_{i=1}^{I} \mathfrak{M}_{i,j} \tag{2.12}$$

Without loss of generality, Figure 2.6 depicts how LOP implements an SDC-Net with three meters.

An issue of symmetric cryptography is the number of keys, which grows quadratically with the number of meters. Thus, we should assume trust between the meters to reduce the number of keys, for instance, the meter one can trust in the meter two and can agree on a key only with the meter two. The protocol works in the same way, but the meter two can reveal measurements from the meter one. The connection between the meters is related with graph theory. Hence, we say that LOP is a fully connected SDC-Net, which enforces privacy, i.e., no one can disclose a single measurement $m_{i,j}$ and only the collusion of $I - 1$ meters can disclose measurements from the unique meter that is not in the collusion, where $i$ is the number of meters in the aggregation. A paranoiac information security analyst might consider the situation of all against one a vulnerability. However, this is a strong result in comparison with AHEPs, which need only the collusion of two parties to disclose measurements $m_{i,j}$. Total enforcement of privacy or everlasting privacy does not exist. Even for a fully connected SDC-Net that is unconditional secure, adversaries can use Table 2.1 to estimate measurements without collusion [1] and collusion of all against one is always effective.

The second protocol presented in this section is a star SDC-Net, which was presented as part of a protocol called iKUP [12]. For simplicity, let us call the iKUP's SDC-Net as iKUP. Thus, we can say that iKUP can behave as an AHEP. Moreover, it is a lower bounder for processing time of encryption function in classical computers [1]. In summary, no other protocol can be faster than iKUP to encrypt consolidated consumptions $c_j$.

Using iKUP, each meter $i$ agrees on a key $k_i$ with the supplier. Afterwards, for each round, the meters encrypt computing:

$$\mathfrak{M}_{i,j} = \mathsf{Enc}(m_{i,j}) = m_{i,j} + \mathsf{H}(k_i \,||\, j) \tag{2.13}$$

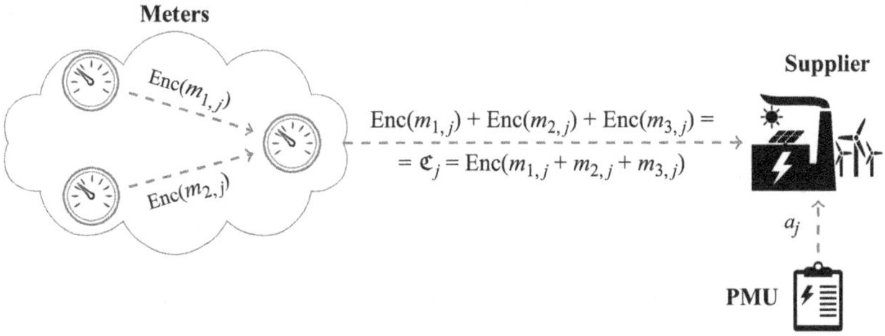


*Figure 2.7 iKUP: a star SDC-Net behaving as AHEPs*

where || denotes string concatenation and H is the hash function chosen for the protocol. Comparing the processing time for the encryption functions, iKUP is $2(I-1)$ times faster than LOP.

The data aggregation happens in a process called in-network aggregation, which is used in protocols for smart grid [17,25,26]. In summary, the meters send the measurements to each other computing:

$$\mathfrak{C}_j = \sum_{i=1}^{I} \mathfrak{M}_{i,j} \tag{2.14}$$

until the last meter, which sends the consolidated consumption $c_j$ to the supplier. Once received $c_j$, the supplier decrypts computing:

$$c_j = \mathsf{Dec}(\mathfrak{C}_j) = \sum_{i=1}^{I} \mathfrak{M}_{i,j} - \mathsf{H}(k_i \,||\, j) \tag{2.15}$$

Figure 2.7 depicts the communication model for iKUP and introduces an aggregated measurement $a_j$, collected by a PMU. With $a_j$ and $c_j$ in a round $j$, the supplier can detect frauds and failures in the smart grid, because discounting the transmission costs $\epsilon_j$, we have $a_j - \epsilon_j = c_j$. Therefore, aggregated measurements $a_j$ represent expected consolidated consumptions $c_j$. This can be seen in Figure 2.1.

Similar to AHEP, we need to assume that the supplier cannot spoof the communication between the meters. In addition, every technique used with AHEP to aggregate can also be used with iKUP. They have the same communication model, which may be seen in Figure 2.4.

**We do not need AHEPs**

iKUP is a star SDC-Net and can behave as AHEPs. Moreover, iKUP is the fastest scheme to encrypt measurements for privacy-preserving data aggregation. Therefore, we do not need AHEPs to add up measurements $m_{i,j}$ over encrypted measurements $\mathfrak{M}_{i,j}$.

On the one hand, a star SDC-Net is faster than an AHEP, which is faster than a fully connected SDC-Net. On the other hand, only fully connected SDC-Net enforces privacy hitherto.

### 2.4.4 Asymmetric DC-Net (ADC-Net)

An ADC-Net was premature introduced in a short paper in 2014 [27]. This section presents an ADC-Net presented posteriorly [1].

Excluding unconditional security, ADC-Nets can have the same properties than SDC-Nets and other properties, e.g., they are non-iterative, customers do not need to be reliant on trusted third party (TTP), meters store permanent keys, each meter store only one key, time complexity is most polynomial, they are scalable without iteration over the number of meters $I$ in the encryption, meters can send the minimum number of messages, which can be signed and sent directly to the supplier. Moreover, ADC-Nets use asymmetric cryptography enabling two important properties: verification as Pedersen commitment can do and enforcement of privacy as fully connected SDC-Net can do.

With enforcement of privacy, meters can sign and send their encrypted measurements $\mathfrak{M}_{i,j}$ directly to the supplier. Thus, the supplier can detect any issue in the information network. Meters can send the minimal number of messages in the communication channel. When using in-network aggregation, meters can increase the number of messages, which can be only signed with homomorphic signatures [11]. Figure 2.8 depicts a communication model for ADC-Nets, where meters send their encrypted measurements $\mathfrak{M}_{i,j} = \mathsf{Enc}(m_{i,j})$ concatenated with the respective digital signatures $\mathfrak{S}_{i,j} = \mathsf{Sign}(\mathfrak{M}_{i,j})$ in a round $j$. As stated, the supplier can compare the aggregated measurement $a_j$ with consolidated consumption $c_j$ to detect issues in the

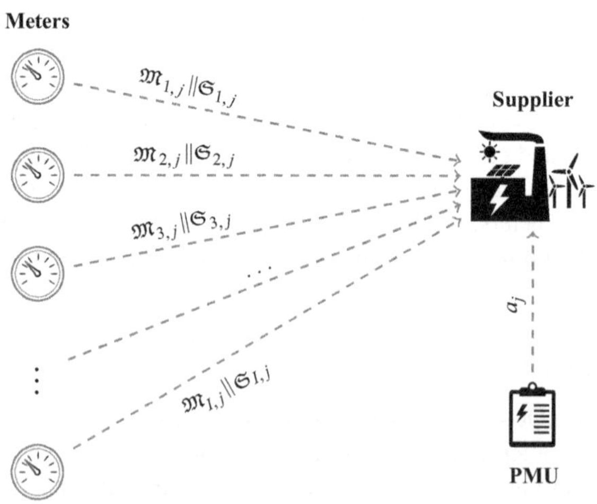


*Figure 2.8 Communication model for ADC-Nets*

information network. Notice that even if a sensor does not measure correctly, this failure will be detected.

Similar to other asymmetric cryptography schemes, the process to set up the keys is a hard part, but we can simplify with the aid of an SDC-Net. To set up, each meter $i$ chooses its private key $k_i$ and the supplier chooses its private key $k_0$. Subsequently, they use an SDC-Net to compute:

$$s = \sum_{i=0}^{I} k_i \tag{2.16}$$

This set-up phase is done only once.

For each round $j$, each meter $i$ encrypts its measurement $m_{i,j}$ with:

$$\begin{aligned} \mathsf{Enc} : \mathbb{Z}_n &\rightarrow \mathbb{Z}_{n^2} \\ \mathsf{Enc}(m_{i,j}) &\mapsto (1+n)^{m_{i,j}} \cdot g^{h_j + k_i} \mod n^2 \end{aligned} \tag{2.17}$$

where $n$ is a product of two safe primes generated in the set-up phase by the meters [28], $h_j = \mathsf{H}(j)$, $\mathsf{H}$ is a hash function and $g$ is a random number invertible over $\mathbb{Z}_n$. In addition, they sign and send their encrypted measurement $\mathfrak{M}_{i,j}$ directly to the supplier.

Meanwhile, the supplier can aggregate the encrypted measurements computing:

$$\mathfrak{C}_j = \prod_{i=1}^{I} \mathfrak{M}_{i,j} \mod n^2 \tag{2.18}$$

If a digital signature $\mathfrak{S}_{i,j}$ is not valid or a message does not come in a time window, the supplier can take measures, for instance, requesting for the meter to re-send the message or excluding the meter from the protocol.

At the end of the data aggregation, the supplier can decrypt computing:

$$\begin{aligned} \mathsf{Dec} : \mathbb{Z}_{n^2} &\rightarrow \mathbb{Z}_n \\ \mathsf{Dec}(\mathfrak{C}_j) &\mapsto \frac{\left(\mathfrak{C}_j \cdot g^{-I \cdot h_l - s + k_0} \mod n^2\right) - 1}{n} \end{aligned} \tag{2.19}$$

where $s = \sum_{i=0}^{I} k_i$ is established in the set-up phase.

At the end of a week, a month or a year, customers and suppliers want to be sure that the bills are correct. To verify, a meter $i$ presents the billed consumption $b_i$ given by

$$b_i = \sum_{j=1}^{J} m_{i,j}$$

or

$$b_i = \sum_{j=1}^{J} m_{i,j} \cdot t_{i,j} \tag{2.20}$$

for dynamic pricing and positive pricing discrimination. In addition, it presents the respective verifier $v_i$ given by

$$v_i = \prod_{j=1}^{J} g^{h_j + k_i} \mod n^2 \tag{2.21}$$

The supplier presents the product of encrypted measurements $\mathfrak{M}_{i,j}$ that belong to the meter $i$, i.e.,

$$\mathfrak{B}_i = \prod_{j=1}^{J} \mathfrak{M}_{i,j} \mod n^2 \tag{2.22}$$

At the end, they can open the committed values to verify whether the function returns true computing:

$$\mathsf{Open}(b_i, v_i, \mathfrak{B}_i) \Rightarrow \mathfrak{B}_i \stackrel{?}{=} (1+n)^{b_i} \cdot v_i \mod n^2 \tag{2.23}$$

The open function returns true iff the values are correct.

## 2.5 An improved ADC-Net

The previous ADC-Net has a limitation, because it assumes that consolidated consumptions $c_j$ are public and adversaries can compute $c_j$ without knowledge of $s$ [1]. However, some suppliers might dislike this assumption, because $c_j$ represents the revenue in a geographic area. For this reason, this improved ADC-Net allows only the supplier to decrypt the consolidated consumptions $c_j$. This issue would be solved replacing $h_j + k_i$ with $h_j \cdot k_i$ in the previous ADC-Net. However, this little change generates performance issues due to the size of the exponent in the modular exponentiation [29]. To have an ADC-Net as efficient as the previous, let us consider the following encryption function:

$$\begin{aligned} \mathsf{Enc} &: \mathbb{Z}_n \longmapsto \mathbb{Z}_{n^2} \\ \mathsf{Enc}(m_{i,j}) &\longmapsto g^{m_{i,j}} \cdot h_j^{k_i} \mod n^2 \end{aligned} \tag{2.24}$$

where $n$ is the product of two safe primes generated by the meters [28], $g$ is a multiple of $n$ chosen by the supplier in the set-up phase, $h_j = \mathsf{H}(j)$ and $\mathsf{H}$ is a hash function.

Similarly, the supplier computes the data aggregation given by

$$\mathfrak{C}_j = \prod_{i=1}^{I} \mathfrak{M}_{i,j} \mod n^2 \tag{2.25}$$

After the aggregation, the supplier computes the decryption function given by

$$\begin{aligned} \mathsf{Dec} &: \mathbb{Z}_{n^2} \mapsto \mathbb{Z}_n \\ \mathsf{Dec}(\mathfrak{C}_j) &\mapsto \mathrm{L}\left(\left(\mathfrak{C}_j \cdot h_j^{n-s+k_0}\right)^{\lambda} \mod n^2\right) \cdot d \mod n \end{aligned} \tag{2.26}$$

where $s$ is the sum of the keys defined in the set-up phase, $\mathrm{L}(u) = (u-1)/n$, $k_0$ is the supplier's private key as well as $d = \mathrm{L}(g^{\lambda} \mod n^2)^{-1}$ and $\lambda = \lambda(n) = \mathsf{lcm}(p-1, q-1)$ that is defined by Carmichael's function.

The verification process is similar to the previous ADC-Net. In addition, the improved ADC-Net has equivalent performance than the previous ADC-Net considering parameter sizes equal. The most costly operation is the modular exponentiation [29]. In both ADC-Nets, the encryption function has two modular exponentiation, but the sizes are different with a slight advantage for the improved ADC-Net. Similarly, the decryption of this improved ADC-Net is slightly slow than the previous ADC-Net.

The construction of this ADC-Net comes from the Paillier encryption, which is an AHEP. Indeed, ADC-Nets are generalisations of AHEP and there is a technique to transform AHEPs in ADC-Nets [1]. Therefore, we can quickly construct at least one ADC-Net for each AHEP.

## 2.6 Comparison with related work

This section presents a theoretic comparison between the improved ADC-Net introduced in Section 2.5 and the state of the art presented in Section 2.4. Six protocols are compared to each other, namely:

1. Paillier [17] is the best representative for AHEPs and is reviewed in Section 2.4.1.
2. Pedersen [22] is an application of Pedersen commitment [21] in smart grids and is reviewed in Section 2.4.2.
3. LOP [24] is the best usage of fully connected SDC-Net and is reviewed in Section 2.4.3.
4. iKUP [12] is the best usage of a star SDC-Net and is reviewed in Section 2.4.3.
5. ADC-Net [1] is the state of the art and is reviewed in Section 2.4.4.
6. This ADC-Net is the improved and innovative ADC-Net introduced in Section 2.5.

To compare, analyses of privacy, communication and performance issues are presented. At the end of this section, we can analyse different properties for the main techniques of privacy-preserving data aggregation, i.e., SDC-Net, AHEP and ADC-Net.

### 2.6.1 Privacy

Let us analyse privacy issues. To start, a third-party aggregator is a single point of failure. Thus, protocols that do not require a third-party aggregator are better than the others. If a supplier colludes with an aggregator, they can disclose all encrypted measurement $\mathfrak{M}_{i,j}$ intended for them with AHEPs. In contrast, DC-Nets and Pedersen require a collusion of all against one to disclose measurements. Only DC-Nets and Pedersen enforce privacy. In addition, only DC-Nets and Pedersen provide verification

*Table 2.2 Comparison of privacy*

| Protocol | Aggregator | Collusion | Verification | Public $c_j$ |
|---|---|---|---|---|
| Paillier [17] | Yes | 2 | No | No |
| Pedersen [22] | No | $I-1$ | Yes | NA |
| LOP [24] | No | $I-1$ | No | No |
| iKUP [12] | Yes | 2 | No | No |
| ADC-Net [1] | No | $I$ | Yes | Yes |
| This ADC-Net | No | $I$ | Yes | No |

for the computation of all billed consumption $b_i$, consolidated consumption $c_j$ and set of measurements $m_{i,j}$. Considering aggregated measurements $a_j$, they can also verify from where incorrect measurements $m_{i,j}$ are coming, but without disclosing other measurements $m_{i,j}$ [1]. An issue in the state of the art motivates the creation of a new ADC-Net. Specifically, consolidated consumptions are public for the ADC-Net [1] but some suppliers might prefer a hidden $c_j$. Hence, this new ADC-Net provides a hidden $c_j$. Both ADC-Nets have equivalent processing time. Therefore, it is more suitable.

Table 2.2 summarises the comparison of privacy issues between the protocols for privacy-preserving data aggregation. Notice that Pedersen does not decrypt. Thus, in its spot is denoted Not Applicable (NA).

## 2.6.2 Communication

Performance is an important aspect of protocols for privacy-preserving data aggregation. Indeed, meters should have a constrained hardware for economic reasons. Let us discuss about performance of data communication. Considering enforcement of privacy with fully connected SDC-Net, DC-Nets request that each meter sends a key to each other. Thus, the communication complexity grows quadratically with respect to the number of meters $I$. The others have linear complexity. Using LOP, each meter has to store the biggest number of keys in comparison to other protocols. Pedersen does not have a key, but each meter $i$ needs to store a verifier $v_i$. ADC-Net needs to store a key and a verifier. Considering that the $PC_i$ introduced in Section 2.4.2 is attached to the meter $i$, only Pedersen and DC-Nets can sign their measurements and send them directly to the supplier. Consequently, meters using the other protocols can disrupt the communication sending a huge number due to failure or malice. To avoid disruption, protocols should have verification and digital signature $\mathfrak{S}_{i,j}$ of encrypted measurements $\mathfrak{M}_{i,j}$.

Table 2.3 summarises the differences between the protocols with respect to data communication. Notice that some of them go beyond performance and presents issues related with security and privacy.

*Table 2.3 Comparison of communication*

| Protocol | Setup | Key | Direct | Disruption |
|---|---|---|---|---|
| Paillier [17] | $O(I)$ | 1 | No | Yes |
| Pedersen [22] | $O(I)$ | NA | Yes | No |
| LOP [24] | $O(I^2)$ | $2(I-1)$ | Yes | Yes |
| iKUP [12] | $O(I)$ | 1 | No | Yes |
| ADC-Net [1] | $O(I^2)$ | 1 | Yes | No |
| This ADC-Net | $O(I^2)$ | 1 | Yes | No |

*Table 2.4 Comparison of processing time*

| Protocol | Encryption (Commit) | Aggregation | Decryption (Open) |
|---|---|---|---|
| Paillier [17] | $O(\log(n))$ | $O(I)$ | $O(\log(n))$ |
| Pedersen [22] | $O(\log(k))$ | $O(J)$ | $O(\log(k))$ |
| LOP [24] | $O(I)$ | NA | $O(I)$ |
| iKUP [12] | $O(1)$ | $O(I)$ | $O(I)$ |
| ADC-Net [1] | $O(\log(k+h_j))$ | $O(I)$ | $O(\log(k))$ |
| This ADC-Net | $O(\log(k))$ | $O(I)$ | $O(\log(n \cdot \lambda))$ |

## 2.6.3 Processing time

To save money and electricity, protocols should be very efficient. They should fulfil their tasks as fast as possible. iKUP has the fastest encryption function. However, iKUP only computes what AHEPs compute, i.e., consolidated consumptions $c_j$. Whereas the keys represented by $k$ are much smaller than the group size represented by $n$, Pedersen and ADC-Nets have the second fastest encryption functions with this ADC-Net slightly faster than ADC-Net to encrypt. LOP needs an interaction per meter in its encryption function. Thus, LOP is not scalable. Notice that the set-up phase of ADC-Nets might use LOP, might use iKUP or other technique to aggregate the keys. Excluding LOP, the aggregation of measurements is equivalent for all protocols, but Pedersen aggregate by rounds $j$ instead of meters $i$. LOP does not have aggregation because it happens along with the decryption function. ADC-Net and Pedersen are faster than Paillier, which is faster than this ADC-Net to decrypt. The performance of iKUP and LOP in the decryption function depends on the number of meters $I$. Although this ADC-Net has the most time-consuming decryption function, this process happens only once per round $j$. Furthermore, the supplier has powerful computers in contrast with constrained meters.

Table 2.4 summarises the time complexity required for protocols to encryption, aggregation and decryption, but for Pedersen, it means commit, aggregation and open.

**The fastest of all**

iKUP has complexity $O(1)$. Thus, we cannot develop a protocol for encryption of consolidated consumption $c_j$ faster than iKUP in classical computers. Perhaps, quantum reality can give better results [30]. However, such technology is still away from ubiquitous computing. Therefore, the processing time of iKUP's encryption function is a lower bound to compare efficient protocols.

### 2.6.4 Techniques

We have seen the comparison between the most important protocols for privacy-preserving data aggregation in a smart grid scenario. However, many other protocols are being created. For this reason, we should compare the three techniques used to construct the protocols, namely: SDC-Net, AHEP and ADC-Net. We have seen that AHEPs are not resistant against collusion, but DC-Nets can be. Furthermore, DC-Nets have a concept of trusted set of meters that enables an SDC-Net to behave as AHEPs. Fully connected SDC-Nets are not scalable, because their processing time depends on the number of meters $I$ and the number of keys grows quadratically with respect to the number of meters $I$. However, the meters only need to store a linear number of keys with respect to the number of meters $I$. AHEPs simplify the management of keys with only one pair. However, the minimal number of keys to enforce privacy is $O(I)$. ADC-Nets differ from other techniques in two important properties: meters cannot disrupt the communication, because they can be detected and the whole process can be verified as commitment schemes do. Therefore, ADC-Net is a technique that overcomes the others in all main points.

Table 2.5 summarises the comparison between the techniques used in many protocols for privacy-preserving data aggregation in a smart grid scenario. Moreover, we can see in Table 2.5 that ADC-Net overcomes independent of application scenario.

*Table 2.5 Comparison between SDC-Nets, AHEPs and ADC-Nets*

| **Properties** | **SDC-Nets** | **AHEPs** | **ADC-Nets** |
|---|---|---|---|
| Resistant against collusion | ✓ | | ✓ |
| Set of trusted measurements | ✓ | | ✓ |
| Messages to the supplier | ✓ | | ✓ |
| Scalable | | ✓ | ✓ |
| Total of keys | $O(I^2)$ | 2 | $O(I)$ |
| Keys stored per meter | $2(I-1)$ | 1 | 1 |
| Meter cannot disrupt | | | ✓ |
| Verification as commitment | | | ✓ |

**ADC-Net is the best technique**

SDC-Nets and AHEPs have been used in many application scenarios [1], e.g., electronic voting, image processing, reputation systems, and electronic money. All of them can profit from ADC-Nets.

## 2.7 Simulations

In this section, we find a description in Section 2.7.1 of the real-world data set used as input, a description of the software and hardware in Section 2.7.2, the parameters chosen in Section 2.7.3 and the simulation results in Section 2.7.4.

### *2.7.1 Real-world data set*

As might be expected, the real-world data set has some inconsistencies. This is quite normal for large amount of data without a verification process, because machines might fail. With real-world data, we see the necessity for verification. If they had collected the measurements with an ADC-Net, they could verify and detect issues. Whereas we cannot ask meters for new measurements, we should remove failures to sanitise the data set [1].

The simulation ran into the sanitised data set, which has $I \cdot J = 165\,546\,810$ measurements collected from $I = 6\,435$ meters in $J = 25\,726$ rounds with an intervals of 30 minutes. Thus, each meter collected 25 726 measurements in almost one and half years.

### *2.7.2 Software and hardware*

The source codes for Paillier [17], LOP [24], iKUP [12], ADC-Net [1] and this ADC-Net were written in C programming language, compiled with GCC version 4.1.2 (Red Hat 4.1.2-55) and linked with the GNU Multiple Precision Arithmetic Library (GMP), the Open Multi Processing (OpenMP) and the open-source toolkit for SSL/TLS (OpenSSL). GMP provides the commands to work with big numbers. OpenMP provides the commands to parallelise the encryption functions, i.e., to run many threads with a code. OpenSSL provides the hash function. Excluding LOP that does not need aggregation, each protocol has an encryption algorithm, an aggregation algorithm and a decryption algorithm implemented without precomputation.

The hardware used was a machine with 32 cores recognised of Intel® Xeon® CPU E5520 of 2.27 GHz and 23 Gigabytes of shared memory.

### *2.7.3 Simulation parameters*

Although the algorithms use the same hash function SHA256 generated by OpenSSL, LOP and iKUP use only 32 bits, when ADC-Nets use 160 bits. Paillier does not use hash but instead it uses a pseudo-random numbers generated by GMP, which also provides the prime numbers of 512 bits. Thus, their product has 1024 bits.

This simulation used the parameters for the smaller security level [1]. Increasing the security level, Paillier will become increasingly slower than the others will.

In opposite to other simulation [1,12,31] that collected the wall clock time – time function in C – of each round, the encryption algorithms run in 30 threads and collected the total processing time – clock function in C – per round in seconds with two decimal digits.

### *2.7.4 Simulation results*

The processing time outputted came as expected in the Section 2.6 and with time equivalent to other simulations [1,12,31]. The time for encryption, aggregation and decryption algorithms are presented separately. However, the processing time for encryption has the highest impact on total processing time. At the end of this section, we find the global comparison.

#### 2.7.4.1 Encryption

Figure 2.9 depicts the boxplots of the processing time collected for the encryption algorithms per round in seconds with two decimal digits. The arithmetic mean time is represented by a grey dot. Excluding Paillier encryption that has many outliers, the others have values close to the mean. The $y$-axis is split to simplify the visualisation, because LOP encryption with time complexity $O(I)$ took much more time than the others did. As expected, iKUP is the fastest with time complexity $O(1)$ followed by this ADC-Net and ADC-Net [1] with time complexities $O(\log(k))$ and $O(\log(k + h_j))$, respectively. They are followed by Paillier with time complexity $O(\log(n))$. As $k$ has 160 bits and $n$ has 1024 bits, the order is theoretically justified.

Table 2.6 shows the collected processing times for encryption algorithms. iKUP is much faster than the others are. Although this ADC-Net and ADC-Net [1] have almost the same time, this ADC-Net is slightly faster. Paillier is almost six times slower than ADC-Nets and could be worse, if the simulation run with better level of

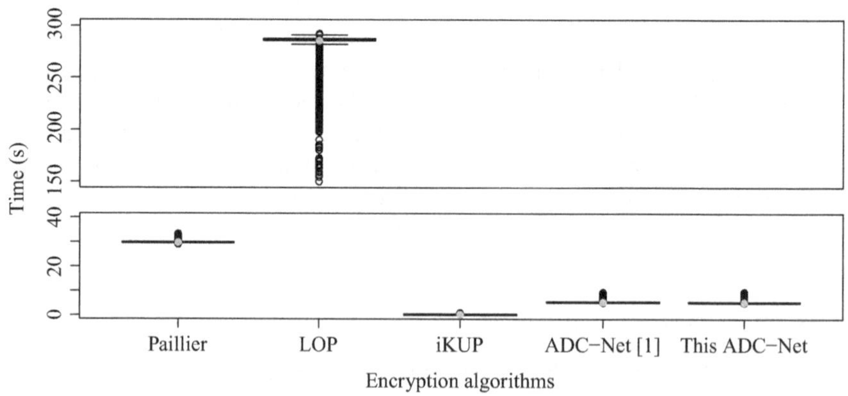


*Figure 2.9 Boxplots of the time collected for the encryption algorithms per round in seconds*

*Table 2.6 Time collected for the encryption algorithms per round in seconds*

| | Paillier | LOP | iKUP | ADC-Net [1] | This ADC-Net |
|---|---|---|---|---|---|
| Minimum | 29.57 | 282.28 | 0.29 | 5.29 | 5.26 |
| Lower quartile | 29.64 | 285.55 | 0.30 | 5.35 | 5.30 |
| Median | 29.66 | 286.90 | 0.30 | 5.37 | 5.31 |
| Mean | 29.69 | 285.61 | 0.31 | 5.38 | 5.32 |
| Upper quartile | 29.69 | 287.73 | 0.31 | 5.39 | 5.33 |
| Maximum | 29.76 | 290.97 | 0.32 | 5.44 | 5.37 |

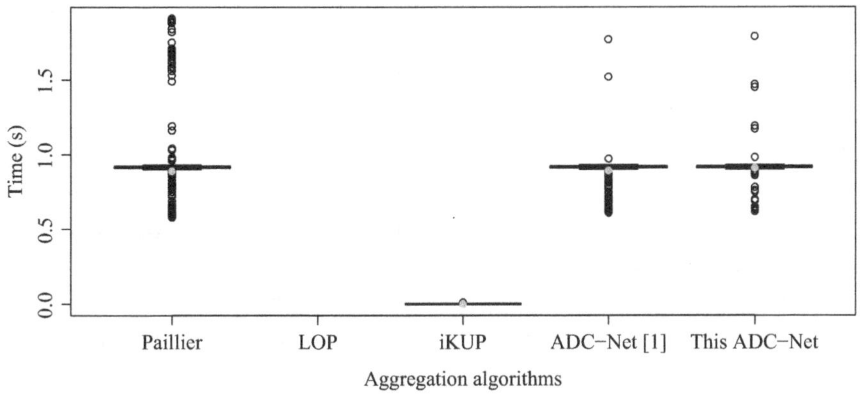


*Figure 2.10 Boxplots of the time collected for the aggregation algorithms per round in seconds*

security. LOP is almost ten times worse than Paillier. LOP can have better time for smaller number of meters and worse for bigger. The processing times for a meter to run other encryption algorithms do not change with the number of meters. On the one hand, the processing time per round for LOP grows quadratically with the number of meters $I$. On the other hand, the time per round for the others is the number of meters $I$ times a constant for the encryption function. The minimum value collected for Paillier was 149.96 s and the maximum was 292.19 s. However, Table 2.6 does not include the outliers in the maximum and minimum.

### 2.7.4.2 Aggregation

Figure 2.10 depicts the boxplots for the collected processing time for aggregation algorithms. We can say that LOP is the fastest, because it does not need any aggregation algorithm. Specifically, its aggregation happens along with the decryption. In the sequence, we have iKUP, which is much faster than ADC-Nets and Paillier, though they have the same time complexity $O(I)$.

Analogously, Table 2.7 shows the collected processing time for aggregation algorithms. While NA represents the absolute zero, two decimal digits was not enough

*Table 2.7 Time collected for the aggregation algorithms per round in seconds*

| | Paillier | LOP | iKUP | ADC-Net [1] | This ADC-Net |
|---|---|---|---|---|---|
| Minimum | 0.90 | NA | 0 | 0.90 | 0.90 |
| Lower quartile | 0.91 | NA | 0 | 0.91 | 0.91 |
| Median | 0.92 | NA | 0 | 0.92 | 0.92 |
| Mean | 0.89 | NA | 0 | 0.89 | 0.91 |
| Upper quartile | 0.92 | NA | 0 | 0.92 | 0.92 |
| Maximum | 0.93 | NA | 0 | 0.93 | 0.93 |

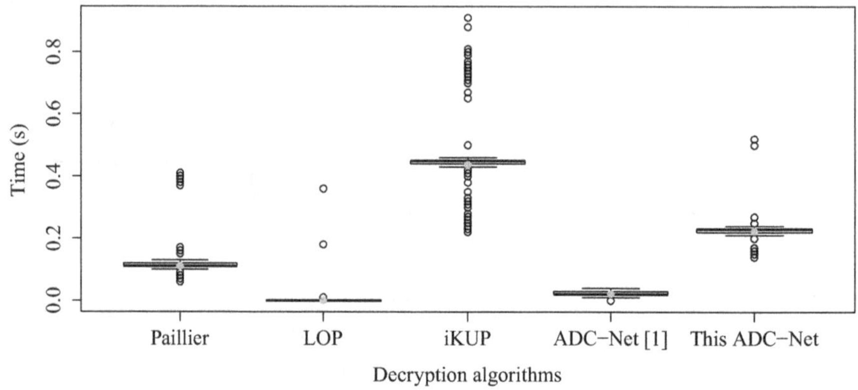


*Figure 2.11 Boxplots of the time collected for the decryption algorithms per round in seconds*

to register the time for iKUP, which was registered as zero. However, iKUP has a maximum outlier of 0.01 s. Indeed, we need microseconds [1], but the clock function used do not have such precision. The time for the others are quite similar to each other with about nine-tenths of a second. Notice that the mean of Paillier is smaller than its minimum, because the maximum and minimum exclude the outliers. The minimum and maximum collected for Paillier were 0.58 s and 1.91 s, respectively.

### 2.7.4.3 Decryption

Similarly, Figure 2.11 depicts the boxplots for the decryption algorithms. As well as the aggregation algorithms, the variation of decryption algorithms are not very high. From the fastest to the slowest, we have: LOP with time complexity $O(I)$, ADC-Net [1] with $O(\log(k))$, Paillier with $O(\log(n))$, this ADC-Net with $O(\log(n \cdot \lambda))$ and iKUP with $O(I)$.

Table 2.8 shows the processing time collected for the decryption algorithms, where LOP was faster than the precision can measure. However, LOP has a maximum outlier with 0.36 s. ADC-Net is about five times faster than Paillier and ten times faster than this ADC-Net, which is two times faster than iKUP.

*Table 2.8 Time collected for the decryption algorithms per round in seconds*

| | Paillier | LOP | iKUP | ADC-Net [1] | This ADC-Net |
|---|---|---|---|---|---|
| Minimum | 0.10 | 0 | 0.43 | 0.01 | 0.21 |
| Lower quartile | 0.11 | 0 | 0.44 | 0.02 | 0.22 |
| Median | 0.11 | 0 | 0.45 | 0.02 | 0.23 |
| Mean | 0.11 | 0 | 0.44 | 0.02 | 0.22 |
| Upper quartile | 0.12 | 0 | 0.45 | 0.03 | 0.23 |
| Maximum | 0.13 | 0 | 0.46 | 0.04 | 0.24 |

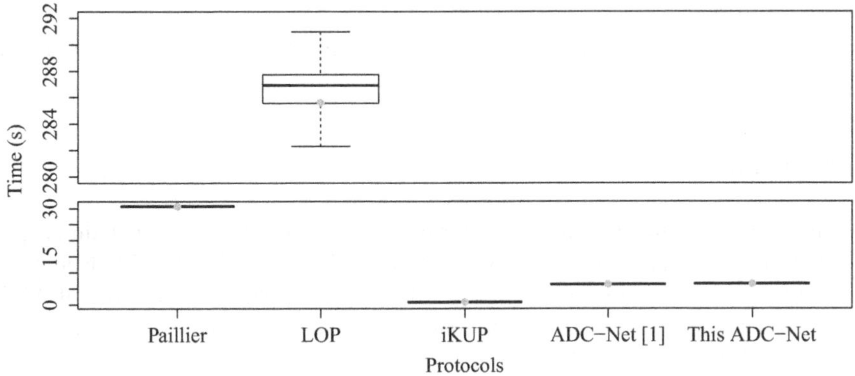


*Figure 2.12 Boxplots of the processing time collected for the protocols per round in seconds*

#### 2.7.4.4 Overall processing time

Figure 2.12 depicts the boxplots of the overall processing time collected for each protocol. Different from the previous boxplots, Figure 2.12 does not present the outliers to simplify the visualisation. Whereas encryption algorithms dominate the processing time, Figure 2.12 is similar to Figure 2.9. Therefore, we have almost the same order of performance, namely: iKUP is the fastest, ADC-Net [1] is the second, this ADC-Net is the third, Paillier is the fourth and LOP is the slowest. Notice that the mean is away from the median indicating outliers for LOP, which spends so much time in comparison with the others.

Table 2.9 shows the processing time collected for the protocols. iKUP is much faster than the others are. Practically, there is no difference in time between ADC-Net [1] and this ADC-Net. The former is faster and the latter has a more interesting propriety for suppliers, i.e., it provides consolidated consumptions $c_j$ only to suppliers. iKUP is about 40 times faster than Paillier, which is about ten times faster than LOP. Again, the maximum and minimum do not include the outliers.

iKUP outperforms all protocols. However, it as well as Paillier and LOP does not have many interesting properties described in Section 2.6. ADC-Nets have a very

*Table 2.9 Time collected for the protocols per round in seconds*

| | Paillier | LOP | iKUP | ADC-Net [1] | This ADC-Net |
|---|---|---|---|---|---|
| Minimum | 30.57 | 282.28 | 0.71 | 6.23 | 6.40 |
| Lower quartile | 30.66 | 285.55 | 0.74 | 6.29 | 6.44 |
| Median | 30.69 | 286.90 | 0.75 | 6.31 | 6.45 |
| Mean | 30.69 | 285.61 | 0.75 | 6.29 | 6.45 |
| Upper quartile | 30.72 | 287.73 | 0.76 | 6.33 | 6.47 |
| Maximum | 30.81 | 290.97 | 0.79 | 6.39 | 6.51 |

good performance and many interesting properties, for instance, verification and enforcement of privacy.

## 2.8 Conclusions

In this chapter, a survey of protocols for privacy-preserving data aggregation is presented. In particular, a powerful cryptographic technique called as ADC-Net is also presented and an improved ADC-Net is developed. Besides all benefits of ADC-Nets the previous state of the art assumed consolidated consumptions $c_j$ as public, this ADC-Net can also hide the consolidated consumptions $c_j$ giving privacy also for suppliers. This work also call the attention that iKUP [12] determines a lower bounder for processing time to compute consolidated consumptions $c_j$. Furthermore, iKUP behaves as AHEPs, which are used in many applications. Hoverer, iKUP and AHEPs do not enforce privacy and do not provide verification. In contrast, ADC-Nets do. In addition, ADC-Nets also can replace iKUP and AHEPs in smart grid scenarios. In fact, ADC-Nets can also be used in many application scenarios beyond smart grids.

The protocols are presented and analysed in theory. In addition, the results of simulations with real-world measurements are presented to validate the theory. Encryption algorithms run in constrained meters and heavily require more processing than aggregation and decryption algorithms. Efficiency is paramount for such protocols, especially for encryption algorithms. Remember that devices should be as cheaper as possible and consume the minimal amount of energy.

Real-time consolidated consumption, billing process and verification are essential for smart grids. To achieve them, we can use ADC-Nets. Since the basis of a smart grid is established, we can go further with new technologies for smart grid scenarios.

## Acknowledgements

The author thanks Hongjian Sun, Nikolaos Hatziargyriou, Laurence Carpanini, Miguel Fornié and Vincent Poor for all their support.

The author thanks the Brazilian National Laboratory of Scientific Computing for providing the hardware used in the simulation and the Irish Social Science Data Archive and the Commission for Energy Regulation, Electricity Customer Behaviour Trial, for providing the Dataset issued by The Research Perspective Ltd on 12 March 2012.

## Bibliography

[1] Borges F. "On privacy-preserving protocols for smart metering systems" [Ph.D. Thesis – Dr.-Ing.]. Technische Universität Darmstadt. Darmstadt; 2015. Available from http://tuprints.ulb.tu-darmstadt.de/4693/. Accessed on 07/27/2016.

[2] McBee KD, Simoes MG. "General smart meter guidelines to accurately assess the aging of distribution transformers". *IEEE Transactions on Smart Grid*. 2014 Nov;5(6):2967–2979.

[3] Lovins AB, Datta EK, Feiler T, *et al. Small Is Profitable*. Rocky Mountain Institute; USA, 2002.

[4] Molina-Markham A, Shenoy P, Fu K, Cecchet E, Irwin D. "Private memoirs of a smart meter". In: *Proceedings of the Second ACM Workshop on Embedded Sensing Systems for Energy-Efficiency in Building (BuildSys'10)*. New York, NY: ACM; 2010. pp. 61–66. Available from http://doi.acm.org/10.1145/1878431.1878446. Accessed on 07/27/2016.

[5] Greveler U, Justus B, Löhr D. "Multimedia content identification through smart meter power usage profiles", translated from "Identifikation von Videoinhalten über granulare Stromverbrauchsdaten". In: Suri N, Waidner M, editors. *Sicherheit 2012: Sicherheit, Schutz und Zuverlässigkeit, Beiträge der 6. Jahrestagung des Fachbereichs Sicherheit der Gesellschaft für Informatik e.V. (GI)*, 7.–9. März 2012 in Darmstadt, Germany, vol. 195 of *LNI*. GI; 2012. pp. 35–45.

[6] Holvast J. "History of privacy". In: Matyáš V, Fischer-Hübner S, Cvrček D, Švenda P, editors. *The Future of Identity in the Information Society*, vol. 298 of *IFIP Advances in Information and Communication Technology*. Berlin: Springer; 2009. pp. 13–42. Available from http://dx.doi.org/10.1007/978-3-642-03315-5_2. Accessed on 07/27/2016.

[7] Wang Z, Zheng G. "Residential appliances identification and monitoring by a nonintrusive method". *IEEE Transactions on Smart Grid*. 2012 Mar; 3(1):80–92.

[8] Batra N, Kelly J, Parson O, *et al.* "NILMTK: an open source toolkit for non-intrusive load monitoring". In: *Proceedings of the Fifth International Conference on Future Energy Systems (e-Energy'14)*. New York, NY: ACM; 2014. pp. 265–276. Available from http://doi.acm.org/10.1145/2602044.2602051. Accessed on 07/27/2016.

[9] Shannon CE. "Communication theory of secrecy systems". *The Bell System Technical Journal*. 1949 Oct;28(4):656–715.

[10] Diffie W, Hellman ME. "New directions in cryptography". *IEEE Trans Information Theory*. 1976;IT-22(6):644–654.

[11] Li F, Luo B. "Preserving data integrity for smart grid data aggregation". In: *Smart Grid Communications (SmartGridComm), 2012 IEEE Third International Conference* on Tainan; 2012. pp. 366–371.

[12] Borges F, Martucci LA. "iKUP keeps users' privacy in the smart grid". In: *IEEE Conference on Communications and Network Security, CNS 2014*, San Francisco, CA, USA, October 29–31, 2014; 2014. pp. 310–318. Available from http://dx.doi.org/10.1109/CNS.2014.6997499. Accessed on 07/27/2016.

[13] Borges F, Volk F, Mühlhäuser M. "Efficient, verifiable, secure, and privacy-friendly computations for the smart grid". In: *2015 IEEE Power and Energy Society Innovative Smart Grid Technologies Conference, ISGT 2015*, Washington, DC, USA, February 18–20, 2015; 2015. pp. 1–5. Available from http://dx.doi.org/10.1109/ISGT.2015.7131862. Accessed on 07/27/2016.

[14] Lei Yang, Xu Chen, Junshan Zhang and H. Vincent Poor, "Cost-Effective and Privacy-Preserving Energy Management for Smart Meters," *IEEE Transactions on Smart Grid*, 6(1), pp. 486–495, January 2015.

[15] Onur Tan, Deniz Gündüz and H. Vincent Poor, "Increasing Smart Meter Privacy Through Energy Harvesting and Storage Devices," *IEEE Journal on Selected Areas in Communications: Smart Grid Communications*, 31(7), pp. 1331–1341, July 2013.

[16] Lalitha Sankar, S. Raj Rajagopalan, Soheil Mohair and H. Vincent Poor, "Smart Meter Privacy: A Theoretical Framework," *IEEE Transactions on Smart Grid*, 4(2), pp. 837–846, June 2013.

[17] Li F, Luo B, Liu P. "Secure information aggregation for smart grids using homomorphic encryption". In: *Smart Grid Communications (SmartGridComm), 2010 First IEEE International Conference* on Gaithersburg, Maryland; 2010. pp. 327–332.

[18] Jawurek M, Kerschbaum F. "Fault-tolerant privacy-preserving statistics". In: Fischer-Hübner S, Wright M, editors. *Privacy Enhancing Technologies*, vol. 7384 of *Lecture Notes in Computer Science*. Berlin: Springer; 2012. pp. 221–238. Available from http://dx.doi.org/10.1007/978-3-642-31680-7_12. Accessed on 07/27/2016.

[19] Ruj S, Nayak A. "A decentralized security framework for data aggregation and access control in smart grids". *IEEE Transactions on Smart Grid*. 2013 Mar;4(1):196–205.

[20] Paillier P. "Public-key cryptosystems based on composite degree residuosity classes". In: *Advances in Cryptology – EUROCRYPT 1999*, vol. 1592 of *Lecture Notes in Computer Science*. Berlin: Springer; 1999. pp. 223–238.

[21] Pedersen TP. "Non-interactive and information-theoretic secure verifiable secret sharing". In: *Proceedings of the 11th Annual International Cryptology Conference on Advances in Cryptology (CRYPTO'91)*. London: Springer-Verlag; 1992. pp. 129–140. Available from http://dl.acm.org/citation.cfm?id=646756.705507. Accessed on 07/27/2016.

[22] Jawurek M, Johns M, Kerschbaum F. "Plug-in privacy for smart metering billing". In: *Privacy Enhancing Technologies*, vol. 6794 of *Lecture Notes in Computer Science*. Berlin: Springer; 2011. pp. 192–210.

[23] Chaum D. "The dining cryptographers problem: unconditional sender and recipient untraceability". *Journal of Cryptology*. 1988 Mar;1(1):65–75. Available from http://dl.acm.org/citation.cfm?id=54235.54239. Accessed on 07/27/2016.

[24] Kursawe K, Danezis G, Kohlweiss M. "Privacy-friendly aggregation for the smart-grid". In: *Privacy Enhancing Technologies*, vol. 6794 of *Lecture Notes in Computer Science*. Berlin: Springer; 2011. pp. 175–191.

[25] Yang L, Li F. "Detecting false data injection in smart grid in-network aggregation". In: *Smart Grid Communications (SmartGridComm), 2013 IEEE International Conference* on Vancouver, Canada; 2013. pp. 408–413.

[26] Borges F, Demirel D, Bock L, Buchmann JA, Mühlhäuser M. "A privacy-enhancing protocol that provides in-network data aggregation and verifiable smart meter billing". In: *IEEE Symposium on Computers and Communications, ISCC 2014*, Funchal, Madeira, Portugal, June 23–26, 2014; 2014. pp. 1–6. Available from http://dx.doi.org/10.1109/ISCC.2014.6912612. Accessed on 07/27/2016.

[27] Borges F, Buchmann JA, Mühlhäuser M. "Introducing asymmetric DC-Nets". In: *IEEE Conference on Communications and Network Security, CNS 2014*, San Francisco, CA, USA, October 29–31, 2014; 2014. pp. 508–509. Available from http://dx.doi.org/10.1109/CNS.2014.6997528. Accessed on 07/27/2016.

[28] Boneh D, Franklin M. "Efficient generation of shared RSA keys". *Journal of the ACM*. 2001 Jul;48(4):702–722. Available from http://doi.acm.org/10.1145/502090.502094. Accessed on 07/27/2016.

[29] Lara PCS, Borges F, Portugal R, Nedjah N. "Parallel modular exponentiation using load balancing without precomputation". *Journal of Computer and System Sciences*. 2012;78(2):575–582. Available from http://dx.doi.org/10.1016/j.jcss.2011.07.002. Accessed on 07/27/2016.

[30] Borges F, Santos RAM, Marquezino FL. "Preserving privacy in a smart grid scenario using quantum mechanics". *Security and Communication Networks*. 2015;8(12):2061–2069. Available from http://dx.doi.org/10.1002/sec.1152. Accessed on 07/27/2016.

[31] Borges F, Mühlhäuser M. "EPPP4SMS: efficient privacy-preserving protocol for smart metering systems and its simulation using real-world data". *IEEE Trans Smart Grid*. 2014;5(6):2701–2708. Available from http://dx.doi.org/10.1109/TSG.2014.2336265. Accessed on 07/27/2016.

# Chapter 3
# Smart price-based scheduling of flexible residential appliances

*Dimitrios Papadaskalopoulos*[1] *and Goran Strbac*[1]

Residential appliances are characterised by a significant flexibility potential. Smart scheduling of these appliances' demand, supported by the envisaged roll-out of smart metering, can greatly improve the economic efficiency of future low-carbon power systems. However, this scheduling task exhibits serious challenges. Traditional centralised scheduling architectures face scalability issues and raise privacy concerns by the residential consumers. Dynamic pricing constitutes a promising scheduling approach that overcomes these limitations. However, as demonstrated in this chapter, naive application of dynamic pricing leads to new demand peaks and inefficient system operation since appliances' demand response is concentrated at the periods with the lowest prices.

This chapter presents, analyses, and compares three different smart measures to avoid such concentration effects and achieve more efficient system operation without centralised knowledge of residential appliances' characteristics. The design of these measures is customised to the specific operating properties of different types of flexible residential appliances, namely appliances with continuously adjustable power levels and appliances with shiftable cycles. Smart-charging electric vehicles (EV) and wet appliances (WA) with delay functionality are used as representative examples of these two types, due to significant level of expected penetration and flexibility potential.

The first measure imposes a relative flexibility restriction on residential appliances. In case of appliances with continuously adjustable power levels, this restriction corresponds to a maximum power limit, preventing each of these appliances from requesting a large proportion of its total energy requirements at the lowest-priced periods. In case of appliances with shiftable cycles, this restriction corresponds to a maximum cycle delay limit, preventing them from synchronising their operation at the lowest-priced periods.

Since imposing flexibility restrictions may be considered by the consumers as a direct intervention in the control of their appliances, an alternative proposed measure replaces the hard flexibility restriction by a soft non-linear price signal, penalising

[1]Department of Electrical and Electronic Engineering, Imperial College London, London SW7 2AZ, UK

the extent of flexibility utilised by the appliances. Specifically, this price penalises the square of the power demand and the duration of cycle delay of continuously adjustable and shiftable cycle appliances, respectively. Regarding the former type, apart from the above acceptability advantage, this non-linear pricing approach is demonstrated to outperform the flexibility restriction approach in flattening the demand profile and thus achieving more efficient operation solutions. Regarding the latter type, a third proposed smart measure randomising the non-linear price signal posted to different appliances is demonstrated to bring significant additional benefits.

These novel concepts presented in this chapter are supported by two sets of case studies. The first one deals with scheduling flexible residential appliances in electricity markets, where the objective lies in minimising the total generation costs. The second one deals with scheduling flexible residential appliances in the context of local distribution networks management, where the objective lies in minimising the cost of interventions required to resolve network constraints and the amount of network losses.

## Nomenclature

| | |
|---|---|
| $t \in T$ | Index and set of time periods of scheduling problem |
| $t^{res}$ | Length of time period in hours |
| $\lambda_t$ | Electricity price at time period $t$ |
| $P_t^{ev}$ | Power demand of EV at time period $t$ |
| $P^{max}$ | Maximum charging rate of EV battery |
| $E_t$ | Energy in EV battery at the end of time period $t$ |
| $E^{min}$ | Minimum energy level in EV battery |
| $E^{max}$ | Maximum energy level in EV battery |
| $E_t^{dr}$ | Driving energy requirements of EV at time period $t$ |
| $\eta^{ch}$ | Charging efficiency of EV battery |
| $T^{gr}$ | Set of time periods that the EV is connected to the grid |
| $\tau$ | Index of time steps of the WA cycle |
| $P_\tau^{wa}$ | Power demand at step $\tau$ of the WA cycle |
| $T^{dur}$ | Duration of WA cycle |
| $t^{act}$ | Activation time period of WA cycle |
| $t^{in}$ | Initiation time period of WA cycle |
| $\delta$ | Maximum delay limit of WA cycle |

## 3.1 Introduction

### 3.1.1 Context – emerging challenges for low-carbon electrical power systems

Energy systems across the world are currently undergoing fundamental changes since a number of diverse factors justify the need for a new technical, commercial and regulatory framework for their future development and operation. First of all, the continuously increasing levels of greenhouse gases emissions in the atmosphere have raised serious environmental and climate change concerns. Numerous governments have taken significant initiatives in response to such concerns. In the United Kingdom (UK), for example, the *2008 Climate Change Act* [1] set a legally binding target of 26% reduction in greenhouse gases emissions by 2026 (with respect to the 1990 baseline), extended to a further ambitious target of 80% reduction by 2050. Apart from the issue of climate change, growing energy security concerns emerge over the dependency of energy systems on fossil fuels exhibiting a continuously reducing availability and a subsequent increase of their prices.

In the context of addressing the above environmental and energy security concerns, energy systems are facing the challenge of decarbonisation. At the generation side, this decarbonisation is under way through the wide deployment of renewable and low-carbon generation sources. The *2008 European Commission (EC) directive* [2] for example, put forward a legally binding target for renewable energy sources to cover 20% of the total energy consumption in the countries of the European Union (EU) by 2020. However, the majority of these sources are inherently characterised by limited predictability and controllability, implying that significant amounts of underutilised conventional generation capacity need to remain in the system to provide security of supply and balance renewable generation variability.

At the demand side, traditional technologies of the transport and heat sectors (internal combustion engines for transport and gas/oil fired technologies for heating) are based on the intense consumption of fossil fuels and the emission of a significant portion of the total greenhouse emissions [3–6]. In combination with the ongoing and future decarbonisation of electricity generation systems, strong motives arise for the electrification of these technologies. Recent technological developments in the automotive and heating sectors have techno-economically enabled this transition with the production and efficient operation of electric vehicles (EV) [4,6] and electric heat pumps (EHP) [5], respectively. Nevertheless, due to the natural energy intensity of transportation vehicles and heating loads, the environmental and energy security potential of this transition is accompanied by the introduction of a considerable amount of new demand in electrical power systems. Going further, the electrification of transport and heat sectors will lead to disproportionately larger demand peaks – and subsequently disproportionately higher generation and network costs – than the increase in the total electrical energy consumption, due to the temporal patterns of users' driving and heating requirements [7].

### 3.1.2 Role of residential demand in addressing emerging challenges

The role of the demand side in the emerging power system setting described above attracts continuously increasing interest. Particular focus has been put on residential demand as it is responsible for the larger proportion of electricity consumption, particularly during system peaks [5]. In the first place, in line with the effort to address environmental and energy security concerns, significant energy efficiency programmes have been developed around the world. These promote the reduction of electrical energy consumption through the development of suitable economic and environmental incentives for the adoption of energy efficiency measures, such as the improved building insulation, the use of efficient residential appliances, and consumers' behavioural changes [8].

However, apart from the above static energy reduction measures, a dynamic flexibility potential is associated with residential demand and can greatly contribute to addressing emerging system challenges. The consumers may shift the operation of some of their appliances in time if they are provided with sufficient incentives, higher than the related inconvenience. Furthermore, some demand technologies incorporate explicit storage components, enabling the temporal decoupling between the acquisition of the required electrical energy and its actual consumption, and thus exhibiting a flexibility potential without affecting the level of service delivered to the consumers. The technologies supporting the electrification of transport and heat sectors constitute significant examples; EV are necessarily accompanied by an electrical energy battery replacing the traditional fuel tank and EHP may be accompanied by heat storage, usually in the form of hot water tanks.

Smart scheduling of such residential appliances can reduce the underutilised, conventional generation capacity required to balance renewable generation variability and limit peak demand levels by redistributing consumers' electricity requirements towards off-peak time periods. As a result, it exhibits a potential of significantly improving the economic efficiency of future low-carbon power systems by reducing generation and network operational and investment costs [9–11].

In order to enable such smart coordination of residential appliances, deployment of advanced information, communication and control infrastructure in domestic premises is necessary. First of all, residential consumers have neither the expertise nor the comfort to commit themselves in the task of optimally scheduling their appliances' operation according to the system objectives; therefore, the deployment of suitable automation technologies would be required to minimise direct human engagement. Furthermore, suitable two-way communication technologies are required to enable the coordination between the residential appliances and system entities (e.g. market operator, network operator). The envisaged roll-out of smart metering constitutes the gateway for the transition to this new paradigm [12–14]. Smart meters will have access to a communication link with system entities, through which they will be communicating the operational characteristics of the residential appliances and they will be receiving control signals for their optimal scheduling.

### *3.1.3 Challenges in scheduling residential appliances*

Approaches examined in the relevant literature to achieve integration of flexible demand in system operation can be generally divided into two categories. The first one involves traditional centralised scheduling architectures [15–18]. The flexible appliances communicate their technical and economic characteristics to a system coordinator (according to the application this could correspond to a national market operator or a local distribution network operator (DNO)) through the dedicated smart meter. This entity then makes decisions on the scheduling of the appliances based on the solution of a global optimisation problem and communicates respective dispatch signals to the appliances through the smart meter.

Under a significant penetration of flexible residential appliances, however, the communication and computational scalability of centralised mechanisms is at least questionable in both technical and economic terms. Transmission of the diverse complex operational constraints and physical parameters of a very large number of residential appliances to the system coordinator will yield information collection and communication problems, while the vast number of decision variables and constraints associated with these appliances will create a massive computational burden to the system coordinator. Furthermore, centralised mechanisms are likely to raise privacy concerns by the consumers, as they are not generally willing to disclose sensitive information (such as habits, preferences and appliances' properties) and be directly controlled by an external entity.

The second approach to integrate flexible demand in system operation involves dynamic pricing schemes [19,20] that avoid the above scalability and privacy limitations of centralised architectures. Without communicating their individual characteristics to the central coordinator, residential appliances are exposed through the smart meter to time-differentiated prices and are thus encouraged to activate their flexibility and modify their demand patterns accordingly in order to reduce the electricity cost incurred by their operation. More specifically, the posted prices are higher during peak demand periods and lower during off-peak demand periods to incentivise appliances to redistribute their energy requirements towards the latter and therefore improve the efficiency of system operation.

However, as the authors demonstrated in a number of publications [21–23], naive application of dynamic pricing in combination with the envisaged automation in appliances' control (Section 3.1.2) leads to serious loss of diversity and demand response concentration phenomena. Specifically, appliances' demand is concentrated at the periods with the lowest prices, yielding significant new demand peaks and thus inefficient system operation.

### *3.1.4 Overview of alternative approaches for smart scheduling of residential appliances*

In this context, this chapter presents, analyses, and compares three different smart measures to avoid such concentration effects and achieve more efficient system operation without centralised knowledge of residential appliances' characteristics. The design of these measures is customised to the specific operating properties of different types of flexible residential appliances, namely appliances with continuously adjustable power levels and appliances with shiftable cycles. Smart-charging EV and WA with delay functionality are used as representative examples of these two types, due to their significant penetration and flexibility potential [14].

The first measure imposes a relative flexibility restriction on residential appliances. In case of appliances with continuously adjustable power levels, this restriction corresponds to a maximum power limit, preventing each of these appliances from requesting a large proportion of its total energy requirements at the lowest-priced periods. In case of appliances that cannot continuously adjust their power levels but can only defer their fixed operation cycles, this restriction corresponds to a maximum cycle delay limit, preventing them from synchronising their operation at the lowest-priced periods.

However, imposing a flexibility restriction may not be deemed acceptable by the consumers as it may be considered as a direct intervention in the control of their appliances. In this context, an alternative proposed measure replaces the hard flexibility restriction by a soft non-linear price signal, penalising the extent of flexibility utilised by the appliances. Specifically, this price penalises the square of the power demand and the duration of cycle delay of continuously adjustable and shiftable cycle appliances, respectively. Regarding the former type, apart from the above acceptability advantage, this non-linear pricing approach is demonstrated to outperform the flexibility restriction approach in flattening the demand profile and thus achieving more efficient solutions. Regarding the latter type, a third proposed smart measure randomising the non-linear price signal posted to different appliances is demonstrated to bring significant additional benefits.

These three measures against demand response concentration are tested and compared in two sets of case studies. The first one deals with scheduling flexible residential appliances in electricity markets, where the objective lies in minimising the total generation costs. The second one deals with scheduling flexible residential appliances in the context of local distribution networks management, where the objective lies in minimising the cost of interventions required to resolve network constraints and the amount of network losses.

The rest of this chapter is organised as follows. Section 3.2 derives operational models of the two examined types of flexible residential appliances and mathematically formulates their price response optimisation problems. Section 3.3 details the measures proposed to prevent response concentration effects. Case studies are presented in Section 3.4. Finally, Section 3.5 discusses conclusions and future extensions of this work.

## 3.2 Modelling operation and price response of flexible residential appliances

Two types of flexible residential appliances are examined, both characterised by flexibility in terms of the specific time period(s) they can acquire the amount of energy required for their operation, as soon as this is done within a specific temporal interval allowed by their users. For the first type, the power demand level can be continuously adjusted up to a maximum rate. Smart-charging EV [24], which need to obtain the energy required for the desired journeys over the interval they are connected to the grid, are employed as a representative example of this type.

The operation of the second type is based on the execution of user-called cycles the power profile of which cannot be altered; their flexibility involves the deferability of these cycles up to a maximum delay limit set by their users. WA (e.g. dishwashers

(DW) and washing machines (WM)) with delay functionality [25] are employed as a representative example of this type.

The following subsections derive flexibility models of these two types of residential appliances and mathematically formulate the optimisation problems expressing their optimal response to a set of time-differentiated prices.

## *3.2.1 Appliances with continuously adjustable power levels – EV with smart charging capability*

### **3.2.1.1 Flexibility modelling**

This type of appliances is characterised by:

(a) flexibility in terms of the specific time period(s) they can acquire the amount of energy required for their operation, as soon as this is done within a specific temporal interval allowed by their users (referred to as scheduling interval in the remainder of this chapter) and
(b) flexibility in terms of their power demand at each time period within this interval, as soon as it is lower than their maximum power limit.

Figure 3.1 illustrates these flexibility characteristics. The solid and the dashed power profile correspond to two different demand patterns. Both of them ensure that the total energy required for the operation of the appliance is acquired within the scheduling interval, but the timing and power level of this energy input is different. It is evident that these appliances' extent of operational flexibility depends on the size of their scheduling interval and the relative size of their maximum power limit with respect to their overall required energy.

EV with smart charging capability are employed as a representative example of this type. Their scheduling interval corresponds to the temporal interval that their users park them and connect them to the grid, and the energy acquired from the grid and stored in the EV battery during the scheduling interval will be later consumed for the journeys desired by the users. The users are indifferent regarding the specific time period(s) that the required energy will be acquired, as soon as their EV batteries have sufficient energy to carry out the desired journeys when they disconnect them from the

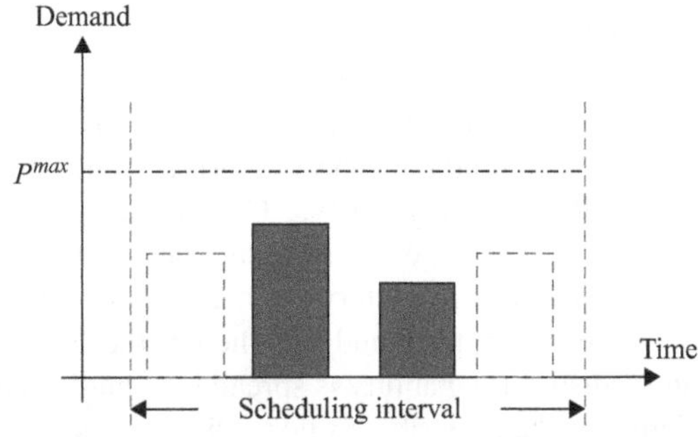


*Figure 3.1 Flexibility of appliances with continuously adjustable power levels*

grid and they start travelling. The power demand level of the EV battery is generally assumed continuously adjustable between zero and its maximum charging rate.

EV with smart charging capability are characterised by significant flexibility, due to their:

(a) Stationary character: According to Strbac *et al.* [14], vehicles are, on average, parked and idle for more than 90%–95% of the time. If EV charging infrastructure is widely deployed, this property will be translated into a large scheduling interval.
(b) Low energy requirements: Analysis of the users' travelling patterns and the technical characteristics of already developed and marketable electric motors and batteries for EV reveal that the driving requirements in terms of electrical energy are modest with respect to the significant energy capacities and power ratings of the batteries [14].

#### 3.2.1.2 Formulation of optimal price response problem

The optimisation problem solved by the control system of the EV charger to determine the optimal response to the time-differentiated price signals communicated through the smart meter is formulated as follows:

$$\textit{Objective function}\text{: } \min \sum_{t \in T} \lambda_t * P_t^{ev} \tag{3.1}$$

$$\textit{Decision variables}\text{: } P_t^{ev} \in R^+, \forall t \in T \tag{3.2}$$

*Constraints*:

$$E_t = \eta^{ch} * P_t^{ev} * t^{res} + E_{t-1} - E_t^{dr}, \quad \forall t \in T \tag{3.3}$$

$$E^{min} \le E_t \le E^{max}, \quad \forall t \in T \tag{3.4}$$

$$0 \le P_t^{ev} \le P_t^{max}, \quad \forall t \in T \tag{3.5}$$

$$P_t^{max} = \begin{cases} P^{max} & \textit{if } t \in T^{gr} \\ 0 & \textit{otherwise} \end{cases} \tag{3.6}$$

$$E_0 = E_{|T|} \tag{3.7}$$

The objective function (3.1) involves the minimisation of the electricity cost associated with the EV charging during the examined temporal horizon, by adjusting the charging power at each time period of this horizon (3.2). Constraint (3.3) expresses the energy balance in the EV battery including the charging losses of the battery and the users' energy requirements for the desired journeys. Constraint (3.4) corresponds to the battery's maximum depth of discharge and state of charge ratings. Constraints (3.5) and (3.6) represent the limit of the battery's power input, which depends on the maximum charging rate of the battery and on whether the EV is connected to the grid. The EV demand redistributing ability is spread beyond the temporal horizon of the examined problem; for the sake of simplicity, the battery energy content at the start and the end of the horizon are assumed equal (3.7).

## *3.2.2 Appliances with shiftable cycles – WA with delay functionality*

### 3.2.2.1 Flexibility modelling

As the type of appliances examined in the previous section, this type is characterised by flexibility in terms of the timing they can acquire their required amount of energy within their scheduling interval. However, they are not flexible in terms of their power demand at each time period. This is due to the fact that these appliances do not incorporate explicit storage components (such as the battery of EV) and their operation is based on the execution of user-called cycles the power profile of which cannot be altered. As a result, their flexibility is only associated with their ability to shift these cycles in time within the scheduling interval allowed by their users.

Figure 3.2 illustrates this flexibility characteristic. The solid and the dashed power profile correspond to two different demand patterns. Both of them ensure that the appliance cycle is executed following the inherent power profile, but the timing that this cycle is executed is different. These appliances' extent of operational flexibility depends only on the size of their scheduling interval.

WA (e.g. DW and WM) with delay functionality are employed as a representative example of this type. The operation of these appliances is based on the execution of user-called cycles, which comprise a sequence of phases or sub-processes occurring at a fixed order with generally fixed duration and fixed electrical power requirements that cannot be altered [25]. Taking a WM as an example, one cycle corresponds to the cleaning process of the laundry loaded in the machine by the users, according to the washing programme and temperature they have selected. This cycle generally consists of a water heating (to the desired temperature) phase, carried out by an electrical resistive heating system, followed by several wash and rinse phases, through the rotation of the machine's drum by an electric motor, and concluded by a water and suds extraction from the load through high-speed rotation of the drum [25].

WA have been traditionally operated in an inflexible fashion; once the user loaded the clothes/dishes and switched the appliance on, the respective cycle began immediately. Technological developments, however, have techno-economically enabled the incorporation of advanced control schemes in these appliances, allowing the delay

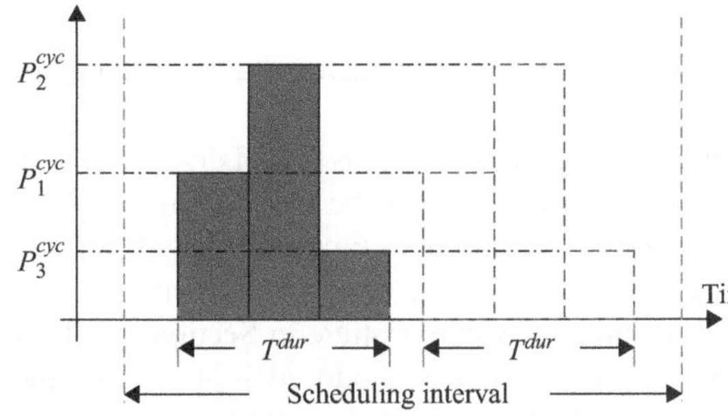


*Figure 3.2 Flexibility of appliances with shiftable cycles*

of their cycle initiation. This time-delay functionality allows the decoupling between the appliance activation time (time when the user switches the appliance on) and the cycle initiation time by deferring the latter in time. Therefore, the users can predetermine the latest time by which the cycle should end (latest desired termination time) and the cycle can be initiated any time after the activation, as soon as it ends before the latest desired termination time. Therefore, their scheduling interval corresponds to the temporal interval between the activation time and the latest desired termination time. The energy requirements of the WA at the period between the activation and the initiation time ('standby' period) are generally deemed negligible [25].

#### 3.2.2.2 Formulation of optimal price response problem

Assuming without loss of generality that the WA is activated by its users once during the examined temporal horizon, the optimisation problem solved by the control system of the WA to determine the optimal response to the time-differentiated price signals communicated through the smart meter is formulated as follows:

$$\textit{Objective function}: \ \min \sum_{\tau=1}^{T^{dur}} \lambda_{t^{in}+\tau-1} * P_{\tau}^{wa} \tag{3.8}$$

$$\textit{Decision variables}: \ t^{in} \in \mathbb{N} \tag{3.9}$$

$$\textit{Constraints}: \ t^{act} \leq t^{in} \leq t^{act} + \delta \tag{3.10}$$

The objective function (3.8) involves the minimisation of the electricity cost associated with the WA operation during the examined temporal horizon, by adjusting the initiation time of its cycle (3.9). Constraint (3.10) expresses the users' requirements; the cycle cannot be initiated before the activation time of the appliance (since users need to load and switch on the appliance before the latter initiates its cycle) or after the time determined by the maximum delay limit $\delta$ set by their users. In order to capture out-of-horizon effects, a periodic continuation is assumed, implying that the control system of the WA assumes $\lambda_{t+|T|} = \lambda_t$ and the demand of WA migrating towards/away from period $t + |T|$ is added to/subtracted from the total demand at period $t$.

## 3.3 Measures against demand response concentration

Application of dynamic pricing leads to serious loss of diversity and demand response concentration phenomena, as the solution of the above optimisation problems involves requesting the largest possible demand at the time periods with the lowest prices. These concentration effects are translated to significant new demand peaks and inefficient solutions, as we will quantitatively demonstrate in Section 3.4. This means that the basic form of dynamic pricing needs to be modified through smart measures avoiding such concentration effects. The following sub-sections present three such measures, which will be quantitatively tested and compared in Section 3.4.

### *3.3.1 Flexibility restriction*

As quantitatively demonstrated in Section 3.4, the size of the demand concentration effect is enhanced when the flexibility of the appliances is larger. Regarding appliances with continuously adjustable power levels, their flexibility is mainly associated with their maximum power limit (Section 3.2.1); a larger maximum power limit enables them to acquire larger proportion of their energy requirements at the lowest-priced periods and thus aggravates the concentration effect. Regarding appliances with shiftable cycles, their flexibility is associated with their maximum cycle delay (Section 3.2.2); a larger maximum cycle delay limit enables them to execute their cycles over a wider time window. Given that different appliances are activated by their users at different periods (due to the diversity of users' preferences), a larger maximum cycle delay limit increases the number of appliances executing their cycles at the lowest-priced time periods and thus aggravates the concentration effect.

Driven by this observation, the first proposed smart measure transmits – in addition to the dynamic prices – a relative flexibility restriction signal $\omega \in (0, 1]$ to the appliances' control systems through the smart meter. The value of this signal represents the fraction of available flexibility that is allowed to be utilised by the appliances.

For appliances with continuously adjustable power levels, $\omega$ is interpreted by their control systems as their allowed maximum power limit as a fraction of the respective nominal one. Application of this measure in the case of EV transforms constraint (3.5) of their optimal price response problem to:

$$0 \leq P_t^{ev} \leq \omega * P_t^{max}, \quad \forall t \in T \tag{3.11}$$

For appliances with shiftable cycles, $\omega$ is interpreted by their control systems as their allowed maximum cycle delay limit as a fraction of the respective limit set by their users. Application of this measure in the case of WA transforms constraint (3.10) of their optimal price response problem to:

$$t^{act} \leq t^{in} \leq t^{act} + \omega * \delta \tag{3.12}$$

### *3.3.2 Non-linear pricing*

Imposing such flexibility restrictions may not be deemed acceptable by the consumers, as they may consider it as a direct intervention of an external entity in the control of their appliances. In this context, an alternative smart measure replaces this hard flexibility restriction by a soft non-linear price signal $\alpha$. The value of this signal represents the penalty set on the extent of flexibility utilised by the appliances.

For appliances with continuously adjustable power levels, $\alpha$ is interpreted by their control systems as a penalty on the square of their power demand and thus deters them from requesting large demand levels at a small number of time periods. Such a quadratic price follows the inclining block rate pricing concept [26], according to which the marginal price increases with the size of the consumed quantity. Application of this measure in the case of EV transforms the objective function (3.1) of their optimal price response problem to:

$$\min \sum_{t \in T} \lambda_t * P_t^{ev} + \alpha * (P_t^{ev})^2 \tag{3.13}$$

For this type of appliances, apart from the above acceptability advantage, this quadratic pricing measure exhibits an optimality advantage over the flexibility restriction approach. With the latter approach, the optimal price response problem remains linear and thus the solution always involves demand equal to the new maximum power limit at the periods with the lowest prices. With the quadratic pricing approach on the other hand, the objective function becomes quadratic, the optimal response can admit a larger number of values in the interior of the feasible operation domain, and thus a better demand flattening effect can be achieved. This advantage will be quantitatively demonstrated in Section 3.4.

For appliances with shiftable cycles, $\alpha$ is interpreted by their control systems as a penalty on the duration of their cycle delay; given that different appliances are activated by their users at different periods, this measure indirectly limits the number of appliances that can execute their cycles at the same periods. Application of this measure in the case of WA transforms the objective function (3.8) of their optimal price response problem to:

$$\min \sum_{\tau=1}^{T^{dur}} \lambda_{t^{in}+\tau-1} * P_\tau^{wa} + \alpha * (t^{in} - t^{act}) \tag{3.14}$$

### 3.3.3 *Randomised pricing*

According to the above discussion, a critical assumption for the effectiveness of both flexibility restriction and non-linear pricing measures in the case of appliances with shiftable cycles is that different appliances exhibit different activation times. As such operating diversity gets lower, the performance of both above measures gets worse. This can be better understood by considering the extreme example where all shiftable cycle appliances have identical operating parameters; irrespective of the value of $\omega$ or $\alpha$, all appliances will initiate their cycle at the same time period and the concentration effect cannot be avoided. Although the operating parameters of all appliances are not identical in reality, some parameters will be more prominent; for example, European consumers surveys' have demonstrated that a large proportion of DW are operated at late afternoon/evening times [25].

Therefore, a third smart measure, diversifying the appliances' responses by posting diversified price signals to their control systems, can bring significant efficiency gains over the uniform flexibility restriction and non-linear pricing measures, in the case of shiftable cycle appliances. Since central information on the appliances' parameters to drive the specifics of such diversification is not available, a randomisation approach is proposed. In order to reduce the complexity and computational requirements of such randomisation, the proposed measure randomises the single non-linear price signal $\alpha$ instead of the multiple linear price signals $\lambda_t$.

The population of flexible appliances is randomly allocated to a number $J$ of groups $j$. A vector $X$ is derived by a random number generator, the size of which is equal to $J$ and its elements $x_j$ take random values following the standard normal distribution (normal distribution with zero mean and unity standard deviation). The randomised non-linear price posted to appliance $i$ belonging to group $j$ is given by (3.15), where $\alpha^*$ and $\sigma$ denote, respectively, the mean value and standard deviation of the normal distribution of non-linear prices.

$$\alpha_i^{rand} = \alpha^* + \sigma * x_j \tag{3.15}$$

### 3.3.4 Tuning the parameters of smart measures

Relatively large values of $\omega$ and relatively small values of $\alpha$ and $\sigma$ may not sufficiently limit the appliances' flexibility to concentrate their demand at the lowest-priced periods, while relatively small values of $\omega$ and relatively large values of $\alpha$ and $\sigma$ may limit excessively their flexibility and thus prevent them from shaving the peaks and filling the off-peak valleys of the inflexible demand profile.

In this context, suitable values of $\omega$, $\alpha$ and $\sigma$ should be employed to achieve an effective trade-off between these two effects, leading to a flatter demand profile and thus more efficient operation solutions. As quantitatively demonstrated in Section 3.4, such suitable values will depend on the correlation between the characteristics of the flexible appliances population (number, nominal flexibility and diversity) and the temporal variation of inflexible demand. For a certain inflexible demand profile, a larger number and flexibility and a lower diversity of the flexible appliances population generally result in a smaller value of the most suitable $\omega$ and a larger value of the most suitable $\alpha$ and $\sigma$.

In order to heuristically search for a suitable value of $\omega$, $\alpha$ or $\sigma$, the system coordinator posts a set $\Omega = \{\omega_k, k = 1, 2, \ldots, K\}$, $A = \{\alpha_k, k = 1, 2, \ldots, K\}$ or $A^{rand} = \{\alpha_k^{rand}, k = 1, 2, \ldots, K\}$ (based on Section 3.3.3, each element of the last set is a vector of randomised prices submitted to the different appliances, and corresponding to a value $\sigma_k$ of the standard deviation), respectively, to the flexible appliances' control systems through the smart meter. The appliances' control systems solve their optimal price response problems for each element of the set, and for their response corresponding to each element, the system coordinator determines $K$ alternative solutions of the coordination problem and realises the most efficient one by transmitting according signals to the flexible appliances. In real implementations, it is envisaged that by employing suitable forecasting and learning algorithms, the system coordinator will be able to determine an efficient solution without the need to try out a large number of $\omega$, $\alpha$ and $\sigma$ values. The development of such techniques is out of the scope of this chapter but constitutes a significant future research challenge, as discussed in Section 3.5.

## 3.4 Case studies

### 3.4.1 Scheduling of flexible residential appliances in electricity markets

#### 3.4.1.1 Description of studies and input data

The first set of case studies deals with the participation of flexible residential appliances in electricity markets. The market operator aims at optimally scheduling the generation and the demand side of the market in order to minimise the total generation costs required for supplying the system's demand. A flatter demand profile results in lower generation costs due to the fact that the marginal cost of the generation side increases with the size of the demand. Therefore, intelligent scheduling of flexible residential appliances can bring significant generation cost savings.

Case studies are carried out on a model of the UK system. A day-ahead horizon with hourly resolution is considered for the market-clearing problem. Data regarding the generation side of the system and a typical winter day (inflexible) demand profile is taken from Reference 27.

Concerning the modelling of EV travelling patterns, each EV is assumed to carry out a journey from users' home- to work-place and a journey in the opposite direction every day. Based on publicly available data [28], the EV fleet is grouped into a set of types, each defined by the combination of the start time, end time and electrical energy requirements of each of its two daily journeys. EV are assumed connected to the grid during the period between the end of their second and the start of their first journey, in line with the home-charging scenario, deemed as the most plausible in the literature. The rest of the EV parameters are given in Table 3.1 and assumed equal for all EV in the system. In the case where smart charging capability is not available (inflexible EV operation), EV are assumed to start charging immediately after being connected to the grid until they are fully charged.

Two different types of WA are considered, namely DW and WM, executing one operational cycle per day. The total number of each type of WA in the UK and data regarding the typical activation times as well as the duration and power demand profile of a typical cycle of each type (Table 3.2) are derived from Reference 25. In the case where delay functionality is not available (inflexible WA operation), the cycle of each WA starts immediately after their users load and activate them, i.e. $\delta = 0$.

*Table 3.1 Values of EV parameters*

| Parameter | Value |
|---|---|
| $P^{max}$ | 3 kW |
| $E^{min}$ | 3 kWh |
| $E^{max}$ | 15 kWh |
| $\eta^{ch}$ | 0.93 |

*Table 3.2 Values of WA parameters*

| Parameter | Value | |
|---|---|---|
| | DW | WM |
| $T^{dur}$ | 2 h | 2 h |
| $P_1^{wa}$ | 0.56 kW | 0.78 kW |
| $P_2^{wa}$ | 0.63 kW | 0.11 kW |

Different scenarios regarding the penetration of EV in the UK system and the flexibility of WA (in terms of their maximum cycle delay limit) are examined. In all examined scenarios, the linear prices $\lambda_t$ posted to the flexible appliances have been assumed equal to their values in the case where appliances' flexibility is neglected and the whole demand is assumed inflexible, following the rationale of existing dynamic pricing schemes. These prices have been calculated as the Lagrangian multipliers associated with the demand–supply balance constraints, at the optimal solution of the generation scheduling problem without flexible demand in the system, following the marginal pricing principle.

According to the discussion in Section 3.3.4, a number of values for the flexibility restriction $\omega$, the non-linear price $\alpha$ and the standard deviation $\sigma$ were inputted into the optimal price response problems of EV and WA. Specifically, the trialled values of $\omega$ ranged from 0.05 to 1 with a step of 0.05, the values of $\alpha$ ranged from 0 to 0.03 with a step of 0.001 and the values of $\sigma$ ranged from 0 to 0.01 with a step of 0.001.

The system demand profile emerging from their response for each of the trialled $\omega$, $\alpha$ and $\sigma$ values, was inputted into a generation scheduling problem in order to determine the most suitable value $\omega^*$, $\alpha^*$ and $\sigma^*$ of the above parameters as the one resulting in the lowest system generation costs.

In the case of randomised non-linear pricing, the most suitable value of the uniform non-linear price was employed as the mean value of the distribution of non-linear prices, and the number of groups receiving differentiated non-linear prices has been set to $J = 100$.

#### 3.4.1.2 Analysis of cases with flexible EV

In this section, all WA in the system are assumed inflexible and different scenarios regarding the penetration of EV as a percentage of the total UK vehicle population are investigated. Figure 3.3 presents the system demand profile for each scenario, in the cases where (a) EV do not exhibit smart charging capability (solid lines) and (b) EV exhibit smart charging capability and are scheduled through dynamic pricing (dashed lines).

In the former case, EV start charging immediately after returning home (Section 3.4.1.1); given that most users return home during late afternoon/evening hours 17–21 when the non-EV demand peak occurs, the system peak demand at these hours

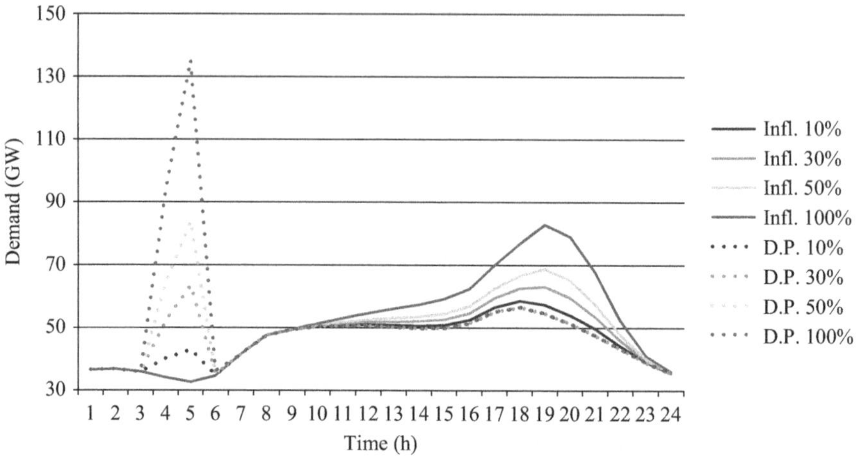


*Figure 3.3 System demand profile under inflexible EV operation and dynamic pricing in different EV penetration scenarios*

*Table 3.3 Optimal tuning and benefits of measures against demand response concentration in different EV penetration scenarios*

| Penetration | Flexibility restriction | | Non-linear pricing | | Randomised pricing | |
|---|---|---|---|---|---|---|
| (%) | $\omega^*$ | Benefit (%) | $\alpha^*$ (£/kW$^2$) | Benefit (%) | $\sigma^*$ (£/kW$^2$) | Benefit (%) |
| 10 | 0.35 | 0.15 | 0.001 | 0.15 | 0 | 0.15 |
| 30 | 0.25 | 10.01 | 0.004 | 10.27 | 0 | 10.27 |
| 50 | 0.2 | 29.17 | 0.008 | 29.72 | 0 | 29.72 |
| 100 | 0.15 | 57.40 | 0.023 | 57.70 | 0 | 57.70 |

is significantly increased. In the latter case, the demand response of smart-charging EV is concentrated and creates a new demand peak at the late night hours of the day (4–5), since the latter exhibit the lowest prices due to their low inflexible demand levels. This concentration effect is enhanced as the number of EV is increased and for an EV penetration higher than 30% the new demand peak is even higher than the evening peak under inflexible EV operation.

In order to address this concentration effect, the three smart measures described in Section 3.3 are employed. Table 3.3 presents the most suitable values $\omega^*$, $\alpha^*$ and $\sigma^*$ of the parameters of the three measures, as well as their benefits in terms of generation cost savings (with respect to the case where no measure against demand response concentration is applied), for each of the examined EV penetration scenarios.

As the penetration of EV is enhanced, the size and the cost implications of the new peaks created in the case without measures are significantly aggravated; as a result,

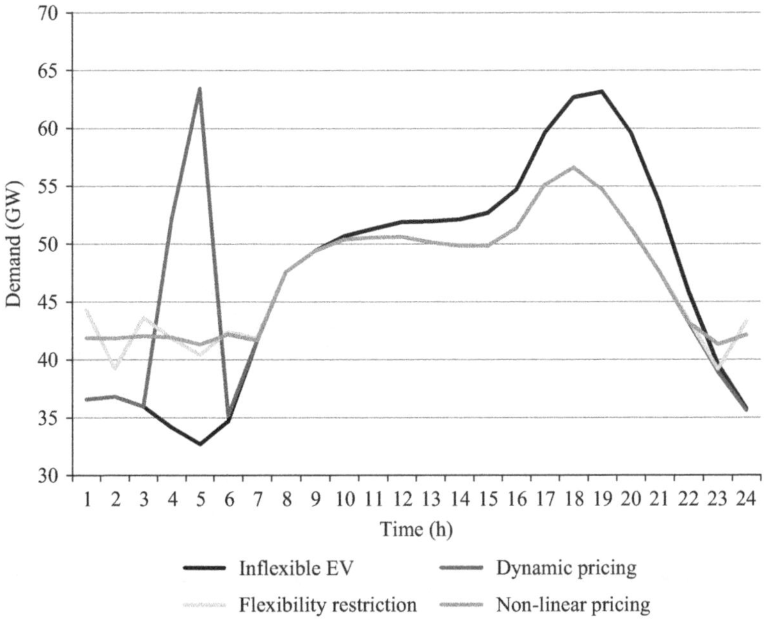


*Figure 3.4 System demand profile under different EV scheduling approaches in 30% EV penetration scenario*

the benefits of flexibility restriction and non-linear pricing are increased. Such benefits get dramatically higher in the 50% and 100% EV penetration scenarios, because the emerging demand peak under pure dynamic pricing cannot be satisfied by the existing generation capacity in the UK and requires expensive demand shedding. Furthermore, $\omega^*$ and $\alpha^*$ are decreased and increased, respectively, as a more significant restriction and penalisation of EV flexibility, respectively, is required to avoid the demand concentration effect. In the case of EV, randomisation of the non-linear prices does not bring additional benefits and the most suitable value of the standard deviation is $\sigma^* = 0$ in all examined scenarios.

Figure 3.4 illustrates the system demand profile in the 30% EV penetration scenario, for the case where EV are inflexible as well as the case where EV are flexible and they are scheduled through (i) pure dynamic pricing (without any measure against demand response concentration), (ii) dynamic pricing along with the flexibility restriction measure (tuned with $\omega^* = 0.25$ according to Table 3.3) and (iii) dynamic pricing along with the non-linear pricing measure (tuned with $\alpha^* = 0.004$ according to Table 3.3). Both flexibility restriction and non-linear pricing measures prevent the concentration effect and efficiently distribute the EV demand across the off-peak valley. For the reason analysed in Section 3.3.2, non-linear pricing yields a flatter demand profile at the off-peak hours (Figure 3.4) and therefore higher generation cost savings (Table 3.3) than the flexibility restriction approach.

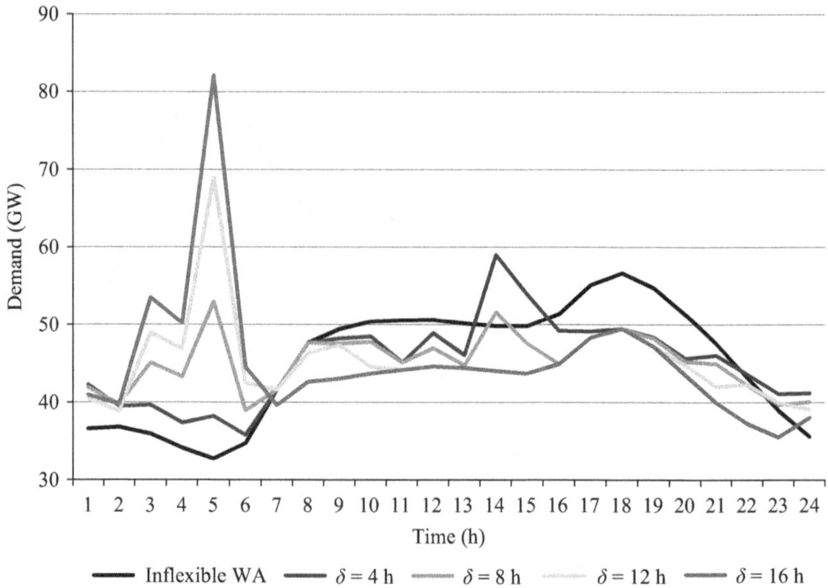


*Figure 3.5 System demand profile under dynamic pricing in different WA flexibility scenarios*

#### 3.4.1.3 Analysis of cases with flexible WA

In this section, it is assumed that the system does not include EV and different scenarios regarding the flexibility of WA in terms of their maximum cycle delay limit are investigated. Figure 3.5 presents the system demand profile for each scenario when the WA are scheduled through dynamic pricing.

In the same fashion with the case of smart-charging EV investigated in the previous section, the demand response of WA with delay functionality is concentrated and creates a new demand peak at the late night hours of the day, since the latter exhibit the lowest prices. This concentration effect is enhanced as the maximum cycle delay of WA is increased, as a larger number of WA are enabled to execute their cycles during these lowest-priced hours (Section 3.3.1), and for a maximum cycle delay of 12 h and 16 h the new demand peak is notably higher than the evening peak under inflexible WA operation.

In order to address this concentration effect, the three smart measures described in Section 3.3 are employed. Table 3.4 presents the most suitable values $\omega^*$, $\alpha^*$ and $\sigma^*$ of the parameters of the three measures, as well as their benefits in terms of generation cost savings (with respect to the case where no measure against demand response concentration is applied), for each of the examined WA flexibility scenarios.

As the WA maximum cycle delay is increased, the size and the cost implications of the new peaks created in the case without measures are significantly aggravated; as a result, the benefits of the three measures are increased. Furthermore, $\omega^*$ and $\alpha^*$

*Table 3.4 Optimal tuning and benefits of measures against demand response concentration in different WA flexibility scenarios*

| Maximum cycle delay (h) | Flexibility restriction | | Non-linear pricing | | Randomised pricing | |
|---|---|---|---|---|---|---|
| | $\omega^*$ | Benefit (%) | $\alpha^*$ (£/h) | Benefit (%) | $\sigma^*$ (£/h) | Benefit (%) |
| 4 | 1 | 0 | 0.001 | 0.34 | 0.001 | 1.29 |
| 8 | 1 | 0 | 0.001 | 0.52 | 0.002 | 2.96 |
| 12 | 0.7 | 1.79 | 0.001 | 0.93 | 0.007 | 8.53 |
| 16 | 0.5 | 5.00 | 0.002 | 4.09 | 0.009 | 19.06 |

are decreased and increased, respectively, as a more significant restriction and penalisation of WA flexibility, respectively, is required to avoid the demand concentration effect. In contrast with the EV cases, the comparison between flexibility restriction and non-linear pricing does not lead to a definite conclusion, as the latter yields higher benefits in the first two scenarios and the former yields higher benefits in the other two.

Figure 3.6 illustrates the system demand profile for the case where WA are inflexible as well as the case where WA are flexible with $\delta = 12\,\text{h}$ and they are scheduled through (i) pure dynamic pricing, (ii) dynamic pricing along with the non-linear pricing measure (tuned with $\alpha^* = 0.001$ according to Table 3.4) and (iii) dynamic pricing along with the randomised non-linear pricing measure (tuned with $\sigma^* = 0.007$ according to Table 3.4). Uniform non-linear pricing partially limits the new peak created by the WA response concentration effect. In contrast with EV cases, and for the reason analysed in Section 3.3.3, randomisation of the non-linear prices yields a much flatter demand profile at the off-peak hours (Figure 3.6) and therefore much higher generation cost savings (Table 3.4). As the maximum cycle delay is increased, the diversity of WA in terms of the time window they can execute their cycles is reduced; as a result, the benefits of randomised over uniform non-linear pricing are enhanced and the most suitable value of the standard deviation $\sigma^*$ is increased (Table 3.4).

## 3.4.2 *Scheduling of flexible residential appliances for management of local distribution networks*

### 3.4.2.1 Description of studies and input data

The second set of case studies deals with the participation of flexible residential appliances in the management of local distribution networks. The DNO aims at optimally scheduling flexible residential appliances in order to (a) satisfy thermal and voltage constraints of the network in a cost-effective manner and (b) minimise the amount of network losses. A flatter demand profile contributes to both objectives, since: (a) demand peaks breaching the constraints of the network and therefore requiring

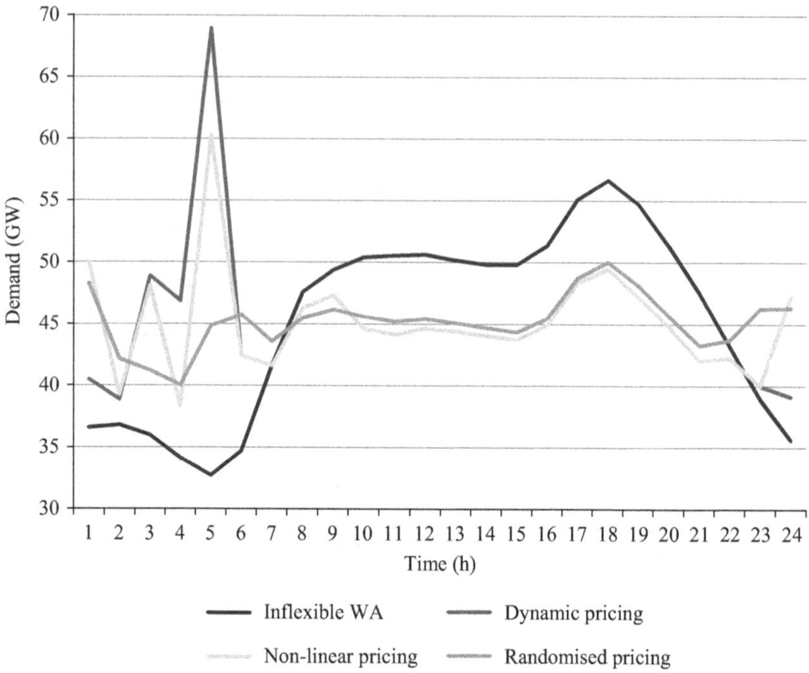


*Figure 3.6 System demand profile under different WA scheduling approaches in $\delta = 12$ h WA flexibility scenario*

expensive demand shedding are limited and (b) network losses are reduced since they are proportional to the square of power demand. Therefore, intelligent scheduling of flexible residential appliances can bring significant benefits to the operation of the local distribution network.

Case studies are carried out on a model of a test distribution feeder in Brixton, London, UK (Figure 3.7). As in the previous set of studies, a day-ahead horizon with hourly resolution is considered for the network operation problem. Data regarding the network and the inflexible demand is taken from Reference 29. Similar data and assumptions with Section 3.4.1.1 are employed to model EV and WA operational characteristics. Furthermore, similar scenarios of EV penetration and WA flexibility are investigated. The numbers of EV at each bus of the feeder under each penetration scenario are given in Table 3.5, while the numbers of WA at each bus of the feeder are given in Table 3.6.

A similar approach with the one described in Section 3.4.1.1 is followed to determine the prices posted to the flexible appliances by the DNO. The latter solves an AC optimal power flow (ACOPF) problem neglecting appliances' flexibility, and the prices are set equal to the values of the resulting Lagrangian multipliers associated with the power balance at each bus. The difference with respect to the studies

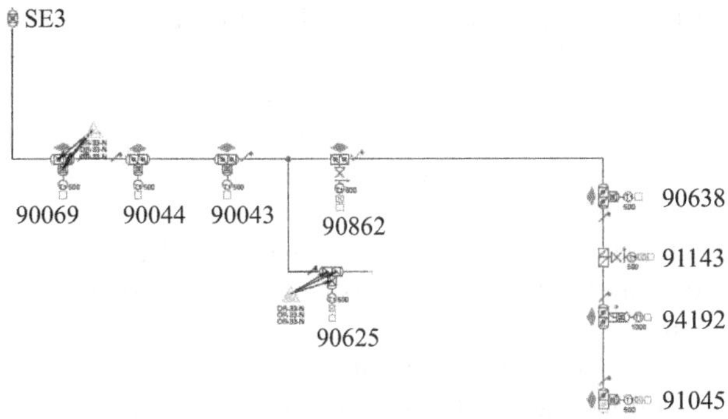


*Figure 3.7 Test distribution feeder*

*Table 3.5 Number of EV per feeder bus*

| **Bus ID** | **Number of EV at each penetration scenario** | | | | |
|---|---|---|---|---|---|
| | **10%** | **20%** | **30%** | **50%** | **100%** |
| 90069 | 18 | 36 | 54 | 90 | 180 |
| 90044 | 5 | 10 | 15 | 25 | 50 |
| 90043 | 6 | 12 | 18 | 30 | 60 |
| 90625 | 13 | 26 | 39 | 65 | 130 |
| 90862 | 2 | 4 | 6 | 10 | 20 |
| 90638 | 12 | 24 | 36 | 60 | 120 |
| 91143 | 18 | 36 | 54 | 90 | 180 |
| 94192 | 52 | 104 | 156 | 260 | 520 |
| 91045 | 190 | 380 | 570 | 950 | 1,900 |

*Table 3.6 Number of WA per feeder bus*

| **Bus ID** | **Dishwashers** | **Washing machines** |
|---|---|---|
| 90069 | 75 | 169 |
| 90044 | 21 | 47 |
| 90043 | 25 | 57 |
| 90625 | 54 | 122 |
| 90862 | 8 | 19 |
| 90638 | 50 | 112 |
| 91143 | 75 | 169 |
| 94192 | 215 | 488 |
| 91045 | 786 | 1,783 |

of Section 3.4.1 lies in the fact that prices are location-specific, reflecting the effect of network congestion and losses.

Furthermore, following the approach discussed in Section 3.4.1.1, a number of values for the flexibility restriction $\omega$, the non-linear price $\alpha$ and the standard deviation $\sigma$ were inputted into the optimal price response problems of EV and WA. The feeder demand profile emerging from their response for each of the trialled $\omega$, $\alpha$ and $\sigma$ values, was inputted into an ACOPF problem in order to determine the most suitable value $\omega^*$, $\alpha^*$ and $\sigma^*$ of the above parameters as the one resulting in the most efficient network operation.

In the case of randomised non-linear pricing, the most suitable value of the uniform non-linear price was employed as the mean value of the distribution of non-linear prices, and each appliance has received differentiated non-linear prices (the number of groups receiving differentiated non-linear prices has been set equal to the number of flexible appliances at each scenario).

### 3.4.2.2 Analysis of cases with flexible EV

In this section, all WA in the network are assumed inflexible and different scenarios regarding the penetration of EV are investigated.

As in the studies of Section 3.4.1.2, when EV do not exhibit smart charging capability, the overall peak demand of the feeder during late afternoon/evening hours is significantly increased. As a result, the power flow at the top feeder section (section between Bus 910 and Bus 90069) breaches its thermal capacity limit in every EV penetration scenario. In order to securely operate the network, the DNO needs to shed some demand. As observed in Figure 3.8, the demand shed is increased with a higher EV penetration.

When EV exhibit smart charging capability and are scheduled through dynamic pricing, their demand response is concentrated and creates a new demand peak at the lowest-priced (late night) hours of the day. This concentration effect is enhanced as the number of EV is increased, and under the 50% and 100% penetration scenarios, the power flow at the top feeder section breaches its thermal capacity limit, leading to demand shedding (Figure 3.8). It is worth observing that under the 100% penetration scenario the amount of demand shed under dynamic pricing is almost equal to the respective amount under inflexible EV operation.

In order to address this concentration effect and avoid expensive demand shedding, the three smart measures described in Section 3.3 are employed. In line with the results of Section 3.4.1.2, randomisation of the non-linear prices does not bring additional benefits (the most suitable value of the standard deviation is $\sigma^* = 0$ in all examined scenarios) and therefore is not examined further in this section. Table 3.7 presents the most suitable values $\omega^*$ and $\alpha^*$ of the parameters of the flexibility restriction and non-linear pricing measures for each of the examined EV penetration scenarios. In line with the results of Section 3.4.1.2, as the EV penetration is increased, $\omega^*$ and $\alpha^*$ are decreased and increased, respectively, as a more significant restriction and penalisation of EV flexibility, respectively, is required to avoid the demand concentration effect.

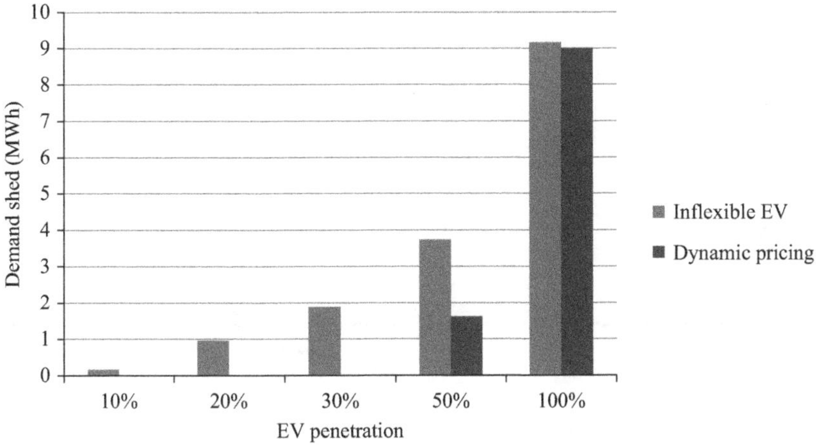


*Figure 3.8 Demand shed in different EV penetration scenarios*

*Table 3.7 Optimal tuning of measures against demand response concentration in different EV penetration scenarios*

| **Parameters** | **EV penetration** | | | | |
|---|---|---|---|---|---|
| | **10%** | **20%** | **30%** | **50%** | **100%** |
| $\omega^*$ | 0.4 | 0.35 | 0.3 | 0.25 | 0.2 |
| $\alpha^*$ (p/kW$^2$) | 0.0005 | 0.001 | 0.0015 | 0.002 | 0.005 |

Figure 3.9 illustrates the power flow at the top feeder section in the 100% EV penetration scenario and Figure 3.10 illustrates the network losses in each EV penetration scenario, for the case where EV are inflexible as well as the case where EV are flexible and they are scheduled through (i) pure dynamic pricing (without any measure against demand response concentration), (ii) dynamic pricing along with the flexibility restriction measure (tuned according to the values of Table 3.7) and (iii) dynamic pricing along with the non-linear pricing measure (tuned according to the values of Table 3.7).

Both flexibility restriction and non-linear pricing measures mitigate the concentration effect and efficiently distribute the EV demand across the off-peak valley. As a result, the thermal constraint of the top feeder section is not breached (Figure 3.9), and therefore demand shedding is not required. Furthermore, as observed in Figure 3.10, the flatter demand profile is translated into a reduction of network losses with respect to pure dynamic pricing, given that losses are proportional to the square of the power

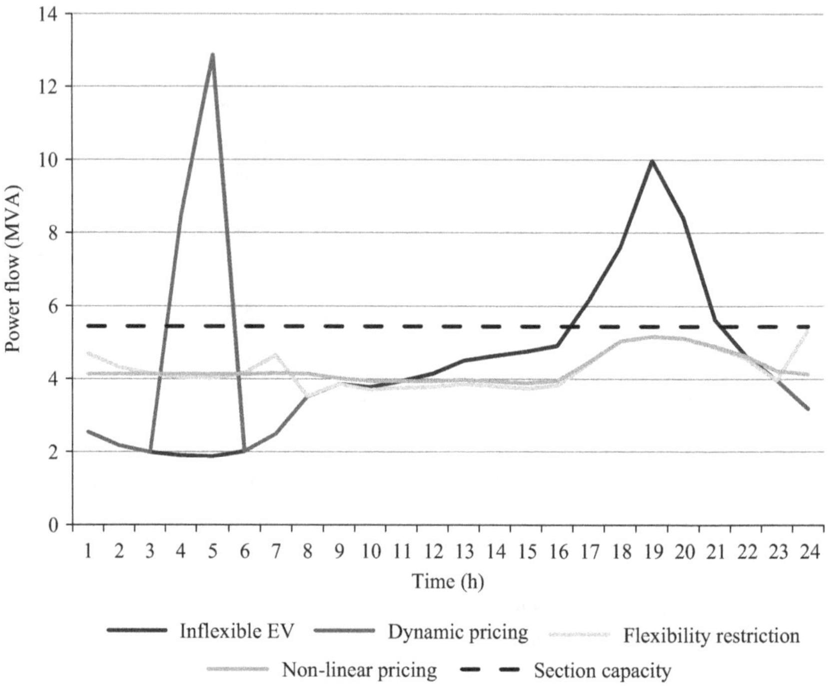


*Figure 3.9 Power flow on top feeder section under different EV scheduling approaches in 100% EV penetration scenario*

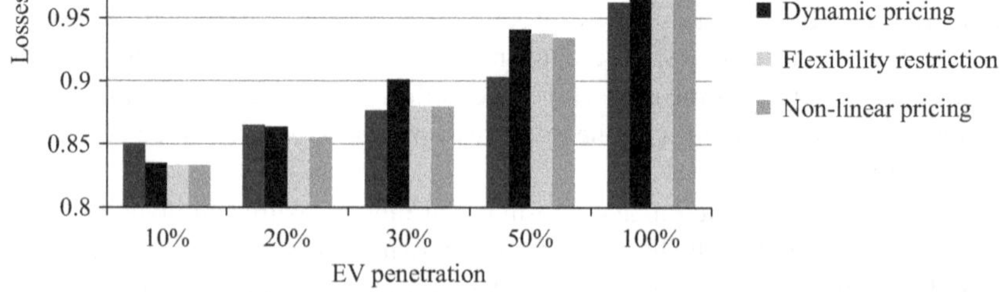


*Figure 3.10 Network losses under different EV scheduling approaches in different EV penetration scenarios*

demand (the 100% EV penetration scenario constitutes an exception as the significant required demand shedding under dynamic pricing reduces losses).

For the reason analysed in Section 3.3.2, non-linear pricing yields a flatter demand profile at off-peak hours (Figure 3.9) and therefore lower network losses (Figure 3.10) than the flexibility restriction approach.

#### 3.4.2.3 Analysis of cases with flexible WA

In this section, a 10% EV penetration is assumed; all EV are assumed inflexible while different scenarios regarding the flexibility of WA are investigated.

When WA do not exhibit delay functionality, the power flow at the top feeder section breaches its thermal capacity limit during late afternoon/evening hours due to EV demand (Section 3.4.2.2), and the DNO needs to shed demand in order to securely operate the network (Figure 3.11). When WA exhibit delay functionality and are scheduled through dynamic pricing, they shift their cycles away from these hours, since the latter exhibit the highest prices due to their high inflexible demand levels. As a result, demand shedding during late afternoon/evening hours is avoided. However, in a similar fashion with the studies of Section 3.4.1.3, the WA demand response is concentrated and creates a new demand peak at the lowest-priced (late night) hours of the day. This concentration effect is enhanced as the maximum cycle delay of WA is increased, and under the $\delta = 20\,\text{h}$ scenario, the power flow at the top feeder section breaches its thermal capacity limit during these hours, leading to demand shedding (Figure 3.11); it is worth observing that this amount of demand shed is much higher than the respective amount under inflexible WA operation.

In order to address this concentration effect and avoid expensive demand shedding, the three smart measures described in Section 3.3 are employed. The performance of flexibility restriction and non-linear pricing measures is similar, so only the latter and randomised pricing are explored in the remainder of this section. Table 3.8 presents the most suitable values $\alpha^*$ and $\sigma^*$ of the parameters of the two

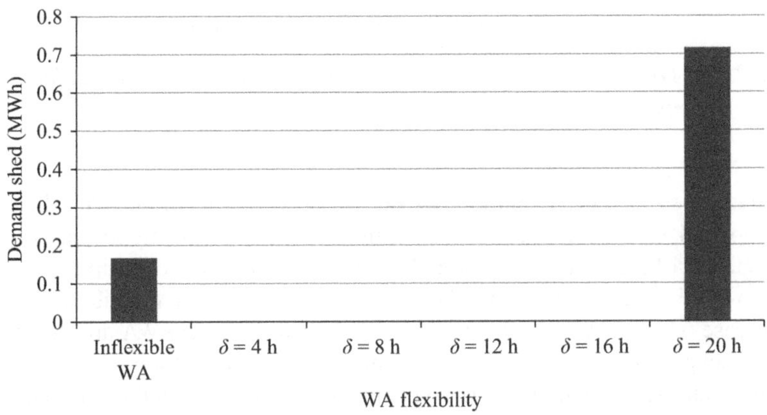


*Figure 3.11 Demand shed under dynamic pricing in different WA flexibility scenarios*

*Table 3.8 Optimal tuning of different measures against demand response concentration in different WA flexibility scenarios*

| **Parameters** | **WA maximum cycle delay** | | | | |
|---|---|---|---|---|---|
| | $\delta = 4$ **h** | $\delta = 8$ **h** | $\delta = 12$ **h** | $\delta = 16$ **h** | $\delta = 20$ **h** |
| $\alpha^*$ (p/h) | 0.0001 | 0.0002 | 0.0002 | 0.0002 | 0.0003 |
| $\sigma^*$ (p/h) | 0.00005 | 0.0002 | 0.0003 | 0.0005 | 0.0008 |

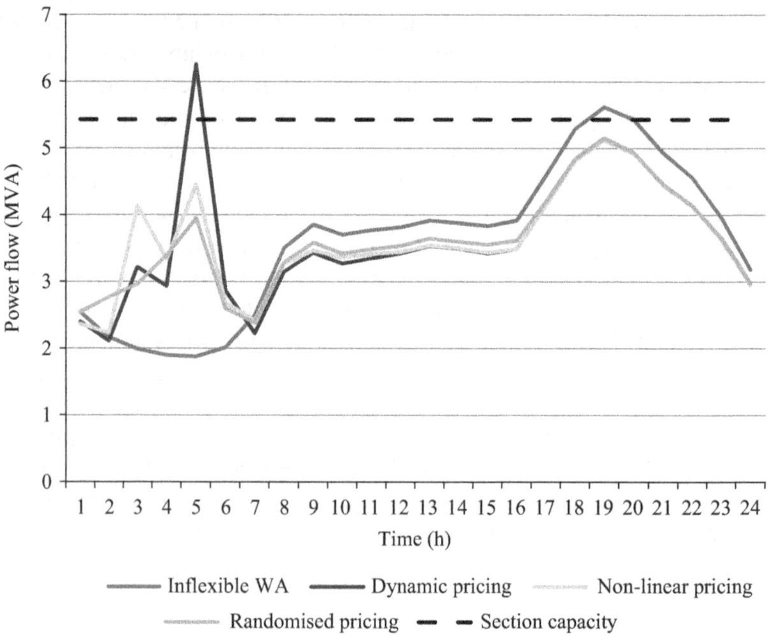


*Figure 3.12 Power flow on top feeder section under different WA scheduling approaches in $\delta = 20$ h WA flexibility scenario*

measures for each of the examined WA flexibility scenarios. In line with the results of Section 3.4.1.3, as the WA flexibility is increased, $\alpha^*$ and $\sigma^*$ are increased.

Figure 3.12 illustrates the power flow at the top feeder section in the $\delta = 20$ h scenario and Figure 3.13 illustrates the network losses in each WA flexibility scenario, for the case where WA are inflexible as well as the case where WA are flexible and they are scheduled through (i) pure dynamic pricing (without any measure against demand response concentration), (ii) dynamic pricing along with the non-linear pricing measure (tuned according to the values of Table 3.8) and (iii) dynamic pricing along with the randomised non-linear pricing measure (tuned according to the values of Table 3.8).

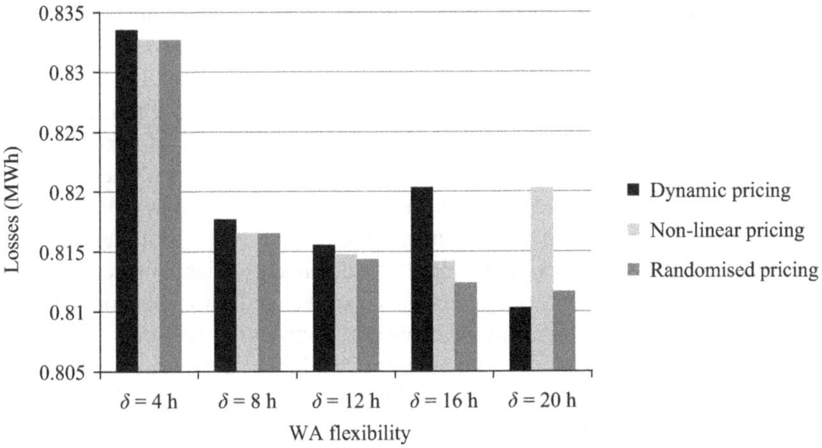


*Figure 3.13 Network losses under different WA scheduling approaches in different WA flexibility scenarios*

Both non-linear pricing and randomised pricing measures mitigate the concentration effect and avoid the need for demand shedding (Figure 3.12). Furthermore, as observed in Figure 3.13, the flatter demand profile is translated into a reduction of network losses with respect to pure dynamic pricing, given that losses are proportional to the square of the power demand (the $\delta = 20$ h scenario constitutes an exception as the significant required demand shedding under dynamic pricing reduces losses).

For the reason analysed in Section 3.3.3, randomisation of the non-linear prices yields a flatter demand profile at off-peak hours (Figure 3.12) and therefore lower network losses (Figure 3.13) than the uniform non-linear pricing measure.

## 3.5 Conclusions and future work

Dynamic pricing, supported by the envisaged roll-out of smart metering, constitutes a promising approach for the realisation of flexible residential appliances' potential in future power systems, due to its scalability and privacy advantages over traditional centralised scheduling architectures. However, this chapter has demonstrated that naive application of dynamic pricing leads to new demand peaks since appliances' response is concentrated at the periods with the lowest prices. In the context of electricity markets, such response concentration effects yield high generation costs, while in the context of local distribution networks management they yield high network losses and potentially expensive demand shedding required to operate the networks within their operating constraints. The size of these concentration effects and their adverse implications in system operation are aggravated as the penetration and flexibility extent of flexible residential appliances is increased, and in certain cases dynamic pricing was demonstrated to yield worse results than totally neglecting appliances flexibility.

This chapter presents, analyses, and compares three different smart measures to avoid such concentration effects and achieve more efficient system operation. The design of these measures is customised to the specific operating properties of different types of flexible residential appliances, namely appliances with continuously adjustable power levels and appliances with shiftable cycles. Smart-charging EV and WA with delay functionality are used as representative examples of these two types, due to their significant penetration and flexibility potential. When the parameters of these smart measures are suitably tuned, the latter are demonstrated to mitigate response concentration effects and yield flatter demand profiles. This results in lower generation costs in the context of electricity markets and lower network losses and demand shedding requirements in the context of distribution networks management. These benefits are demonstrated to increase with a higher penetration and flexibility extent of flexible residential appliances, highlighting the significance of these smart measures in future power systems.

The first measure imposes a relative flexibility restriction on residential appliances. In case of appliances with continuously adjustable power levels, this restriction corresponds to a maximum power limit, preventing each of these appliances from requesting a large proportion of its total energy requirements at the lowest-priced periods. In case of appliances with shiftable cycles, this restriction corresponds to a maximum cycle delay limit, preventing them from synchronising their operation at the lowest-priced periods.

Since imposing flexibility restrictions may be considered by the consumers as a direct intervention in the control of their appliances, an alternative proposed measure replaces the hard flexibility restriction by a soft non-linear price signal, penalising the extent of flexibility utilised by the appliances. Specifically, this price penalises the square of the power demand and the duration of cycle delay of continuously adjustable and shiftable cycle appliances, respectively. Regarding the former type, apart from the above acceptability advantage, this non-linear pricing approach is demonstrated to outperform the flexibility restriction approach in flattening the demand profile and thus achieving more efficient operation solutions. Regarding the latter type, a third proposed smart measure randomising the non-linear price signal posted to different appliances is demonstrated to bring significant additional benefits.

The parameters of these smart measures need to be suitably tuned in order to achieve an effective trade-off between avoiding demand response concentration while still enabling the benefits of demand flexibility. As demonstrated in this chapter, the most suitable values of these parameters depend on the correlation between the characteristics of the flexible appliances population (number, nominal flexibility and diversity) and the temporal variation of inflexible demand. For a certain inflexible demand profile, a larger number and flexibility of the flexible appliances population generally result in a smaller value of the most suitable relative flexibility restriction $\omega$ and a larger value of the most suitable non-linear price $\alpha$, since a more significant restriction and penalisation of appliances' flexibility, respectively, is required to avoid the demand concentration effect. Furthermore, a larger flexibility of shiftable cycle appliances results in a larger value of the most suitable standard deviation $\sigma$ associated with the randomisation of non-linear prices, since the diversity of appliances in terms of the time window they can execute their cycles is reduced.

In this chapter, this tuning has been carried out by heuristically trying out a number of different values. Future work aims at developing forecasting and learning algorithms in order to achieve efficient tuning of these parameters without the need to try out a large number of $\omega$, $\alpha$ and $\sigma$ values. This constitutes a challenging task given that central information on the characteristics of the flexible appliances population is not available.

## Bibliography


[1] *Building a Low Carbon Economy – The UK's Contribution to Tackling Climate Change*, Committee on Climate Change, London, UK, 2008.

[2] *Directive of the European Parliament and of the Council on the Promotion of the Use of Energy from Renewable Sources*, Commission of the European Communities, Brussels, 2008.

[3] *Meeting the Energy Challenge: A White Paper on Energy*, Department of Trade and Industry, UK, May 2007.

[4] *Investigation into the Scope for the Transport Sector to Switch to Electric Vehicles and Plug-in Hybrid Vehicles*, Department for Transport, UK, October 2008.

[5] *2050 Pathways Analysis*, Department of Energy and Climate Change, UK, July 2010.

[6] *Technology Roadmap: Electric and Plug-in Hybrid Electric Vehicles*, International Energy Agency, June 2011.

[7] C.K. Gan, M. Aunedi, V. Stanojevic, G. Strbac and D. Openshaw, "Investigation of the impact of electrifying transport and heat sectors on the UK distribution networks," in *2011 CIRED Conference*, Frankfurt, Germany, June 2011.

[8] A. Sallam and O. Malik, "Demand Side Management and Energy Efficiency," *Electric Distribution Systems*, Hoboken, NJ, USA, pp. 433–468, Wiley-IEEE Press, 2011.

[9] D.S. Kirschen, "Demand-side view of electricity markets," *IEEE Transactions Power Systems*, vol. 18, no. 2, pp. 520–527, May 2003.

[10] G. Strbac, "Demand side management: Benefits and challenges," *Energy Policy*, vol. 36, no. 12, pp. 4419–4426, December 2008.

[11] M.H. Albadi and E.F. El-Saadany, "A summary of demand response in electricity markets," *Electric Power Systems Research*, vol. 78, no. 11, pp. 1989–1996, November 2008.

[12] M. Venables, "Smart meters make smart consumers," *Engineering & Technology*, vol. 2, no. 4, p. 23, April 2007.

[13] S. Davies, "Smart meters," *Engineering & Technology*, vol. 7, no. 6, pp. 48–49, July 2012.

[14] G. Strbac, C.K. Gan, M. Aunedi, *et al.*, *Benefits of Advanced Smart Metering for Demand Response Based Control of Distribution Networks*, Summary

report, Centre for Sustainable Electricity and Distributed Generation, April 2010.

[15] D.S. Kirschen, G. Strbac, P. Cumperayot and D. de Paiva Mendes, "Factoring the elasticity of demand in electricity prices," *IEEE Transactions Power Systems*, vol. 15, no. 2, pp. 612–617, May 2000.

[16] C.-L. Su and D. Kirschen, "Quantifying the effect of demand response in electricity markets," *IEEE Transactions Power Systems*, vol. 24, no. 3, pp. 1199–1207, August 2009.

[17] K. Singh, N.P. Padhy and J. Sharma, "Influence of price responsive demand shifting bidding on congestion and LMP in pool-based day-ahead electricity markets," *IEEE Transactions Power Systems*, vol. 26, no. 2, pp. 886–896, May 2011.

[18] A. Khodaei, M. Shahidehpour and S. Bahramirad, "SCUC with hourly demand response considering intertemporal load characteristics," *IEEE Transactions Smart Grid*, vol. 2, no. 3, pp. 564–571, September 2011.

[19] A.P. Sanghvi, "Flexible strategies for load/demand management using dynamic pricing," *IEEE Transactions Power Systems*, vol. 4, no. 1, pp. 83–93, February 1989.

[20] A.J. Svoboda and S.S. Oren, "Integrating price-based resources in short-term scheduling of electric power systems," *IEEE Transactions Energy Conversion*, vol. 9, no. 4, pp. 760–769, December 1994.

[21] D. Papadaskalopoulos and G. Strbac, "Decentralized participation of flexible demand in electricity markets – Part I: Market mechanism," *IEEE Transactions Power Systems*, vol. 28, no. 4, pp. 3658–3666, November 2013.

[22] D. Papadaskalopoulos, D. Pudjianto and G. Strbac, "Decentralized coordination of microgrids with flexible demand and energy storage," *IEEE Transactions Sustainable Energy*, vol. 5, no. 4, pp. 1406–1414, October 2014.

[23] D. Papadaskalopoulos and G. Strbac, "Non-linear and randomized pricing for distributed management of flexible loads," *IEEE Transactions Smart Grid*, vol. 7, no. 2 pp. 1137–1146, March 2016.

[24] J.A.P. Lopes, F.J. Soares and P.M.R. Almeida, "Integration of electric vehicles in the electric power system," *IEEE Proceedings*, vol. 99, no. 1, pp. 168–183, January 2011.

[25] R. Stamminger (ed.), *Synergy Potential of Smart Domestic Appliances in Renewable Energy Systems*. University of Bohn, Shaker Verlag, Aachen, 2009.

[26] A.H. Mohsenian-Rad and A. Leon-Garcia, "Optimal residential load control with price prediction in real-time electricity pricing environments," *IEEE Transactions Smart Grid*, vol. 1, no. 2, pp. 120–133, September 2010.

[27] *National Grid Website*, UK [Online]. Available from www.nationalgrid.com/uk. Accessed on October 2012.

[28] *2008 National Travel Survey*, Department for Transport, UK [Online]. Available from http://www.dft.gov.uk. Accessed on October 2012.

[29] I. Konstantelos, D. Papadaskalopoulos, D. Pudjianto, M. Woolf and G. Strbac, "Novel commercial arrangements for smart distribution networks," *Report D5 for the Low Carbon London LCNF project*, Imperial College London, 2014.

# Chapter 4
# Smart tariffs for demand response from smart metering platform

*Chenghong Gu*[1], *Zhimin Wang*[2] *and Furong Li*[1]

## 4.1 Introduction

The rollout of the smart metering platform provides a good opportunity for electricity customers to participate in demand side response (DSR). They can reduce, shift or increase electricity use in response to economic signals, incentives or technical control signals. The customer behaviour change could bring many benefits and flexibilities to the electricity networks. Reference 1 has identified a range of potential benefits of DSR on improving distribution network investment efficiency:

- deferring new network investment;
- increasing the amount of distributed generation that can be connected to the existing distribution network infrastructure;
- relieving voltage-constrained power transfer problems;
- relieving congestion in distribution substations;
- simplifying outage management and enhancing the quality and security of supply to critical-load customers;
- providing corresponding carbon reduction.

The role of DSR has been highly recognised by the government and industry because of its remarkable impacts on economic benefit savings and carbon emission reductions. According to the UK statistics, if 10% of its peak demand is shifted to off-peak periods, the maximum daily energy cost reduction could reach £1.7 m, annual network investment cost saving is £28 m, and 2,550 t of $CO_2$ emission reduction can be realised.

Because of its advantages, DSR has been identified as an important area of future innovation enabled by smart metering and forms part of the Department of Energy and Climate Change (DECC) and the Office of Gas and Electricity Markets' (Ofgem's) broader strategies for the energy market. In April 2014, Ofgem launched a project– 'Smarter Markets Programme' to realise the opportunities of settling half-hourly tariffs for domestic and smaller non-domestic electricity customers [2].

[1]Department of Electronic and Electrical Engineering, University of Bath, Bath, BA2 7AY, UK
[2]JiBei Electric Power Company Limited, Beijing, 100053, China

They have also directed the Balancing and Settlement Code Panel to consult on the implementation date of a modification that would mandate larger non-domestic electricity consumers to be settled using half-hourly consumption data.

A smart meter records the time and day that a customer uses electricity and it has two key features: time-based measurement and two-way communication [3]. It allows frequent data and information exchange between the meter and the utility, and at the same time allows customers to have timely and easily accessible information about their usage, enabling time-based pricing and other types of demand response.

In order to accommodate this new environment, the design of smart tariff for DSR and study of their effectiveness in enabling DSR are becoming fundamentally important. In an Ofgem's report on DSR [4], it stresses that

> The rollout of smart meters will enable electricity suppliers to deploy new time-of-use (ToU) based approaches to pricing electricity. These have the potential to stimulate customers to shift demand from peak to off-peak periods or simply to reduce peak demand, in both cases reducing system costs and improving system efficiency.

Currently, however, the vast majority of domestic electricity customers are on flat or average rates that do not vary by time of day or season, no matter how much the cost to generate or deliver electricity fluctuates in respond to demand/generation variations.

This chapter first discusses the current tariff products used by the industry and some desirable features of tariffs needed for facilitating DSR from domestic customers. Thereafter, two ToU design approaches are proposed to convert variable tariffs into ToU tariffs, where the key information of unit prices and tariff variation patterns are maintained properly. The first method determines ToU tariff time windows according to the distribution of real-time pricing (RTP) prices throughout a year with equal interval grouping. The settlement periods are classified into different groups based on a price variation envelope and the time window can be determined without the perturbations of critical peak or trough energy prices. The second method, which employs hierarchical clustering, is able to integrate the settlement periods with similar prices into a cluster. Meanwhile, this approach provides a solid theoretical foundation for ToU pricing. The results show that the RTP tariffs, under all the scenarios, can be represented by ToUs with three tariff rates and no more than eight-time intervals per day.

## 4.2 Electricity tariff review

In facilitating DSR in a deregulated market, price signals are a key factor to affect customers' decision in electricity use. The price is made up of a number of key cost elements of electricity generation and transportation, including energy cost in the wholesale market, transmission and distribution cost in networks, the environmental costs, etc. An example of electricity price composition in the UK in Figure 4.1 [5] is

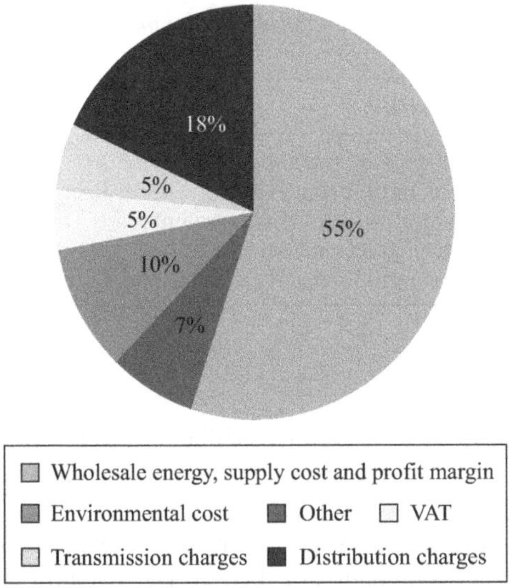


*Figure 4.1 Breakdown for electricity price*

selected to show its breakdown elements. These percentages are obtained by averaging annual costs across all the incumbent suppliers and payment methods across the UK. Obviously, generation cost accounts for over half of the total cost and the proportion of transmission and distribution network charges reaches 23%, where distribution network charge takes up approximately 18%. The tariff composition indicates that the wholesale energy cost and distribution cost need to be carefully considered to retain the cost reflectiveness.

### *4.2.1 Current energy tariff products*

Currently, a range of energy products based on flat-rate tariffs are supplied by large-energy retail companies to domestic consumers. In the UK, flat-rate tariffs are the most common electricity prices products for small customers with a capacity below 100 kW. Conventional flat-rate household tariffs were developed in the 1960s [6] and they reflect the total cost of energy generation, transmission, distribution and supply. Currently, the vast majority of consumers purchase their electricity from suppliers on flat-rate tariffs with no price variations throughout a calendar day or a year. Two types of commonly available flat-rate tariffs for domestic consumers are: standing-charge tariffs and two-tier tariffs.

- Standing-charge tariff: Standing tariff is a fixed annual amount of cost paid by customers to suppliers. It includes all the costs of meter reading, maintenance, network connection, etc. The annual standing charge is averaged at £54.35 across UK's six electricity suppliers [7].

*Table 4.1 Example of tariff products provided by the UK suppliers*

| Retail companies | Sample tariffs |
|---|---|
| Southern Electric | Better Plan, Go Direct, Fix Priced Product |
| E.ON | Standard Tariff, Fixed Price, Go Green, Energy Saver Capped Product, Standard Tariff, Energy Discount 5, Fix Online, Energy Saver Capped Product, Track and Save |
| British Gas | Standard Tariff, Websaver 4 and 5, Online Saver |

- Two-tier tariff: Consumers under this tariff are subject to two-tier unit prices and the fixed cost is built into the unit rate: (i) Tier 1 unit price is applied to the first block of consumers' energy use, recovering suppliers' fixed cost; (ii) Tier 2 unit price is applied to electricity usage at and above the first tier of consumption, recovering suppliers' total operational costs. Tier 1 unit price is normally higher than that of Tier 2. The average Tier 1 and Tier 2 unit charges are approximately 17.06p and 12.46p, respectively [7].

In addition to flat-rate tariffs, all UK suppliers offer Economy 7 and Economy 10 [8] tariffs, which have significantly less customer volume, around 9.7% of total UK customers. These tariffs introduce cheaper night or afternoon rates, and the '7'/'10' means seven/ten hours of lower rate electricity. In practice, the time intervals of the economic rates may slightly differ from one supplier to another. Economy 7 and Economy 10 tariffs represent the simplest form of variable tariffs and attract domestic consumers with electrical storage heaters to take advantage of cheap energy overnight. Table 4.1 provides some sample tariff products by three big electricity suppliers in the UK, Southern Electric, E.ON and British Gas [9].

The main purpose of the most existing flat-rate tariffs is to attract more customers to buy their products rather than to encourage them to dynamically follow the conditions of the energy market. Although energy generation prices and congestions costs change over time, the price variation information is passed on only to larger consumers but not small domestic customers. It impedes mass domestic consumers from participating in DSR schemes.

### *4.2.2 Variable electricity tariffs*

Apart from the flat-rate tariffs, here are a number of advanced tariffs that are better placed for encouraging flexible load shifting. The typical representatives are ToU, critical peak pricing (CPP) and RTP tariffs, each with its own benefits and drawbacks. Some of them have been widely tested in practice under smart grid demonstration projects. For example, in the project – SoLa Bristol jointly conducted by Western Power Distribution and the University of Bath, it studies 'if DNOs and customers could share battery storage on DC networks with a variable tariff, then the mutual benefits may make battery storage financially viable' [10].

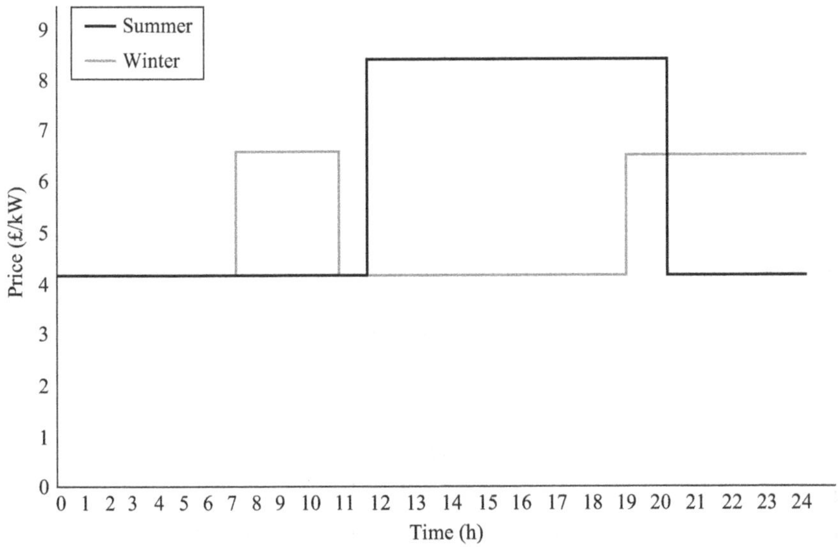


*Figure 4.2 Example of ToU tariffs*

#### 4.2.2.1 ToU tariff

This type of tariffs provides a number of pre-defined peak periods with an intention to reduce peak demand thus peak energy costs. It becomes possible because of the enabling technologies such as smart metering and smart reading technologies. Figure 4.2 illustrates a daily ToU price profile. For both winter and summer, the price profiles are segmented into several peak and off-peak periods, where the unit prices vary significantly [11, 12]. The total costs in a settlement day under a ToU tariff is decided by the electricity consumption in each hour multiplied by the unit price of the hour. Currently, ToU tariff is the most commonly adopted in smart pricing schemes and the time-based rates have been applied by a large number of domestic customers in the US and Canada.

#### 4.2.2.2 Critical peak pricing tariff

CPP tariffs are improved ToU tariffs that trace critical supply periods dynamically in the system [13]. The critical peak periods, always associated with extremely high unit prices, change from 1 h to another and the periods are notified to consumers, at least, one day ahead. To illustrate the difference between ToU and CPP tariff structures, an example of the two tariffs are depicted in Figure 4.3. The highest price in CPP tariffs can last up to 4–6 h within a day [14–15]. CPP emphasises the critically important hours of a year by introducing an extremely high rate so that the demand is attempted to be limited. Both ToU and CPP tariffs are effective on peak shaving and valley filling in energy consumption management considering their variations with time.

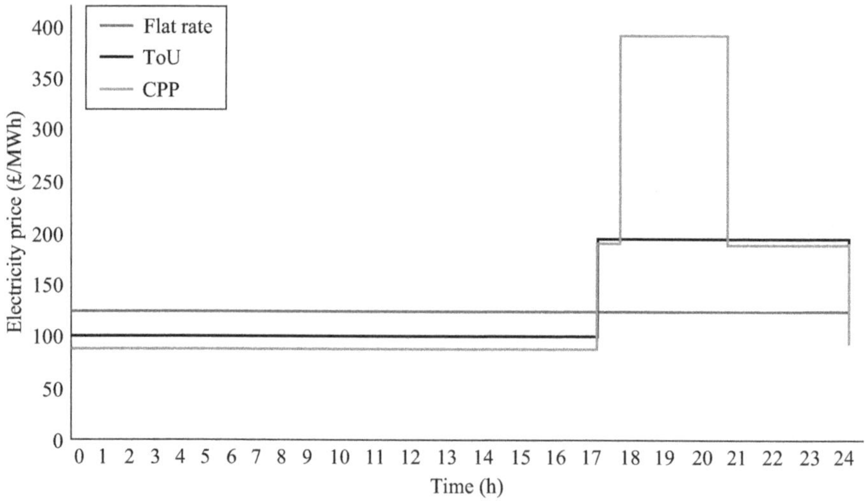


*Figure 4.3 An example of CPP tariff structure*

#### 4.2.2.3 Real-time pricing tariff

RTP tariffs provide the most direct way to reflect the dynamics and variations of wholesale prices throughout a day or throughout a year [16, 17]. Accordingly, this type of tariffs generally varies hourly or half-hourly complying with the wholesale market. Customers can very flexibly adjust their electricity usage in response to RTP tariffs, but demand control requires frequent operation linked with the dynamic price variations. Advanced information and communications technologies (ICTs) are essential here to enable RTP and thus by now, RTP is mostly applied to large commercial and industrial customers.

#### 4.2.2.4 Direct load control tariff

With load control tariffs, a lower unit price is provided to customers in exchange for the control of some appliance electricity use from time to time [18].

Variable tariffs provide customers flexible incentives to manage their electricity use more efficiently in order to reduce the energy cost. Smart meters can easily facilitate two-way communication to facilitate domestic customer DSR. Considering that the cost of the smart metering system is relatively low compared to electricity cost savings over the lifespan, it is beneficial for customers to install. It should be noted that the degree of DSR and its impact on demand reduction is highly uncertain, which are influenced by many economic, technical or social factors, such as income, education, tolerance to price rise and demand elasticity.

## 4.3 Variable ToU tariff design

### *4.3.1 Introduction*

In order to convey the representative costs of electricity generation and transportation to end customers to affect their electricity use behaviours, new types of smart tariffs are essential, particularly considering the increasing significant low-carbon flexible demand technologies. Generally, the challenge in ToU tariff design lies in two aspects: (i) quantifying tariff rates, i.e. price within each window and (ii) determining tariff shape, i.e. duration of each window. Normally, the new tariffs should persist the following key features [6]:

- considering economic efficiency, costing resources in terms of fuel, conversion costs and effects on the environment;
- reflecting the costs of generation, transmission and distribution;
- maintaining equity between diverse consumers;
- ensuring simple and transparent tariffs to customers.

A number of previous studies have been conducted to design ToU tariffs for retail markets. A computable equilibrium model is proposed in Reference 19, which can be utilised to determine ex-ante ToU tariffs in the electricity market. By analysing mixed complementarity programming models of equilibrium, the response speed between customers and suppliers is accelerated. This model considers the time-differentiated pricing concept and closely adheres to the marginal pricing principle. Reference 20 introduces a variational inequality model to determine ToU tariffs of different electricity market structures. This method can be used to forecast ToU tariff and determine the changing electricity market welfare. In Reference 21, the authors consider data mining and load profiles clustering techniques to implement ToU tariffs in partially deregulated systems. K-means algorithm and Silhouette strategy are applied to cluster load profiles at primary distribution level to three sectors, industrial, residential and commercial. After decomposing the load in primary feeders to these three sectors, the typical profiles of these sectors are used to design the ToU tariff. Stochastic optimisation techniques and quadratically constrained quadratic programming are proposed in Reference 22 for ToU tariff design. This method can mitigate the uncertainty in price-elasticated demand and assess various aspects in tariff design based on regulator/regulated utility. It should be noted that in some previous work, the number of price categories and time intervals for pricing are predefined and no sufficient justifications have been provided for the proposed rates settings in ToU pricing schemes. In practice, flat-rate tariffs have been implemented by suppliers for long history with comprehensive consideration of network operation and investment costs. They are good indicators of cost drivers in tariffs, but the problem is that flat-rate tariffs include no price variation information.

In this chapter, two novel approaches: equal interval grouping and hierarchical clustering are proposed for designing ToU tariffs with the most significant price elements and price variations captured. The RTP tariffs obtained from the wholesale market prices are first studied. The profile patterns of RTP tariffs are then inherited in designing new smart variable tariffs because they can reflect the tariff variation with time to reflect actual generation cost. The two approaches divide a settlement day into several short-time intervals and each interval has a constant unit price. The ToU tariff rates of all intervals are determined by maintaining the total daily electricity bills for a typical domestic load profile to be equal with those charged under RTP.

The wholesale energy cost covers more than half of the electricity bills paid by domestic customers according to the survey by conducted by Ofgem [23]. It is mainly because it consists of the costs of fuel procurement, generation operation and investment. Into the future, the wholesale energy cost is still likely to be a major part of electricity bills for domestic customers [7] and thus significant cost savings might be realised if they can respond to the price variations of wholesale energy costs.

In order to convey the information of energy price variations to end customers, variable tariffs need to inherit the time-varying features of wholesale energy prices. The real-time dynamic pricing in the wholesale market makes a closer alignment of price with generation cost. However, it varies very quickly, for example, half hourly in the UK wholesale electricity market [24], thus too complex for small customers to respond if no automatic technologies are in place. By contrast, they are more appropriate for large consumers signing pre-established peak load reduction agreements [25].

ToU and CPP tariffs are less complex for customers to respond because their variation intervals are much longer compared to RTP tariffs. Their intervals have various unit rates and, therefore, they can differentiate the cost drivers. The critical peak periods of CPP are often associated with extremely high unit prices to reflect the unusually high generation costs or network congestion costs or the both [26]. ToU divides a settlement day into several certain fixed time periods to roughly capture the variations of real-time energy prices. Generally, ToU is one of the simplest variable tariffs for customers to understand and respond. It is thus designed in this chapter and its impact is extensively investigated.

## *4.3.2 Rationale of proposed tariff design*

### 4.3.2.1 Characteristics of energy price variation

Before conducting tariff design, it is essential to understand the variation characteristics of wholesale energy prices. The energy prices in each settlement period throughout 2010 in Figure 4.4 [27] are taken to explain. Clearly, dramatic changes can be observed and the prices are extremely volatile during wintertime. The price distribution is further summarised in Figure 4.5 and two key characteristics are seen:

- The first major characteristic is that the price varies dramatically throughout all seasons. Clearly, the settlement periods with a price higher than 70 £/MWh account for more than 5% time in January. By contrast, in August, less than 1% of the prices are higher than 70 £/MWh.
- The second characteristic is that the peak energy price and peak demand usually coincide with each other within the same settlement periods.

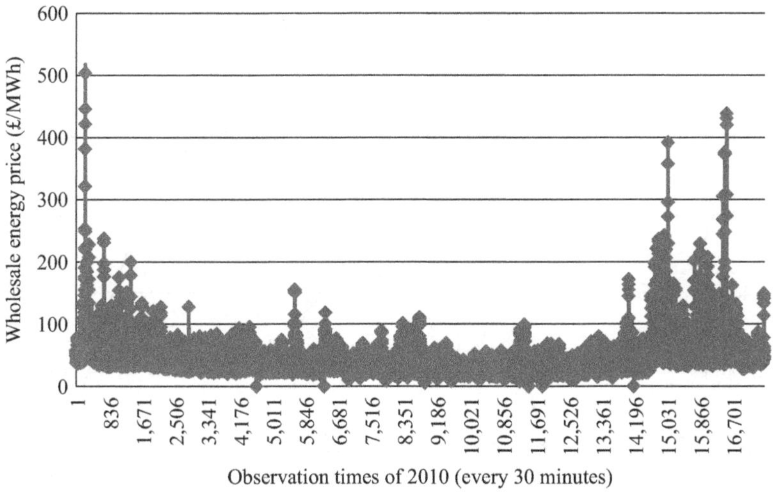


*Figure 4.4 Wholesale energy price variation of 2010*

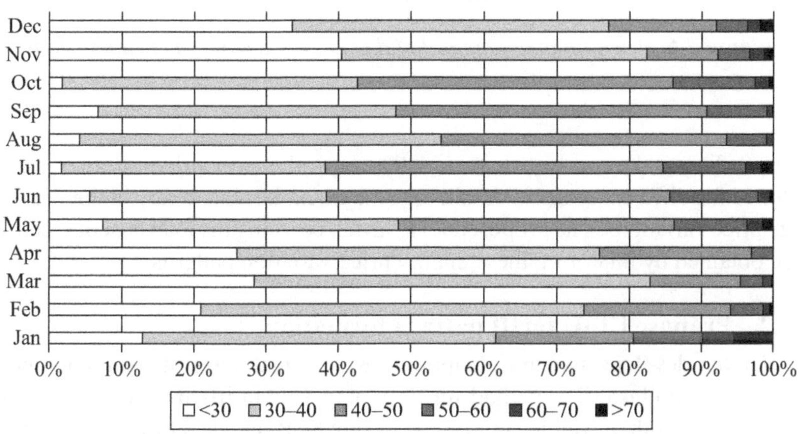


*Figure 4.5 Energy price distribution in a whole year*

The energy price and demand curve in a typical day is provided in Figure 4.6 to further demonstrate their alignment. As seen, demand increases dramatically from the beginning of the 10th settlement period and the morning peak demand of 2,608 MW occurs at the 19th settlement period. After settlement period 14, load saturates at around 2,650 MW until another increment appears period 33. The peak demand of the sample day occurs at the 35th settlement period. On the other hand, energy price gradually increases from around 30 to 46 £/MWh. After a slight drop at the 31st settlement period, the price peaks at 55 £/MWh in the 36th period and falls ultimately back to 30 £/MWh.

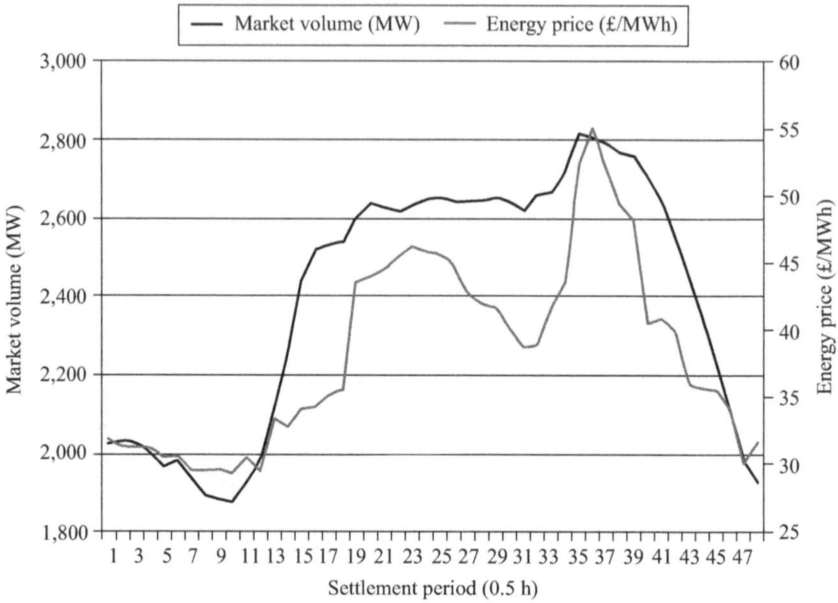


*Figure 4.6 Variations of demand and energy price on 3 February 2010*

According to the two key characteristics, typical energy price variation patterns developed for different seasons should to some extent be able to reflect the wholesale energy price variations. The proposed RTP tariffs, covering the total cost of electricity, can be obtained by inheriting these energy price variation patterns.

#### 4.3.2.2 Proposed ToU tariff pattern formation

This chapter develops two novel approaches – equal interval grouping method and hierarchical clustering method for designing ToU tariffs in order to encourage domestic DSR, both of which work by converting RTP signals into ToU without compromising much precision. Their biggest difference lies in whether the interval number is determined based upon previous experience or by mathematical approaches.

The unit tariff rate is derived by maintaining the total electricity bill unchanged, i.e., for a given load profile, the total bill is the same under the RTP and ToU. The implementation procedures for the proposed ToU pricing schemes are summarised as follows and the detailed flow chart is illustrated in Figure 4.7:

- The wholesale energy prices through a calendar year are represented by eight typical energy price variation patterns, i.e. weekdays and weekends in all four seasons.
- The energy price variation patterns are converted to eight RTP tariffs.
- For each RTP tariff, 48 settlement periods are grouped into appropriate price categories by equal interval grouping method and hierarchical clustering method.

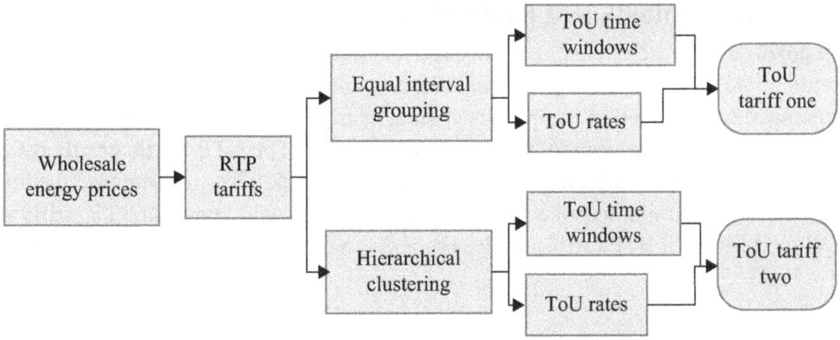


*Figure 4.7 Flow chart of the investigation process in smart variable tariff design*

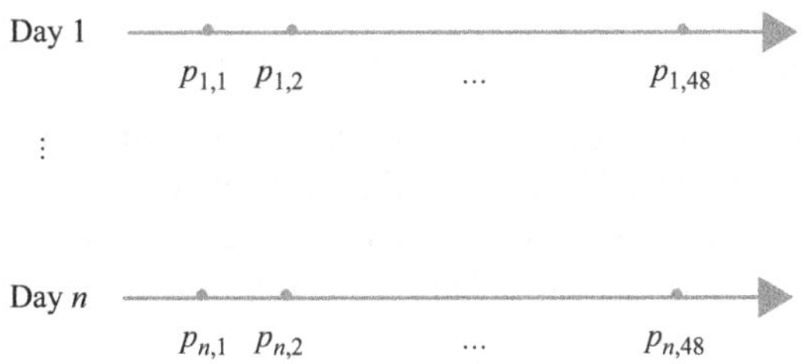


*Figure 4.8 Energy prices collected from settlement periods belonging to a day type*

#### 4.3.2.3 Typical energy price variations

Since the information of wholesale energy prices over a year is essential for ToU tariff design, the energy price collected from each half-hour period in any of eight-day types can be represented by $p_{j,i}$ as shown in Figure 4.8. Here, $p_{j,i}$ is the energy price level of the $i$th settlement period in the $j$th day and $n$ stands for the number of days in one day type.

In order to identify an appropriate price $p_{0,i}$ to represent generic energy price in the $i$th settlement period of a specific day type, it is essentially obtained by minimising the sum of the squares of the distances between the estimated price and each single known price. The objective function is therefore formulated as follows:

$$\text{Min}(p_{0,i} - p_{1,i})^2 + (p_{0,i} - p_{2,i})^2 + \cdots + (p_{0,i} - p_{n,i})^2 \tag{4.1}$$
$$\text{s.t.: } 0 < \min(p_{1,i}, p_{2,i}, \ldots, p_{n,i}) < p_{0,i} < \max(p_{1,i}, p_{2,i}, \ldots, p_{n,i})$$

Once the energy price in each settlement day is estimated, the energy price variation patterns for any weekdays or weekends in all seasons are easy to obtain, which could be defined as $(p_{0,1}, p_{0,2}, \ldots, p_{0,48})$. This is the price used for deriving ToU tariff rates.

#### 4.3.2.4 Determination of RTP tariffs

As known, wholesale energy cost only reflects part of the total energy supply costs, around 55%. Other cost components, such as transmission and distribution network costs, need to be added in the proposed smart variable tariffs to capture all makeup costs of electricity generation and transportation [5]. The RTP tariffs are therefore scaled up based on the obtained energy price variation patterns to recapture the total electricity cost. Similar to price variation pattern design, eight types of RTP tariffs are developed for each RTP tariff profile. For a given tariff, its rate in the $i$th settlement period is

$$p_{R,i} = \frac{p_{0,i}}{\alpha} \tag{4.2}$$

where $p_{R,i}$ is the rate in the $i$th settlement period and $\alpha$ is the proportion of energy generation cost in the total electricity cost.

### *4.3.3 ToU tariff design by equal interval grouping*

Once eight RTP tariff profiles by seasons and day types are determined, the follow-up work is then to form ToU tariffs by converting RTP tariffs, where the eight scenarios continue to be used. The first approach employs the quantity analysis of annual RTP prices, where the ToU tariffs are decided by the following three elements:

- ToU price category (number of price blocks)
- ToU time window (width of each price block)
- ToU rate (height of price block)

#### 4.3.3.1 Determination of ToU time windows

The most common ToU tariff schemes are developed with three price categories – peak, shoulder and off-peak [28]. This type of pattern has the advantage of not only simple form, but also a better use of peak and trough periods of price variations. A calendar day could be divided into several time intervals, each belonging to one of the three defined categories. In this proposed approach, the peak, shoulder and off-peak price periods are determined according to RTP tariff distributions.

In the eight typical scenarios/days, each settlement period in a RTP tariff is assigned to peak, shoulder or off-peak category. The grouping process is as follows:

(i) To identify group number (price block) of ToU tariffs. As discussed above, three-rate tariffs with peak, shoulder and off-peak periods are selected.
(ii) To determine group interval. In all RTP tariffs, a range at 95% confidence level is selected as the area for grouping, defined as an envelope for price variations. Within this envelope, the group interval is set to be equal, given by

$$L = \frac{E_{\max} - E_{\min}}{g} \tag{4.3}$$

where $L$ is group interval, $E_{\max}$ and $E_{\min}$ are the maximum and minimum prices within the envelope and $g$ is group number.

*Table 4.2 Price range determinations in grouping*

| | Off-peak | Shoulder | Peak |
|---|---|---|---|
| Initial grouping (£/MWh) | $E_{min,1}$–$E_{max,1}$ | $E_{min,2}$–$E_{max,2}$ | $E_{min,3}$–$E_{max,3}$ |
| Improved grouping (£/MWh) | 0–$E_{max,1}$ | $E_{min,2}$–$E_{max,2}$ | $E_{min,3}$–$\infty$ |

(iii) To identify price range of each group. With the increment of prices, the first group corresponds to off-peak price and the peak price is in the last group. For the $k$th group, its maximum and minimum prices are obtained by

$$E_{max,k} = E_{max} - (g - k) \cdot L \tag{4.4}$$

$$E_{min,k} = E_{min} + (k - 1) \cdot L \tag{4.5}$$

(iv) To assign RTP prices of each settlement period to one of the three groups, i.e. the corresponding settlement periods are assigned to peak, shoulder or off-peak periods.

The initial price range of each group within the price variation envelope is listed in the second row of Table 4.2. In order to accommodate the diversity of RTP prices, the results are further improved given in the third row, where the minimum price for off-peak periods can be as low as zero and the maximum value for peak periods can be as high as infinite.

#### 4.3.3.2 Determination of ToU rates

The time window determination decides ToU price shape based on the price variation trend of RTP tariffs, but the targeted ToU should also include the unit rate, i.e. the height of each block. In order to calculate each rate, the most convenient way is to use the average RTP rates within the group, but this approach cannot capture the cost variance. In addition, by using the average price within a group from each settlement period with the same weight would compromise the ToU's representativeness.

The rate for each price category in one of the eight-day types is determined by (4.6). The concept is to maintain the total cost during the settlement periods within a group unchanged, whatever charged by ToU or RTP. The rate is

$$TR_k = \frac{\sum_{t \in \mathbf{K}} e_{t,k} \cdot v_{t,k}}{\sum_{t \in \mathbf{K}} v_{t,k}} \tag{4.6}$$

where $e_{t,k}$ is the RTP price of the $t$th settlement period in group $k$ and $v_{t,k}$ is the energy consumption during that period. $\boldsymbol{K}$ is the set of settlement period whose RTP prices belong to group $k$. $TR_k$ is the final obtained peak, shoulder or off-peak unit tariffs.

### 4.3.4 ToU tariff development by hierarchical clustering

In equal interval grouping method, the number of price categories is predefined and the time windows are determined by grouping RTP prices. Here, an alternative ToU designing approach of determining price blocks is introduced by using hierarchical clustering. The price category number, i.e. the number of clusters, is obtained by an optimisation model considering the accuracy and feasibility of implementation [29]. All price elements grouped in a cluster have the least price rate difference.

By employing the clustering approach, a settlement day is divided into several time intervals, each belonging to one cluster. This process is more complex and the flow chart is provided in Table 4.8. Similarly, the ToU tariffs of different seasons and day types are also obtained in a similar way. The number of clusters and the assignment of settlement periods largely depend on the patterns of initial RTP tariffs. For example, a steady RTP can be well represented by a flat rate while a very dynamic RTP tariff would require many different tariff rates and time intervals to represent.

In order to derive appropriate clusters out from RTP tariffs without pre-knowledge, hierarchical clustering is adopted by calculating the distance between prices in each settlement period, which indicates the similarity between periods. At the initial stage, each settlement period is classified as its own cluster and they are then merged according to the similarity until a single cluster is formed in the end. The implementation steps of this hierarchical clustering are listed as follows:

(i) For different numbers of clusters, the within-group dissimilarity is calculated, which is expected to decrease with increasing cluster number. After a certain number, say $N$, the decreasing rate will drop significantly, indicating much less impacts of further partition. The number of clusters can be tested from 1 to the total individual settlement periods $T$.
(ii) Each settlement period is classified into one of the $N$ clusters so that the RTP tariffs of a day are divided into several time intervals.
(iii) Unit rates of the ToU tariffs are determined by ensuring the same total costs for a typical load profile (Figure 4.9).

#### 4.3.4.1 Number of price categories and time window

For any day type, a RTP tariff profile can be denoted by a $1 \times 48$ vector, say $(p_{0,1}, p_{0,2}, \ldots, p_{0,48})$. Hierarchical clustering is adopted to partition the rates of all settlement periods (hourly or half hourly) into clusters. The distance $dis_{h,f}$ between two prices $p_{0,h}$ and $p_{0,f}$ is determined by

$$DIS_{h,f} = \|p_{0,h} - p_{0,f}\| = \sqrt{(p_{0,h} - p_{0,f})^2} \tag{4.7}$$

where $p_{0,h}$ and $p_{0,f}$ are the prices in the $h$th and $f$th settlement period of the selected RTP tariff.

Thereafter, the settlement periods are gathered into a hierarchical cluster tree by merging together those with the smallest price distance and the merged settlement

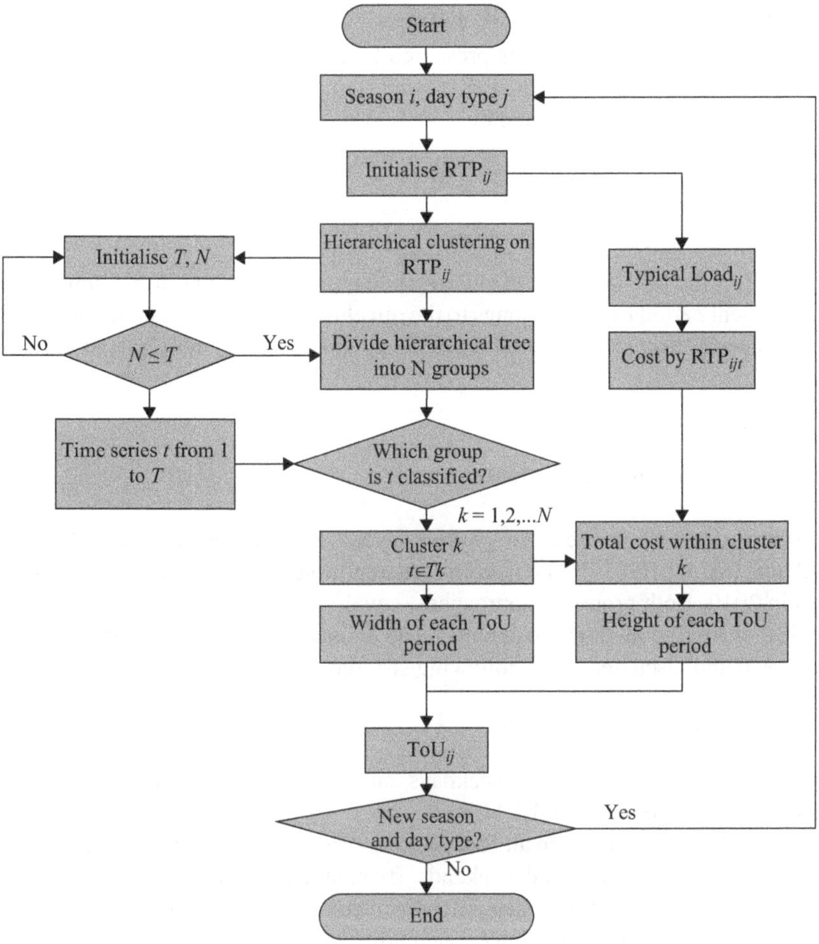


*Figure 4.9 Flow chart of developing ToU by hierarchical clustering*

periods form a cluster. Then, a new distance is calculated between existing clusters to form new clusters and the process of forming new clusters is repeated until only one cluster remains [30]. The distance between clusters is calculated by the Ward distance in (4.8), assuming matrices $\boldsymbol{A}$ and $\boldsymbol{B}$ are two clusters:

$$DIS_{A,B} = \frac{1}{|\mathbf{A}||\mathbf{B}|}\sqrt{\sum_{p_a \in \mathbf{A}} \sum_{p_b \in \mathbf{B}} (p_a - p_b)^2} \tag{4.8}$$

where $p_a$ and $p_b$ are unit prices in settlement periods clustered into cluster $\boldsymbol{A}$ and $\boldsymbol{B}$ in the previous step.

#### 4.3.4.2 Determination of ToU rates

The tariff rate for each cluster is presented by the quotient of the total costs under RTP tariffs and the corresponding energy consumption during all settlement periods within the cluster. It is calculated by

$$TR_g = \frac{\sum_{t\in\mathbf{G}} e_{t,g} \cdot v_{t,g}}{\sum_{t\in\mathbf{G}} v_{t,g}} \tag{4.9}$$

where $TR_g$ is the ToU tariff rate of cluster $g$ and $v_{t,g}$ represents the energy consumption in settlement period $i$ which belongs to the $g$th cluster. The settlement period numbers within the $g$th cluster are gathered in set $\boldsymbol{G}$. $e_{t,g}$ stands for RTP tariff rate in settlement period $t$.

## 4.4 Results and discussion

For demonstration purposes, energy prices over a calendar year are adopted to design RTP and ToU tariffs [27]. The data reflects real-time energy prices of all settlement days in 2010. Both proposed approaches, equal interval grouping method and hierarchical clustering method are adopted. The derived tariffs together with RTP tariffs are shown and discussed in the following sections.

### *4.4.1 Results of RTP tariffs*

The resultant RTP tariffs for weekdays and weekends in different seasons are illustrated in Figures 4.10 and 4.11.

Clearly, the RTP tariffs in all four seasons vary remarkably. Peak prices generally appear in winter weekdays and weekends. By contrast, the peak price rate in summer weekdays is 76% of that in winter weekdays, and its peak price in weekends is only 66% of the highest price in winter weekends. Overall, the RTP prices in weekends are higher than those in weekdays.

### *4.4.2 ToU tariffs by equal interval grouping*

This subsection provides the results of time windows and tariff rates for peak, shoulder and off-peak periods developed by equal interval grouping approach.

#### 4.4.2.1 Time windows

With statistical analysis, the RTP tariffs in Figures 4.7 and 4.8 show that 95% of price rates are within the range of 60–120 £/MWh. This range is therefore defined as the envelope for price variations. The price range of each group within the envelope is listed in the second row of Table 4.3 and the improved results are in the third row by extending the ranges of peak and off-peak prices. Eventually, the settlement periods whose RTP prices are lower than 80 £/MWh are grouped into off-peak price category and the settlement periods with over 100 £/MWh prices are assigned to peak periods.

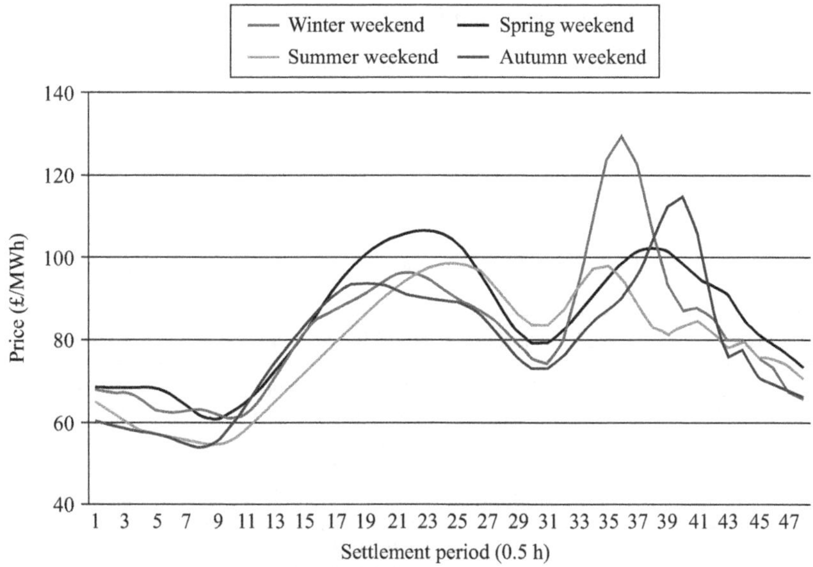


*Figure 4.10 RTP tariffs for weekdays in different seasons*

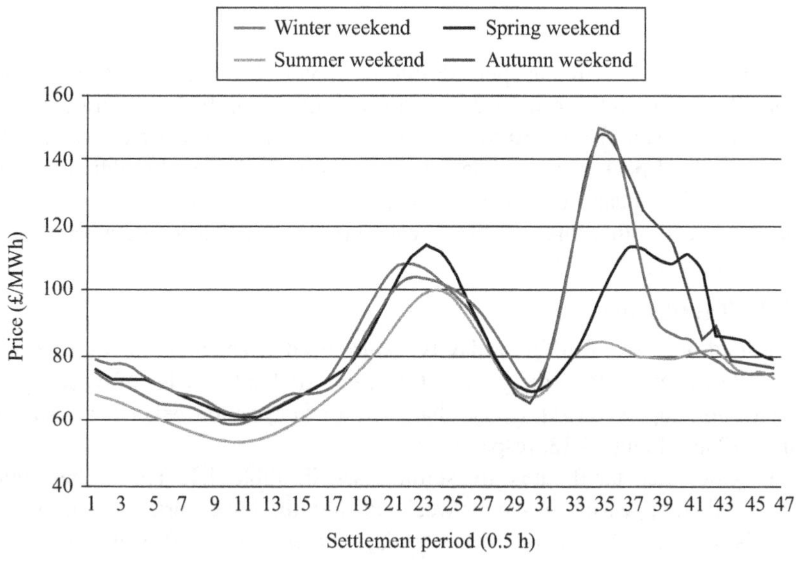


*Figure 4.11 RTP tariffs for weekends in different seasons*

*Table 4.3 Price range determination in the case study*

| | Off-peak | Shoulder | Peak |
|---|---|---|---|
| Initial grouping (£/MWh) | 60–80 | 80–100 | 100–120 |
| Improved grouping (£/MWh) | 0–80 | 80–100 | 100–∞ |

*Table 4.4 ToU tariff time windows obtained by equal interval grouping*

| Scenarios | Peak period | Shoulder period | Off-peak period |
|---|---|---|---|
| Winter weekday | 16:30–19:00 | 6:30–16:30, 19:00–22:30 | 22:30–6:30 |
| Spring weekday | 9:00–12:30, 18:00–19:30 | 6:30–9:00, 12:30–18:00, 19:30–23:30 | 23:30–6:30 |
| Summer weekday | NA | 7:30–23:00 | 23:00–7:30 |
| Autumn weekday | 18:30–20:30 | 6:30–18:30, 20:30–22:00 | 22:00–6:30 |
| Winter weekend | 10:30–12:30, 16:30–19:30 | 9:00–10:30, 12:30–16:30, 19:30–1:30 | 1:30–9:00 |
| Spring weekend | 10:30–13:00, 18:00–21:30 | 9:00–10:30, 13:00–14:30, 16:30–18:00, 21:30–0:30 | 0:30–9:00, 14:30–16:30 |
| Summer weekend | NA | 9:30–14:00, 16:30–23:30 | 14:00–16:30, 23:30–9:30 |
| Autumn weekend | 10:00–12:30, 16:00–21:00 | 8:30–10:00, 12:30–14:30, 21:00–0:00 | 0:00–8:30, 14:30–16:00 |

By grouping the settlement periods into eight typical scenarios/days, peak, shoulder and off-peak periods for each day type are summarised in Table 4.4. In all eight-day types, time intervals from 1:30 to 6:30 am are assigned to off-peak periods. There are additional off-peak periods in late evenings of weekdays and middle afternoons of weekends. The peak periods do not appear in summer due to flatter RTP tariff variations, but for other three seasons, peak periods are either in late mornings or in early evenings.

#### 4.4.2.2 Tariff rates

Typical load profiles of different day types are used to calculate total energy costs and ToU tariff rates. Here, the generic GB domestic load profiles of weekdays and weekends are employed [31], whose shapes in weekdays and weekends are shown in Figure 4.12 and Figure 4.13, respectively.

The calculated tariff rates are summarised in Table 4.5. The highest rate is 121.89 £/MWh appearing in winter weekends and the lowest price during off-peak periods is 63.55 £/MWh in summer weekdays. Overall, both off-peak and shoulder prices for the eight-day types vary insignificantly. All off-peak rates are within the range of 63–70 £/MWh and shoulder prices vary between 82 and 89 £/MWh. The peak prices in winter are generally much higher than those in spring and autumn, but on the contrary, there are no peak rates for summer weekdays and weekends.

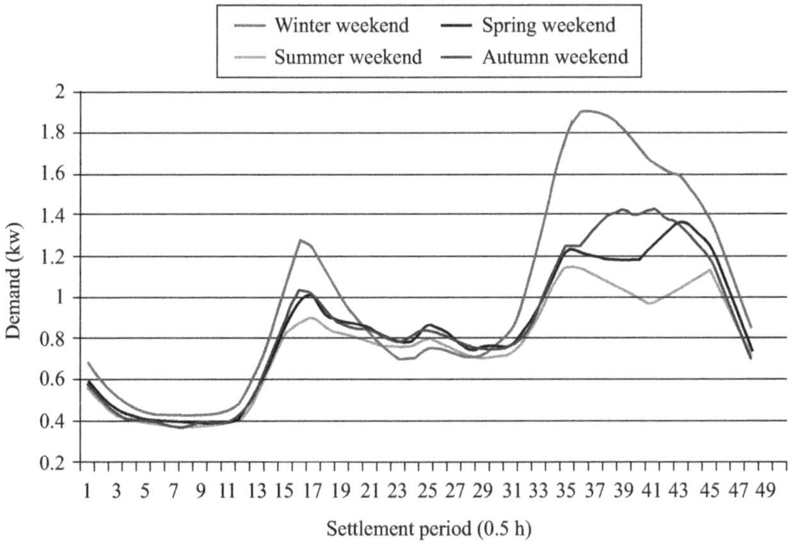


*Figure 4.12 Typical domestic individual load profiles for weekdays*

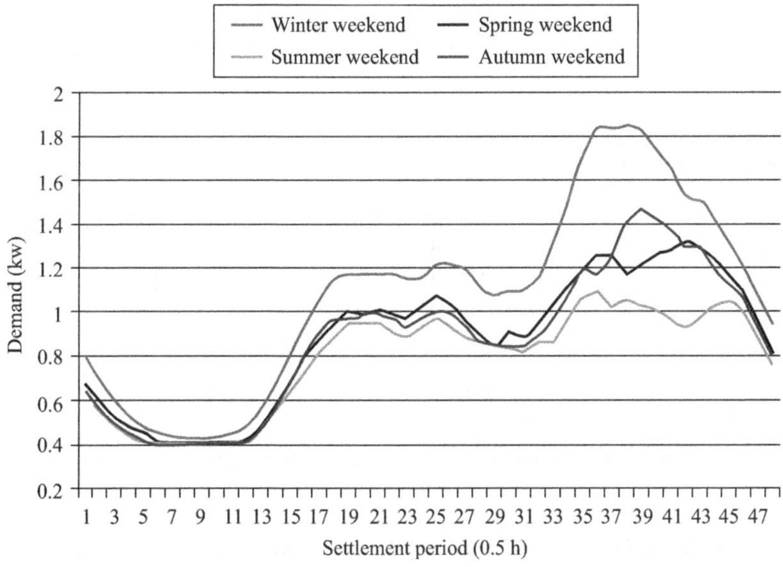


*Figure 4.13 Typical domestic individual load profiles for weekends*

*Table 4.5 ToU tariff rates obtained by equal interval grouping*

| ToU rate (£/MWh) | Off-peak rate | Shoulder rate | Peak rate |
|---|---|---|---|
| Winter weekday | 66.5 | 86.0 | 118.7 |
| Spring weekday | 67.5 | 88.5 | 103.3 |
| Summer weekday | 63.6 | 86.9 | – |
| Autumn weekday | 65.0 | 86.3 | 109.0 |
| Winter weekend | 68.4 | 82.8 | 121.9 |
| Spring weekend | 69.3 | 85.1 | 109.5 |
| Summer weekend | 65.0 | 83.8 | – |
| Autumn weekend | 67.3 | 84.3 | 119.0 |

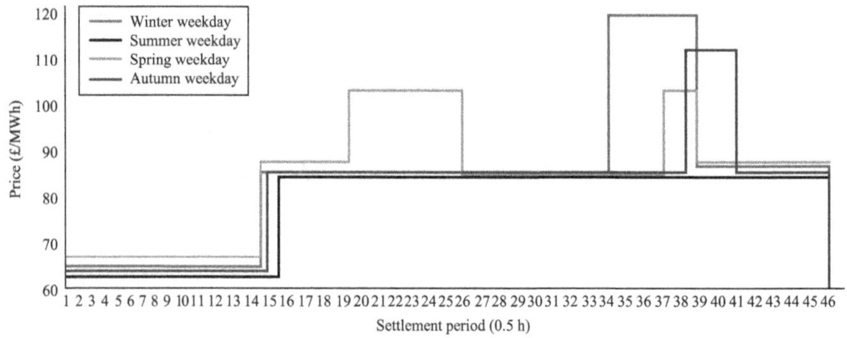


*Figure 4.14 ToU tariffs for weekdays by equal interval grouping*

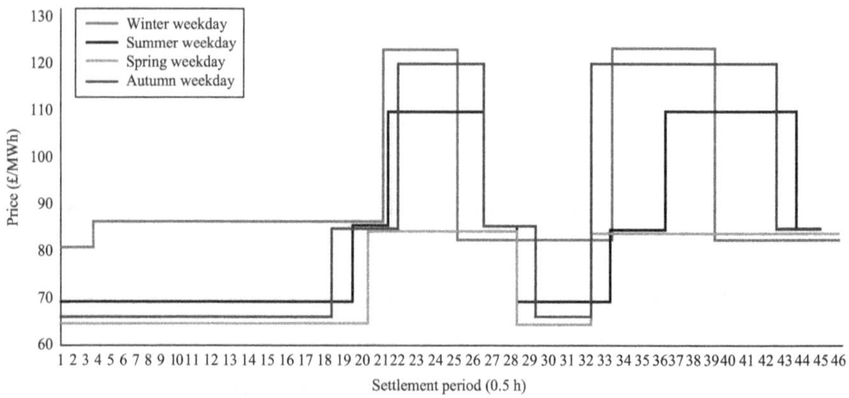


*Figure 4.15 ToU tariffs for weekends by equal interval grouping*

#### 4.4.2.3 ToU tariff profiles

Once tariff rates for peak, shoulder and off-peak periods in each settlement day are calculated, the tariff results for different seasons are plotted in Figures 4.14 and 4.15.

*Table 4.6 Average within-group distance for different cluster numbers*

| Cluster number | Average within-group distance (£/MWh) |
|---|---|
| 1 | 47 |
| 2 | 17.6 |
| 3 | 5.6 |
| 4 | 3.3 |
| 5 | 2.2 |
| 6 | 2.4 |

It can be observed that the off-peak price rates and their durations in weekdays are roughly the same for all eight scenarios/days. However, the case is totally different for peak periods. Two-time intervals make up peak periods in a typical spring weekday, but none of the settlement periods in a summer weekday are assigned to peak categories. Even though both winter and autumn have a time interval with peak price, the degree of price levels and durations in winter are much higher/longer than those in autumn. Compared to ToU tariffs in weekdays, the tariffs in weekends have longer peak periods. Besides, two peak time intervals appear in the morning and evening, respectively, in all typical weekends except in summer.

## *4.4.3 ToU tariffs by hierarchical clustering*

### 4.4.3.1 Number of clusters

The lower average within-group distance indicates a higher similarity within the group. The distance variations with respect to the number of clusters within winter weekdays are listed in Table 4.6 for demonstration.

For a typical winter weekday, when the prices of all settlement periods are grouped in one cluster, the intra-group distance is very high at 47 £/MWh. It starts to decrease when similar prices are categorised into more clusters, decreasing quickly to 5.61 £/MWh with three clusters and further partition producing slight improvement. For other day types, the decreasing rates become much slower when the cluster number is larger than three. Therefore, in designing ToU tariffs, three price categories are sufficient for representing all eight typical scenarios/days.

### 4.4.3.2 Time windows

By taking winter weekdays as an example, each settlement period in a day can be assigned into one of the three categories, i.e. peak, shoulder and off-peak. As shown by the dendrogram in Figure 4.16, forty-eight settlement periods are clustered according to the price distance. Three clusters are partitioned as shown in the grey boxes.

Once the clustering approach determines the number of clusters and their corresponding time intervals, they can form the overall shape of ToU tariffs following the variation trends of RTP tariffs. Accordingly, the ToU time windows obtained are

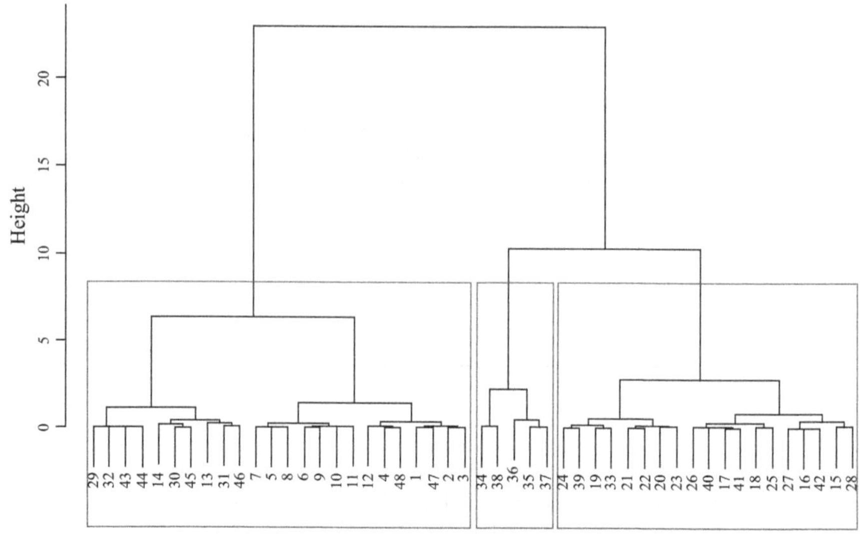


*Figure 4.16 Winter weekday clustering dendrogram*

*Table 4.7 ToU tariff time windows obtained by hierarchical clustering*

| | **Peak period** | **Shoulder period** | **Off-peak period** |
|---|---|---|---|
| Winter weekday | 16:30–19:00 | 7:00–14:00, 19:00–21:00 | 21:00–7:00, 14:00–16:30 |
| Spring weekday | 8:00–13:30, 16:30–21:30 | 6:00–8:00, 13:30–16:30, 21:30–0:00 | 0:00–6:00 |
| Summer weekday | 10:00–13:30, 16:00–18:00 | 6:30–10:00, 13:30–16:00, 18:00–0:00 | 0:00–6:30 |
| Autumn weekday | 7:00–13:30, 16:30–21:00 | 5:00–7:00, 13:30–16:30, 21:00–0:00 | 0:00–5:00 |
| Winter weekend | 16:30–19:30 | 10:00–13:30 | 13:30–16:30, 19:30–10:00 |
| Spring weekend | 10:30–13:00, 18:00–21:30 | 9:00–10:30, 13:00–14:00, 16:30–18:00, 1:30–0:00 | 0:00–9:00, 14:00–16:30 |
| Summer weekend | 9:30–14:00, 16:30–22:30 | 7:30–9:30, 14:00–16:30, 22:30–2:00 | 2:00–7:30 |
| Autumn weekend | 17:00–19:00 | 9:00–14:00, 16:00–17:00, 19:00–22:00 | 14:00–16:00, 22:00–9:00 |

summarised in Table 4.7, each RTP tariff converted into a ToU pattern with three price steps.

As seen, by employing the clustering method, the overnight settlement periods are still assigned to off-peak blocks. In the RTP tariffs, prices in winter weekdays

*Table 4.8 ToU tariff rates obtained by hierarchical clustering*

| ToU rate (£/MWh) | Off-peak rate | Shoulder rate | Peak rate |
|---|---|---|---|
| Winter weekday | 73.3 | 89.2 | 118.7 |
| Spring weekday | 66.1 | 81.1 | 98.4 |
| Summer weekday | 59.2 | 80.8 | 96.1 |
| Autumn weekday | 57.4 | 73.7 | 94.4 |
| Winter weekend | 77.3 | 100.2 | 129.9 |
| Spring weekend | 69.9 | 86.1 | 109.5 |
| Summer weekend | 56.6 | 70.0 | 84.7 |
| Autumn weekend | 70.7 | 102.5 | 141.2 |

and weekends are generally higher than those in other seasons but the durations of peak periods in winter are not as long as those in other three seasons. The reason is that the settlement periods with critical high RTP prices mainly contribute to the peak periods in ToU tariffs. Due to the large gap between critical high prices and other prices, all settlement periods with non-critical prices are assigned to shoulder or off-peak periods.

#### 4.4.3.3 Tariff rates

The calculated tariff rates are listed in Table 4.8. Different from the results in Table 4.5, the tariff rates in the eight scenarios/days vary dramatically. In detail, the difference between off-peak rates in winter weekends and summer weekends is 20.3 £/MWh. Meanwhile, the shoulder price rate in autumn weekends is 1.4 times of that in autumn weekdays, and the peak rate in summer weekends is only 65% of that in winter weekends.

#### 4.4.3.4 ToU tariff profiles

Similarly, the ToU tariffs developed based on hierarchical clustering method with eight price profiles are illustrated in Figures 4.17 and 4.18 for weekdays and weekends. Obviously, the ToU tariffs are different from each other in terms of time window duration and tariff rate. For the peak, shoulder and off-peak prices, the lowest rate is in summer weekends and the highest appears in winter weekends.

## 4.5 Impact analysis of ToU tariffs

This section employs the developed ToU tariffs on the typical UK domestic customers to understand their impact and quantifies the corresponding economic benefits. Beforehand, it is essential to classify load according to their sensitivity to economic signals.

For typical GB domestic customers, electricity is majorly used for heating, cooking and lighting. Ideally, each composition of the consumption could be flexibly

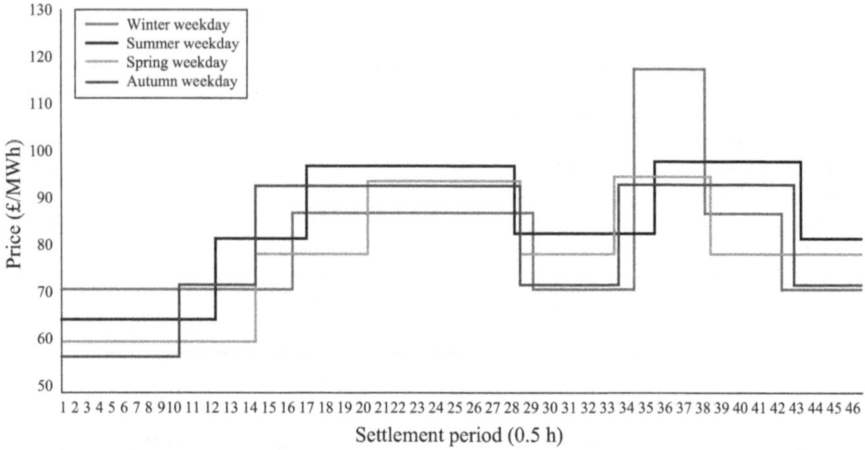


*Figure 4.17 ToU tariffs for weekdays by hierarchical clustering*

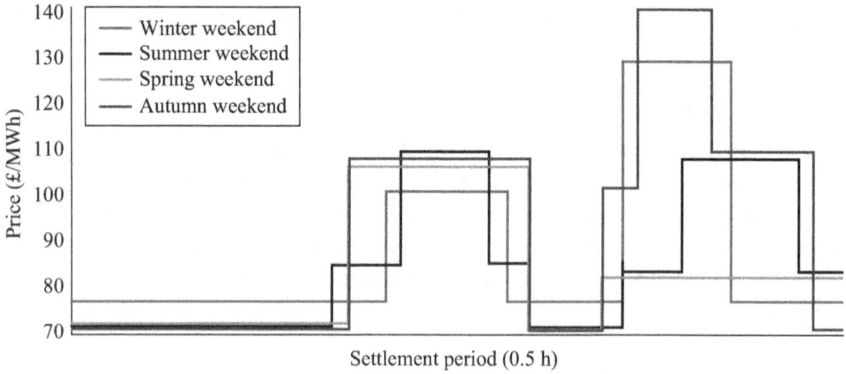


*Figure 4.18 ToU tariffs for weekends by hierarchical clustering*

shifted to alter load profiles, but the reality is that the scope of domestic DSR is rather narrow. The flexibility and capability of household appliances are shown in Figure 4.19. Because heating, cooking and lighting are essential for daily life, their consumptions can hardly be shifted for general users and thus are regarded as inflexible load. By contrast, the shifting of appliances, such as dishwashers, washing machines and tumble dryers in domestic households, is less likely to result in large disruptions to daily activities. This type of appliances is defined as wet appliances, which consume about 15% of the total electricity of a typical household [32]. As flexible load, the consumption of wet appliances can be effectively shifted in response to economic signals. By contrast, the inflexible load, which is insensitive to economic signals can be managed by energy storage systems to increase the flexibility.

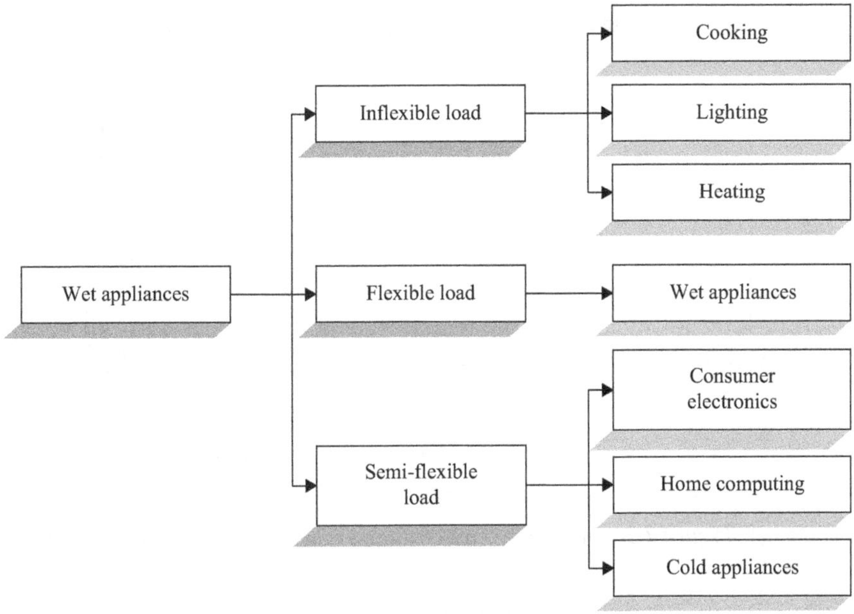


*Figure 4.19 Flexibility of household appliances*

## 4.5.1 Flexible load modelling

In order to reflect the diversity of energy consumption, three typical household profiles, reflecting different start running time of wet appliances, are selected for modelling here [33, 34]. The first type represents customers who largely use wet appliances in evenings, and the second and third types represent the cases that wet appliances are mainly in mornings and evenings. The profiles for the flexibilities of different loads are shown in Figure 4.20, Figure 4.21 and Figure 4.22 respectively. Generally, they possess two distinctive features:

- Wet appliances are usually used after lunch or supper.
- Tumble dryers are always used following the use of washing machines.

For each type of households, the flexible consumption of wet appliances is expected to be shifted to off-peak periods in response to ToU tariffs. The operation of washing machines and dryers needs special attention in load shifting considering their coordination. Normally, dryers run after washing machines according to Figures 4.20–4.22. Therefore, washing machines should be run early enough to guarantee the start time of dryers to be before 12 a.m. In addition, the off-peak period right before midnight is expected to be fully utilised for running washing machines.

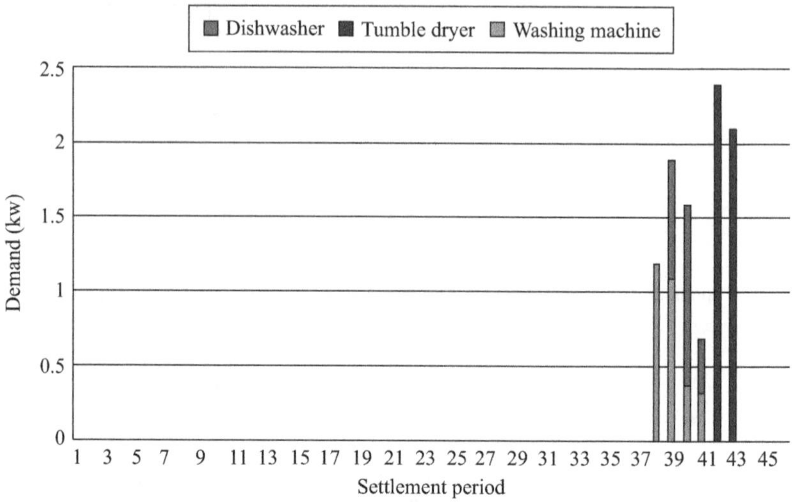


*Figure 4.20 Flexible load profiles for household type one*

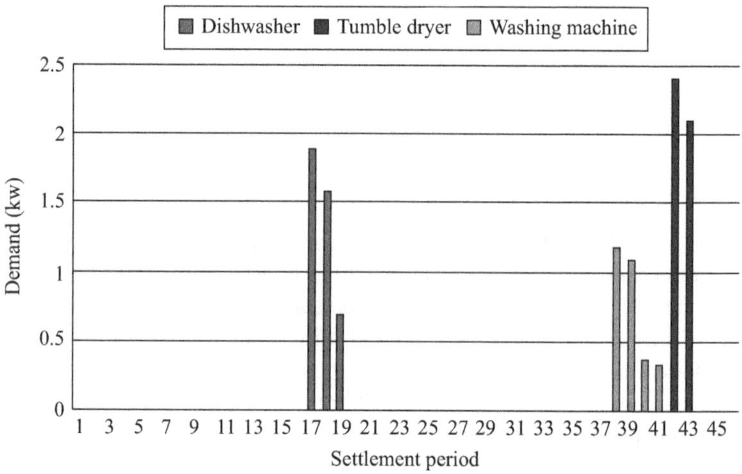


*Figure 4.21 Flexible load profiles for household type two*

### *4.5.2 Impact analysis of designed ToU tariffs*

For domestic consumers, half-hourly varying tariffs could hardly effectively guide DSR if no automatic operation is in place due to the high price complexity and volatility. Therefore, the developed ToU tariffs with the three-block pattern are much more practical in triggering DSR considering its simplicity but at the same time without loss price variations. Based on the eight ToU tariffs developed for weekdays and weekends in four seasons, the periods for accommodating shifted loads are determined by the following steps and the detailed flow chart is shown in Figure 4.23.

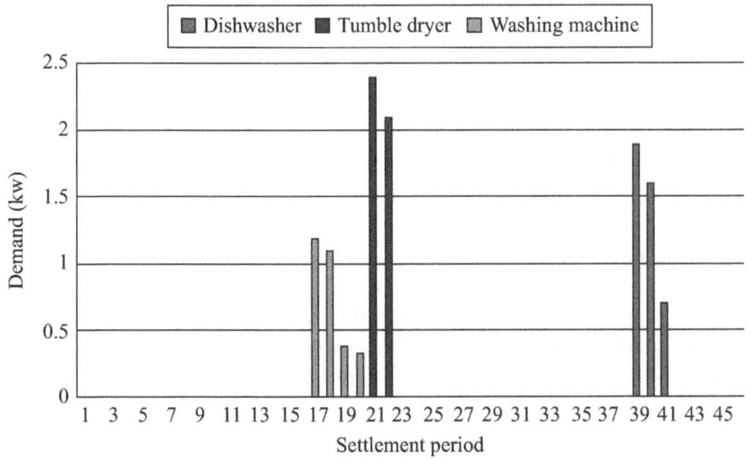


*Figure 4.22 Flexible load profiles for household type three*

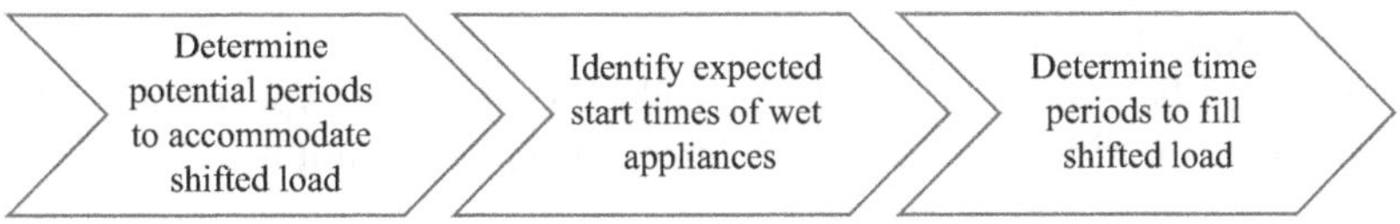


*Figure 4.23 Flow chart for determining periods to accommodate shifted demand*

(i) Quantify potential periods to accommodate shifted load from other periods in response to price signals. The candidate periods are the off-peak periods in the employed ToU tariffs.

(ii) Identify the expected start time of wet appliances to realise effective load shifting, considering customers' daily life habits and energy price variations. For every off-peak period, its duration should be compared with the operation of wash/dry cycle of each wet appliance. If the off-peak period lasts longer, its start time is set as the expected start time for the wet appliance; otherwise, the flexible load is shifted to the following off-peak period.
Within a day, the longest off-peak period lasts from late evening to early morning. If wet appliances can start no later than midnight, the operation is considered to be feasible. Therefore, for general wet appliances, the proposed start time is 12 a.m. if the beginning of the longest off-peak period in ToU tariff is later than midnight. Otherwise, their start is the beginning time point of the overnight off-peak period.

(iii) Determine the periods to accommodate the shifted load. For wet appliances, the proposed periods for shifting can be decided from the expected start time of a wet appliance to the end of one operation cycle. For more flexible customers, the load shifting strategies could be much more flexible.

### 4.5.3 Benefit quantification

When customers change their energy consumption by shifting wet appliances to off-peak price periods in response to ToU tariffs, the financial benefits in terms of electricity bill savings are quantified for a typical day, which is then extrapolated to a whole year. For the *i*th type of appliances in the *j*th day type, the daily financial cost for a customer is

$$C_{old,i\_j} = \mathbf{E} \cdot \mathbf{L}_{h\_old,i\_j} \cdot t \tag{4.10}$$

where $\boldsymbol{E}$ is a 1×48 price matrix, representing the ToU tariffs over a settlement day. $\boldsymbol{L}_{h\_old,i\_j}$ is daily demand profile of the *i*th type of appliance in the *j*th scenario, which can be represented by a 48×1 matrix containing half-hourly demand. $t$ is the length of each settlement period.

Similarly, after flexible load shifting, the new cost is

$$C_{new,i\_j} = \mathbf{E} \cdot \mathbf{L}_{h\_new,i\_j} \cdot t \tag{4.11}$$

where $L_{h\_new,i\_j}$ is the new daily demand profile of the *i*th type of appliance in the *j*th day after flexible load shifting.

The daily benefit in terms of cost reduction for each type of appliance is determined by the difference before and after wet appliance shifting:

$$B_{i\_j} = \mathbf{C}_{new,i\_j} - \mathbf{C}_{old,i\_j} \tag{4.12}$$

The daily benefit is regarded as the benefits from one running cycle of a wet appliance. In order to quantify the annual benefits, the annual running cycles under any one of the eight scenarios/days are:

$$cy_{i\_j} = \frac{AD_i}{D_i} \cdot \frac{d_j}{365} \tag{4.13}$$

where $AD_i$ represents the annual electricity consumption of the *i*th appliance and $D_i$ is its energy consumption in one running cycle. $d_j$ is the number of days in the *j*th scenario/day type.

Therefore, when wet appliances are shifted in response to ToU tariffs, the annual benefits $ANB$ can be determined by

$$ANB = \sum_{j=1}^{8} \sum_{i=1}^{3} B_{i\_j} \cdot cy_{i\_j} \tag{4.14}$$

### 4.5.4 Cooperation with energy storage

The benefits from DSR could be further extended by shifting inflexible loads with the assistance of battery storage. Here, the benefits from DSR enabled by energy storage are compared with those harvested from economic demand shifting.

#### 4.5.4.1 Benefit comparison with energy storage

The effect of DSR enabled by household demand shifting is first compared with that enabled by battery storage. The equivalent capacity of storage that could achieve the same benefit as load shifting is then quantified for each household. It aims to interconnect the DSR from both technical and economic aspects.

As there are three price steps in the developed ToU tariffs, off-peak periods are defined as the periods for battery charging and by contrast, batteries are discharged during shoulder periods in summer and peak periods in other seasons. In these charging and discharging periods, it is assumed that charging and discharging power is constant and there is only one charging cycle within a settlement day. Following this principle, the annual financial benefits obtained from using per kWh storage is approximated by

$$B_{st,un} = \sum_{j=1}^{8} \mathbf{E} \cdot (\mathbf{L}_{st,dis,j} - \mathbf{L}_{st,ch,j}) \cdot t \cdot d_j \tag{4.15}$$

where $\boldsymbol{L}_{st,dis,j}$ and $\boldsymbol{L}_{st,ch,j}$ are $48 \times 1$ matrixes, representing half-hourly charging and discharging demand stemmed from a unit storage in the $j$th scenario/day types.

Once the annual benefit from using one unit kWh storage battery is obtained, it could be compared with that from shifting load. The quotient of annual household benefit from load shifting and from unit kWh storage is defined as the equivalent storage capacity:

$$Eq_{st} = \frac{ANB}{B_{st,un}} \tag{4.16}$$

#### 4.5.4.2 Cooperation with energy storage

The study could be further extended to quantify benefits when both energy storage and household demand shifting are mobilised in response to ToU tariffs. The consequential total benefit for a household is

$$ANB_{total} = ANB + B_{st,un} \cdot Ca \tag{4.17}$$

where $ANB_{total}$ is the annual household benefit from the proposed cooperation. $Ca$ represents the capacity battery for a household.

### *4.5.5 Case study*

In order to test the proposed methodology, typical customers chosen from the UK system are used and the ToU tariffs selected for load shifting are those developed by using the equal interval grouping method described in the previous sector. Power factor and predicted load growth rate together with coincidence factor, annuity factor and discount rated are chosen as 0.95, 2%, 0.8, 0.074 and 5.6%, respectively, and the parameters of storage are given in Table 4.9 [35].

The level of wet appliance ownership in the UK is derived from National Statistics [36], shown in the fourth column of Table 4.10. In addition, for each type of wet appliance, their energy consumption in one cycle and one calendar year are summarised in the second and the third columns of Table 4.9.

*Table 4.9 Parameters of lithium-ion storage battery*

| | **Unit** |
|---|---|
| Battery capacity | 2.4 kWh |
| Depth of discharge limited | 2 kWh (80% depth of discharge) |
| Battery charging current limit | <20% of rated AmpHours |
| Battery discharging current limit | <20% of rated AmpHours |

*Table 4.10 Wet appliances in modelling and the parameters*

| | **Energy consumption per cycle (kWh)** | **Household electricity consumption (kWh/year)** | **Ownership level (%)** |
|---|---|---|---|
| Washing machine | 1.5 | 205 | 96 |
| Tumble dryer | 2.2 | 427 | 57 |
| Dishwasher | 1.3 | 306 | 40 |

*Table 4.11 Annual benefit for DSR from demand shifting and storage*

| | **Per unit household/storage benefit (£)** | **Equivalent storage capacity (kWh)** |
|---|---|---|
| Household demand shifting | 53.5 | 1.4 |
| Per unit storage battery application | 37.8 | 1.0 |

By utilising the proposed benefit quantification algorithm, the annual benefit achieved from shifting flexible wet load can reach £13,750 in the test system. If it is averaged among all grid-connected households, the financial cost saving is £53.5 per dwelling.

The annual benefit obtained from per kWh storage battery, in this case, is £37.8. If a storage battery is expected to generate the same benefit of £53.5, its equivalent capacity is quantified as 1.4 kWh. The annual benefits for DSRs from shifting household demand and applying storage batteries are summarised in Table 4.11.

This chapter proposes a RTP tariff design approach and two methods to develop ToU tariffs from the obtained RTP tariffs using equal interval grouping and hierarchical clustering techniques. The demonstration results prove that the developed ToUs can reflect the variation of real-time energy prices in the wholesale market with fewer price rates. Generally, the smart variable tariffs designed based on RTP tariffs are expected to encourage customers to shift their loads in response to the price variations.

All the RTP and ToU tariffs are achieved depending on a significant factor, i.e. the percentage of energy cost in the total electricity bill. In practice, this value may not be appropriate for all the situations in GB. Therefore, future smart variable tariffs can be improved by scaling up or down the obtained results to accommodate the real situation.

## 4.6 Impact of networks on tariff design

By far, the introduced ToU tariff design only considers the energy side, i.e. mainly capturing the variation in cost of generating electricity. Actually, DSR can bring many benefits to our energy sector, which is extensively discussed in Reference 37. Considering the key cost elements of bills to domestic customers, they can reduce their electricity bills to some extent if they could respond to the corresponding cost drivers. As known that network cost takes up nearly 25% of the total costs, if customers can respond to network condition, the potential network investment might be reduced or deferred, leading to lower bills. This section thus discusses the approach for quantifying DSR impact on network investment and some considerations for designing ToU tariffs considering network costs.

### *4.6.1 Quantification of DSR on network investment*

One approach to quantify the demand reduction during peak time is proposed in Reference 38, which works by measuring the change in network's time to reinforce. The change is then translated into change in present value of future investment.

For a given load growth $r$, the investment horizon is the time taking load to grow from the current level to full loading level:

$$C = D \cdot (1 + r)^n \tag{4.18}$$

where $n$ is time horizon, $C$ is the capacity of a component and $D$ is its current loading level.

Rearranging 4.18 and taking the logarithm of it produces:

$$n = \frac{\log C - \log D}{\log(1 + r)} \tag{4.19}$$

If customers reduce their peak demand by $\Delta p$, the new time horizon, $n_{new}$ can be obtained by replacing $\log D$ with $\log(D - \Delta p)$.

The present value ($PV$) of a component is

$$PV = \frac{Asset\ Cost}{(1 + d)^n} \tag{4.20}$$

The change in $PV$ with a discount rate $d$ is

$$\Delta PV = Asset\ Cost \cdot \left( \frac{1}{(1 + d)^{n_{new}}} - \frac{1}{(1 + d)^n} \right) \tag{4.21}$$

The above equation reflects the saving from demand shifting by $\Delta p$ load on network investment.

### *4.6.2 Tariff design in response to network conditions*

The shortcoming of most existing tariff design is that they do not properly consider the impact of networks, and thus are not able to reflect the costs incurred by practical network operation and investment. The study in this chapter designs tariffs by scaling RTP tariffs with a certain ratio to reflect network costs, but the method is static and rather rudimentary. In addition, the tariff design mainly focuses on transforming flat rate into ToU tariffs or directly designing according to wholesale market prices so that tariffs have time variation features. This concept actually does not quantify the real costs of generating and transporting, and then allocating them to customers depending on their sizes and locations.

In reality, customer behaviour change affects both required generation output and network transfer capacity. Electricity network conditions are decided by the balance between the supply and demand, i.e. generation and load, both of which however fluctuate frequently in real time. The periods of high-energy costs and congested network conditions do not necessarily coincide with each other particularly at distribution level due to local generation. If they reduce or shift demand from peak time, network investment could be deferred, but on the other hand, if they increase consumption during this period, network investment might be brought forward. It justifies that tariffs should reflect not only energy costs, but also network operation and investment costs. Thus, the economic signals should be able to encourage customers to avoid using electricity not only during energy peak period, but also during network-congested periods.

Currently, this type of network investment cost is recovered from customers in terms of use-of-system charges in the UK, for example, Investment Cost-Related Pricing (ICRP) [39] at transmission network, Long-Run Incremental Cost Pricing (LRIC) [40, 41] and Forward Cost Pricing (FCP) [42] at higher-voltage distribution networks, Distribution Reinforcement Model (DRM) at lower-voltage distribution networks. The disadvantage is that network charges generated by these pricing approaches are static and do not change with network conditions, thus unable to directly and dynamically reflect the impact of DSR on network planning and operation.

Once demand is shifted during system peak time, it can reduce either system investment cost or operation cost. When the system peak demand is below a branch capacity, there is no congestion. In this case, demand shifting or reduction can only defer network investment. The actual cost saving in investment deferral is determined through the change in investment horizon introduced in the previous section. On the other hand, when system peak demand along branches are above their capacity, and then network operators need to decide whether to go with network investment or generation/demand curtailment. Then the choice is normally decided by the difference between investment cost and operation cost.

Figure 4.24 illustrates a conceptual relation between network investment cost and operation cost. Normally, investment cost is recovered long time and in each

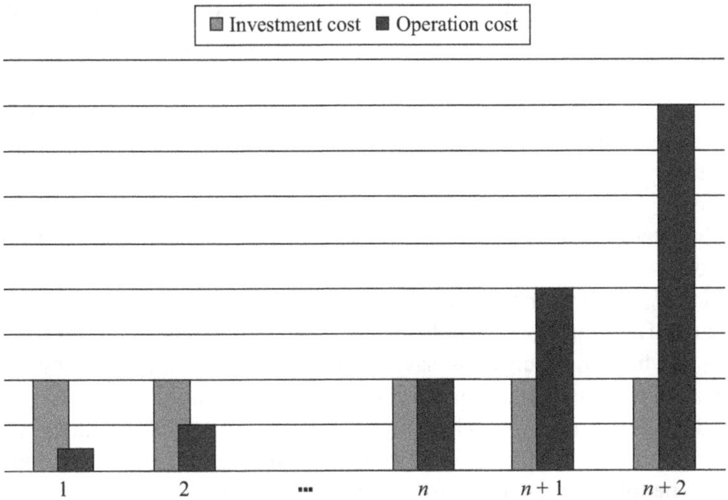


*Figure 4.24 Network investment cost vs. operation cost*

year the recovery is the same, but by contrast, the operation cost would increase on a yearly basis. Thus, at the initial stage, it could be much more economical to conduct generation/demand curtailment, and in the later years, it would be more economical to conduct network investment. Therefore, in designing ToU tariff for reflecting network costs, this trade-off between the two costs has to be dealt with very carefully.

## 4.7 Discussion and conclusion

### *4.7.1 Discussion*

Based on the study and literature review in this chapter, we have identified the following key research directions in tariff design and the application to facilitate domestic DSR for benefiting different parties.

- One key aspect is to tariff the design tariffs considering network costs. As discussed in the previous sector, network cost is the second largest element in electricity bills for domestic customers and if they manage electricity properly in response to network conditions, huge investment or operation costs might be saved. Currently, however, customers have no such motivations due to that the tariffs they see do not reflect the actual network operation practices. We have provided some preliminary concept on this topic, i.e. this part of tariff design should balance the relationship between network investment cost and operation costs.
- In order to test the proposed energy algorithms in the domestic sector, generic GB load profiles designed by Elexon are employed to represent the demand variations

of individual domestic customers. However, when applied into practice, they may be representative enough for all domestic customers in order to conduct DSR due to uncertain demand spikes and uncertainties in real load profiles. Therefore, a key task in the future is to investigate the impact of the proposed method on practical individual load profiles, which are not smooth at all. This can be facilitated by the installed smart metering systems, which have the capabilities to provide high granularity customer electricity use information. Big data techniques are thus very fit for digging out essential demand consumption from the real-time data.

- In this chapter, only the benefits for domestic customers in terms of bill savings are quantified. Further study also needs to investigate the benefit assessment in terms of wholesale energy cost savings and network investment saving to identify the feasibility of the proposed method in practice. Possibly, a whole-system study would be ideal but it is very challenging. We also explore DSR with the assistance of energy storage, but the battery control algorithms and charging envelope design need to be improved to accommodate real situations.
- In the case study, we assume that the appliances at customer ends can be roughly classified into two categories, where wet appliances can be flexibly shifted in response to ToU tariffs. This classification is rather simple and whether customers are willing to respond has not been considered. From economic perspective, some studies have employed demand elasticity to measure the degree of customers' willingness to respond to economic signals. It should be noted that this problem is affected by many factors such as technical, economic, psychosocial and social. Thus, cross-disciplinary research is necessary to combine different expertise together to investigate the three-dimensional challenges of time, location and amount of domestic demand response.
- We can also foresee that into the future with the development of ICTs, dynamic pricing could be feasible. It can deliver the real-time information of costs of both electricity generation and transportation. By then, new tariff schemes can dynamically reflect the costs and engage customers very close.

### 4.7.2 Conclusion

Smart tariffs, as incentives to trigger DSR, play a critical role in the energy sector to exploit the huge resources from the customer side. Load shifting in response to appropriate pricing signals could produce energy cost savings and network investment deferral in addition to other benefits. This chapter investigates smart tariff design enabled by smart metering platform and its application to end customers. The variable ToU tariffs are designed based on energy price variations, which are then scaled to include network transportation costs.

In detail, RTP tariffs are developed based on annual energy price variations, where two novel approaches are proposed to convert the tariffs into ToU tariffs. The two new approaches are equal interval grouping and hierarchical clustering. The developed tariffs by both approaches are for eight scenarios/days, i.e. weekdays and weekends in four seasons, to reflect the tariff diversity throughout a calendar year. The DSR to these tariffs are therefore able to reduce energy cost by moving part of energy consumption from periods with expensive energy, where demand is met by expensive generation units. Each approach has its own characteristics and strength:

- The ToU tariffs achieved by statistically grouping settlement periods of RTP tariffs consider annual price variations as a whole. The time windows can be identified without perturbing critical peak or trough energy prices.
- In the second approach, the ToU tariffs are obtained by clustering settlement periods with similar prices into a cluster. Therefore, these tariff profiles inherit the shape of original RTP tariffs. The number of clusters is obtained through an optimisation model with the consideration of both accuracy and feasibility in implementation.

As illustrated by the demonstration results, the RTP tariffs developed for different days and reasons vary considerably dynamically. Peak prices normally appear in winter and the prices in summer are relatively flatter. The peak price rate in summer weekday is 76% of that in winter weekday, and its peak price in a weekend is only 66% of the highest price in winter weekend. The ToU results achieved from the two proposed methods shows that:

- The RTP prices, which reflect energy price variations, can be represented by ToUs with no more than three price steps. The off-peak periods of ToU tariffs usually appears between midnight and 6 a.m. in the morning, and the peak periods are either in the early evening or in late morning.
- In the ToU tariffs from the equal interval grouping method, the off-peak price rates and durations for weekdays are roughly the same. Two-time intervals are considered as peak periods in a typical spring weekday, but no peak exists for a summer weekday. Even though both winter and autumn have time intervals with peak price, the unit prices and durations in winter are much higher than those in autumn. Two peak periods, which occur in the morning and evening, respectively, are designed for all weekends except those in summer.
- In the ToU results obtained by the hierarchical clustering method, although its prices are generally higher, the durations of peak-price periods in winter are not as long as those in other seasons. The reason is that the settlement periods with critical high prices in winter RTP tariffs are mainly clustered to form peak periods. In addition, the price rates of eight scenarios/days vary dramatically, whatever for peak, shoulder or off-peak periods.

We also insist that ToU tariffs should include other key cost components, particularly those network operation and investment. In this way, the tariffs designed for customers can also benefit networks, which are normally seen as a bottleneck for transporting increasing renewables to end customers. In turn, it can further reduce electricity bills of customers and at the same time enables a low-cost transition of the energy sector to low-carbon infrastructure.

## Bibliography

[1] Strbac G. Demand side management: Benefits and challenges. *Energy Policy*. 2008;36(12):4419–4426.

[2] The Office of Gas and Electricity Market (Ofgem). Electricity settlement reform – moving to half-hourly settlement. 2016. https://www.ofgem.gov.uk/publications-and-updates/electricity-settlement-%E2%80%93-moving-half-hourly-settlement

[3] Environmental and Energy Study Institute (EESI). Smart meters, demand response, and low-income customers. Available from http://wwweesiorg/files/drsg_low_income_111308pdf. 2015. Accessed on May 2016.

[4] Thumim J. Investigating the potential impacts ofTime of Use (TOU) tariffs on domestic electricity customers, Centre for Sustainable Energy. https://www.ofgem.gov.uk/ofgem-publications/87361/toutariffsandclustering-reportvfinal 160414-pdf. 2014

[5] Ofgem. Updated household energy bills explained. Available from https://wwwofgemgovuk/sites/default/files/docs/2013/01/household_energy_bills_explained_udjuly2013_webpdf. 2013. Accessed on May 2016.

[6] Ofgem. Structure of electricity distribution charges: Consultation on the longer term charging framework. Available from https://wwwofgemgovuk/sites/default/files/docs/2005/05/10763-13505pdf. 2005. Accessed on May 2016.

[7] Prandini A. Good, BETTA, best? The role of industry structure in electricity reform in Scotland. *Energy Policy*. 2007;35(3):1628–1642.

[8] Owen. G, Ward J. Smart tariffs and household demand response for Great Britain. *Sustainability First*, 2010. https://www.cse.org.uk/pdf/Sustainability_First_2010_smart_tariffs_and_household_demand.pdf

[9] Wang Z, Li F, editors. Developing trend of domestic electricity tariffs in Great Britain. *Innovative Smart Grid Technologies (ISGT Europe), 2011 Second IEEE PES International Conference and Exhibition on*; 5–7 December 2011.

[10] WPD. SoLa Bristol 2015. Available from http://www.westernpowerinnovation.co.uk/Projects/SoLa-Bristol.aspx. Accessed on May 2016.

[11] Tang Y, Song H, Hu F, Zou Y, editors. *Investigation on TOU pricing principles* 2005: IEEE/PES Transmission & Distribution Conference & Exposition: Asia and Pacific, Dalian, 2005, 1–9.

[12] Na Y, Ji-Lai Y, editors. Optimal TOU decision considering demand response model. *Power System Technology, 2006 PowerCon 2006 International Conference on*; 22–26 October 2006.

[13] Faruqui A, Wood L. Quantifying the benefits of dynamic pricing In the mass market. 2008, The Brattle Group, http://www.edisonfoundation.net/iee/documents/quantifyingbenefitsdynamicpricing.pdf.

[14] Wolak FA. An experimental comparison of critical peak and hourly pricing: The PowerCents DC program. Department of Economics Stanford University. 2010. http://pesd.fsi.stanford.edu/file/218030/download?token=Sw4QDoI6.

[15] Zhang Q, Wang X, Fu M, editors. *Optimal implementation strategies for critical peak pricing* 2009 6th International Conference on the European Energy Market: IEEE, page:1–6.

[16] Demand response program evaluation – Final report Quantum Consulting Inc. and Summit Blue Consulting. LLC. Working Group 2 Measurement and Evaluation Committee, and California Edison Company, 2005.

[17] Yun Z, Quan Z, Caixin S, Shaolan L, Yuming L, Yang S. RBF neural network and ANFIS-based short-term load forecasting approach in real-time price environment. *Power Systems, IEEE Transactions on*. 2008;23(3):853–858.

[18] Schiller P. The control of the domestic load. *Electrical Engineers – Part II: Power Engineering, Journal of the Institution of*. 1941;88(5):373–389.

[19] Celebi E, Fuller JD. A Model for efficient consumer pricing schemes in electricity markets. *Power Systems, IEEE Transactions on*. 2007;22(1):60–67.

[20] Celebi E, Fuller JD. Time-of-Use pricing in electricity markets under different market structures. *Power Systems, IEEE Transactions on*. 2012;27(3): 1170–1181.

[21] Zala HN, Abhyankar A, editors. A novel approach to design time of use tariff using load profiling and decomposition. *Power Electronics, Drives and Energy Systems (PEDES), 2014 IEEE International Conference on*; 2014: IEEE.

[22] de Sa Ferreira R, Barroso LA, Rochinha Lino P, Carvalho MM, Valenzuela P. Time-of-Use tariff design under uncertainty in price-elasticities of electricity demand: A stochastic optimisation approach. *Smart Grid, IEEE Transactions on*. 2013;4(4):2285–2295.

[23] The Office of Gas and Electricity Market (Ofgem). Updated household energy bills explained. 2013. https://www.ofgem.gov.uk/ofgem-publications/64006/householdenergybillsexplainedudjuly2013web.pdf

[24] Barbose G, Goldman C, Neenan B. A survey of utility experience with real time pricing. 2004. Lawrence Berkeley National Laboratory. https://emp.lbl.gov/sites/all/files/REPORT%20lbnl%20-%2054238.pdf

[25] Albadi M, El-Saadany E. A summary of demand response in electricity markets. *Electric Power Systems Research*. 2008;78(11):1989–1996.

[26] Faruqui A, George SS. The value of dynamic pricing in mass markets. *The Electricity Journal*. 2002;15(6):45–55.

[27] UK power market result [Internet]. 2010. Available from http://www.apxendex.com/index.php?id=49. Accessed on May 2016.

[28] Sheen J-N, Chen C-S, Yang J-K. Time-of-Use pricing for load management programs in Taiwan Power Company. *Power Systems, IEEE Transactions on*. 1994;9(1):388–396.

[29] Li R, Wang Z, Gu C, Li F, Wu H. A novel time-of-Use tariff design based on Gaussian mixture model. *Applied Energy*. 2016;162:1530–1536.

[30] Freitas AA. *Data mining and knowledge discovery with evolutionary algorithms*. Berlin: Springer; 2002.

[31] Elexon. Load profiles and their use in electricity settlement. 2013. https://www.elexon.co.uk/wp-content/uploads/2013/11/load_profiles_v2.0_cgi.pdf

[32] The Office of Gas and Electricity Market (Ofgem). Creating the right environment for demand-side response: next steps. 2013. https://www.ofgem.gov.uk/ofgem-publications/85129/creatingtherightenvironmentfordemandsideresponsenextsteps.pdf

[33] Gottwalt S, Ketter W, Block C, Collins J, Weinhardt C. Demand side management – A simulation of household behaviour under variable prices. *Energy Policy*. 2011;39(12):8163–8174.

[34] Knight I, Ribberink H. European and Canadian non-HVAC electric and DHW load profiles for use in simulating the performance of residential cogeneration systems. Natural Resources Canada; 2007.

[35] ENA. Common distribution charging methodology 2010. Available from http://www.energynetworks.org/electricity/regulation/structure-of-charges-cdcm/common-distribution-charging-methodology.html. Accessed on May 2016.

[36] Office for National Statistics. General lifestyle survey – Trends in household size: 1971 to 2009. 2010. http://www.ons.gov.uk/ons/rel/ghs/general-lifestyle-survey/2010/general-lifestyle-survey-overview-report-2010.pdf.

[37] Bradley P, Leach M, Torriti J. A review of current and future costs and benefits of demand response for electricity. University of Surrey; 2011. Available from https://www.surrey.ac.uk/ces/files/pdf/1011_WP_Bradley_et_al_DemandResponse_3.pdf. Accessed on May 2016.

[38] Gu C, Li F, editors. Quantifying the long-term benefits of interruptible load scheme for distribution network investment. *Power and Energy Society General Meeting*, 2011 IEEE; 24–29 July 2011.

[39] Li J, Zhang Z, Gu C, Li F, editors. Long-run incremental pricing based transmission charging method distinguishing demand and generation technologies. *PES General Meeting – Conference and Exposition*, 2014 IEEE; 27–31 July 2014.

[40] Gu C, Li F. Long-run marginal cost pricing based on analytical method for revenue reconciliation. *Power Systems, IEEE Transactions on*. 2011;26(1): 103–110.

[41] Gu C, Wu J, Li F. Reliability-based distribution network pricing. *Power Systems, IEEE Transactions on*. 2012;27(3):1646–1655.

[42] Gu C, Li F, editors. New development in distribution network pricing for revenue recovery in the Great Britain. *Power and Energy Society General Meeting*, 2011 IEEE; 24–29 July 2011.

*Part II*

# Smart grid modeling, control and optimization

*Chapter 5*

# Decentralized models for real-time renewable integration in future grid

*Kiyoshi Nakayama*[1]

## 5.1 Introduction to future smart grid

Future grid will likely support flexible bidirectional power flows and include energy production from multiple, disparate, and uncontrollable sources due to a high penetration of distributed renewable energy resources [1–3]. It is substantiated in Reference 4 that battery systems save a large amount of generation cost without hurting customers' utility. Therefore, it is required to maximize the use and efficiency of renewable resources and realize optimal demand and power production responses that can complement renewable intermittency by incorporating storage systems [5–7].

For this reason, the following important issues need to be addressed:

- balanced allocation of distributed energy resources (DERs);
- production of highly reliable and sustainable operation at all times;
- preventing blackouts caused by sudden load peaks and imbalanced allocation of dispersed renewable energies;
- reducing the adverse environmental impact (e.g. $CO_2$ emissions) and the cost of excessive power generation by fossil fuels, nuclear powers, etc.

To compensate the real time and random nature of renewable generation, decentralized control has become a new direction in creating architectures of future power networks [8,9]. In this chapter, we describe a decentralized model of future grid, formulate a problem to maximize real-time renewable integration, and introduce approaches for solving the problem in a decentralized manner.

## 5.2 Hybrid model of centralized resource management and decentralized grid control

In a future grid, more and more devices are connected to the network(s) and controlled by remote systems. Many vendors are developing smart power devices these days that

[1]NEC Laboratories America, Inc., Cupertino, CA, USA

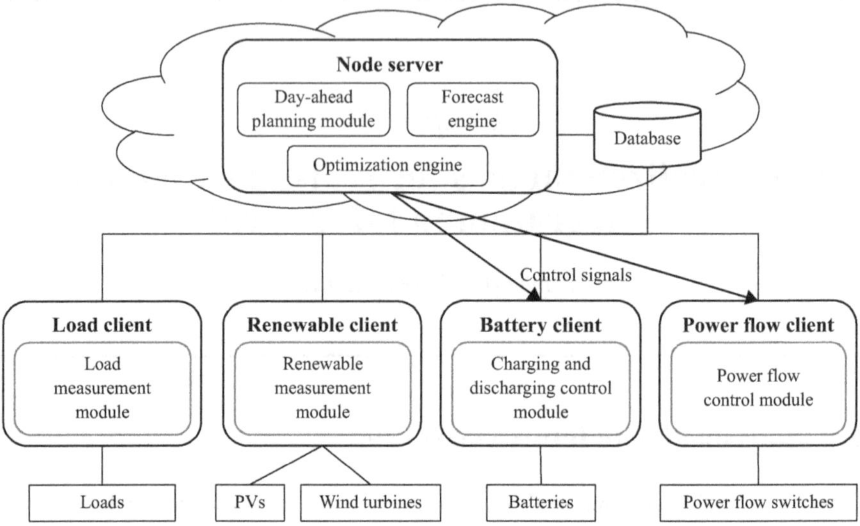


*Figure 5.1 Example of cloud platform at a grid node*

can be controlled through a network by providing access to the devices. Therefore, the increasing number of energy systems and devices are connected with each other and exchange their state information and measurement data such as load profile and local generation. With the significant increase in the number of real-time smart devices, it is necessary to set up a centralized server to aggregate multiple types of energy devices as in Figure 5.1. However, it is not realistic for just one centralized server to aggregate and control all the devices exist in an entire grid since a grid may become quite large such as national transmission grid in the US. Even within a certain area of a country or state, numerous smart devices may exist so that the amount of real-time data that can constantly be collected from energy devices becomes huge and its management gets complicated.

Therefore, both centralized resource management and decentralized grid control are required to achieve efficient future grid operation. In particular, the use cases of the hybrid model of centralized management and decentralized control are described in Sections 5.2.1 and 5.2.2.

### *5.2.1 Centralized resource management*

Centralized management is needed to aggregate the resources that exist in certain geographical area. In particular, one Micro Grid is usually controlled by a centralized energy management system for reliable integration of DERs [10,11].

The multiple resource types include residential devices such as photovoltaics (PVs), electric vehicles (EVs), HVAC (heating, ventilating, and air conditioning) devices, and water pumps. The resources in grid-tied virtual power plants and micro grid may also be aggregated by their energy management system. In this case, larger-size stationary batteries, wind turbines will be considered. The forecasting technologies sometimes are applied not only to individual devices but also to aggregated resources. The typical model of aggregated centralized management is shown in Figure 5.1 where each client talks to individual devices to collect measurement data or issue a command to control them. For instance, the battery client collects the data of state of charge (SOC) of batteries and controls charging and discharging processes of them. The measurement data collected by the clients can be stored in the server's database so that the data could be used in optimizing the resources and the grid control.

### 5.2.1.1 Cloud computing for smart grid

Cloud computing is becoming a way to make the power grid scalable by providing a variety of platforms as well as reduce operational costs and the environmental impact [12,13]. A model for data management with cloud computing platform is discussed in Reference 14, which includes real-time distributed data management, parallel processing of information retrieval. Amazon Web Services (AWS) [15] as a representative platform offers reliable, scalable, and inexpensive cloud computing resources and services for both Windows and UNIX environments. Therefore, a sever can be set up on top of a node or bus in the grid to aggregate data collected from a number of resources such as smart meters. In this way, one cloud server can support a number of real-time devices whose data will be utilized for maximizing the renewable integration as cloud storage also provides efficient data management schemes. Figure 5.1 depicts the case where a server at a node or bus is installed in the cloud platform. The server and clients can communicate over the Internet by using Hypertext Transfer Protocol (HTTP) communications. If the clients are also installed in the cloud platform, a gateway to communicate with local devices should be required to support the energy systems without cloud-based access point. A centralized database (cloud storage) serves as the core part of the platform. A cloud storage also provides multiple access interfaces such as Java Database Connection (JDBC) so that the collected measurement data can be stored in the database in real-time manner.

### 5.2.1.2 Internet of Things devices

The technology based on "Internet of Things" (IoT) [16] has been enabling a lot of network devices to be controlled remotely without manually changing operation settings for achieving autonomous network management. Smart grid is one of the domains where the concept and technologies of IoT are intensively applied [17]. Now a lot of power devices especially for residential areas feature remote access mechanisms such as local application programming interface (API) and cloud-based open API. Therefore, more and more devices are becoming programmable and easy to access to obtain the data of load consumption, renewable generation, and energy level of a storage device [18] so that the aggregated computation can be realized by installing a server at each node in a grid and constantly collecting data from the real-time IoT devices.

The examples of the smart IoT devices are NEST thermostats and EVs sold by Greenlots supporting cloud-based API. On the basis of cloud API mechanism, a remote server can go through the cloud platform provided by the vendor and directly access the device. The security mechanism utilized to support cloud API, among others, is OAuth that is an open standard for authorization.

### *5.2.2 Decentralized grid control*

The power grid is adapting decentralized approaches to control bidirectional flow incurred by renewable energy sources more efficiently at consumption sectors. Intelligent grid automation makes the transition possible by integrating decentralized power management as well as communication technologies that enable smart grid modernization. Autonomous distribution networks have been designed in Reference 19 to take scalability into account by dividing the legacy distribution network into a set of subnetworks. Decentralized control based on the utilization of overlapping information on load frequency is discussed in Reference 20 to introduce distributed power generations and minimize the cost function of load frequency control problem. Decentralized control is also conducted in different applications in a smart/micro grid such as managing the change in local grid frequency and the active power generation within a micro grid [21] and controlling charging demands from EV customers [22,23].

Therefore, control for an entire grid should be decentralized also because setting up a centralized system for a large-scale grid control will be inefficient especially in terms of resiliency. In addition, local autonomy is required due to communication times that are longer than those required to communicate grid perturbations and actuate hardware to prevent widespread outages. Local authority regimes also contain disparate policy and market frameworks in which the optimization and control infrastructure must operate. For this reason, development of an autonomous distributed architecture that can realize optimum allocation of complementary demand and power response together with dispersed renewable energy resources is inevitably required to produce highly efficient and reliable operation of future grid.

This chapter focuses on the decentralized control of an entire future grid to maximize the renewable integration.

## 5.3 Graph modeling

Topological studies have been conducted in terms of power line communications [24,25]. The topology of today's traditional grids must be modified to integrate and manage DERs when creating future grids as current grid topology employs a "tree" structure in the distribution system. However, the tree topology will not hold anymore when DER will be integrated into the grid since the topology is the worst in communication delay and power loss as the number of hops between nodes grows in the order of the size of the network [24]. In addition, recent attention has been focused on mesh topologies for development of new power network infrastructure because

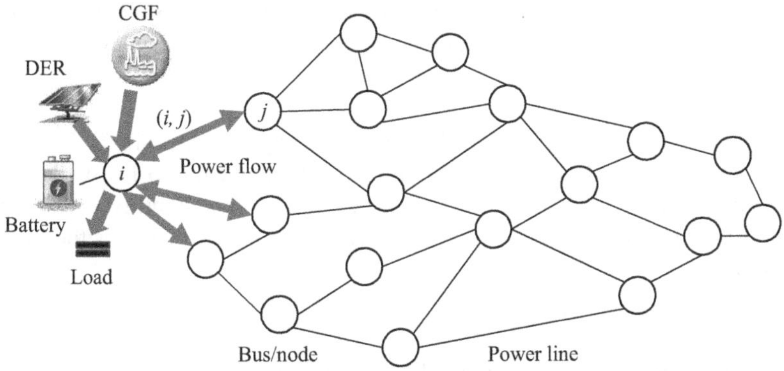


*Figure 5.2 Graph modeling of a future grid*

of the increased flexibility, efficiency, and resiliency they provide and the familiarity, good understanding and technology associated with the analogous information and communications network of the Internet [26].

The grid model used in this chapter is assumed to be a connected network that supports DC power flows such as a transmission grid so that the concept could be applied to a variety of topologies. A connected grid can be expressed using a graph as $G = (\mathscr{V}, \mathscr{E})$ with a set of nodes $\mathscr{V} = \{1, 2, \ldots, n\}$ and a set of edges $(i,j) \in \mathscr{E}$ representing power lines or power routes among nodes where $\mathscr{E} \subseteq \mathscr{V} \times \mathscr{V}$. Each node can be a System Operating Point (SOP) that has the task of integrating growing amounts of renewable power, load processes, incoming and outgoing power flows, etc. The SOP makes decisions to balance the supply and load of electric power at every instant. As forecasts of these variables may have errors, a real-time approach to improve their accuracy was proposed on the basis of accumulating observations when supply and load are balanced [27]. Therefore, as in Figure 5.2, one node aggregates the process of loads, renewable generation, energy storage(s), and incoming and outgoing power flows by setting up a server. A grid has centralized power plant(s) that generate energy using traditional resources. Assume there is a facility that aggregates power plants using the generation by traditional resources defined as *Centralized Generation Facility (CGF)*. An SOP also controls the generation by CGF.

## 5.4 Maximizing real-time renewable integration

Real-time renewable integration becomes the key to achieve future smart grid since there are a lot of economical and social needs to reduce power generation based on the traditional energy resources. The traditional resources include fossil fuels (e.g. coal, gas) and nuclear resources except renewable energies. One of the important factors in integrating renewable energies is minimizing the power supply from CGF.

For any two vertices $i,j$ $(i \neq j)$, let $f_{ij}(t)$ be a real power flow over an edge $(i,j)$ at time $t$. The real power flow from node $i$ to node $j$ in a power network at time $t$ is

$$f_{ij}(t) = V_i V_j Y_{ij}(\theta_i(t) - \theta_j(t)), \quad (i,j) \in \mathscr{E}, \tag{5.1}$$

Assume that $|\theta_i(t) - \theta_j(t)|$ is small enough to approximate $\sin(\theta_i(t) - \theta_j(t))$ by $|\theta_i(t) - \theta_j(t)|$. As voltages must be constantly maintained close to 1 (usually between 0.95 and 1.05), we consider $V_i = 1$ at each node $i$. As a stability constraint, the angle difference $\theta_i(t) - \theta_j(t)$ between two neighboring nodes satisfies the following condition:

$$-\frac{\pi}{2} \leq \theta_i(t) - \theta_j(t) \leq \frac{\pi}{2}. \tag{5.2}$$

Due to a capacity and/or a thermal safety limit, a power flow $f_{ij}(t)$ is bounded by the capacity $c_{ij}(t) > 0$ as follows:

$$-c_{ij}(t) \leq f_{ij}(t) \leq c_{ij}(t), \quad (i,j) \in \mathscr{E}. \tag{5.3}$$

Even though future power networks become enough flexible such as information networks today, a capacity of a branch should still be considered.

Here, let $b_i(t)$ be the battery energy level of node $i \in \mathscr{V}$ at time $t$ indexed by $t \in \mathscr{T} := \{1, 2, \ldots\}$. $b_i(t)$ is properly scaled according to each time step, so that in this case it is not needed to distinguish between energy and power.

An aggregated load $l_i(t)$ by a server of node $i$ at time $t$ is assumed to be an uncontrollable process.

Each node also has an aggregation function for distributed generation (DG) systems that exploit renewable energy resources. The aggregated amount of power derived from renewable generation devices is defined as $r_i(t)$ at time $t$, which is also an uncontrollable intermittent process. Then, *net demand* $d_i(t)$ is defined as follows:

$$d_i(t) := l_i(t) - r_i(t). \tag{5.4}$$

If the renewable generation is less than the load, node $i$ demands power at time $t$; otherwise $d_i(t)$ is charged to a battery.

The generation by CGF $g_i(t)$, which is an controllable process, is bounded by

$$0 \leq g_i(t) \leq \overline{g}_i, \quad i \in \mathscr{V}, \tag{5.5}$$

where $\overline{g}_i$ is a CGF generation capacity at node $i$.

From Kirchhoff's laws, the *net power export* from node $i$ to adjacent nodes at time $t$ is given by

$$F_i(t) := \sum_{j \in \mathscr{V}} V_i V_j Y_{ij}(\theta_i(t) - \theta_j(t)), \quad i \in \mathscr{V}. \tag{5.6}$$

If the net power export at node $i$ is positive, $i$ supplies power at time $t$; otherwise $i$ demands power.

Then, the battery energy level $b_i(t)$ at node $i$ at time $t$ follows the dynamics as

$$b_i(t) = b_i(t-1) - d_i(t) + g_i(t) - F_i(t), \quad i \in \mathscr{V}, \tag{5.7}$$

with a given initial energy level $b_i(0) \geq 0$. The battery energy level $b_i(t)$ is bounded by a battery capacity where $0 \leq b_i(t) \leq \overline{b}_i$.

Some of the factors are determined in a day-ahead planning phase. For instance, the battery may have to be charged while electricity price is cheap and discharged during peak-price time. In addition, there are load peaks especially in summer season so that batteries should be discharged to smooth the peaks. As this chapter focuses on real-time control, a day-ahead planning scheme of optimal SOC for batteries is not discussed. Instead, *Lower Energy Limit (LEL)* $e_i(t)$ is defined in case that a battery needs to satisfy a certain SOC at time $t$. Then, the boundary condition of a battery at node $i$ at time $t$ is

$$e_i(t) \le b_i(t) \le \overline{b}_i, \quad i \in \mathcal{V}, \tag{5.8}$$

with a given initial LEL $e_i(0)$. $\Delta e_i(t) = \overline{b}_i - e_i(t)$ is preserved for unexpected renewable generation peak when $r_i(t)$ is larger than the load process $l_i(t)$. In this case, the net demand becomes negative as $d_i(t) < 0$, so that $|d_i(t)|$ is stored in the battery at node $i$.

Given an initial battery energy level $b_i(0)$, battery capacity $\overline{b}_i$, LEL $e_i(t)$, load $l_i(t)$, and renewable generation $r_i(t)$ at each time step $t$ for each $i \in \mathcal{V}$, and a link capacity $c_{ij}(t)$ and admittance $Y_{ij}$ for each $(i,j) \in \mathcal{E}$, the objective function to maximize the renewable integration at each time step $t \in \mathcal{T}$ is

$$\begin{aligned} \text{Minimize} \quad & \sum_{i \in \mathcal{V}} g_i(t), \\ \text{Over} \quad & g, f, F, b, \theta, \\ \text{Subject to} \quad & (5.1)\text{–}(5.8). \end{aligned} \tag{5.9}$$

## 5.5 General decentralized approaches

The principle that underlies decentralized algorithms is that each node tries to solve a problem by exchanging messages with other peripheral nodes. All the basics of distributed algorithms are described in Reference 28. The application of distributed approaches can also be achieved by setting up HTTP servers and clients. For example, once the HTTP methods and functions are implemented in the server of a node, a node can talk to other nodes with HTTP client functions over the Internet. Communications among the nodes can therefore be implemented in multiple ways.

There are two typical scenarios when conducting decentralized control for a power grid modeled by a graph:

- Communicating with adjacent nodes only (nodal approach)
- Clustering a grid into a set of segments (clustering approach)

The nodal approach is a typical way to realize distributed control within a grid and its mechanism often becomes the foundation of a lot of other decentralized approaches. A message that is exchanged among the nodes can be a notification signal or a message that contains state information, measurement data of devices, etc. For example, if a communications scheme exploits HTTP protocols, simple GET and POST, Extensible

Markup Language (XML), JavaScript Object Notation (JSON), among others, can be supported to transfer those data.

The clustering approach based on graph modeling divides a network into set of segments where the segment can be a tree, cut sets, loops, among others. Sometimes clustering controls demonstrate better performance than nodal approaches in terms of failure detection or flow controls [29].

## 5.6 Distributed nodal approach

Distributed nodal approach (DNA) is a representative approach to solve distributed problems.

In this scenario to solve the formulated problem, the message includes the value of the function that evaluates a node with certain predefined criteria. When deciding the process priority for overlapping resources shared by several adjacent nodes, each node $i$ exchanges the value of the *Nodal Evaluation Function (NEF)* denoted as $\Phi_i(t)$ with its adjacent nodes $\mathscr{A}_i$. NEF is subject to objective functions, constraints, characteristics of a grid, etc. Here, for each node $i \in \mathscr{V}$, a *Battery Penalty Cost (BPC)* $H_i(t)$ is defined as

$$H_i(t) := \begin{cases} 0, & \text{if } b_i(t) \geq e_i(t), \\ \alpha_i(e_i(t) - b_i(t)), & \text{if } b_i(t) < e_i(t), \end{cases} \tag{5.10}$$

with some penalty coefficient $\alpha_i > 0$. If BPC is utilized for NEF, $\Phi_i(t) = H_i(t)$. When the process priority cannot be determined by NEF with BPC, node $i$ can use precomputed value or predefined ID such as Multiple Access Control (MAC) address.

*Nodal Flag (NF)* denoted as $\zeta_i$ is used to distinguish a node in process from a stand-by node once the process priority is decided using NEF. The definition of NF is as follows:

- $\zeta_i = 0$: Node $i$ is Stand-By
- $\zeta_i = 1$: Node $i$ is In Process

When $\zeta_i = 1$, all the adjacent nodes $\mathscr{A}_i$ must stand by setting its NF as $\zeta_i = 0$ until node $i$ finishes its procedure. Parallel optimization among nodes is feasible with the concept of NF.

A *Nodal Evaluation Function Message (NEFM)* is utilized when exchanging the value of NEF $\Phi_i(t)$ with adjacent nodes $\mathscr{A}_i$. A *Nodal Flag Signal (NF Signal)* is used for exchanging the binary value of NF of node $i$. DNA exchanges NEFM asynchronously without any global clock. Flowchart of DNA conducted in each node $i$ is described in Figure 5.3. The procedure of DNA is mainly consists of *Initialize*, *Send*, *Receive*, *Optimize*, *Notify*, *Confirm*, and *StandBy*.

### *5.6.1 Initialize*

In *Initialize*, NEF and NF of node $i$ are set as $\Phi_i(t) = 0$ and $\zeta_i = 0$, respectively. Net demand is calculated as $d_i(t) = l_i(t) - r_i(t)$ based on the measurement data of load

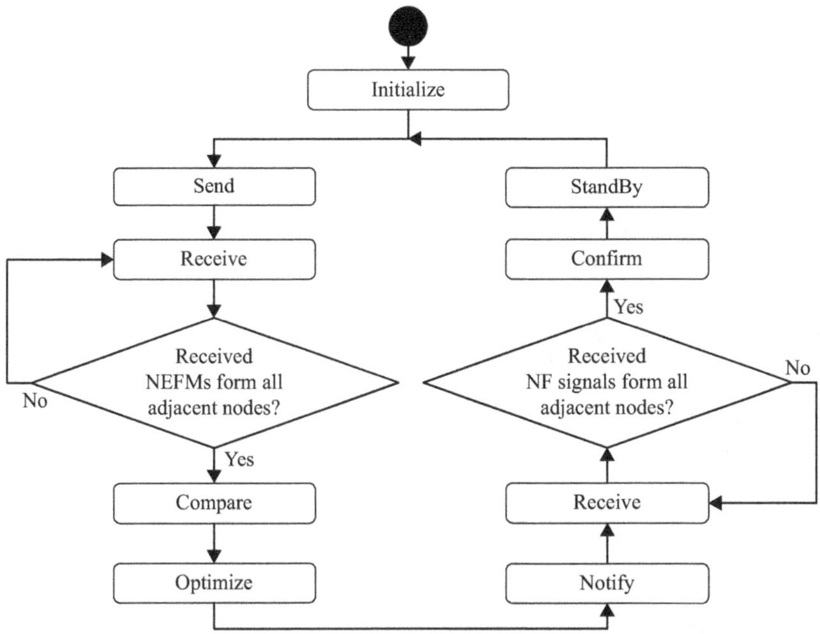


*Figure 5.3 Flowchart of distributed nodal approach (DNA)*

and renewable generation. In addition, set $g_i(t)$, $\theta_i(t)$, and $F_i(t)$ to 0 so that at this moment of initialization, the battery energy level is $b_i(t) = b_i(t-1) - d_i(t)$.

## 5.6.2 Send

The value of NEF $\Phi_i(t)$ is calculated based upon the measurement data of node $i$ at time $t$. After calculating $\Phi_i(t)$, node $i$ writes its value of NEF in NEFM. Then, node $i$ sends the NEFMs to all the adjacent nodes $\mathscr{A}_i$.

## 5.6.3 Receive

Receive is called when node $i$ receives a message (e.g. NEFM, NF Signal) from an adjacent node $j \in \mathscr{A}_i$. In case that one of the other procedures of DNA is not completed, the message is temporarily stored in node $i$. Until node $i$ receives a message from all the adjacent nodes $\mathscr{A}_i$, node $i$ stands by to receive another NEFM. Once $i$ received all the messages from $\mathscr{A}_i$, it moves on to the next procedure.

## 5.6.4 Compare

After receiving all the NEFMs form $\mathscr{A}_i$, node $i$ compares its value of $\Phi_i(t)$ with the value of $\Phi_j(t)$ of each adjacent node $j \in \mathscr{A}_i$. If the value of $\Phi_i(t)$ is the largest among those of all the adjacent nodes, node $i$ sets its NF as $\zeta_i = 1$, otherwise $\zeta_i = 0$. If the value of $\Phi_i(t)$ is the same as the value of $\Phi_j(t)$ of node $j$, node $i$ uses another NEF to decide the procedure priority such as a precomputed value $\Phi_i^p(t)$ that is unique for each node $i$.

**Algorithm 1** Compare

Set NF as $\zeta_i = 1$.
**for** each $j \in \mathscr{A}_i$ **do**
  **if** $H_i(t) < H_j(t)$ **then**
    Set NF as $\zeta_i = 0$.
  **else if** $H_i(t) = H_j(t)$ **then**
    **if** $\Phi_i^p(t) < \Phi_j^p(t)$ **then**
      Set NF as $\zeta_i = 0$.
    **end if**
  **end if**
**end for**

### 5.6.5 *Optimize*

The most common way to solve the problem on maximizing renewable integration (5.9) would be exploiting the notion of Linear Programming that has been applied to many linear-optimization problems. As we are solving (5.9) in a decentralized manner, Distributed Linear Programming (DLP) approach is formulated in Algorithm 2 where each node $i$ tries to minimize the amount of power generated at its CGF $g_i(t)$ under the constraints (5.1)–(5.8) by communicating with adjacent nodes $\mathscr{A}_i$.

If $\zeta_i = 1$, node $i$ conducts Algorithm 2. After conducting DLP, node $i$ sets its NF as $\zeta_i = 0$.

### 5.6.6 *Notify*

In *Notify*, node $i$ sends NF Signal to each adjacent node $j \in \mathscr{A}_i$ to notify that $\zeta_i = 0$.

### 5.6.7 *Confirm*

*Confirm* is called when node $i$ receives NF Signals from an adjacent node $j \in \mathscr{A}_i$ after *Optimize*. Until node $i$ receives NF Signals from all the adjacent nodes $\mathscr{A}_i$, node $i$ waits for another NF Signal. After receiving all the NF Signals form $\mathscr{A}_i$, node $i$ confirms that each NF of $j \in \mathscr{A}_i$ is $\zeta(L_j) = 0$.

### 5.6.8 *StandBy*

Let $\Delta t$ be a communication interval among adjacent nodes. In StandBy, node $i$ waits for $\Delta t$ minute(s), and then proceed to the next round of DNA as in Figure 5.3.

The procedure of DNA can be terminated if a node receives a termination signal.

**Algorithm 2** Distributed Linear Programming (DLP) Algorithm

**if** NF $\zeta_i = 1$ **then**
  Set $flag_{opt} = 1$.
  **for each** $j \in \mathscr{A}_i$ **do**
    **if** $b_j(t) < e_j(t)$ **then**
      Set $flag_{opt} = 0$.
    **end if**
  **end for**
  **if** $flag_{opt} = 1$ **then**
    $\theta_{min} \leftarrow +\infty$.
    **for each** $j \in \mathscr{A}_i$ **do**
      $\theta_{tmp} := -\frac{b_j(t) - e_j(t)}{V_i V_j Y_{ij}}$.
      **if** $|\theta_{tmp}| < |\theta_{min}|$ **then**
        $\theta_{min} \leftarrow \theta_{tmp}$.
      **end if**
    **end for**
    **if** $|F_i(t)| > e_i(t) - b_i(t)$ **then**
      $\theta_i(t) \leftarrow -\frac{e_i(t) - b_i(t)}{\sum_{j \in \mathscr{A}_i} V_i V_j Y_{ij}}$.
      $g_i(t) = 0$.
      $F_i(t) = \sum_{j \in \mathscr{A}_i} V_i V_j Y_{ij} \theta_i(t)$.
    **else**
      $\theta_i(t) \leftarrow \theta_{min}$.
      $F_i(t) = \sum_{j \in \mathscr{A}_i} V_i V_j Y_{ij} \theta_i(t)$.
      $g_i(t) \leftarrow e_i(t) - b_i(t)$.
    **end if**
  **else**
    $g_i(t) \leftarrow e_i(t) - b_i(t)$.
    $\theta_i(t) = 0$.
    $F_i(t) = 0$.
  **end if**
  Set NF as $\zeta_i = 0$.
**else**
  Skip *Optimize*.
**end if**

## 5.7 Distributed clustering approach

Graph clustering or partitioning approaches have often utilized when dividing a network into a set of segments. For most of the cases, the partitioning is conducted in a centralized manner so that there is a master node that grasp the entire topology of the partition.

Although there are many partitioning approaches have been provided, the concept of "cut-set" and "tie-set" has been used with the Kirchhoff's laws (e.g. Kirchhoff's Current Law (KCL) and Kirchhoff's Voltage Law (KVL)). Refer to the literature [29–31] for how to configure the cut-set and tie-set structure in a network.

As one of the clustering approaches on maximizing renewable integration, we introduce an approach based on Tie-Set Graph Theory.

## *5.7.1 Tie-set graph theory and its application to distributed systems*

The concept of tie-sets has been applied to core fields of circuits and systems in formulating, solving, and characterizing properties of various problems from reliability, resiliency, and optimization perspectives. The tie-set graph theory has been described in References 31–33 and applied to many areas of smart grid applications [34,35]. Here, we introduce the basis for the unfamiliar reader.

### 5.7.1.1 Fundamental system of tie-sets

For a given bi-connected graph $G = (\mathscr{V}, \mathscr{E})$ with a set of vertices $\mathscr{V} = \{1, 2, \ldots, n\}$ and a set of edges $\mathscr{E} = \{e_1, e_2, \ldots, e_m\}$, let $L_i = \{e_1^i, e_2^i, \ldots\}$ be a set of all edges that constitutes a loop in $G$. The set of edges $L_i$ is called a *tie-set* [30]. Let $T$ and $\overline{T}$, respectively, be a spanning tree and a co-tree of $G$, where $\overline{T} = \mathscr{E} - T$. For $l \in \overline{T}$, $T \cup \{l\}$ includes one tie-set. Focusing on a subgraph $G_T = (\mathscr{V}, T)$ of $G$ and an edge $l = (a, b) \in \overline{T}$, there exists only one elementary path $P_T$ whose origin is $b$ and terminal is $a$ in $G_T$. Then, a *fundamental tie-set* that consists of the path $P_T$ and the edge $l$ is uniquely determined as $L(l) = \{l\} \cup P_T(b, a)$. It is known that $\mu = |\overline{T}|$ fundamental tie-sets exist in $G$, and they are called a *fundamental system of tie-sets* $\mathbb{L}_B = \{L_1, L_2, \ldots, L_\mu\}$. A fundamental system of tie-sets guarantees that it covers all the vertices and edges (Figure 5.4(a)) even in a bi-connected and non-planar graph $G$ (Figure 5.5).

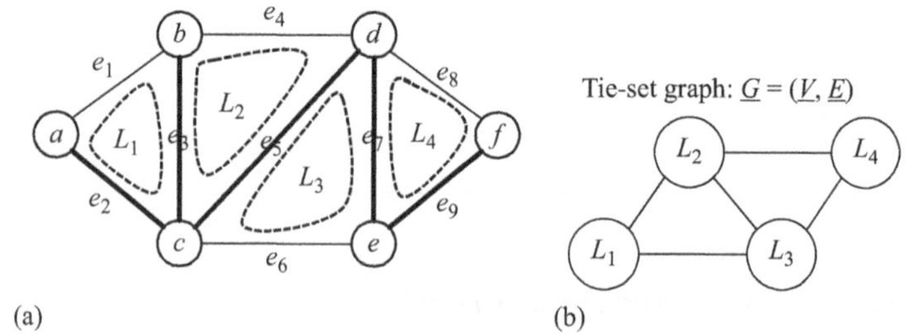


*Figure 5.4 An example of a fundamental system of tie-sets and its tie-set graph. (a) A fundamental system of tie-sets $\{L_1, L_2, L_3, L_4\}$ and (b) the tie-set graph of panel (a)*

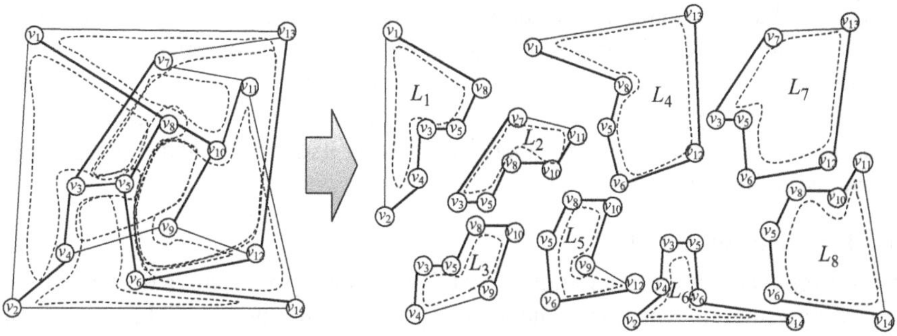

*Figure 5.5 A fundamental system of tie-sets in a non-planar graph*

Fundamental tie-sets are independent of each other; any fundamental tie-set cannot be obtained by calculus $\oplus$[2] among other tie-sets.

#### 5.7.1.2 Tie-set graph

A graph $\underline{G} = (\underline{\mathcal{V}}, \underline{\mathcal{E}})$ is defined as a *tie-set graph*, where a set of vertices $\underline{\mathcal{V}}$ corresponds to a fundamental system of tie-sets $\{L_1, L_2, \ldots, L_\mu\}$, and a set of edges $\underline{\mathcal{E}}$ corresponds to a set of edges $\{\underline{e}(L_i, L_j)\}, (i \neq j)$, which represent the connections among tie-sets. $\underline{e}(L_i, L_j)$ is determined by the set of common vertices of $L_i$ and $L_j$. Let $\mathcal{V}(L_i)$ be a set of all the vertices included in a tie-set $L_i$. If $\mathcal{V}(L_i) \cap \mathcal{V}(L_j) \neq \emptyset$, $L_i$ and $L_j$ have an edge $\underline{e}(L_i, L_j)$.

Each fundamental tie-set of a given graph $G$ is uniquely mapped to the specific tie-set graph $\underline{G}$ as shown in Figure 5.4(b).

#### 5.7.1.3 Tie-set Information $\mathbb{L}_i$

Tie-set Information is the State Information of tie-set(s) to which the node $i$ belongs. When node $i$ belongs to $L_i$ where $i \in \mathcal{V}(L_i)$, $i$ has Tie-set Information of $L_i$. The node $c$ in Figure 5.4(a), for example, has information of fundamental tie-sets $\{L_1, L_2, L_3\}$ as Tie-set Information.

#### 5.7.1.4 Communications among tie-sets

After constructing Tie-set Information at each node, a leader node $v_l^i$ is selected in each tie-set $L_i$ by exchanging state information of the nodes within $L_i$. A leader node of a tie-set determines the optimal control for distributed problems and solves them. A leader also decides how to use distributed resources allocated to each node in a tie-set domain. The criteria to decide a leader of a tie-set are magnitude relation of node ID (physical addresses), the number of adjacent nodes (incident links), the number of tie-sets to which a node belongs, among others [36]. A leader node $v_l^i$ of

[2]The definition of $\oplus$ for a set $A$ and a set $B$ is defined as follows:
$A \oplus B = (A - B) \cup (B - A) = (A \cup B) - (A \cap B)$.

a tie-set $L_i$ has information of *Adjacent Tie-sets* $\mathbb{L}_i^a = \{L_1^i, L_2^i, \ldots\}$ so that a leader $v_l^i$ of $L_i$ communicates with leaders of adjacent tie-sets $\mathbb{L}_i^a$. An adjacent tie-set $L_j$ of $L_i$ is determined according to the relation of connection $\underline{e}(L_i, L_j) \in \underline{\mathscr{E}}$ of $\underline{G}$. For example, adjacent tie-sets of $L_1$ in Figure 5.4(a) are $\mathbb{L}_1^a = \{L_2, L_3\}$ so that the leader node of $L_1$ constantly communicates with leaders of $\mathbb{L}_1^a$ using its precomputed routing table as explained in Reference 36.

#### 5.7.1.5 Tie-set Agent

*Tie-set Agent (TA)* is an autonomous agent that constantly navigates a tie-set $L_i$ around and brings state information of nodes in $L_i$ to its leader node $v_l^i$. The concept of a distributed intelligent agent system with autonomous agents are explained in Reference 37. TA contains a measurement vector $y_i(t)$ that provides the value of current loads $l_i(t)$ and renewable power generations $r_i(t)$, where $i \in \mathscr{V}(L_i)$. The leader node of a tie-set $L_i$, which is considered to be an SOP in a tie-set, receives a measurement vector $y_i(t)$ and decides the procedure in $L_i$ based upon $y_i(t)$. TA constantly brings the measurement data $y_i(t)$ of $L_i$ to its leader node with certain time interval.

#### 5.7.1.6 Tie-set Evaluation Function

*Tie-set Evaluation Function (TEF)* is the function that evaluates a tie-set with certain predefined criteria. When deciding the process priority for overlapping resources shared by several adjacent tie-sets, each tie-set exchanges the value of TEF denoted as $\Phi(L_i, t)$ with its adjacent tie-sets $\mathbb{L}_i^a$. TEF $\Phi(L_i, t)$ is calculated based upon the current measurement vector $y_i(t)$.

Here, TEF is defined using BPC as follows:

$$\Phi(L_i, t) = \sum_{i \in \mathscr{V}(L_i)} H_i(t). \tag{5.11}$$

For instance, as BPC is used for TEF, the TEF values of tie-sets in the upper network in Figure 5.6 are

$$\begin{aligned}
\Phi(L_1, t) &= H_a(t) + H_b(t) + H_c(t) = 25.0, \\
\Phi(L_2, t) &= H_b(t) + H_c(t) + H_d(t) = 12.0, \\
\Phi(L_3, t) &= H_c(t) + H_d(t) + H_e(t) + H_f(t) = 24.0, \\
\Phi(L_4, t) &= H_e(t) + H_f(t) + H_g(t) + H_h(t) = 28.0,
\end{aligned} \tag{5.12}$$

as LELs of all nodes in the example network is set as $e_i(t) = 40.0$. Therefore, the NFs of $\{\zeta(L_1), \zeta(L_2), \zeta(L_3), \zeta(L_4)\}$ are set as $\{1, 0, 0, 1\}$ so that $L_1$ and $L_4$ are in process in Figure 5.6.

### *5.7.2 Tie-set Based Optimization Algorithm*

Tie-set Based Optimization (TBO) Algorithm is described in Algorithm 3 when an tie-set $L_i$ optimizes its resources. In a tie-set graph $\underline{G}$, each node represents a tie-set itself. Therefore, if we consider a tie-set as a node, the DNA can be directly applied to the parallel computation among tie-sets where each tie-set communicates with

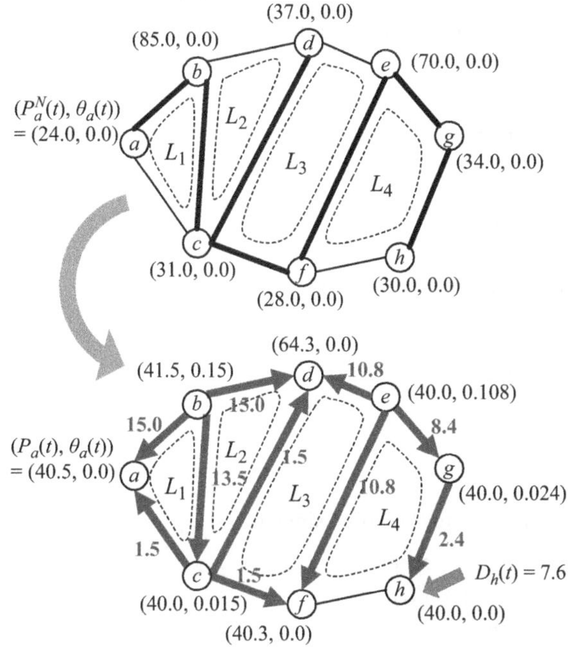


*Figure 5.6 Parallel optimization in $L_1$ and $L_4$*

adjacent tie-set(s) in order to decide which tie-set conducts optimization. The tie-set that conducts optimization calculates the power flows $f_{ij}(t)$ that realize the balanced renewable energy sharing as well as decide the amount of CGF generation $g_i(t)$.

*STEP 0:* Step 0 is an initialization procedure. For each node $i$ in a tie-set $L_i$, its angle $\theta_i(t)$ and CGF generation $g_i(t)$ are set as 0. Battery energy levels in $L_i$ from a previous time step $b_i(t-1)$ have been preserved in a leader node $v^i_l$ of $L_i$ at time $t$. As information of loads $l_i(t)$ and renewables $r_i(t)$ in $L_i$ are provided by TA, the net demand is decided as $d_i(t) := l_i(t) - r_i(t)$. Then, the net battery energy level is calculated as $b^d_i(t) := b_i(t-1) - d_i(t)$.

*STEP 1:* In Step 1, the angles of nodes in $L_i$ are calculated, and thus power flows on the edges are decided. We define a *flag* $= \{0, 1\}$ to decide whether or not $L_i$ conducts the **while** loop of Algorithm 3. If there is a node in $L_i$ that has energy that can be distributed to other nodes, that is, $b^d_i(t) > e_i(t)$, then *flag* $= 0$. Otherwise, $L_i$ goes to Step 2 without distributing energy of nodes. In the **while** loop, phase angles are calculated in order to balance the battery energy levels. The node with maximum energy that meets $b^d_i(t) > e_i(t)$ distributes its power to other nodes until it reaches LEL $e_i(t)$. The **while** procedure stops if all the nodes in $L_i$ store energy greater than or equal to LEL where $b^d_i(t) \geq e_i(t)$ is satisfied. The **while** procedure also stops if energy of the nodes that meet $b^d_i(t) > e_i(t)$ converges on its LEL $e_i(t)$, where a certain small value $\varepsilon$ is used to check the convergence.

**Algorithm 3** Tie-set Based Optimization (TBO) Algorithm

**STEP 0:** Initialization
**for** each $i \in \mathscr{V}(L_i)$ **do**
  $\theta_i(t) = 0, g_i(t) = 0.$
  $d_i(t) := l_i(t) - r_i(t).$
  $b_i^d(t) := b_i(t-1) - d_i(t).$
**end for**

**STEP 1:** Deciding Angles for Power Flows
Set *flag* $= 1.$
**for** each $i \in \mathscr{V}(L_i)$ **do**
  **if** $b_i^d(t) > e_i(t)$ **then**
    *flag* $= 0.$
  **end if**
**end for**
**while** *flag* $\neq 1$ **do**
  Select $i \in \mathscr{V}(L_i)$, where $b_i^d(t)$ is maximum.
  Calculate $\theta_i'(t) = \frac{b_i^d(t) - e_i(t)}{\sum_{j \in \mathscr{A}_i} V_i V_j Y_{ij}}.$
  Update $\theta_i(t) \leftarrow \theta_i(t) + \theta_i'(t).$
  **for** each $j \in \mathscr{A}_i$ **do**
    Update $b_i^d(t) \leftarrow b_i^d(t) - V_i V_j Y_{ij} \theta_i'(t).$
    Update $b_j^d(t) \leftarrow b_j^d(t) + V_i V_j Y_{ij} \theta_i'(t).$
  **end for**
  Set *flag* $= 1.$
  **for** each $i \in \mathscr{V}(L_i)$ **do**
    **if** $b_i^d(t) < e_i(t)$ **then**
      Set *flag* $= 0.$
    **end if**
  **end for**
  $b_{max} := \max_{i \in \mathscr{V}(L_i)} \{b_i^d(t)\}.$
  **if** $b_{max} - e_i(t) \leq \varepsilon$ **then**
    *flag* $= 1.$
  **end if**
**end while**
Set angles $\theta_i(t)$ for all nodes $i \in \mathscr{V}(L_i)$ with (5.2).
Distribute power flows in $L_i$ according to (5.1) and (5.3).

**STEP 2:** Deciding Power Supply from CGF
**for** each $i \in \mathscr{V}(L_i)$ **do**
  **if** $b_i(t) < e_i(t)$ **then**
    $g_i(t) = e_i(t) - b_i(t).$
    CGF supplies power to $i$ with the amount of $g_i(t)$.
  **end if**
**end for**

After calculating the angles, the leader $v_l^i$ notifies the nodes in $L_i$ of the magnitude of angles to distribute power flows.

*STEP 2:* Step 2 decides the amount of power generated at CGF $g_i(t)$ at nodes in $L_i$. With the LEL $e_i(t)$ at each node in a tie-set, if node $i$ does not have minimum level of power as in $b_i(t) < e_i(t)$, then CGF supplies power to the node $i$ with the amount of $g_i(t) = e_i(t) - b_i(t)$.

## 5.8 Case study of decentralized grid control

In order to validate the behavior of the decentralized model facing with different levels of uncertainty that are seen in load processes and renewable generation, case studies are discussed on the performance of nodes $i$ and $j$ of the network in Figure 5.2, which are shown in Figure 5.7(a) and 5.7(b). The DNA is exploited to validate the behavior of the processes.

At time $t = 0$, the battery energy level and LEL of nodes $i$ and $j$ are given with $b_i(0) = b_j(0) = 60$ and $e_i(0) = e_j(0) = 20$, respectively. As in the case studies, nodes $i$ and $j$ increase their LEL $e_i(t)$ and $e_j(t)$ to cope with sudden load peaks. For instance, at $t = 30$ and 135, $e_i(30)$ and $e_i(135)$ at node $i$ are updated as there are sudden load peaks expected in control planning phase. This procedure demonstrates that those nodes can deal with the uncertainty of load processes by changing their LEL in order to prevent the nodes from running out of power.

There also are sudden renewable peaks as indicated in Figure 5.7(a) and 5.7(b). For example, when $t = 45$, 75, 105, and 180 at node $i$, the renewable generation is higher than the load process. In this case, some amount of the sudden renewable

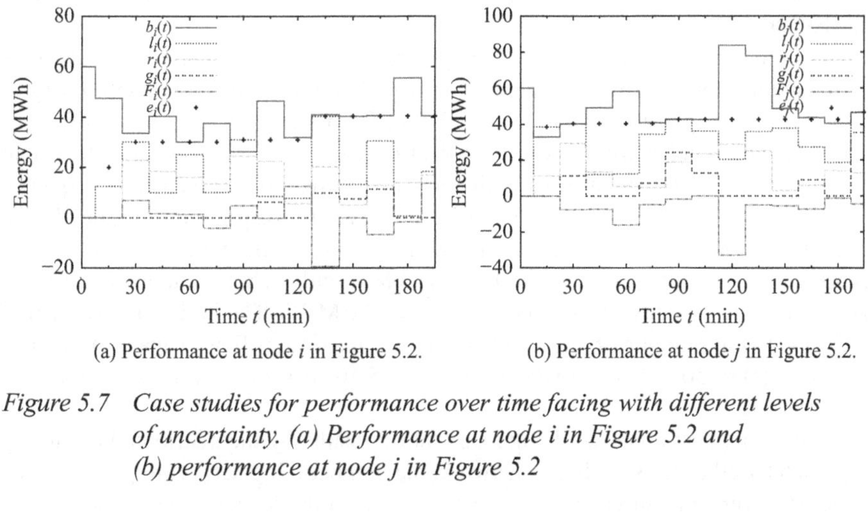


(a) Performance at node $i$ in Figure 5.2.

(b) Performance at node $j$ in Figure 5.2.

*Figure 5.7 Case studies for performance over time facing with different levels of uncertainty. (a) Performance at node i in Figure 5.2 and (b) performance at node j in Figure 5.2* generation is stored in the battery of node $i$, and then shared to the adjacent nodes at the next time step. This also demonstrates the effectiveness of having preserved space $\Delta e_i(t) = \overline{b}_i - e_i(t)$ for the renewable peak as the excess amount of renewable generation can be temporarily stored at the node, and then delivered to the adjacent nodes that require power to satisfy their net demands.

Power flows between the nodes $i$ and $j$ on the case studies also demonstrate the framework that each node distributes power to the other peripheral nodes when a node has enough power, that is, $b_i(t) > e_i(t)$. At the same time, the node in short of energy ($b_i(t) < e_i(t)$) receives incoming power flows with negative value of $F_i(t)(<0)$ from the nodes that have the excess amount of energy. For example, as node $i$ had enough energy from the previous time step $t = 15$, $i$ shares its energy to the adjacent nodes at $t = 30$. The portion of the energy from $i$ is sent to node $j$ as seen in the incoming power flow $F_j(30)$, which satisfies the net demand $d_j(30) = l_j(30) - r_j(30)$ together with supplementary CGF generation $g_j(30)$. Thereby, the node $j$ makes up for the first load peak that happened from $t = 15$ to 30.

The process of CGF constantly compensates the net demand that has been partially satisfied by incoming power flows. As in the case studies, although node $i$ obtains incoming power flows at $t = 135$, $i$ still needs more power to meet $b_i(135) \geq e_i(135)$. In order to satisfy the demand, CGF supplies power to $i$ with the amount of $g_i(135)$.

## 5.9 Simulation and experiments

Let us introduce the simulation results based on the decentralized control models described in the previous sections. In a bi-connected power network, power flows can pass bidirectionally. Each node employs common buffering method and polling mechanism.

A simulation network $G = (\mathscr{V}, \mathscr{E})$ is created with 100 nodes and 190 links with random connections. Aggregated data of loads $l_i(t)$ and renewable energies $r_i(t)$ at each node are available every 15 min from smart meters. The profiles of loads and renewables are based on the data provided in References 38–40 so that each node has different profiles with random nature. For a given time sequence ($\mathscr{T} = \{0, 15, 30, \ldots\}$) at each node $i$, $l_i(t)$ and $r_i(t)$ are, respectively, obtained. The communication interval $\Delta t$ in StandBy of Figure 5.3 is defined as 1 min. In this experiment, the voltages are set to be $V_i = 1\,\text{MV}$, and each admittance is set to be $Y_{ij} = 100$. The capacity of each battery is $\overline{b}_i = 100\,\text{MWh}$. The initial energy $b_i(0)$ of each node is assigned as $b_i(0) = 60\,\text{MWh}$. The maximum flow (capacity) on each edge is $c_{ij}(t) \equiv 50\,\text{MW}$. $\alpha_i$ of BPC defined in (5.10) is set as 1. $\varepsilon$ in Algorithm 3 is set to be 0.1.

Following sections describe the simulation results by the distributed clustering approach based on tie-sets. In this network, the number of tie-sets is uniquely decided to be 91. Every measurement vector $y_i(t)$ is sent to the leader node of each tie-set constantly using a TA.

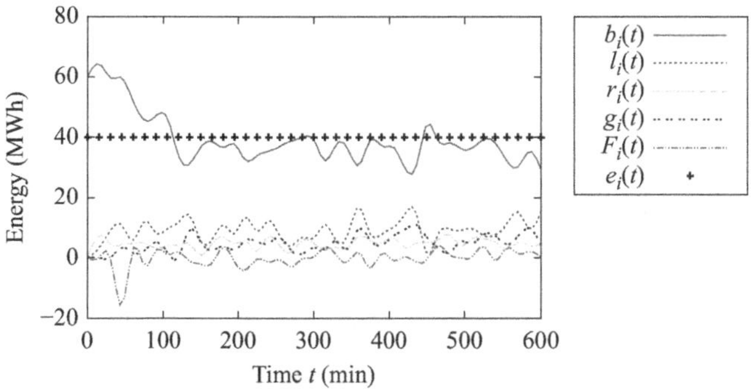


*Figure 5.8* *Energy stimulus response of $b_i(t)$, $l_i(t)$, $r_i(t)$, $g_i(t)$, and $F_i(t)$ from $t = 0$ to 600 where load and renewable generation are given with $0 \leq l_i(t) \leq 20$ and $0 \leq r_i(t) \leq 10$ MWh, respectively*

## 5.9.1 Energy stimulus response

In this experiment, we look at the simulated behavior of energy stimulus response by picking up several nodes with different load and renewable profiles. Initial LELs are set as $e_i(0) = 40$ MWh.

Figure 5.8 shows the changing process of energy stimulus response of $b_i(t)$, $l_i(t)$, $r_i(t)$, $g_i(t)$, and $F_i(t)$ from $t = 0, 15, 30, \ldots, 600$ (min) at node $i$. At this node, the load input and renewable generation are randomly given between 0–20 and 0–10 MWh, respectively. As the decentralized approach based on tie-sets is applied to controlling $g_i(t)$ and $F_i(t)$, the value of $b_i(t)$ keeps to be around 40 MWh from $t \approx 100$ min. The value of $e_i(t)$ is always 40 MWh as the load is bounded by 20 MWh. It is also indicated that $i$ distributes power flows to adjacent nodes when it has an excess amount of battery energy. When the size of the battery is enough large compared with the load and renewable generation, the stimulus response for battery energy level $b_i(t)$ becomes stable. Therefore, it realizes sustainable control against the randomness of loads and renewables in case of installing large-size energy storages.

Next, we look at the node where the magnitude of load and renewable generation is large against the size of energy storage. Figure 5.9 is the energy stimulus response at another node $i$ where the load input is assigned with the range of 0–80 MWh, and the renewable generation is given between 0 and 60 MWh. There are many sudden peak load processes that make convergence of battery energy level at $i$ unstable. However, the iteration of decentralized optimization constantly tries to cope with the uncertainty of load processes so that the energy at node $i$ never runs out. The behavior of LEL at this node shows dynamic change according to the load peaks, which is also the key control point to prevent a node from running out of energy. The experiments shown in Figures 5.8 and 5.9 utilize larger load bound than maximum renewable generation. Therefore, we analyze the node that exploit larger renewable generation than the process of load.

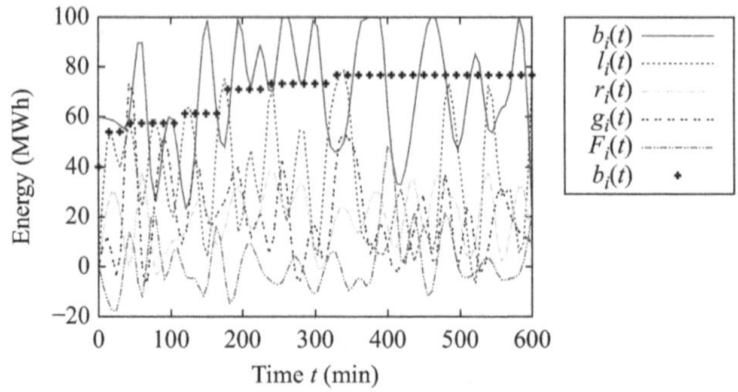


*Figure 5.9 Energy stimulus response where load and renewable generation are given with $0 \leq l_i(t) \leq 80$ and $0 \leq r_i(t) \leq 60$ MWh, respectively*

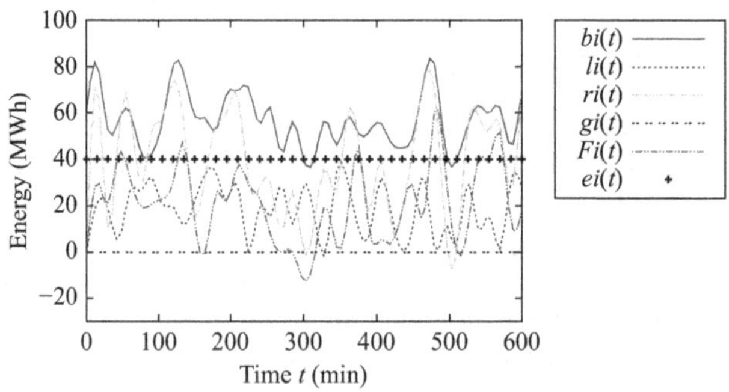


*Figure 5.10 Energy stimulus response where load and renewable generation are given with $0 \leq l_i(t) \leq 40$ and $0 \leq r_i(t) \leq 80$ MWh, respectively*

At a node at which the maximum load and renewable generation are, respectively, bounded by 40 and 80 MWh as shown in Figure 5.10, the energy level is generally greater than the LEL at almost all times unless net demand is positive and large. As $\Delta e_i(t) = \overline{b}_i - e_i(t)$ is preserved for unexpected renewable peaks, the preserved space is effectively used to temporally store excess renewable generation. It is also indicated that when the node cannot store the excess amount of renewable, it distributes the energy to the other nodes that require it. The process of this node contributes to minimizing total amount of energy generated at CGFs as this node constantly distributes the DER stored in its battery.

From the results shown in Figures 5.8–5.10, it is demonstrated that the decentralized model copes with unexpected load peaks and renewable generation by dynamically controlling power flows and CGF generation.

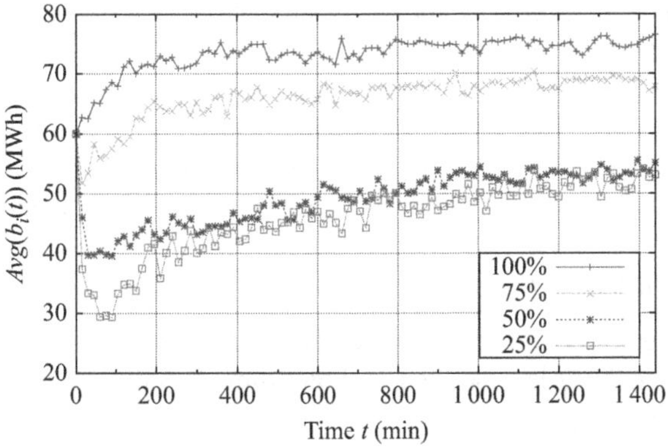


*Figure 5.11 Convergence behavior of the average battery energy level $Avg(b_i(t)) = \overline{B}(t)$ with different renewable penetration rates (RPRs) (100%, 75%, 50%, 25%) in a 100-node power network*

### 5.9.2 Convergence with different renewable penetration rates

We conducted simulation with different renewable penetration rates (RPRs) as it is an important issue to be considered when integrating renewable energies. Here, the RPR is adjusted by changing the boundary of renewable generation capacity. In this simulation, load input is random bounded by 60 MWh at each node $i$ at each time $t$. When expecting RPRs to be 100%, 75%, 50%, and 25% as in Figure 5.11, the random renewable input is bounded by 60, 45, 30, and 15 MWh at each node $i$ at each time $t$, respectively. Initial LELs are set as $e_i(0) = 20$ MWh.

Simulation data of $b_i(t)$ of all the nodes taken at $t = 0, 15, 30, \ldots, 1\,440$ min (24 h) are averaged. Hence, $Avg(b_i(t)) = \overline{B}(t)$, which stands for the average battery energy level at time $t$ defined as follows, is used to see the convergence behavior of entire grid energy level.

$$\overline{B}(t) = \frac{\sum_{i \in \mathscr{V}} b_i(t)}{|\mathscr{V}|} \tag{5.13}$$

Experiments are conducted 20 times for each case of RPR to average the results.

Figure 5.11 shows the simulated behavior of the average battery energy level $\overline{B}(t)$ in a 100-node power network with different RPRs. When RPR is 100%, the average battery energy level $\overline{B}(t)$ converges around 75 MWh since each node stores the renewable energy in case the amount of renewable generation is larger than the load process. After the average battery energy level draws to 75 MWh, power flows among nodes become stable by distributing stored energy to the nodes in short of energy with minimum energy supply from CGF. The similar behavior can be seen when RPR is 75% where $\overline{B}(t)$ converges around 70 MWh. In case of 50% and 25% RPRs, the average powers $\overline{B}(t)$ first drops since the initial LELs are $e_i(0) = 20$ MWh at all the nodes. The values of $e_i(t)$ are dynamically updated according to the peak load profile at each node over the time range. Thus, the average energy gradually converge to 60 MWh with the updated LELs $e_i(t)$.

It has been verified that the convergence behavior of the average battery energy level in 200-node to 500-node networks with the different RPRs from 25% to 100% is similar to the behavior in the 100-node network. In addition, it is confirmed that the convergence process does not depend on the initial energy at each node.

### 5.9.3 Comparison of TBO and DLP

In this experiment, the optimality of the TBO and DLP is assessed. The mechanism of both DLP and TBO is described in Algorithms 2 and 3, respectively. Then, we compare the following factors:

- Changing process of the average battery energy level: $Avg(b_i(t)) = \overline{B}(t)$.
- Amount of total energy generated at CGF: $Total(g_i(t)) = \sum_{i \in \mathcal{V}} g_i(t)$.

Although any optimization technique cannot excel in the solution of the problem, it is important to analyze how the result of those approaches are close to the theoretical limitation in terms of convergence of battery energy levels and the energy supply from CGF. The simulation network is the same as the 100-node network in the previous section. The RPR is intended to be 50% by giving load and renewable generation with the range of 0–60 and 0–30 MWh, respectively. Initial LELs are $e_i(0) = 20$ MWh. Simulation data of $\sum_{i \in \mathcal{V}} g_i(t)$ are taken at $t = 0, 15, 30, \ldots,$ 600 (min). Experiments are conducted 20 times and the average is calculated.

Figure 5.12 demonstrates the average battery energy level $\overline{B}(t) = Agv(b_i(t))$ within the time from $t = 0$ to 600 min that compares DLP and TBO (Tie-set) with the optimal $\overline{B}^*(t)$. The average battery energy level by TBO is closer to the theoretical solution than DLP since TBO can calculate power flows that balance the battery

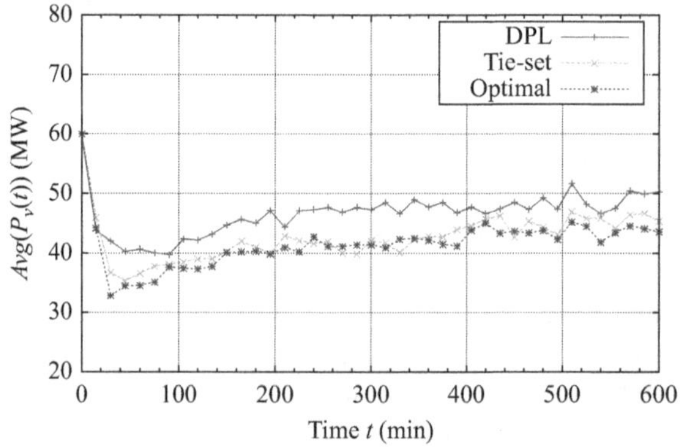


*Figure 5.12 Comparison on the average battery energy level $\overline{B}(t) = Agv(b_i(t))$ of Distributed Linear Programming (DLP), Tie-set Based Optimization (TBO), and the optimal $\overline{B}^*(t)$ when RPR is 50%*

energy levels within individual tie-sets. DLP requires more energy than TBO almost at each time step since the optimization is based on individual nodes and energy is exchanged only with adjacent nodes, which restricts the scale of optimization, even though the process of DLP is completely decentralized by communicating with other peripheral nodes with limited state information. Therefore, CGF generation at each node is not minimized as it must satisfy $b_i(t) \geq e_i(t)$ to compensate the demand that has not been met by its net power export $F_i(t)$.

Figure 5.13 shows the changing process of $\sum_{i \in \mathcal{V}} g_i(t)$ by DLP and TBO as well as the theoretical solution (Optimal). As indicated in Figure 5.13, the optimal line shows minimum allocation of energy generated at CGFs at each time. The result by TBO is similar to the theoretical solution as both TBO and optimal line converge between 1 400 and 1 600 MWh after $t = 60$ min. DLP makes up for the demand not sufficiently satisfied by power flows at each node, which leads to excessive energy supply from CGF. That is why the simulated behavior of the average battery energy level by DLP in Figure 5.12 is constantly higher than that of TBO and theoretical solution.

Table 5.1 shows comparison of the total CGF power provision from $t = 0$ to 600 at all the nodes in $\mathcal{V}$ calculated as $\sum_{t=0}^{600} \sum_{i \in \mathcal{V}} g_i(t)$ with different RPRs. The result

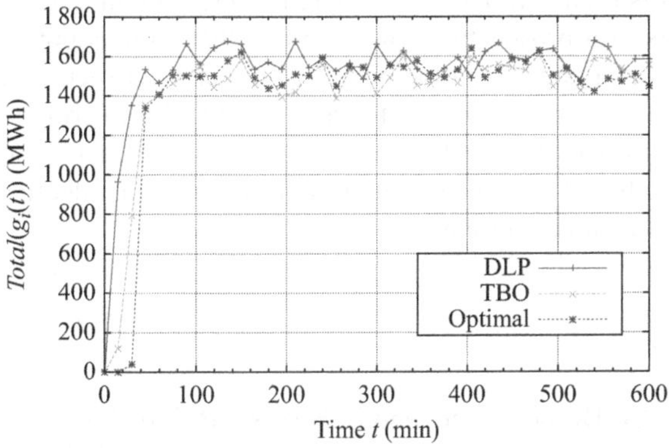


*Figure 5.13 Comparison on total CGF energy supply $Total(g_i(t)) = \sum_{i \in \mathcal{V}} g_i(t)$ of DLP, TBO, and the optimal supply from CGFs when RPR is 50%*

*Table 5.1 Comparison of total CGF energy supply $\sum_{t=0}^{600} \sum_{i \in \mathcal{V}} g_i(t)$ (MWh) with different renewable penetration rates*

| **RPR** | **25%** | **50%** | **75%** | **100%** |
|---|---|---|---|---|
| DLP | 86 708.70 | 62 315.36 | 38 134.25 | 28 874.94 |
| Tie-set | 85 637.09 | 57 945.28 | 36 460.37 | 25 686.16 |
| Optimal | 85 007.26 | 57 535.15 | 34 474.77 | 25 302.25 |

indicates that the total CGF supply by TBO is close to the theoretical solution with modest increase, whereas the amount of CGF supply by DLP is larger than TBO and the optimal amount. Therefore, we could reduce the cost of using fossil fuels and nuclear plants by fully taking advantage of renewable generation on the basis of the nature of iterative TBO where the optimization in a system of $\mu$ independent tie-sets quickly leads to global optimization [32,41].

## 5.10 Summary

This chapter has discussed a decentralized model for maximizing renewable integration in future grid and described how to realize real-time energy sharing and power production response by incorporating batteries and many end-use smart devices connected to a local network or through the Internet. The uncontrollability of loads and renewables is dealt with by the real-time decentralized algorithms such as DNA with DLP and clustering approach with TBO. The simulation results show that balanced energy sharing of renewable energies constantly conducted with decentralized control models realize a sustainable power network complimented by dynamic power supplies from centralized generation facilities even if the load process and renewable generation is highly variable and unpredictable. The discussed models extract essential aspects of a prospective power grid so that various optimization theories and algorithms being studied in the boundary domain between power systems and distributed network systems can be integrated into the future grid. A variety of latest technologies and forecasting techniques on load and renewable generation can also be utilized for both planning and real-time control phases in order to create a reliable, resilient, and sustainable future grid.

## Bibliography


[1] C. Batlle, A. Dòria-Cerezo, and E. Fossas, "Bidirectional power flow control of a power converter using passive Hamiltonian techniques," *International Journal of Circuit Theory and Applications*, vol. 36, pp. 769–788, October 2008.

[2] T. Takuno, M. Koyama, and T. Hikihara, "In-home power distribution systems by circuit switching and power packet dispatching," in *IEEE International Conference on Smart Grid Communications (SmartGridComm)*, Gaithersburg, Maryland, USA, pp. 427–430, 2010.

[3] K. Tashiro, R. Takahashi, and T. Hikihara, "Feasibility of power packet dispatching at in-home dc distribution network," in *IEEE International Conference on Smart Grid Communications (SmartGridComm)*, Tainan City, Taiwan, pp. 401–405, 2012.

[4] N. Li, L. Chen, and S. Low, "Optimal demand response based on utility maximization in power networks," in *2011 IEEE Power and Energy Society General Meeting*, Detroit Michigan, USA, pp. 1–8, 2011.

[5] T. Yau, L. N. Walker, H. Graham, A. Gupta, and R. Raithel, "Effects of battery storage devices on power system dispatch," *IEEE Transactions on Power Apparatus and Systems*, vol. PAS-100, no. 1, pp. 375–383, 1981.

[6] E. Sortomme and M. El-Sharkawi, "Optimal power flow for a system of microgrids with controllable loads and battery storage," in *Power Systems Conference and Exposition, 2009. PSCE'09. IEEE/PES*, Seattle, WA, USA, pp. 1–5, 2009.

[7] C. Hill, M. Such, D. Chen, J. Gonzalez, and W. Grady, "Battery energy storage for enabling integration of distributed solar power generation," *IEEE Transactions on Smart Grid*, vol. 3, no. 2, pp. 850–857, 2012.

[8] P. Yang, P. Chavali, and A. Nehorai, "Parallel autonomous optimization of demand response with renewable distributed generators," in *2012 IEEE Third International Conference on Smart Grid Communications (SmartGridComm)*, Tainan City, Taiwan, pp. 55–60, 2012.

[9] Y. Zhang, N. Gatsis, and G. Giannakis, "Robust distributed energy management for microgrids with renewables," in *2012 IEEE Third International Conference on Smart Grid Communications (SmartGridComm)*, Tainan City, Taiwan, pp. 510–515, 2012.

[10] D. Olivares, C. Canizares, and M. Kazerani, "A centralized optimal energy management system for microgrids," in *IEEE Power and Energy Society General Meeting*, Detroit Michigan, USA, pp. 1–6, July 2011.

[11] A. Tsikalakis and N. Hatziargyriou, "Centralized control for optimizing microgrids operation," in *IEEE Power and Energy Society General Meeting*, Detroit Michigan, USA, pp. 1–8, July 2011.

[12] A. Beloglazov, J. Abawajy, and R. Buyya, "Energy-aware resource allocation heuristics for efficient management of data centers for cloud computing," *Future Generation Computer Systems*, vol. 28, no. 5, pp. 755–768, 2012. Special Section: Energy efficiency in large-scale distributed systems.

[13] J. Popeangă, "Cloud computing and smart grids," *ERP and E-Business Application Deployment in Open Source Distributed Cloud Systems*, Vol. 3, p. 57–71, 2012.

[14] S. Rusitschka, K. Eger, and C. Gerdes, "Smart grid data cloud: A model for utilizing cloud computing in the smart grid domain," in *IEEE International Conference on Smart Grid Communications (SmartGridComm)*, Gaithersburg, Maryland, USA, pp. 483–488, October 2010.

[15] F. P. Miller, A. F. Vandome, and J. McBrewster, *Amazon Web Services*. Alpha Press, London, UK, 2010.

[16] R. H. Weber and R. Weber, *Internet of Things*. Berlin: Springer, 2010.

[17] M. Yun and B. Yuxin, "Research on the architecture and key technology of Internet of Things (IoT) applied on smart grid," in *International Conference on Advances in Energy Engineering (ICAEE)*, Beijing, China, pp. 69–72, June 2010.

[18] S. Karnouskos, "The cooperative Internet of Things enabled smart grid," in *IEEE International Symposium on Consumer Electronics (ISCE2010)*, Braunschweig, Germany, pp. 7–10, June, 2010.

[19] C.-H. Lo and N. Ansari, "Decentralized controls and communications for autonomous distribution networks in smart grid," *IEEE Transactions on Smart Grid*, vol. 4, pp. 66–77, March 2013.

[20] T. Suehiro and T. Namerikawa, "Decentralized control of smart grid by using overlapping information," in *Proceedings of SICE Annual Conference (SICE)*, Akita, Japan, pp. 125–130, August 2012.

[21] J. Shah, B. Wollenberg, and N. Mohan, "Decentralized power flow control for a smart micro-grid," in *IEEE Power and Energy Society General Meeting*, Detroit Michigan, USA, pp. 1–6, July 2011.

[22] H. Liu, Z. Hu, Y. Song, and J. Lin, "Decentralized vehicle-to-grid control for primary frequency regulation considering charging demands," *IEEE Transactions on Power Systems*, vol. 28, pp. 3480–3489, August 2013.

[23] C. Ahn, C.-T. Li, and H. Peng, "Decentralized charging algorithm for electrified vehicles connected to smart grid," in *American Control Conference (ACC), 2011*, San Francisco, USA, pp. 3924–3929, June 2011.

[24] S. Galli, A. Scaglione, and Z. Wang, "For the grid and through the grid: The role of power line communications in the smart grid," *Proceedings of the IEEE*, vol. 99, no. 6, pp. 998–1027, 2011.

[25] Z. Wang, A. Scaglione, and R. Thomas, "Generating statistically correct random topologies for testing smart grid communication and control networks," *IEEE Transactions on Smart Grid*, vol. 1, no. 1, pp. 28–39, 2010.

[26] P. C. Mosaic, "Mesh networks is communications winner in utility survey," 2009. Available from : http://www.smartgridnews.com/story/mesh-networks-communications-winner-utility-survey/2009-12-15

[27] R. Rajagopal, E. Bitar, P. Varaiya, and F. Wu, "Risk-limiting dispatch for integrating renewable power," *International Journal of Electrical Power and Energy Systems*, vol. 44, no. 1, pp. 615–628, 2013.

[28] N. A. Lynch, *Distributed Algorithms*. San Francisco, CA: Morgan Kaufmann Publishers Inc., 1996.

[29] K. Nakayama, N. Shinomiya, and H. Watanabe, "An autonomous distributed control method for link failure based on tie-set graph theory," *IEEE Transactions on Circuits and Systems I: Regular Papers*, vol. 59, no. 11, pp. 2727–2737, 2012.

[30] M. Iri, I. Shirakawa, Y. Kajitani, and S. Shinoda, *Graph Theory with Exercises*. Tokyo: CORONA Publishing, 1983.

[31] N. Shinomiya, T. Koide, and H. Watanabe, "A theory of tie-set graph and its application to information network management," *International Journal of Circuit Theory and Applications*, vol. 29, no. 4, pp. 367–379, 2001.

[32] T. Koide, H. Kubo, and H. Watanabe, "A study on the tie-set graph theory and network flow optimization problems," *International Journal of Circuit Theory and Applications*, vol. 32, no. 6, pp. 447–470, 2004.

[33] J. Malinowski, "A new efficient algorithm for generating all minimal tie-sets connecting selected nodes in a mesh-structured network," *IEEE Transactions on Reliability*, vol. 59, no. 1, pp. 203–211, 2010.

[34] K. Nakayama, K. Benson, L. Bic, and M. Dillencourt, "Complete automation of future grid for optimal real-time distribution of renewables," in *2012 IEEE Third International Conference on Smart Grid Communications (SmartGridComm)*, Tainan City, Taiwan, pp. 418–423, 2012.

[35] K. Nakayama, K. E. Benson, V. Avagyan, M. B. Dillencourt, L. Bic, and N. Venkatasubramanian, "Tie-set based fault tolerance for autonomous recovery of double-link failures," in *IEEE International Symposium on Computers and Communications (ISCC)*, Split, Croatia, pp. 391–397, 2013.

[36] K. Nakayama, N. Shinomiya, and H. Watanabe, "An autonomous distributed control method based on tie-set graph theory in future grid," *International Journal of Circuit Theory and Applications*, vol. 41, no. 11, pp. 1154–1174, 2013.

[37] L. Bic, "Distributed computing using autonomous objects," in *Proceedings of the Fifth IEEE Computer Society Workshop on Future Trends of Distributed Computing Systems*, Cheju Island, pp. 160–168, 1995.

[38] B. Tarroja, F. Mueller, J. D. Eichman, J. Brouwer, and S. Samuelsen, "Spatial and temporal analysis of electric wind generation intermittency and dynamics," *Renewable Energy*, vol. 36, no. 12, pp. 3424–3432, 2011.

[39] "Irvine smart grid demonstrate project," Tech. rep., EDISON and EPRI, November 2010. Available from http://www.sustainability.uci.edu/Resources1/ISGDOverview.pdf. Accessed on 29 July 2016.

[40] J.-P. Maton, L. Zhao, and J. Brouwer, "Dynamic modeling of compressed gas energy storage to complement renewable wind power intermittency," *International Journal of Hydrogen Energy*, vol. 38, no. 19, pp. 7867–7880, 2013.

[41] Y. Sakai, K. Nakayama, and N. Shinomiya, "A node-weight equalization problem with circuit-based computations," in *IEEE International Symposium on Circuits and Systems (ISCAS)*, Beijing, China, pp. 2525–2528, 2013.

# Chapter 6
# Distributed and decentralized control in future power systems

*Emmanouil Loukarakis*[1] *and Chris Dent*[2]

## 6.1 Introduction

Control in power systems consists of a sequence of energy markets optimization problems [1] followed by fast local device controls [2]. The former determines the set points and operating modes of all devices connected to the system, while the latter, based on local frequency and voltage measurements, controls individual devices for the purpose of maintaining system stability. While currently the aforementioned problems largely concern large generating units' power output, with the advent of smart grids (once smart meters are in place and the necessary communications have been established) large numbers of electric vehicles, small-scale storage and other flexible devices will also become available to control. In order to achieve optimal operation, the latter would have to be taken into account in both optimization and control design. As a consequence, the way associated problems are formulated and solved has to change. Given the potentially huge number of available controls – several devices in each individual building, millions across a whole country – this implies dealing with problems of an unprecedented for the energy sector, large scale. Obtaining a solution in a timely manner is not a trivial matter, and quite probably – especially in terms of optimization – the traditional centralized approaches may no longer be viable. Decentralized solutions to energy management problems would be necessary.

This chapter reviews the traditional power systems control structure and discusses how it could evolve to incorporate all the new controllable devices which will be available in the future, as well as cope with the potential difficulties these may create. Given the scale of these problems, we place particular emphasis on distributed solutions. We provide an overview of relevant approaches and put them into perspective with regards to when and how they could be applied. Rather than a rigorous mathematical treatment of the presented methods, we use simple, indicative, power system based formulations and examples to illustrate how they work, establish similarities between them, and identify the challenges – both in mathematics and

[1]School of Electrical and Electronic Engineering, University of Manchester, Manchester, UK
[2]School of Engineering & Computing Sciences, Durham University, Durham, UK

implementation – that still lie ahead. For the reader interested in mathematical proofs or extended results regarding a particular method, we provide an adequate number of relevant bibliographic references.

## 6.2 A look into current power systems control

The basic power system physical structure is illustrated in Figure 6.1. Achieving optimal operation implies meeting at any given point in time the requirements of demand, at minimum possible cost while satisfying certain reliability conditions. This is largely achieved by controlling the operating points of generating units (mostly located at the transmission level) but also a number of devices (e.g. transformers) which modify the characteristics of the network itself. When it comes to determining the optimal operating points, there are three fundamental factors affecting the problem formulation and solution approach:

- *Device response times:* These indicate when a control decision has to be made. One representative example is that of conventional generators start-up and shut-down times, which may take up to several hours. As a consequence, the operating status of these generators has to be determined at least as many hours ahead of real time. Given the periodic pattern of demand during a day, in relevant literature this is often assumed to happen in a day-ahead context through the solution of the so-called unit commitment (UC) problem.
- *Uncertainty*: When optimizing ahead of time (as in UC), it is not possible to know exactly the demand or renewables generation output, or even device availability.

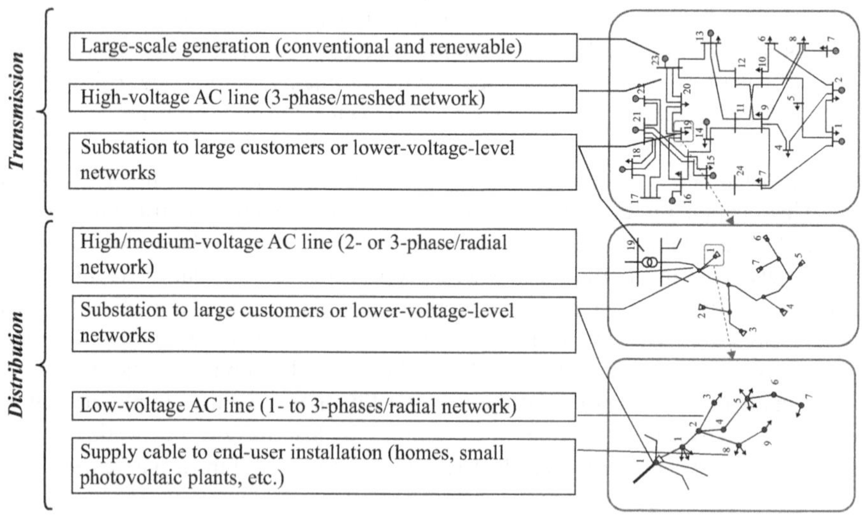


*Figure 6.1 An indicative overview of basic power systems structure (circles indicate generating plants; arrows indicate demand)*

Managing uncertainty has significant implications for power systems operation. First and foremost, modelling uncertainty in various optimization problems is a significant challenge and it is not always clear what the best approach would be, e.g. there does not yet appear to be a consensus regarding UC formulation and a variety of options appear in the current literature [3]. Second, given that the UC decision will have to be adjusted closer to real time to reflect the actual realizations of uncertain quantities, a variety of additional optimization problems are solved throughout the day. Perhaps the most prominent of these problems is the so-called economic dispatch (ED) which is the last system-wide optimization carried out before real time. Both UC and ED involve a number of contingency constraints to account for the possibility of component outages or other major unforeseen events and ensure that sufficient power reserves are kept.

- *Solution speed/tractability*: UC is a mixed-integer multi-time-step optimization problem which unavoidably involves detailed generator models (in terms of operating states), but a simplified network model in order to enable faster solutions. In contrast, ED, given that it is the last problem to be solved ahead of real time, involves a detailed network representation. However, to ensure tractability it is often formulated as a largely continuous problem (also in terms of generator representation). Finally, closer to real time there is not sufficient time to solve optimization problems. Individual device controls, based on locally measured values of frequency and voltage, are used to counter any deviations from expected quantities and maintain system stability.

The three considerations mentioned above lead to today's power system control structure, which roughly involves three successive problems or control mechanisms [1,2]:

- *Unit commitment (UC):* a centralized mixed-integer optimization problem, covering a time period of several hours, which determines the operating status of large conventional generators [1,4]. Different versions of this problem may be solved closer to real time for the specific purpose of scheduling smaller fast-start generators.
- *Economic dispatch (ED):* a centralized non-linear and non-convex optimization problem, covering a short period in time [5] and solved several minutes ahead of real time. While in practice there may be a number of discrete devices involved (e.g. capacitor banks or transformers), ED in research literature is commonly assumed to involve only continuous constraints and controls [6].
- *Fast local controls:* these involve the controllers of individual devices acting instantly based on local signals of frequency or voltage. Their set points and operation mode would typically be determined through the ED mechanism.

This structure strikes a good balance between management of uncertainty and individual problems tractability. UC and ED are the energy management mechanisms, which try to achieve optimal (in terms of economic efficiency and reliability) operation, while the fast local controls are there to ensure system stability. While in the former prices and associated market rules are the defining parameters, in the latter frequency and voltage are all that matters.

## 6.3 Identifying the role of distributed methods

Current UC and ED formulations focus on optimizing the operation of a mix of large-scale conventional and renewable generation, subject to transmission network constraints, given aggregate demand estimates at the transmission bus level. As such individual end-user devices and the distribution networks have never been considered in detail. This was a reasonable approach given that end users have been largely inflexible in their energy requirements, and as a consequence distribution networks were built to cover any foreseeable amount of demand. As a result, their constraints in terms of energy management could simply be ignored. However, recent trends in power systems indicate a high number of flexible demand devices (e.g. electric vehicles [7], small-scale storage [8], or simply controllable versions of today's appliances [9]) in the near future. The most dominant characteristic of flexible devices in terms of energy management is their time-linking characteristics, i.e. instead of meeting a power requirement at a given instance of time (as with today's inflexible demand), energy requirements have to be met over a period of time. In addition, most of these devices will be connected at the distribution level. These considerations raise three basic issues which are not fully addressed through current energy management mechanisms:

- *Demand coordination over time*: Towards achieving efficient operation the use of these flexible devices at the distribution level has to be coordinated with generation at the transmission level, e.g. shift demand to hours when energy is cheap (either due to the fact that demand was in the first place low, or due to increased renewable energy availability). However, this is not covered by today's formulations as ED has at most a very limited look-ahead period in time [5]. At the same time, UC, which does optimize over long periods of time, is typically solved far in advance and would not fully allow micromanaging individual devices.
- *Distribution network constraints*: One important point that has to be considered is the simple fact that the new developments might imply significant strain for current distribution networks [10]. Comparing, for example, the typical household demand in Reference 9 peaking at about 5–14 kW (depending on household size) and the typical electric vehicle charging power at 3.5–6.6 kW (even up to 40 kW for fast charging) [11], could indicate a significant increase in peak demand and maximum network loading. This would especially be the case if users exhibit a high degree of correlation in terms of energy usage (e.g. if most tend to charge their electric vehicles at about the same time). Thus, a second important question pertains to how distribution network limits would be taken into account in energy management problems. The only alternative would imply significant investments in distribution network equipment to increase capacity. This again is something that traditionally has never been considered in either UC or ED.

- *Controls assignment and role*: One further evident question is when a system operator would assign and adjust set points or control modes to these new flexible devices. The nature of the signals that would be exchanged between them would also need to be clarified. In addition, if at any part of the system a large number of often discrete-state (e.g. on/off type) flexible devices try to respond to frequency and voltage signals, another important issue is ensuring that their response is stable. As an example, consider the simple case where a large number of small flexible discrete devices disconnect simultaneously due to a frequency drop. This might create a significant over-frequency event, and vice versa if these devices were allowed to reconnect automatically.

Without question, the basic energy management structure – as presented in the previous section – should evolve to account for these three issues. Only then would procurement of energy at the minimum possible cost and environmental impact be possible. A change in the way UC and ED problems are formulated and solved is unavoidable. Of course, the most straightforward solution would be to simply extent their formulations to include the requirements and limitations of each individual device and network component. This however raises again the question of tractability, i.e. whether or not these problems may be solved fast enough for the answer to be of use for power system operations. In addition, there is the question of a suitable communications structure. In a centralized solution, one operator would have to communicate every few seconds with millions of devices and it is uncertain whether a sufficiently fast and reliable solution in terms of communications infrastructure would be possible at a reasonable cost. Finally, there is the rather subjective issue of privacy. Sending a full schedule of one's use of energy and private activities to one central coordinating authority, might be a problem for several individuals. The same is true with respect to passing full control of the devices (e.g. an electric vehicle) one owns. Distributed solution methods can potentially provide an answer to all three problems.

## 6.4 Distributed optimization definitions and scope

Considering a given optimization problem (such as UC or ED), in terms of tractability – or in other words solution speed – there are three possibilities for improvements: simplifying the problem, improving the solver, or parallelizing the necessary computations [12]. The latter is something to be pursued on several levels, e.g. the parallel execution of basic numerical operations in a processor (a core part of modern computing systems), or the parallel solution of algebraic problems within the solver by exploiting the structure of the associated mathematical method, or the parallel solution of parts of the original problem. This latter higher-level parallelization is the type of distributed approach that may also lead to fully decentralized solution. It involves breaking apart the original problem into a set of smaller subproblems and coordinating iteratively their solution through an appropriate mathematical decomposition approach. The latter should ideally achieve this coordination in a way that is mathematically proven to converge to the optimum of the original problem (global if the problem is convex, possibly local if the problem is non-convex) in a limited number of iterations.

Depending on how the coordination of the subproblems is achieved, part of the solution approach may still be centralized – where even though subproblems are solved in a distributed manner in separate computing systems, coordination is achieved via the communication with one central controller – or decentralized – where there is no need for a central controller and coordination is achieved through the communication between individual subproblems. While a centralized part should not affect a distributed method's ability to cope with tractability problems, the aforementioned issues of communications and privacy point towards more decentralized approaches. One further distinction could be made between synchronous and asynchronous methods. The former require that subproblem solutions wait at specific points for the arrival of certain data while the latter have no such requirements but require a more rigorous assessment of their convergence properties.

### *6.4.1 Distributed optimization fundamentals*

Consider one large power system separated into two independent (not electrically connected) areas. For optimization purposes each area is defined by: a vector $\mathbf{x}$ whose elements are the optimization control variables (e.g. power generated by individual generators, voltages, curtailable demand states); a function $f(\mathbf{x})$ which reflects costs (or negative utility) associated with a given value of $\mathbf{x}$ (e.g. the running costs of generators); a set of constraints $C = \{\mathbf{h}(\mathbf{x}) \leq 0\}$ which models the devices included within the area (e.g. network, equipment capacity limits). The centralized version of the problem would require solving:

$$\min_{\substack{\mathbf{x}_A \in C_A \\ \mathbf{x}_B \in C_B}} \{f_A(\mathbf{x}_A) + f_B(\mathbf{x}_B)\} \tag{6.1}$$

Note that both the objective function and constraints are separable over $\mathbf{x}_A, \mathbf{x}_B$. As such instead of solving (6.1), one could reach the same answer by solving the two areas separately, i.e.:

$$\min_{\mathbf{x}_A \in C_A} \{f_A(\mathbf{x}_A)\} \qquad \min_{\mathbf{x}_B \in C_B} \{f_B(\mathbf{x}_B)\} \tag{6.2}$$

These subproblems not only are simpler but may be solved in parallel. Consequently, the computational burden and time would be significantly reduced. Consider now a new instance of this problem where, as shown in Figure 6.2, the two areas are electrically connected. The optimization problem now becomes:

$$\min_{\substack{\mathbf{x}_A \in C_A \\ \mathbf{x}_B \in C_B}} \{f_A(\mathbf{x}_A) + f_B(\mathbf{x}_B) \colon \tilde{\mathbf{c}}_A \mathbf{x}_A = \tilde{\mathbf{c}}_B \mathbf{x}_B\} \tag{6.3}$$

It should be clear that the coupling introduced through the linear constraints does not allow a direct decomposition as in (6.1). This is a typical case of complicating constraints.

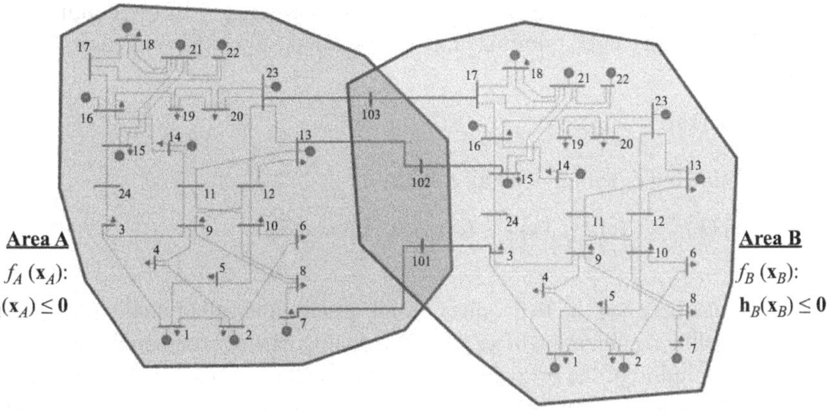


*Figure 6.2 A simple interconnected power systems areas example. Buses 101, 102, and 103 are represented once in each area. As such their associated quantities (voltage, power) exist in duplicates and an additional set of linear constraints $\mathbf{c}_A\tilde{\mathbf{x}}_A = \mathbf{c}_B\tilde{\mathbf{x}}_B$ (or equivalently $\tilde{\mathbf{c}}_A\mathbf{x}_A = \tilde{\mathbf{c}}_B\mathbf{x}_B$) establishes the necessary equalities among elements of $\mathbf{x}_A$ and $\mathbf{x}_B$, with $\mathbf{c}_A, \mathbf{c}_B$ being matrices of 1s and 0s, while $\tilde{\mathbf{x}}_A, \tilde{\mathbf{x}}_B$ are the parts of $\mathbf{x}_A, \mathbf{x}_B$, respectively, which are associated with the common buses*

Consider now the case where a couple of contingency scenarios (a modified set of network constraints reflecting the loss of network components, where there has to be a feasible solution) have to be included in the constraints of area A, i.e.:

$$\min_{\substack{\mathbf{x}_A \in C_A \\ \mathbf{x}_A, \mathbf{x}_{A1} \in C_{A1} \\ \mathbf{x}_A, \mathbf{x}_{A2} \in C_{A2}}} \{f_A(\mathbf{x}_A) + f_{A1}(\mathbf{x}_A + \mathbf{x}_{A1}) + f_{A2}(\mathbf{x}_A + \mathbf{x}_{A2})\} \tag{6.4}$$

Given that $\mathbf{x}_A$ is involved in all constraint sets again decomposition of different scenarios is not possible. This is an example of complicating variables.

Distributed optimization methods – also known as decomposition methods – are algorithms which attempt to bring optimization problems in the form of (6.3) or (6.4) closer to that of (6.1). Subsequently, this can allow an iterative coordinated solution of a number of subproblems similar to those in (6.2), leading to the optimum of the original problem in a mathematically proven way. Note that while the equations above and in the following paragraphs include only two subproblems, they may be easily generalized to accommodate any number of them.

## 6.4.2 Simple price-based decomposition

Considering the problem of (6.3) its dual function may be written as

$$\varphi(\boldsymbol{\lambda}) = \min_{\substack{\mathbf{x}_A \in C_A \\ \mathbf{x}_B \in C_B}} \left\{ \overbrace{f_A(\mathbf{x}_A) + f_B(\mathbf{x}_B) + \boldsymbol{\lambda}^T(\mathbf{c}_A\tilde{\mathbf{x}}_A - \mathbf{c}_B\tilde{\mathbf{x}}_B)}^{\mathcal{L}} \right\} \tag{6.5}$$

where $\boldsymbol{\lambda}$ is the vector of Lagrange multipliers associated with the coupling constraints The quantity to be minimized is the so-called Lagrangian function. The dual problem is

$$\max_{\boldsymbol{\lambda}}\{\varphi(\boldsymbol{\lambda})\} \tag{6.6}$$

The following equation generally holds:

$$\sup_{\boldsymbol{\lambda}}\{\varphi(\boldsymbol{\lambda})\} \leq \inf_{\substack{\mathbf{x}_A \in C_A \\ \mathbf{x}_B \in C_B}} \{f_A(\mathbf{x}_A) + f_B(\mathbf{x}_B) : \mathbf{c}_A\tilde{\mathbf{x}}_A = \mathbf{c}_B\tilde{\mathbf{x}}_B\} \tag{6.7}$$

The difference between the two values in this equation is called duality gap [13]. If for some feasible combination of $\mathbf{x}_A, \mathbf{x}_B, \boldsymbol{\lambda}$ the duality gap is zero, then these $\mathbf{x}_A, \mathbf{x}_B$ are the optimal solution of the primal problem and $\boldsymbol{\lambda}$ is the optimal solution of the dual. Consequently it is possible instead of solving (6.3) directly, to solve the problem (6.6) [14]. Given that the latter may be hard to solve directly, it may be solved through the Lagrangian Relaxation (LR) algorithm:

0. *Initialization:* Select initial values $\boldsymbol{\lambda}^0$ and iteration count $k = 0$.
1. *Subproblem solution:* $\{\mathbf{x}_A^{k+1}, \mathbf{x}_B^{k+1}\} = \text{argmin}_{\mathbf{x}_A \in C_A, \mathbf{x}_B \in C_B}\{f_A(\mathbf{x}_A) + f_B(\mathbf{x}_B) + \boldsymbol{\lambda}^{k^T} \cdot (\mathbf{c}_A\tilde{\mathbf{x}}_A - \mathbf{c}_B\tilde{\mathbf{x}}_B)\}$.
2. *Multipliers (subgradient) update:* Set $\boldsymbol{\lambda}^{k+1} = \boldsymbol{\lambda}^k + a^k(\mathbf{c}_A\tilde{\mathbf{x}}_A^{k+1} - \mathbf{c}_B\tilde{\mathbf{x}}_B^{k+1})$, where $a^k$ a scalar which is updated at every iteration in such a way that $\lim_{k\to\infty} a^k = 0$ and $\sum_{k=1}^{\infty} a^k = \infty$
3. *Convergence check:* If $\max\left|\boldsymbol{\lambda}^{k+1} - \boldsymbol{\lambda}^k\right|$ is less than a certain tolerance then convergence has been achieved, else update iteration count and go to 1.

Comparing the optimization problem at step 1 with the generic decomposable problem of (6.1) it should be clear that they are quite similar. As such, two smaller subproblems could be solved in parallel. Regarding step 2, on the assumption that the problem is convex, the subgradient method is simple to implement and computationally light method. As long as the two subproblems exchange power schedules at their connecting buses this step is also parallelizable and can be carried out in a decentralized manner. Note that the only link between the optimization subproblems is the Lagrangian multiplier (or price term) in their objective function. LR is the most basic decomposition method. Examples of its application in power systems may be found e.g. in References 15,16. Both papers use linear approximations of the ac network, while the latter additionally takes into account the reserve considerations.

### 6.4.3 *From optimization to control using prices*

The concept of prices is very often used in current power systems literature, and especially in papers related to smart grids, where it is commonly assumed that a set of prices is sent to the individual electric vehicle, household, etc., which subsequently optimizes its energy use. While this assumption is a neat way to look into a smaller parts of the energy management problem, when it comes to system-wide energy management, assuming prices by themselves will work, is rather naive. The reason is that in general price-based schemes may easily fail to converge. To illustrate this,

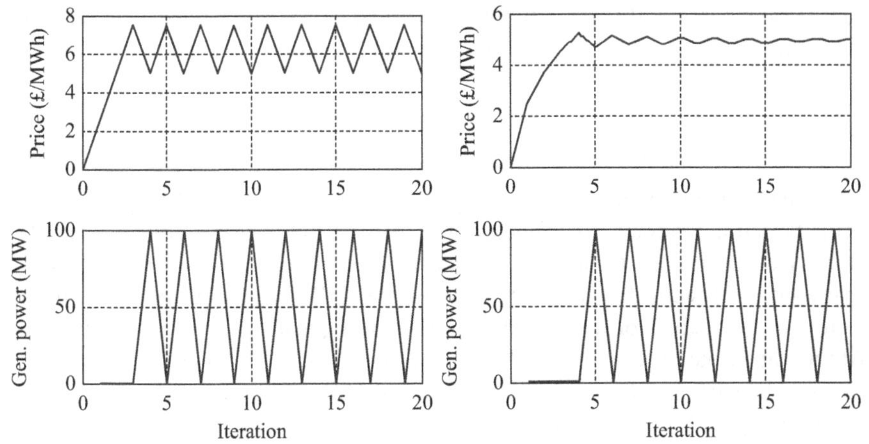


*Figure 6.3 Simple examples illustrating the convergence difficulties of price-based decomposition schemes. On the left the parameter $a^k$ is constant at each iteration (the conditions outlined in step 2 of the LR algorithm are not satisfied), on the right $a^k$ is decreasing at each iteration*

consider as an example in (6.3) a case where area A includes a generator with a linear cost function of 5 £/MWh and a technical maximum of 100 MW, and area B includes a fixed demand equal to 50 MW. As may be seen in Figure 6.3 the price will eventually converge if the method is appropriately setup, but the speed of convergence itself critically depends on how fast the step coefficient $a^k$ is decreasing. A very fast decrease however could potentially result in very slow price convergence. In any case the output of the generator does not converge to the desired result at all. These convergence issues may also appear in more complex formulations of the problem, e.g. when trying to enforce angle equality at the connecting buses of the two areas, or when attempting to schedule flexible demand over time (where power and price oscillations between different time-steps will appear). Such issues have been pointed out in various papers, e.g. References 17–19.

Another problem of interest relates to an approach suggested in a small number of papers, which propose a kind of continuous price-based control: a price is transmitted to users, and the ensuing scheduling decision is directly applied. The result can be an erratic and rather unstable behaviour which can result in price spikes [20] or inconsistent system operation. Due to the unavoidable power imbalances voltage and frequency would suffer, which in turn would generate the requirement for additional resources to be available to balance the system. Despite the fact that the example of Figure 6.3 is a bit extreme, it does illustrate both of the aforementioned problems. It is worth mentioning, however, that price-based approaches have found practical use in telecommunications [21]. However, in that case, different conditions apply: the problem is formulated appropriately through a modification of the user's utility function; there are no frequency and voltage quantities to be concerned about; and there can be safeguards in that if anything goes wrong convergence-wise, data rate could be uniformly reduced for everyone and the process resets itself. Such considerations may not be directly transferred to power systems.

### *6.4.4 Making prices work*

Going back to the simple example of Figure 6.3 it should be noted that if the generator actually had a quadratic cost function the method would easily converge, as there would have been a one-to-one equivalence between output power and price. However, many power system devices do not have such well-defined utility functions. For example, renewables or hydro generation would typically have zero variable costs. At the same time for demand, the differences in actual utility gained from supplying various devices may be hard to define and there is no reason why utility would significantly vary from one device to the next. As such, when network losses are not included in the formulation there might be an infinite number of possible solutions for any given energy management problem. On the other hand, if losses are considered, end node customers will always be the last ones to be served and the first ones to be curtailed. In order to avoid these problems there are two possible alternatives: modification of the original problem objective function; or the use of more advanced decomposition techniques.

In terms of objective function modification, the underlying idea is to associate with each device an appropriate utility function so that the original problem will have a unique solution for any given set of prices. One such class of papers is inspired by similar concepts used in rate control in telecommunications [21]. References 22–24 use a logarithmic utility function of the form $w \log(x)$ where $w$ a parameter that represents the user's willingness to pay and $x$ the decision variable or operating point of the device. Using such a utility function is said to lead to proportional fairness among network users. Note that these papers in order to reach the solution use a control-based scheme which differs from the standard LR, but seen from broad perspective remains quite similar to it, i.e.:

1. *Estimate price:* Calculate energy price (i.e. the control signal) based on a given cost function. Transmit that signal to individual users.
2. *Energy schedule update:* Solve user subproblems, transmit energy schedules to the central controller, and go back to the previous step.

In order to ensure stability, changes in the control signal from one iteration to the next are limited. Device energy schedule updates may be directly applied, rather than waiting for the algorithm's convergence. However, as indicated earlier this can raise concerns with regards to any requirements in terms of additional balancing resources or even system protection settings. Similar in principle utility function modifications – but within the context of centralized optimization – are used in Reference 25 which proposes scaling of electric vehicles utility functions based on their state of charge so that it is ensured that end-node customers get served. A much more elaborate utility function is used in Reference 26. This involves a quadratic function, which depends on the state-of-charge and a variety of other parameters related to the vehicle battery charging requirements. In any case, it should be noted that the new utility functions do not necessarily reflect real physical costs. As a consequence, when these artificial costs are put next to actual generating costs, the end result may be a solution that deviates from the optimum of the original problem. Care should be taken with regards to the definition of the utility functions and the context within which they are used. Note also that all the above papers use single time-step problem formulations. Within a multi time-step optimization, the role of such utility functions is harder to define.

## 6.5 Decomposition methods

In terms of improving the often poor convergence of subgradient-based LR a variety of other techniques and decomposition methods are available. In this section, we briefly present methods that improve the price-update mechanism within LR, or altogether different methods which guarantee faster and reliable convergence, and discuss their application in power systems problems. In the following the index $k$ indicates the $k$th iteration of a method.

### 6.5.1 *Improving price-updates*

A variety of methods are available as an alternative to the simple subgradient update. These methods typically try to build an approximation of the dual function and based on that, find the optimum Lagrange multipliers. The cutting plane method is one representative example. In order to determine the prices at step 2 of the LR algorithm it would solve:

$$\boldsymbol{\lambda}^{k+1} = \underset{z,\lambda}{\operatorname{argmax}}\{z : z \leq \varphi(\boldsymbol{\lambda}^i) + (\mathbf{c}_A\tilde{\mathbf{x}}_A^i - \mathbf{c}_B\tilde{\mathbf{x}}_B^i)^T(\boldsymbol{\lambda} - \boldsymbol{\lambda}^i), \quad i \in \{1, \ldots, k\}\} \tag{6.8}$$

Note however that the size of this price-update subproblem increases at higher degrees of decomposition (i.e. larger number of subproblems) and also with the number of iterations. To cope with the latter, a trust-region variant of the method maintains only a predetermined number of linear approximations which are close to the current Lagrangian multipliers point. One additional modification to this approach comes through the so-called bundle method which adds the term $a^k \|\boldsymbol{\lambda} - \tilde{\boldsymbol{\lambda}}^k\|_2^2$ where $a^k$ is a scalar and $\tilde{\boldsymbol{\lambda}}$ an appropriately updated parameter [27]. This additional term can prevent the oscillating behaviour of LR during convergence. Note that in all these cases the price-update optimization subproblem needs to be solved in a centralized manner.

Examples of more advanced price-update schemes applied to power systems may be found e.g. in Reference 28 which compares two different cutting plane techniques in terms of convergence performance. Tests are carried out in a six-bus system including a number of 4,000 network user subproblems. Also in Reference 29 a different price-update method is used based on a Newton method. Using the full ac network equations, a five-bus system is decomposed to individual generator/demand blocks.

Price updates use derivative information (derived through market players' previous responses) to define a set of equilibrium constraints as a function of price, i.e. building indirectly an approximate model of the user's response to prices.

### 6.5.2 Decomposing an augmented Lagrangian

An alternative way of stabilizing the behaviour of LR with regards to price convergence is by using an augmented Lagrangian of the form:

$$\{\mathbf{x}_A^{k+1}, \mathbf{x}_B^{k+1}\} = \underset{\substack{\mathbf{x}_A \in C_A \\ \mathbf{x}_B \in C_B}}{\operatorname{argmin}} \left\{ \overbrace{f_A(\mathbf{x}_A) + f_B(\mathbf{x}_B) + \boldsymbol{\lambda}^{k^T}(\mathbf{c}_A\tilde{\mathbf{x}}_A - \mathbf{c}_B\tilde{\mathbf{x}}_B) + \frac{\rho}{2}\|\mathbf{c}_A\tilde{\mathbf{x}}_A - \mathbf{c}_B\tilde{\mathbf{x}}_B\|_2^2}^{\mathcal{L}_\rho} \right\} \tag{6.9}$$

where $\rho$ is a penalty factor, which should be sufficiently small so that the problem does not become ill conditioned [14,30]. At the optimal point, the last term is equal to zero and as a result has no impact on the final solution of the problem. However, it makes the problem strongly convex (at least in a region around the optimum) and as a consequence offers improved convergence. The disadvantage of the augmented Lagrangian is that the objective function is no longer separable due to the quadratic penalty term.

An approach suitable for the decomposition of problems with separable constraints but non-separable objective function such as (6.9) is the Auxiliary Problem Principle (APP) method first proposed in References 31,32. The underlying idea behind it is that if for some function $g(\mathbf{x})$, the relation $\nabla g(\mathbf{x}^*) = \epsilon \nabla \mathcal{L}_\rho(\mathbf{x}^*), \epsilon > 0$ holds and $\mathbf{x}^* = \operatorname{argmin}_\mathbf{x} g(\mathbf{x})$, then $\mathbf{x}^* = \operatorname{argmin}_\mathbf{x} \mathcal{L}_\rho(\mathbf{x})$. This means that instead of optimizing $\mathcal{L}_\rho$ it would be possible to optimize an appropriately selected $g$. As indicated in References 33–35 under certain conditions the method generates subproblems with objective functions including a linear Lagrange multiplier term, a separable quadratic proximal term and a linearized component of the augmented Lagrangian quadratic term. With the exception of the last term, the method is quite similar to the proximal methods discussed in the next section, but involves a larger number of parameters. The effect of their tuning on convergence is studied in References 36,37. In Reference 38, a general implementation background is presented and the authors claim that the algorithm can be implemented both in a synchronous and asynchronous fashion. Further details may be found in Reference 39. The results indicate that the asynchronous solutions may produce results faster than synchronous ones, even though no mathematical proof is provided regarding convergence in the former case. However, it should be noted that the results presented in Reference 40 indicate that for certain parameter values the algorithm may fail to converge. Overall, this method requires the tuning of a large number of parameters. Without any loss in terms of convergence performance [33], proximal methods offer similar performance with a limited number of parameters.

### 6.5.3 Proximal decomposition methods

Proximal methods generate problems which have the following general form:

$$\{\mathbf{x}_A^{k+1}, \mathbf{x}_B^{k+1}\} = \underset{\substack{\mathbf{x}_A \in C_A \\ \mathbf{x}_B \in C_B}}{\text{argmin}} \Big\{ f_A(\mathbf{x}_A) + f_B(\mathbf{x}_B) + \boldsymbol{\lambda}^{k^T}(\mathbf{c}_A\tilde{\mathbf{x}}_A - \mathbf{c}_B\tilde{\mathbf{x}}_B) + \frac{\rho}{2}\|\tilde{\mathbf{x}}_A - \mathbf{z}_A^k\|_2^2 + \frac{\rho}{2}\|\tilde{\mathbf{x}}_B - \mathbf{z}_B^k\|_2^2 \Big\} \quad (6.10)$$

The variables $\mathbf{z}_A^k, \mathbf{z}_B^k$ have to be selected so that the sequence of points generated by the iterative solution of (6.10) converges at the optimum. One possibility is to set them equal to $\tilde{\mathbf{x}}_A^k, \tilde{\mathbf{x}}_B^k$. This is an approach used e.g. in Reference 41 which solves a network unconstrained electric vehicle management problem, using the alternative price-update structure discussed in Section 6.4.4. Rather than imposing limits directly to changes in price, a sufficiently high penalty value has the same effect indirectly. In References 42,43, the proximal term penalizes the deviation of a vehicle's power schedule from the mean of an electric vehicle fleet. However, this particular approach might not work well in cases where the electric vehicles exhibit significantly different characteristics from one another. Other variants of this proximal approach are presented in References 44–46. Note that these approaches assume the knowledge of a simple function of system price given the aggregate demand and do not include network constraints. Another more refined method, able to manage any type of additional constraints is the so-called Alternating Direction Method of Multipliers (ADMM).

The ADMM method first reformulates the original problem as follows:

$$\min_{\substack{\mathbf{x}_A \in C_A \\ \mathbf{x}_B \in C_B}} \{f_A(\mathbf{x}_A) + f_B(\mathbf{x}_B) : \tilde{\mathbf{x}}_A = \mathbf{z}_A, \tilde{\mathbf{x}}_B = \mathbf{z}_B, \mathbf{c}_A\mathbf{z}_A - \mathbf{c}_B\mathbf{z}_B = 0\} \quad (6.11)$$

Consider now the augmented Lagrangian of this problem:

$$\mathcal{L}_{\rho,z} = f_A(\mathbf{x}_A) + f_B(\mathbf{x}_B) + \boldsymbol{\lambda}_A(\tilde{\mathbf{x}}_A - \mathbf{z}_A) + \boldsymbol{\lambda}_B(\tilde{\mathbf{x}}_B - \mathbf{z}_B) + \frac{\rho}{2}\|\tilde{\mathbf{x}}_A - \mathbf{z}_A\|_2^2 + \frac{\rho}{2}\|\tilde{\mathbf{x}}_B - \mathbf{z}_B\|_2^2 \quad (6.12)$$

Due to the quadratic terms this is inseparable. To enable the decomposition of (6.4) the method first optimizes $\mathbf{x}_A, \mathbf{x}_B$ with $\mathbf{z}_A, \mathbf{z}_B$ fixed, and then $\mathbf{z}_A, \mathbf{z}_B$ with $\mathbf{x}_A, \mathbf{x}_B$ fixed. Hence the name alternating. More specifically it involves the following steps:

0. Select initial values for $\mathbf{x}_A^0, \mathbf{x}_B^0, \mathbf{z}_A^0, \mathbf{z}_B^0, \boldsymbol{\lambda}_A^0, \boldsymbol{\lambda}_B^0$ and set the iteration count $k = 0$.
1. Solve $[\mathbf{x}_A^{k+1}, \mathbf{x}_B^{k+1}] = \text{argmin}_{\substack{\mathbf{x}_A \in C_A \\ \mathbf{x}_B \in C_B}} \mathcal{L}_{\rho,z}(\mathbf{x}_A, \mathbf{x}_B, \mathbf{z}_A^k, \mathbf{z}_B^k, \boldsymbol{\lambda}_A^k, \boldsymbol{\lambda}_B^k)$.
2. Solve $[\mathbf{z}_A^{k+1}, \mathbf{z}_B^{k+1}] = \text{argmin}_{\mathbf{z}_A, \mathbf{z}_B}\{\mathcal{L}_{\rho,z}(\mathbf{x}_A^{k+1}, \mathbf{x}_B^{k+1}, \mathbf{z}_A, \mathbf{z}_B, \boldsymbol{\lambda}_A^k, \boldsymbol{\lambda}_B^k) : \mathbf{c}_A\mathbf{z}_A - \mathbf{c}_B\mathbf{z}_B = 0\}$.
3. Update $\boldsymbol{\lambda}_A^{k+1} = \boldsymbol{\lambda}_A^k + \rho(\mathbf{x}_A^{k+1} - \mathbf{z}_A^{k+1})$ and $\boldsymbol{\lambda}_B^{k+1} = \boldsymbol{\lambda}_B^k + \rho(\mathbf{x}_B^{k+1} - \mathbf{z}_B^{k+1})$.
4. If not converged update iteration count and go to step 1.

Going back to the simple example of Section 6.4.4, the benefits of the strongly convex terms in (6.12) become clear in the results illustrated in Figure 6.4. Examples of ADMM application in power systems include [35] where a serial implementation of

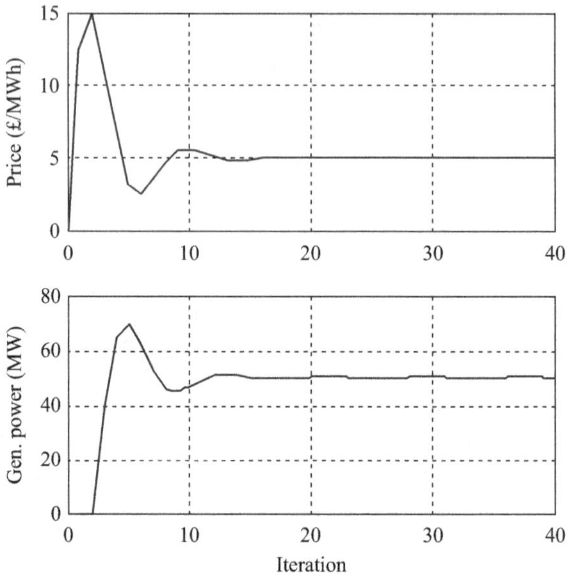


*Figure 6.4 Simple example of the ADMM method application. Compared to LR methods convergence in terms of both prices and power is achieved*

the method was used. In Reference 47, a variant of the method is applied to a linear model of the three-area IEEE Reliability Test System (RTS). Another application may be found in Reference 48 where the initial problem is decomposed down to the individual component level. Tests are carried out in randomly generated networks of up to $3 \times 10^4$ buses. The scaling results indicated by the authors look impressive; however, Kirchhoff's laws are not taken into account. Anese *et al.* [49] use the ADMM method for the decomposition of a semi-definite formulation of the unbalanced Optimal Power Flow (OPF) problem in distribution networks. Finally Loukarakis *et al.* [50] investigate its scalability in a variety of decomposition structures in the full balanced AC OPF problem. A more complex variant of ADMM appears in the form of the Predictor Corrector Proximal Multipliers (PCPMs) method which involves one additional multiplier step. Further details may be found in Reference 51. It should be noted, however, that it is unclear what benefits in terms of convergence performance the additional parameters bring.

### *6.5.4 Optimality Condition Decomposition*

In contrast to proximal methods which (compared to LR) involve a simple modification of the subproblems objective function, there are also mathematical decomposition methods which may involve modifications to subproblem constraints. One popular example is the so-called Optimality Condition Decomposition (OCD). This approach, also known as approximate Newton direction method, was first presented in Reference 52. Its basic advantage is that it does not seem to require any assumptions on the convexity of the problem. The underlying idea behind the method is that a centralized solution of (6.3) would require solving the Karush-Kuhn-Tucker (KKT) optimality conditions. The latter, ignoring the sets $C_A, C_B$, which are already separable and including two copies of the coupling constraints, are

$$\begin{bmatrix} \nabla_{\mathbf{x}_A} f_A + \tilde{\mathbf{c}}_A \boldsymbol{\lambda}_A + \tilde{\mathbf{c}}_A \boldsymbol{\lambda}_B \\ \nabla_{\mathbf{x}_B} f_B - \tilde{\mathbf{c}}_B \boldsymbol{\lambda}_A - \tilde{\mathbf{c}}_B \boldsymbol{\lambda}_B \\ \tilde{\mathbf{c}}_A \mathbf{x}_A - \tilde{\mathbf{c}}_B \mathbf{x}_B \\ \tilde{\mathbf{c}}_A \mathbf{x}_A - \tilde{\mathbf{c}}_B \mathbf{x}_B \end{bmatrix} = 0 \tag{6.13}$$

Consider taking a Newton step towards the solution of this set of equations starting from $\mathbf{x}_A^k, \mathbf{x}_B^k, \boldsymbol{\lambda}_A^k, \boldsymbol{\lambda}_B^k$:

$$\overbrace{\begin{bmatrix} \nabla_{\mathbf{x}_A \mathbf{x}_A} f_A & \mathbf{0} & \tilde{\mathbf{c}}_A & \tilde{\mathbf{c}}_A \\ \mathbf{0} & \nabla_{\mathbf{x}_B \mathbf{x}_B} f_B & -\tilde{\mathbf{c}}_B & -\tilde{\mathbf{c}}_B \\ \tilde{\mathbf{c}}_A & -\tilde{\mathbf{c}}_B & \mathbf{0} & \mathbf{0} \\ \tilde{\mathbf{c}}_A & -\tilde{\mathbf{c}}_B & \mathbf{0} & \mathbf{0} \end{bmatrix}}^{K} \begin{bmatrix} \Delta \mathbf{x}_A \\ \Delta \mathbf{x}_B \\ \Delta \boldsymbol{\lambda}_A \\ \Delta \boldsymbol{\lambda}_B \end{bmatrix} = - \begin{bmatrix} \nabla_{\mathbf{x}_A} f_A + \tilde{\mathbf{c}}_A \boldsymbol{\lambda}_A^k + \tilde{\mathbf{c}}_A \boldsymbol{\lambda}_B^k \\ \nabla_{\mathbf{x}_B} f_B - \tilde{\mathbf{c}}_B \boldsymbol{\lambda}_A^k - \tilde{\mathbf{c}}_B \boldsymbol{\lambda}_B^k \\ \tilde{\mathbf{c}}_A \mathbf{x}_A^k - \tilde{\mathbf{c}}_B \mathbf{x}_B^k \\ \tilde{\mathbf{c}}_A \mathbf{x}_A^k - \tilde{\mathbf{c}}_B \mathbf{x}_B^k \end{bmatrix} \tag{6.14}$$

The method then approximates these equations with the decomposable set:

$$\overbrace{\begin{bmatrix} \nabla_{\mathbf{x}_A \mathbf{x}_A} f_A & \mathbf{0} & \tilde{\mathbf{c}}_A & \mathbf{0} \\ \mathbf{0} & \nabla_{\mathbf{x}_B \mathbf{x}_B} f_B & \mathbf{0} & -\tilde{\mathbf{c}}_B \\ \tilde{\mathbf{c}}_A & \mathbf{0} & \mathbf{0} & \mathbf{0} \\ \mathbf{0} & -\tilde{\mathbf{c}}_B & \mathbf{0} & \mathbf{0} \end{bmatrix}}^{\tilde{K}} \begin{bmatrix} \Delta \mathbf{x}_A \\ \Delta \mathbf{x}_B \\ \Delta \boldsymbol{\lambda}_A \\ \Delta \boldsymbol{\lambda}_B \end{bmatrix} = - \begin{bmatrix} \nabla_{\mathbf{x}_A} f_A + \tilde{\mathbf{c}}_A \boldsymbol{\lambda}_A^k + \tilde{\mathbf{c}}_A \boldsymbol{\lambda}_B^k \\ \nabla_{\mathbf{x}_B} f_B - \tilde{\mathbf{c}}_B \boldsymbol{\lambda}_A^k - \tilde{\mathbf{c}}_B \boldsymbol{\lambda}_B^k \\ \tilde{\mathbf{c}}_A \mathbf{x}_A^k - \tilde{\mathbf{c}}_B \mathbf{x}_B^k \\ \tilde{\mathbf{c}}_A \mathbf{x}_A^k - \tilde{\mathbf{c}}_B \mathbf{x}_B^k \end{bmatrix} \tag{6.15}$$

Working backwards this yields at each iteration two optimization subproblems:

$$\min_{\mathbf{x}_A \in C_A} \{f_A(\mathbf{x}_A) + \tilde{\mathbf{c}}_A \boldsymbol{\lambda}_B^k \mathbf{x}_A : \tilde{\mathbf{c}}_A \mathbf{x}_A = \tilde{\mathbf{c}}_B \mathbf{x}_B^k\}, \ \min_{\mathbf{x}_B \in C_B} \{f_B(\mathbf{x}_B) + \tilde{\mathbf{c}}_B \boldsymbol{\lambda}_A^k \mathbf{x}_B : \tilde{\mathbf{c}}_A \mathbf{x}_A^k = \tilde{\mathbf{c}}_B \mathbf{x}_B\} \tag{6.16}$$

These individual optimization subproblems may or may not be solved to optimality [53], in both cases giving an update for the involved variables. One basic issue is that due to the fixed variables these subproblems may be infeasible. Consequently, the linear equality constraints may need to be relaxed into the objective function (e.g. using a barrier method). Furthermore, for the algorithm to converge, the solution of (6.15) should be well defined and in addition $\rho_s\{\mathbf{I} - \tilde{\mathbf{K}}^{-1}\mathbf{K}\} \leq 1$, where $\rho_s$ denotes the spectral radius of the corresponding matrix. In case this condition does not hold then some preconditioning method may be used, which should adjust the variables and multiplier updates to suitable values. Despite the potentially better performance of OCD compared to dual decomposition techniques, this method might not be such a good candidate for decentralized solutions. Evaluating whether or not the convergence criterion holds, and performing the required preconditioning, might not be easy in a decentralized context.

The OCD method has been successfully applied in power systems multi-area OPF problems. In Reference 54, the method is applied to a variety of test systems of up to 708 buses, separated in up to six areas. The distributed algorithm is shown to converge within a few tens of iterations. Bakirtzis and Biskas [55] combine this approach with linear load flow equations, carrying out tests in systems of up to six areas. Biskas *et al.* [56] further discuss some implementation considerations of this approach. Finally Biskas and Bakirtzis [57] extend the method to full AC equations. Convergence seems to be achieved in a few tens of iterations, however the significant effect of tolerance values used to check convergence is also illustrated. In Reference 58, the OCD method is extended and applied in a power system decomposed to several overlapping areas. The latter are selected independently of each other based on the control effects of various Flexible AC Transmission System (FACTS) devices. Results in Reference 59 seem to indicate faster convergence for this method compared to LR-based approaches for small degrees of decomposition. A technique for improving the convergence speed of the former is also presented (based on a master–slave principle) but no mathematical proof of convergence is provided. Huang *et al.* [60] propose an asynchronous implementation of the OCD method applied to DC equations. The method is tested on the three-area IEEE RTS system, and is shown to converge faster than the synchronous implementation. Finally Lai *et al.* [61] propose a modified form of the method, where a linear approximation for prices as a function of power exchanged between two areas is used, rather than the standard static Lagrange multipliers.

### *6.5.5 On other distributed methods*

A variety of other approaches have also been used in power systems as means of achieving distributed solutions to large-scale problems. One such method is the Dantzig–Wolfe decomposition mostly used in optimal reactive power dispatch problems [62,63], and also in Reference 64 which uses a non-linear version of the method for the solution of security constrained OPF problem. With respect to this last problem a very popular alternative is Bender's decomposition, e.g. in Reference 65 which includes a thorough review of different implementations of the latter for that particular problem. Most commonly, the master problem contains all constraints related with the base (no contingency) case, along with the Benders' cuts derived from the subproblems related to various contingencies. Phan and Kalagnanam [66] include an adaptive variant of the method to circumvent the potential issue of excluding optimum solutions in non-convex problems. Both Dantzig–Wolfe and Bender's decomposition generate one 'master' problem which typically increases in size at each iteration, and a set of smaller distributed subproblems. Overall, they involve a more complex process compared to the methods presented in the previous subsections and they do not offer an as clear interpretation of prices in the subproblems.

One alternative to traditional decomposition methods may be bi-level programming approaches such as the one proposed in Reference 67. At the higher level, the transmission system including an approximate representation of electric vehicle/demand aggregators is optimized. At the lower level, each individual aggregator solves an optimization subproblem that determines individual vehicle schedules while including a penalty for deviations from the higher-level optimization problem solution. It should be noted that this is not exactly a full decomposition method as the higher-level problem includes some representation of the lower-level problem. Finally, other distributed solutions to power systems problems include heuristics, such as the agent-based approach in Reference 68, or control-based approaches such as the max-sum algorithm presented in Reference 69 or the consensus algorithms in References 70,71. However, these approaches have significant limitations in terms of network representation.

## 6.6 OPF insights

Before discussing further optimization within the context of either UC or ED, it is of interest to discuss distributed solutions for the OPF problem. This problem is the basis of any power system optimization application and can provide significant insight to the applicability of distributed optimization methods in network constrained energy management problems (ED included) and the major considerations involved. For the time being we ignore any time linkages that might exist e.g. on the demand side and focus on this single time-step optimization problem. OPF is typically associated with transmission and subsequently balanced system operation. Considering our problem (6.3) for area A $\mathbf{x}_A = [\boldsymbol{\delta}_A \;\; \mathbf{V}_A \;\; \mathbf{P}_A \;\; \mathbf{Q}_A]^T$ and $f_A = \mathbf{P}_A^T \mathbf{c}_{A2p} \mathbf{P}_A + \mathbf{c}_{A1p} \mathbf{P}_A$ where $\boldsymbol{\delta}_A$ are the bus voltage angles for area A buses; $c_{A2p}$ is the quadratic term cost coefficients diagonal matrix; $\mathbf{c}_{A1p}$ is the linear-term cost coefficients vector; $\mathbf{V}_A$ the bus voltages vector; and $\mathbf{P}_A, \mathbf{Q}_A$ the active and reactive power injection vectors of individual generators or demand blocks. The equations describing the network in their standard ac form are well-documented [2] and, along with individual device constraints, may be written in complex number notation as

$$C_A = \left\{ \begin{array}{c} c_{Ap}\mathbf{P}_A + j\mathbf{c}_{Aq}\mathbf{Q}_A = diag\{\mathbf{V}_A\}(\mathbf{Y}_A\mathbf{V}_A)^* \\ \underline{\mathbf{V}_A} \leq |\mathbf{V}_A| \leq \overline{\mathbf{V}_A} \\ |\mathbf{Y}_{At}\mathbf{V}_A| \leq \overline{I_{At}} \\ \underline{\mathbf{P}_A} \leq |\mathbf{P}_A| \leq \overline{\mathbf{P}_A} \\ \underline{\mathbf{Q}_A} \leq |\mathbf{Q}_A| \leq \overline{\mathbf{Q}_A} \end{array} \right\} \tag{6.17}$$

where $\boldsymbol{c}_{Ap}, \mathbf{c}_{Aq}$ matrices linking individual power injections with nodes, i.e. $\boldsymbol{c}_{Ap(i,j)} = 1$ if device $j$ is connected to bus $i$; $\mathbf{Y}_A, \mathbf{Y}_{At}$ the bus and line current admittance matrices for area A; and $\overline{\mathbf{I}_{At}}$ the transmission lines maximum current capacities vector for area A. The constraints for area B are similar. Figure 6.5 illustrates the test network used, and the overall convergence process for the ADMM method for this particular example. For any implementation of a distributed optimization method, there are three basic factors affecting the solution speed:

- *Decomposition method parameters:* These can significantly affect convergence but typically there are no standard rules for setting parameter values for any of the decomposition methods presented in this chapter. For example, with regards to

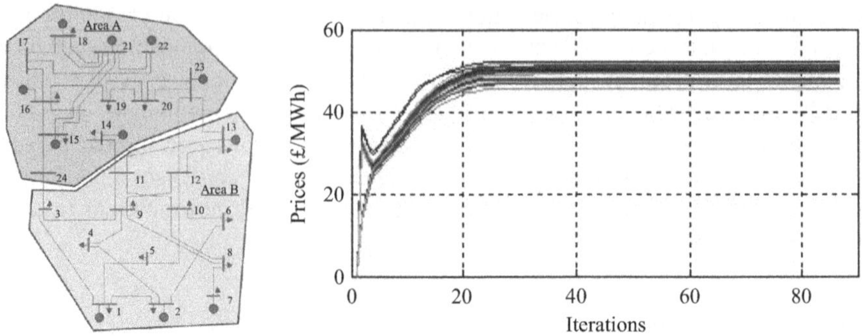


*Figure 6.5 Distributed optimization example. Left: Schematic representation of the test network visualizing the two areas. Right: Prices convergence for the ADMM method*

ADMM further information may be found in References 50,72. Some methods suggest a dynamic adjustment of the penalty factor, however, their efficiency in power systems problems has not been verified [72]. On the other hand, APP may fail to converge if the parameters are not appropriately selected. The OCD approach appears to have a significant advantage in that it simply does not involve any parameters at all, albeit it requires appropriate conditioning.

- *Subproblem solvers:* Given that several iterations would need to be carried out until convergence is achieved, fast centralized solvers could make a significant difference in total convergence time. Methods significantly benefiting from warm starting would be particularly effective. Another important aspect is how various inequality constraints are managed from one iteration to the next. Given that in most cases at the optimum a limited number of inequality constraints are active, an effective constraints management method could significantly improve speed. It should be noted here that the OCD method however, may potentially work just by carrying out a single Newton iteration in the subproblems rather than solving a complete optimization subproblem.
- *Decomposition structure:* The number of subproblems and the constraints they contain, depending on the selected algorithm, can significantly affect performance. Typically, the higher the number of network subproblems, the more iterations to convergence are required. This is further explained in the next section and clearly illustrated in Figure 6.7.

### 6.6.1 Decomposition structure considerations

Consider a network with a significant number of individual flexible devices (rather than the classic flexible demand), each of which has to be assigned to an individual state. Looking back at the simple example of Figure 6.4, the most straightforward way to do this is by adding one decision variable for each individual device in the optimization formulation. This would imply that instead of e.g. a 40 MW aggregate demand on transmission bus 5, we now have a few thousands of devices (a few kilowatt each). It follows that the OPF problem quickly grows in scale. If we were

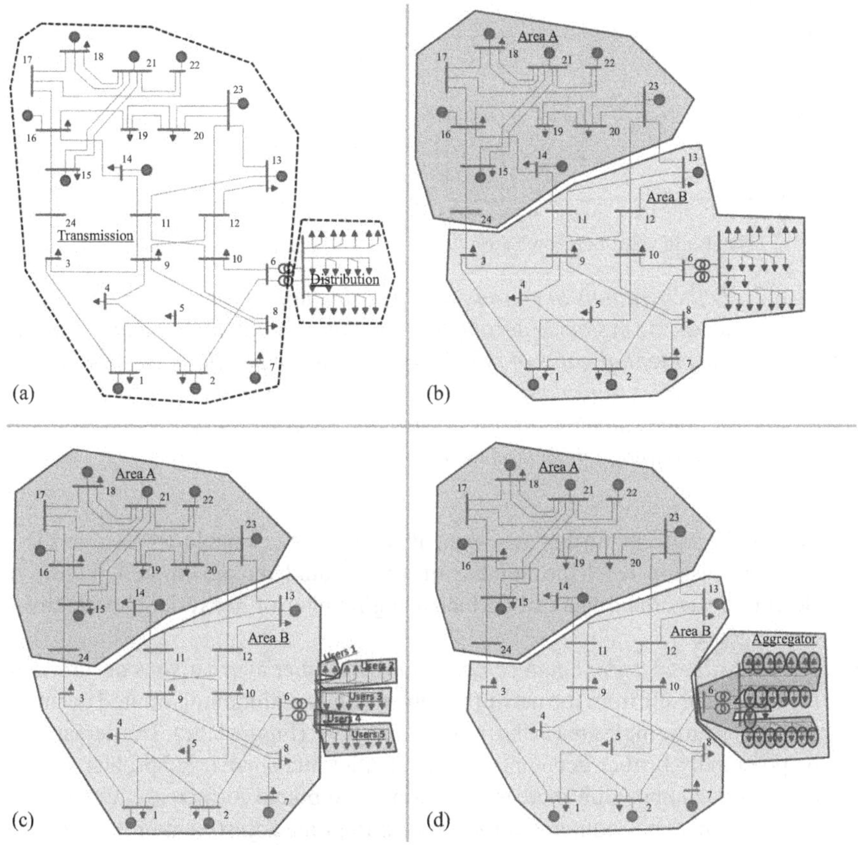


*Figure 6.6 Different decomposition structure possibilities: (a) the test system with the distribution network at node 6 expanded; (b) network areas decomposition; (c) transmission–distribution decomposition; and (d) hierarchical decomposition*

to resort to distributed techniques to manage any scalability issues, then there is the question of how the subproblems would be determined. Following are various alternative considerations:

- *Network areas decomposition:* One possibility would be to simply break down the network into an increasing number of areas (Figure 6.6b). An increase of the number of subproblems may be expected to be followed by an increased number of iterations to convergence, as it would take longer for the prices to be propagated around the system [50]. Eventually after a point, depending on the communications latency, even though the computational burden per subproblem might be very small, convergence in a sufficiently fast way might no longer be possible. One further difficulty with network decomposition arises due to contingency constraints. These introduce additional coupling between subproblems and a significantly larger number of prices and power schedules would have to be transmitted. While

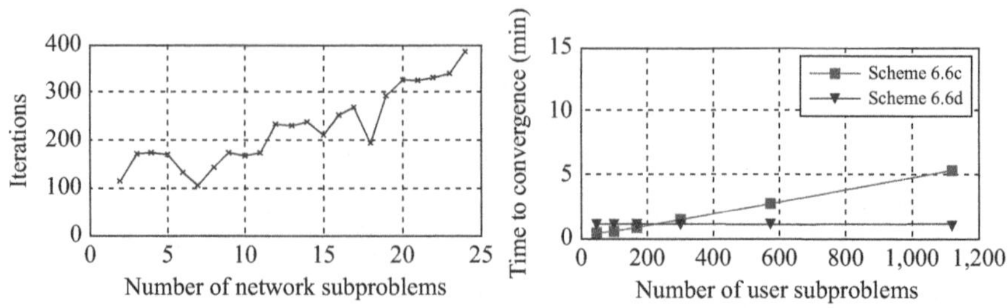


*Figure 6.7 Left: Iterations to convergence as a function of the number of network areas the system is decomposed to. Right: Indicative example showing the potential gains of the hierarchical decomposition compared to the direct transmission–distribution decomposition assuming an average communications delay of 0.1 s (based on the data and methodology of Reference 49)*

this might warrant further research, it might be reasonable to assume that at the transmission level the number of areas would remain limited. Figure 6.7 (left) indicates the impact of an increasing number of network subproblems on convergence.

- *Transmission–distribution decomposition:* One other alternative is decomposing the demand itself from the network problem while maintaining limited decomposition or even a single area at the transmission level (Figure 6.6c). The demand may be potentially further decomposed into smaller and smaller subproblems. However, as decomposition increases iterations are bound to increase again. While this effect may be countered through an appropriate adjustment of the decomposition method's parameters, this is not an easy thing to do. The optimal parameters values cannot be known in advance.
- *Hierarchical decomposition:* One other alternative to reduce computational burden is to carry out a decomposition within the decomposition in a two-level scheme (Figure 6.6d). This is a case where the transmission problem is solved with a standard centralized approach, while the demand/distribution-level subproblem is solved again in a distributed manner. This approach was first proposed in Reference 50 and is based on the simple fact that all loads at a given bus are connected through a single point to the transmission network. That is a point where demand may be aggregated, and potentially different more efficient algorithms may be used (as in the example of Figure 6.7) that take into account the peculiarities of individual demand blocks.

Note that apart from the differences in convergence time, the different structures have direct implications with regards to the communications that are actually required. In the first scheme, all devices in an area would have to pass information to the subproblem optimizer, which might be a significant burden. On the other hand, in the last scheme only the devices served through a given transmission bus need to communicate with the aggregator, which might be an easier thing to achieve.

### *6.6.2 Practical application considerations*

The decomposition methods described in the previous section may be applied to any power systems optimization problem involving continuous variables and mathematically guarantee convergence as long as the problem is convex. Practical power systems problems do not necessarily conform to these requirements however:

- *Convexity:* While the mathematical proofs of convergence for most methods assume convex problems, that does not mean that these methods will not work in non-convex cases [50,73]. In terms of optimality, similar to a centralized mathematical solver, these methods could be considered as local optimizers with their performance dependent on the initial conditions [72,73]. Nevertheless, the formulation of individual subproblems has to be carefully considered. Recent attempts to convexify OPF problems [74] could in the future help provide the necessary convergence guarantees in terms of formulation.
- *Discrete controls:* These often imply non-convex cost functions and lack of an equilibrium based on prices. As is discussed in Section 6.7, distributed methods would need to be complemented with appropriate, likely centralized, mechanisms that set the values of these discrete variables.

Barring these mathematical considerations there are two further important issues:

- *Solution process standards:* It should be clear that, if subproblem parameters change during the distributed algorithm's execution, to reflect either changing conditions in the system or a user/market participant attempting to game the system to increase profits, convergence might be delayed or might even not be possible at all. As such, appropriate standards should govern how the subproblem computations are carried out and how parameters are passed on to them.
- *Communications speed and reliability:* For each iteration taken, the time required for passing messages between subproblems will be added to the time required for computations. While the latter will decrease as the degree of decomposition in a system increases (as subproblems grow smaller in size), the opposite might be true for the time cumulatively required for communications (as iterations to convergence may be expected to increase). The overall distributed solution has to strike a balance between the two. Furthermore, handling communications errors has to be carefully considered, even though it may vary in importance depending on the power actually exchanged between two subproblems (e.g. reliable communications are of critical importance for two transmission areas exchanging several Gigawatt ($10^3$ MW), but of much less importance between users exchanging a few kilowatt).

## 6.7 The UC time frame

UC is a problem which could readily accommodate flexible demand characteristics (e.g. scheduling an electric vehicle or storage unit over time) given its typical multi-period formulation. However, it is characterized by significant uncertainty and it also is already a challenging mixed integer problem when the objective is to schedule a few (relative to the number of small-scale flexible devices) slow-to-start generators. Incorporating the constraints of individual devices or attempting to include the details of distribution networks would not only make the problem unnecessarily large, but also probably be of little practical value. The reason for that being that it is unlikely that demand at the individual end-user level could be reliably forecasted. Consequently, aggregate demand models appear to be a better option. This is in agreement with approaches published in recent relevant literature, e.g. References 75–77 which in essence attempt to build appropriate reduced-order models or determine bids for a collection of flexible devices (most commonly electric vehicle fleets). The latter is assumed to be managed by a single entity, the aggregator.

In terms of UC formulation there are typically three basic approaches [3]: deterministic which uses point estimates for the uncertain parameters; stochastic which uses a reduced set of scenarios; and interval which uses a further reduced scenario set (i.e. a central forecast and upper and lower bounds on it). Applying a decomposition method to any stochastic formulation can be a difficult task. Going back to problem (6.3) consider the case where area A has significant renewables penetration and fluctuations in daily demand. A simple stochastic formulation of the problem which involves minimization of the expected cost over a set of $n_s$ scenarios would be:

$$\min_{\substack{\mathbf{x}_{A,i}\in C_{A,i}\\ \mathbf{x}_{B,i}\in C_{B,i}}} \left\{ \sum_i (f_A(\mathbf{x}_{A,i}) + f_B(\mathbf{x}_{B,i})) : \mathbf{c}_A\tilde{\mathbf{x}}_{A,i} = \mathbf{c}_B\tilde{\mathbf{x}}_{B,i} \right\}, \quad i \in \{1,2,\ldots,n_s\} \tag{6.18}$$

Considering any of the decomposition methods above, e.g. LR, a distributed solution of this problem would presuppose at least the exchange of $n_s$ prices and as many energy schedules. Furthermore, if optimization was carried out over $n_t$ time steps then the number of prices (and schedules) increases to $n_s \times n_t$. This would represent a significant increase in terms of data transfer requirements and in addition might make convergence much more challenging. While in the past LR techniques were commonly used in UC, e.g. Reference 18 or Reference 78, they were applied to deterministic versions of the problem. Of course, given that this is a mixed integer non-convex problem a set of heuristics were additionally used to set unit operating states or adjust prices. For example, in Reference 79, a few simple rules are combined with the standard LR approach: e.g. the start-up cost normalized by the number of operating hours is added to the hourly variable cost to determine whether or not a unit turns on, or units with similar characteristics are committed as a group and then decommitted one by one as long as demand and reserve requirements are satisfied. A similar approach is followed in Reference 80 where a variety of empirical rules are used for updating the Lagrange multipliers depending on whether or not reserve and active power balance constraints are satisfied. In terms of managing discrete demand variables [17] proposes a simple heuristic, which limits the maximum power consumption per hour for individual consumers. In References 81,82, a bi-level optimization approach is proposed, where coordination is achieved through one higher-level master problem. It should be clear however that at least one step of any of these solution approaches remains centralized. Depending on the workings of the centralized mechanism handling price updates and discrete controls states, optimality cannot be always guaranteed.

Another concept of interest that appears in UC is that of local uncertainty management. Several papers deal with the determination of bids, which would be supplied by an independent market participant to the system operator. For example in Reference 83 a simplified approach is presented for constructing bids for an electric vehicle aggregator, where electric vehicles are grouped in three basic categories, based on their expected way of charging. The aggregator builds a forecast for each category and solves an optimization problem that attempts to minimize costs and expected deviation penalties (i.e. differences of power dispatched in ED compared to power initially scheduled in UC). A similar approach of grouping vehicles, this time based on similarities in their usage patterns, is proposed in Reference 84, and also in Reference 85 where the aggregator simply generates a given number of scenarios by sampling distributions for all Electric Vehicle (EV) parameters (e.g. connection time and state of charge) and given forecasts for the energy prices solves a two-stage optimization problem to determine energy bids. In Reference 86 bids for an aggregator are derived based on a stochastic dynamic programming approximation, assuming that the aggregator has knowledge of the resulting probability distribution of energy and regulation prices. EVs are grouped based on their departure times and a penalty is applied if a vehicle departs without being charged at the desired level. The same may be applied to generating systems, either conventional [87] or renewable [88]. Aggregating end-user devices could significantly simplify the problem formulation and limit its scale.

Today end users are not typically directly exposed to fluctuating energy market prices. Assuming this trend continues in the future, through appropriate aggregation of their requirements and flexibility the scale of the UC problem might not dramatically change. In addition, for stochastic formulations decomposition can be hard. As a consequence, there does not, at present, appear to be a particularly strong motivation to move from today's branch and bound methods towards decentralized solutions. Given that UC is unlikely to manage in detail distribution networks and individual devices, this burden will fall to ED. The latter's time frame is where the most drastic changes may be expected to come. Nevertheless, with regards to UC it is possible to make two interesting observations. The first is that managing uncertainty locally at the aggregator level through the submission of aggregate bids or other simplified models appears to be a common enough and practical approach, which can greatly simplify the problem's solution. The second relates to the simple fact that, as may be expected, the aggregation process presupposes an additional dispatch step for individual devices closer to real time. As we will discuss in the following these are ideas that could also be transferred to the ED time frame to allow decentralized solutions.

## 6.8 The ED time frame

The ED problem at its simplest form may be thought of as an OPF problem solved a few minutes ahead of real time. To account for the variability of wind and the ramp rates of conventional generators it might also include a short look-ahead period [5]. Nevertheless, this might not be adequate for scheduling demand over time, and a possible inclusion of distribution network constraints would significantly increase the

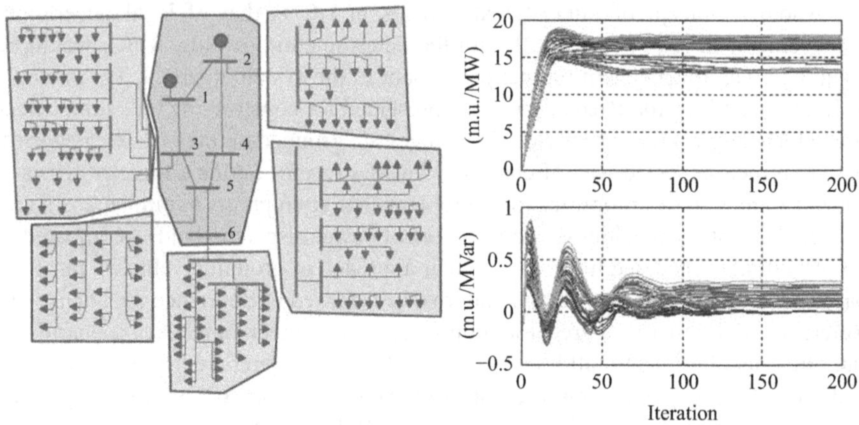


*Figure 6.8 Simple test case. Left: Network schematic including a single transmission area and extended distribution networks at 5 of its buses. Right: Indicative distributed optimization convergence results for active and reactive power prices at the transmission level covering a time period of 12 h separated in as many time steps. The problem is decomposed to one transmission and five distribution subproblems*

size of the problem. Consider for example the simple test network in Reference 89 illustrated in Figure 6.8. A simple transmission level OPF for this network would involve six buses, ten decision variables associated with generation, and five additional variables if demand is assumed curtailable. If medium/high-voltage distribution networks were included that would involve 286 additional nodes, and the demand variables would increase to 195. If optimization over a time period was to be carried out, then the total of the associated constraints and variables would have to be multiplied by the number of time steps. For larger systems, the problem would quickly become intractable. It should also be noted that the inclusion of a look-ahead period brings up questions with regards to representation and management of uncertainty. However, these changes are exactly what is required to schedule flexible devices over time subject to distribution network constraints.

While it is not often that current literature directly links multi-period energy management formulations with the ED time frame, several papers could be directly associated with this problem. For example, Yao *et al.* [67] present a hierarchical multi-period formulation solved via a bi-level programming technique. The decomposed problem structure is close to Figure 6.6(c) but with a single transmission area. Another multi-time-step formulation using a similar decomposition structure may be found in Reference 28 which uses a simplified network representation and assumes exact knowledge of expected demand energy usage. Rather than using aggregators for groups of devices, demand is decomposed to the individual device level. Note, however, that none of these papers considers the associated uncertainties. Taking them into account, however, would bring up the issues associated with (6.17).

In Reference 90, a two-level decomposition scheme similar to Figure 6.6(d) is used based on ADMM. The paper includes considerations with regard to handling uncertainty with regards to problem formulation. The core idea is that while a distributed optimization technique could enable the close-to-real-time solution of a multi-time-step distribution network inclusive ED to large degrees of decomposition, the latter should be carried out up to the level that reliable forecasts on aggregate power can be reasonably accurate. For example, it might be hard to accurately predict the power injections a single electric vehicle could give (due to uncertainty in connection times), but it might be straightforward to do the same for a few tens of vehicles [83]. Approximate aggregate models could be used for subproblems within which uncertainty is high (e.g. an aggregator managing a low voltage network or a generator managing a set of renewable energy sources). This concept does not differ much from certain approaches to UC as discussed in the previous section, or the reasoning of building aggregator models as discussed e.g. in References 75,91,92. The implication is that an additional step would have to be taken even closer to real time to determine the set points of individual devices. Given that this new dispatch step would be associated with a smaller part of the network we call this microgrid dispatch.

## 6.9 Closer to real time

Considering the period in-between two successive ED solutions, it should be clear that system-wide optimization is no longer possible. However, solving smaller problems of a local scale is. An example, already applied in certain power systems, is various secondary voltage control schemes [93]. In these schemes, the transmission network is separated into several control areas. Following an ED and based on phase measurement unit information a local (within each control area) optimization of available reactive power reserves is performed to ensure stable operation as system conditions change. The same principle may be applied to distribution networks or grid-connected microgrids for the purpose of managing individual devices, which might not be represented in detail in ED. In this case, rather than having a decomposition algorithm converging over several iterations, the individual microgrids/distribution networks solve their own local problems and coordination is achieved through the obligation to follow the market reference power set point. This is a sort of coordination over time, as opposed to coordination towards the solution of a specific optimization problem at a given instance of time. In principle, this is similar to what market penalties imposed to market participants for deviating from their promised schedules in UD or ED try to achieve. It should be noted that at this level the subproblems may no longer necessarily be solved within the context of a market and energy prices would not necessarily be a key quantity, while reaching a global optimum in any conceivable formulation of the problem might not be as important. This allows a considerably wider range of solution methods:

- *Centralized approaches:* At this level, the problem might be of a sufficiently small scale to allow for a fast centralized solution where the microgrid operator gathers all relevant data, solves the problem centrally, and broadcasts back the decision.

Optimization approaches applied to distribution network solutions are applicable here. For example, Palma-Behnke *et al.* [94] focus on the control of small microgrids considering both discrete and continuous controls of generation (conventional and photovoltaic) and demand in a multi-period mixed integer programming solution approach solved via branch and bound. A similar approach, but more detailed in terms of network representation, may be found in Reference 95. In cases where the problem is still too complex for branch and bound then a variety of heuristics could allow for a sufficiently fast and reliable solution as e.g. in Reference 96 which offers a comparison of such heuristics, or References 97,98 which focus on building appropriate priority lists. These types of approaches require only a single round of communications and allow for significant control over how the problem is solved. Furthermore, they allow handling discrete controls.

- *Distributed approaches*: Similar to the centralized approaches these methods attempt to solve some problem to optimality or at least recover a good-quality feasible solution. One solution approach is given in References 99,100 where the authors use two distributed optimization heuristics, both with dynamic programming origins for managing flexible users, based on LR. A Lagrangian-based approach which includes an additional heuristic for fixing discrete variables may be found in Reference 101. Proximal methods are also particularly popular e.g. in References 42,44,45,102–104 and also in References 105,106 which take into consideration the discrete nature of certain controls. A consensus-based approach for EV charging is used in Reference 107. This approach, however, is applicable to continuous variables problems only and cannot possibly take into account any network constraints.
- *Control-based approaches*: These methods take one step further from the distributed approaches of the previous paragraph and rather than waiting for convergence, they apply the solution directly. For example, the methods are discussed in References 22,23. The applicability of such approaches however would also depend on the speed and magnitude of changes that take place in the network, i.e. they might not be sufficiently fast to alleviate in time significant voltage deviations or branch current limit violations.

Note that none of the above approaches can be really fully decentralized as in all cases communication with one central operator – who also has knowledge of the network and its current state – is required. As such, carefully considering the scale of the problem and the actual end-user privacy requirements is important to determine whether or not a distributed solution is really needed. It should also be noted that when moving even closer to real time (e.g. when the objective is load shedding due to significant frequency events, or managing an overvoltage to avoid disconnection of a renewable resource) these schemes are not applicable as there is little available time. In these cases, only local frequency and voltage signals may be used and the key consideration is applying appropriate settings to the individual devices so that the overall system is stable. In a more traditional power systems perspective the classic generator droop control is an example of this. In a smart grid setting this could potentially involve new methods which would guarantee that the aggregate response of a large number of devices would give a stable response. As an example authors in References 108 or 109 investigate frequency and voltage control parameter settings to ensure stable system behaviour.

## 6.10 Conclusions

Following the preceding discussions, Figure 6.9 illustrates one possible control structure emphasizing the points of potential applications of distributed methods. The first and most important consideration, at all stages of control, is identifying the timescale of the available controls. This would give a clear indication of how far ahead of real time a control problem has to be solved, the involved uncertainty, and the time available for a solution. Following that, the way uncertainty is managed and any involved security considerations will provide a clear indication of whether a distributed solution is applicable or not. At the same time the emerging aggregator and network operator structures would indicate the decomposition structure, i.e. the subproblems

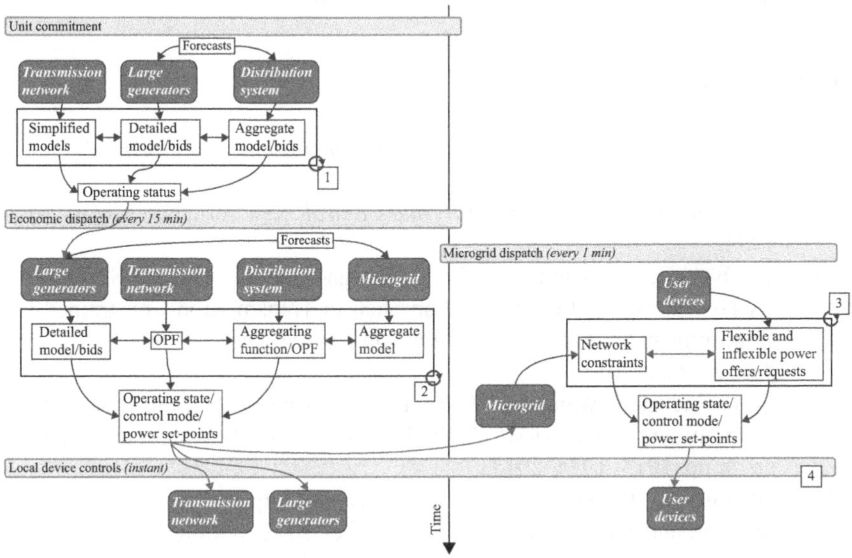


*Figure 6.9 A generic power system control framework. The numbers designate possible applications of distributed solutions. Numbers 1 and 2 are typically price-based decomposition methods that try to reach the optimal solution of the corresponding optimization problem. Number 3 may utilize prices as a control signal, but potentially a suboptimal yet feasible solution might be adequate. In Number 4, stability is the key factor. Coordination is achieved through appropriate controller tuning ahead of time, rather than in real time. The time frames mentioned for economic and microgrid dispatch mechanisms are purely indicative*

in the overall distributed control context. Last but not the least, the communications infrastructure and standards associated with the operation of the available distributed controls are also considerations of significant importance and would play a decisive role when it comes to power systems control in the context of markets. All the aforementioned aspects are interrelated and affect one another. What we need to stress is that in terms of market problems, optimizing subject-to-network constraints is the actual big challenge. In terms of controls, the challenge lies in achieving a stable frequency or voltage control while taking advantage of the flexibility offered by the multitude of end-user devices. Potentially, this could allow maintaining current system operating reliability levels, while requiring less resources (e.g. conventional generation reserves). With the exception of the fast real-time controls, it should be understood that a fully decentralized solution is probably not possible. Power at any given instance in time and energy requirements over a period of time have to be balanced at a system-wide level. The same is true with respect to the overall system reliability and quality of service requirements, as well as enforcement of market rules. There would probably always be a central controller overseeing the smooth solution of the involved optimization problems. In any case, appropriately designed distributed methods could give scalable solutions to most energy management problems.

## Bibliography


[1] Z. Zhu, *Optimization of Power System Operation*. New York, NY: IEEE–Wiley, 2009.

[2] P. Kundur, *Power System Stability and Control*. New York, NY: McGraw-Hill, 1994.

[3] Y. Dvorkin, H. Pandzic, M. A. Ortega-Vazquez and D. S. Kirschen, "A Hybrid Stochastic/Interval Approach to Transmission-Constrained Unit Commitment," *IEEE Transactions on Power Systems*, vol. 30, no. 2, pp. 621–631, 2015.

[4] Q. P. Zheng, J. Wang and A. L. Liu, "Stochastic Optimization for Unit Commitment – A Review," *IEEE Transactions on Power Systems*, vol. 30, no. 4, pp. 1913–1924, 2015.

[5] GE Energy, "Review of Industry Practice and Experience in the Integration of Wind and Solar Generation," November 2012. [Online]. Available from http://www.pjm.com. Accessed on 25-July-2016.

[6] M. B. Cain, R. P. O'Neill and A. Castillo, "History of Optimal Power Flow and Formulations," 2012. [Online]. Available from http://www.ferc.gov/. Accessed on 25-July-2016.

[7] A. T. Al-Alwami and E. Sortomme, "Coordinating Vehicle-to-Grid Services With Energy Trading," *IEEE Transactions on Smart Grid*, vol. 3, no. 1, pp. 453–458, 2012.

[8] L. H. Macedo, J. F. Franco, M. J. Rider and R. Romero, "Optimal operation of distribution networks considering energy storage devices," *IEEE Transactions on Smart Grid*, vol. 6, no. 6, pp. 2825–2836, 2015.

[9] M. Pipattanasomporn, M. Kuzlu, S. Rahman and Y. Teklu, "Load Profiles of Selected Major Household Appliances and Their Demand Response Opportunities," *IEEE Transactions on Smart Grid*, vol. 5, no. 2, pp. 742–750, 2014.

[10] L. P. Fernandez, T. G. S. Roman, R. Cossent, C. M. Domingo and P. Frias, "Assessment of the Impact of Plug-in Electric Vehicles on Distribution Networks," *IEEE Transactions on Power Systems*, vol. 26, no. 1, pp. 206–213, 2011.

[11] P. Zhang, K. Qian, C. Zhou, B. G. Stewart and D. M. Hepburn, "A Methodology for Optimization of Power Systems Demand Due to Electric Vehicle Charging Load," *IEEE Transactions on Power Systems*, vol. 27, no. 3, pp. 1628–1636, 2012.

[12] D. P. Bertsekas and J. N. Tsitsiklis, *Parallel and Distributed Computation: Numerical Methods*. Belmont, MA: Athena Scientific, 1997.

[13] C. Lemarechal, "The Omnipresence of Lagrange," *Annals of Operation Research*, vol. 153, no. 1, pp. 9–27, 2007.

[14] D. P. Bertsekas, *Nonlinear Programming*. Belmont, MA: Athena Scientific, 1999.

[15] A. J. Conejo and J. A. Aguado, "Multi Area Coordinated Decentralized DC Optimal Power Flow," *IEEE Transactions on Power Systems*, vol. 13, no. 4, pp. 1272–1278, Nov. 1998.

[16] A. G. Bakirtzis, "Joint Energy and Reserve Dispatch in a Competitive Pool Using Lagrangian Relaxation," *IEEE Power Engineering Review*, vol. 18, no. 11, pp. 60–62, 1998.

[17] D. Papadaskalopoulos and G. Strbac, "Decentralized Participation of Flexible Demand in Electricity Markets – Part I: Market Mechanism," *IEEE Transactions on Power Systems*, vol. 28, no. 4, pp. 3658–3666, 2013.

[18] S. Dekrajangpetch, G. B. Sheble and A. J. Conejo, "Auction Implementation Problems Using Lagrangian Relaxation," *IEEE Transactions on Power Systems*, vol. 14, no. 1, pp. 82–88, 1999.

[19] J. Batut and A. Renaud, "Daily Generation Scheduling Optimization with Transmission Constraints: A New Class of Algorithms," *Transactions on Power Systems*, vol. 7, no. 3, pp. 982–989, 1992.

[20] M. Roozbehani, M. A. Dahleh and S. Mitter, "Volatility of Power Grids Under Real-Time Pricing," *IEEE Transactions on Power Systems*, vol. 27, no. 4, pp. 1926–1940, 2012.

[21] F. P. Kelly, A. K. Maulloo and D. K. H. Tan, "Rate Control for Communication Networks: Shadow Prices, Proportional Fairness and Stability," *Journal of Operational Research Society*, vol. 49, no. 3, pp. 237–252, 1998.

[22] Z. Fan, "A Distributed Demand Response Algorithm and Its Application to PHEV Charging in Smart Grids," *IEEE Transactions on Smart Grid*, vol. 3, no. 3, pp. 1280–1290, 2012.

[23] Q. R. Hamid and J. A. Barria, "Distributed Recharging Rate Control for Energy Demand Management of Electric Vehicles," *IEEE Transactions on Power Systems*, vol. 28, no. 3, pp. 2688–2699, 2013.

[24] O. Ardakanian, S. Keshav and C. Rosenberg, "Real-Time Distributed Control for Smart Electric Vehicle Chargers: From a Static to a Dynamic Study," *IEEE Transactions on Smart Grid*, vol. 5, no. 5, pp. 2295–2305, 2014.

[25] P. Richardson, D. Flynn and A. Keane, "Local Versus Centralized Charging Strategies for Electric Vehicles in Low Voltage Distribution Systems," *IEEE Transactions on Smart Grid*, vol. 3, no. 2, pp. 1020–1028, 2012.

[26] M. D. Galus, R. A. Waraich, K. S. F. Noembrini, *et al.*, "Integrating Power Systems, Transport Systems and Vehicle Technology for Electric Mobility Impact Assessment and Efficient Control," *IEEE Transactions on Smart Grid*, vol. 3, no. 2, pp. 934–949, 2012.

[27] A. J. Conejo, E. Castillo, R. Minguez and R. Garcia-Bertrand, *Decomposition Techniques in Mathematical Programming: Engineering and Science Applications*. Berlin: Springer, 2006.

[28] N. Gatsis and G. B. Giannakis, "Decomposition Algorithms for Market Clearing with Large-Scale Demand Response," *IEEE Transactions on Smart Grid*, vol. 4, no. 4, pp. 1976–1987, 2013.

[29] A. L. Motto, F. D. Galiana, A. J. Conejo and M. Huneault, "On Walrasian Equilibrium for Pool-Based Markets," *IEEE Transactions on Power Systems*, vol. 17, no. 3, pp. 774–781, 2002.

[30] D. P. Bertsekas, *Constrained Optimization and Lagrange Multiplier Methods*. Belmont, MA: Athena Scientific, 1996.

[31] G. Cohen, "Optimization by Decomposition and Coordination: A Unified Approach," *IEEE Transactions on Automatic Control*, vol. 23, no. 2, pp. 222–232, April 1978.

[32] G. Cohen, "Auxiliary Problem Principle and Decomposition of Optimization Problems," *Journal of Optimization Theory and Applications*, vol. 32, no. 3, pp. 277–305, Nov. 1980.

[33] B. H. Kim and R. Baldick, "Coarse-Grained Distributed Optimal Power Flow," *IEEE Transactions on Power Systems*, vol. 12, no. 2, pp. 932–939, May 1997.

[34] R. Baldick, B. H. Kim, C. Chase and Y. Luo, "A Fast Distributed Implementation of Optimal Power Flow," *IEEE Transactions on Power Systems*, vol. 14, no. 3, pp. 858–864, Aug. 1999.

[35] B. H. Kim and R. Baldick, "A Comparison of Distributed Optimal Power Flow Algorithms," *IEEE Transactions on Power Systems*, vol. 15, no. 2, pp. 599–604, May 2000.

[36] D. Hur, J. K. Park and B. Kim, "Evaluation of Convergence Rate in the Auxiliary Problem Principle for Distributed Optimal Power Flow," *IEE Generation, Transmission and Distribution*, vol. 149, no. 5, pp. 525–532, 2002.

[37] D. Hur, J. K. Park and B. H. Kim, "On the Convergence Rate Improvement of Mathematical Decomposition Technique on Distributed Optimal Power Flow," *Electrical Power and Energy Systems*, vol. 25, no. 1, pp. 31–39, 2003.

[38] J. Contreras, A. Losi, M. Russo and F. F. Wu, "DistOpt: A Software Framework for Modeling and Evaluating Optimization Problem Solutions in Distributed Environments," *Journal of Parallel and Distributed Computing*, vol. 60, no. 6, pp. 741–763, Jun. 2000.

[39] J. Contreras, A. Losi, M. Russo and F. F. Wu, "Simulation and Evaluation of Optimization Problem Solutions in Distributed Energy Management Systems," *IEEE Transactions on Power Systems*, vol. 17, no. 1, pp. 57–61, 2002.

[40] A. Losi and M. Russo, "On the Application of the Auxiliary Problem Principle," *Journal of Optimization Theory and Applications*, vol. 117, no. 2, pp. 377–396, May 2003.

[41] L. Gan, U. Topcu and S. H. Low, "Optimal Decentralized Protocol for Electric Vehicle Charging," *IEEE Transactions on Power Systems*, vol. 28, no. 2, pp. 940–951, 2013.

[42] E. L. Karfopoulos and N. D. Hatziargyriou, "A Multi-Agent System for Controlled Charging of a Large Population of Electric Vehicles," *IEEE Transactions on Power Systems.*

[43] Z. Ma, D. S. Callaway and I. A. Hiskens, "Decentralized Charging Control of Large Populations of Plug-in Electric Vehicles," *IEEE Transactions on Control Systems Technology*, vol. 21, no. 1, pp. 67–78, 2013.

[44] I. Atzeni, L. G. Ordonez, G. Scutari, D. P. Palomar and J. R. Fonollosa, "Demand-Side Management via Distributed Energy Generation and Storage Optimization," *IEEE Transactions on Smart Grid.*

[45] A. H. Mohsenian-Rad, V. W. S. Wong, J. Jatskevich, R. Schober and A. Leon-Garcia, "Autonomous Demand-Side Management Based on Game-Theoretic Energy Consumption Scheduling for the Future Smart Grid," *IEEE Transactions on Smart Grid*, vol. 1, no. 3, pp. 320–331, 2010.

[46] M. H. K. Tushar, C. Assi, M. Maier and M. F. Uddin, "Smart Microgrids: Optimal Joint Scheduling for Electric Vehicles and Home Appliances," *IEEE Transactions on Smart Grid*, vol. 5, no. 1, pp. 239–250, 2014.

[47] X. Wang, Y. H. Song and Q. Lu, "Lagrangian Decomposition Approach to Active Power Congestion Management Across Interconnected Regions," *IEE Proc. Generation, Transmission and Distribution*, vol. 148, no. 5, pp. 497–503, 2001.

[48] M. Kraning, E. Chu, J. Lavaei and S. Boyd, "Message Passing for Dynamic Network Energy Management," *Foundations and Trends in Optimization*, vol. 1, no. 2, pp. 70–122, 2013.

[49] E. D. Anese, H. Zhu and G. B. Giannakis, "Distributed Optimal Power Flow for Smart Grids," *IEEE Transactions on Smart Grid*, vol. 4, no. 3, pp. 1464–1475, 2013.

[50] E. Loukarakis, J. W. Bialek and C. J. Dent, "Investigation of Maximum Possible OPF Problem Decomposition Degree for Decentralized Energy Markets," *IEEE Transactions on Power Systems*, vol. 30, no. 5, pp. 2566–2578, 2015.

[51] B. H. Kim, "An Efficient Implementation of Decentralized Optimal Power Flow," *Journal of Electrical Engineering & Technology*, vol. 2, no. 3, pp. 335–341, 2007.

[52] F. J. Nogales, F. J. Prieto and A. J. Conejo, "Multi-Area AC Optimal Power Flow: A New Decomposition Approach," in *Power Systems Computation Conference (PSCC)*, Trondheim, Norway, 1999.

[53] A. J. Conejo, F. J. Nogales and F. J. Prieto, "A Decomposition Procedure Based on Approximate Newton Directions," *Mathematical Programming*, vol. 93, no. 3, pp. 495–515, Dec. 2002.

[54] F. J. Nogales, F. J. Prieto and A. J. Conejo, "A Decomposition Methodology Applied to the Multi-Area Optimal Power Flow Problem," *Annals of Operations Research*, vol. 120, no. 1, pp. 99–116, 2003.

[55] A. Bakirtzis and P. N. Biskas, "A Decentralized Solution to the DC OPF of Interconnected Power Systems," *IEEE Transactions on Power Systems*, vol. 18, no. 3, pp. 1007–1013, Aug. 2003.

[56] P. N. Biskas, A. G. Bakirtzis, N. I. Macheras and N. K. Pasialis, "A Decentralized Implementation of DC Optimal Power Flow on a Network of Computers," *IEEE Transactions on Power Systems*, vol. 20, no. 1, pp. 25–33, Feb. 2005.

[57] P. N. Biskas and A. G. Bakirtzis, "Decentralised OPF of Large Multiarea Power Systems," *IEE Proceedings Generation, Transmission & Distribution*, vol. 153, no. 1, pp. 99–105, 2006.

[58] G. H. Glanzmann and G. Andersson, "Decentralized Optimal Power Flow Control for Overlapping Areas in Power Systems," *IEEE Transactions on Power Systems*, vol. 24, no. 1, pp. 327–336, Feb. 2009.

[59] M. Arnold, S. Knopfli and G. Andersson, "Improvement of OPF Decomposition Methods Applied to Multi Area Power Systems," in *IEEE PowerTech*, Lausanne, Switzerland, 2007.

[60] A. Huang, S. K. Joo, K. B. Song, J. H. Kim and K. Lee, "Asynchronous Decentralized Method for Interconnected Electricity Markets," *Electrical Power & Energy Systems*, vol. 30, no. 4, pp. 283–290, 2008.

[61] X. Lai, L. Xie, Q. Xia, H. Zhong and C. Kang, "Decentralized Multi-Area Economic Dispath via Dynamic Multiplier-Based Lagrangian Relaxation," *IEEE Transactions on Power Systems*, vol. 30, no. 6, pp. 3225–3233, 2015.

[62] J. C. Lopez, M. Granada and J. R. S. Mantovani, "Multi Area Decentralized Optimal VAr Planning Using the Dantzig–Wolfe Decomposition Principle," in *IEEE Transmission and Distribution Conference and Exposition (T&D-LA)*, Sao Paulo, Brazil, 2010.

[63] N. Deeb and S. M. Shahidehpour, "Linear Reactive Power Optimization in a Large Power Network Using the Decomposition Approach," *IEEE Transactions on Power Systems*, vol. 5, no. 2, pp. 428–438, 1990.

[64] M. Aganagic and S. Mokhtari, "Security Constrained Economic Dispatch Using Nonlinear Dantzig–Wolfe Decomposition," *IEEE Transactions on Power Systems*, vol. 12, no. 1, pp. 105–112, 1997.

[65] S. Cvijic and J. Xiong, "Security Constrained Unit Commitment and Economic Dispatch Through Benders Decomposition: A Comparative Study," in *IEEE Power and Energy Society General Meeting*, San Diego, CA, USA, 2011.

[66] D. Phan and J. Kalagnanam, "Distributed Methods for Solving the Security Constrained Optimal Power Flow Problem," in *IEEE Innovative Smart Grid Technologies (ISGT)*, Washington, DC, USA, 2012.

[67] W. Yao, J. Zhao, F. Wen, Y. Xue and G. Ledwich, "A Hierarchical Decomposition Approach for Coordinated Dispatch of Plug-in Electric Vehicles," *IEEE Transactions on Power Systems*, vol. 28, no. 3, pp. 2768–2778, 2013.

[68] F. Ren, M. Zhang and D. Sutanto, "Multi-Agent Solution to Distribution System Management by Considering Distributed Generation," *IEEE Transactions on Power Systems*, vol. 28, no. 2, pp. 1442–1451, 2013.

[69] S. Miller, S. D. Ramchurn and A. Rogers, "Optimal Decentralized Dispatch of Embedded Generation in the Smart Grid," in *International Conference on Autonomous Agents and Multiagent Systems (AAMAS)*, Valencia, Spain, 2012.

[70] Y. Xu, W. Liu and J. Gong, "Stable Multi-Agent-Based Load Shedding Algorithm for Power Systems," *IEEE Transactions on Power Systems*, vol. 26, no. 4, pp. 2006–2014, 2011.

[71] S. Yang, S. Tan and J. X. Xu, "Consensus Based Approach for Economic Dispatch Problem in Smart Grid," *IEEE Transactions on Power Systems*, vol. 28, no. 4, pp. 4416–4426, 2013.

[72] S. Boyd, N. Parikh, E. Chu, B. Peleato and J. Eckstein, "Distributed Optimization and Statistical Learning via the Alternating Direction Method of Multipliers," *Foundations and Trends in Machine Learning*, vol. 3, no. 1, pp. 1–122, 2010.

[73] T. Erseghe, "Distributed Optimal Power Flow Using ADMM," *IEEE Transactions on Power Systems*, vol. 29, no. 5, pp. 2370–2380, 2014.

[74] J. Lavaei and S. H. Low, "Zero Duality Gap in Optimal Power Flow Problem," *IEEE Transactions on Power Systems*, vol. 27, no. 1, pp. 92–107, 2012.

[75] M. G. Vaya and G. Andersson, "Optimal Bidding Strategy of a Plug-in Electric Vehicle Aggregator in Day-Ahead Electricity Markets Under Uncertainty," *IEEE Transactions on Power Systems.*

[76] S. Vandael, B. Claessens, M. Hommelberg, T. Holvoet and G. Deconinck, "A Scalable Three-Step Approach for Demand Side Management of Plug-in Hybrid Vehicles," *IEEE Transactions on Smart Grid*, vol. 4, no. 2, pp. 720–728, 2013.

[77] N. Rotering and M. Ilic, "Optimal Charge Control of Plug-in Hybrid Electric Vehicles in Deregulated Electricity Markets," *IEEE Transactions on Power Systems*, vol. 26, no. 3, pp. 1021–1029, 2011.

[78] N. J. Redondo and A. J. Conejo, "Short-Term Hydro-Thermal Coordination by Lagrangian Relaxation: Solution of the Dual Problem," *IEEE Transactions on Power Systems*, vol. 14, no. 1, pp. 89–95, 1999.

[79] W. Ongsakul and N. Petcharaks, "Unit Commitment by Enhanced Adaptive Lagrangian Relaxation," *IEEE Transactions on Power Systems*, vol. 19, no. 1, pp. 620–628, 2004.

[80] A. Merlin and P. Sandrin, "A New Method for Unit Commitment at Electricite De France," *IEEE Transactions on Power Apparatus & Systems*, vol. 102, no. 5, pp. 1218–1225, 1983.

[81] A. Kargarian and Y. Fu, "System of Systems Based Security-Constrained Unit Commitment Incorporating Active Distribution Grids," *IEEE Transactions on Power Systems*, vol. 29, no. 5, pp. 2489–2498, 2014.

[82] A. Kargarian, Y. Fu and Z. Li, "Distributed Security-Constrained Unit Commitment for Large-Scale Power Systems," *IEEE Transactions on Power Systems*, vol. 30, no. 4, pp. 1925–1936, 2015.

[83] R. J. Bessa, M. A. Matos, F. J. Soares and J. A. P. Lopes, "Optimized Bidding of a EV Aggregation Agent in the Electricity Market," *IEEE Transactions on Smart Grid*, vol. 3, no. 1, pp. 443–452, 2012.

[84] M. Pantos, "Exploitation of Electric-Drive Vehicles in Electricity Markets," *IEEE Transactions on Power Systems*, vol. 27, no. 2, pp. 682–694, 2012.

[85] S. I. Vagropoulos and A. G. Bakirtzis, "Optimal Bidding Strategy for Electric Vehicle Aggregators in Electricity Markets," *IEEE Transactions on Power Systems*, vol. 28, no. 4, pp. 4031–4041, 2014.

[86] J. M. Foster and M. C. Caramanis, "Optimal Power Market Participation of Plug-in Electric Vehicles Pooled by Distribution Feeder," *IEEE Transactions on Power Systems*, vol. 28, no. 3, pp. 2065–2076, 2013.

[87] A. J. Conejo, F. J. Nogales and J. M. Arroyo, "Price-Taker Bidding Strategy Under Price Uncertainty," *IEEE Transactions on Power Systems*, vol. 17, no. 4, pp. 1081–1088, 2002.

[88] C. J. Dent, J. W. Bialek and B. F. Hobbs, "Opportunity Cost Bidding by Wind Generators in Forward Markets: Analytical Results," *IEEE Transactions on Power Systems*, vol. 26, no. 3, pp. 1600–1608, Aug. 2011.

[89] R. Billinton and S. Jonnavithula, "A Test System for Teaching Overall Power System Reliability Assessment," *IEEE Transactions on Power Systems*, vol. 11, no. 4, pp. 1670–1676, 1996.

[90] E. Loukarakis, C. J. Dent and J. W. Bialek, "Decentralized Multi-Period Economic Dispatch for Real-Time Flexible Demand Management," *IEEE Transactions on Power Systems*, in press, 2015.

[91] D. Wu, D. C. Aliprantis and L. Ying, "Load Scheduling and Dispatch for Aggregators of Plug-in Electric Vehicles," *IEEE Transactions on Smart Grid*, vol. 3, no. 1, pp. 368–376, 2012.

[92] M. Alizadeh, A. Scaglione, A. Applebaum, G. Kesidis and K. Levitt, "Reduced-Order Load Models for Large Populations of Flexible Appliances," *IEEE Transactions on Power Systems*, in print.

[93] V. Alimisis and P. C. Taylor, "Zoning Evaluation for Improved Coordinated Automatic Voltage Control," *IEEE Transactions on Power Systems*, vol. 30, no. 5, pp. 2736–2746, 2015.

[94] R. Palma-Behnke, C. Benavides, F. Lanas, *et al.*, "A Microgrid Energy Management System Based on the Rolling Horizon Strategy," *IEEE Transactions on Smart Grid*, vol. 4, no. 2, pp. 996–1006, 2013.

[95] F. Capitanescu, I. Bilibin and E. R. Ramos, "A Comprehensive Centralized Approach for Voltage Constraints Management in Active Distribution Grid," *IEEE Transactions on Power Systems*, vol. 29, no. 2, pp. 933–942, 2014.

[96] F. Capitanescu and L. Wehenkel, "Sensitivity-Based Approaches for Handling Discrete Variables in Optimal Power Flow Computations," *IEEE Transactions on Power Systems*, vol. 25, no. 4, pp. 1780–1789, 2010.

[97] S. Deilami, A. S. Masoum, P. S. Moses and M. A. S. Masoum, "Real-Time Coordination of Plug-in Electric Vehicles Charging in Smart Grids to Minimize Power Losses and Improve Voltage Profile," *IEEE Transactions on Smart Grid*, vol. 2, no. 3, pp. 456–467, 2011.

[98] A. S. Masoum, S. Deilami, P. S. Moses, M. A. S. Masoum and A. Abu-Siada, "Smart Load Management of Plug-in Electric Vehicles in Distribution and Residential Networks with Charging Stations for Peak Shaving and Loss Minimization Considering Voltage Regulation," *IET Generation, Transmission & Distribution*, vol. 5, no. 8, pp. 877–888, 2011.

[99] R. N. Anderson, A. Boulanger, W. B. Powell and W. Scott, "Adaptive Stochastic Control for the Smart Grid," *IEEE Proceedings*, vol. 99, no. 6, pp. 1098–1115, 2011.

[100] Y. Liang, L. He, X. Cao and Z. J. Shen, "Stochastic Control for Smart Grid Users with Flexible Demand," *IEEE Transactions on Smart Grid*, vol. 4, no. 4, pp. 2296–2308, 2013.

[101] D. Papadaskalopoulos, D. Pudjianto and G. Strbac, "Decentralized Coordination of Microgrids with Flexible Demand and Energy Storage," *IEEE Transactions on Sustainable Energy*, vol. 5, no. 4, pp. 1406–1414, 2014.

[102] X. Xi and R. Sioshansi, "Using Price-Based Signals to Control Plug-in Electric Vehicle Fleet Charging," *IEEE Transactions on Smart Grid.*

[103] Z. Tan, P. Yang and A. Nehorai, "An Optimal and Distributed Demand Response Strategy with Electric Vehicles in the Smart Grid," *IEEE Transactions on Smart Grid*, vol. 5, no. 2, pp. 861–869, 2014.

[104] B. Zhang, A. Y. S. Lam, A. D. Dominguez-Garcia and D. Tse, "An Optimal and Distributed Method for Voltage Regulation in Power Distribution Systems," *IEEE Transactions on Power Systems.*

[105] P. Chavali, P. Yang and A. Nehorai, "A Distributed Algorithm of Appliance Scheduling for Home Energy Management System," *IEEE Transactions on Smart Grid*, vol. 5, no. 1, pp. 282–290, 2014.

[106] C. K. Wen, J. C. Chen, J. H. Teng and P. Ting, "Decentralized Plug-in Electric Vehicle Charging Selection Algorithm in Power Systems," *IEEE Transactions on Smart Grid*, vol. 3, no. 4, pp. 1779–1789, 2012.

[107] Y. Xu, "Optimal Distributed Charging Rate Control of Plug-in Electric Vehicles for Demand Management," *IEEE Transactions on Power Systems.*

[108] F. Katiraei and M. R. Iravani, "Power Management Strategies for a Microgrid with Multiple Distributed Generation Units," *IEEE Transactions on Power Systems*, vol. 21, no. 4, pp. 1821–1831, 2006.

[109] S. M. Ashabani and Y. A. I. Mohamed, "New Family of Microgrid Control and Management Strategies in Smart Distribution Grids – Analysis, Comparison and Testing," *IEEE Transactions on Power Systems*, vol. 29, no. 5, pp. 2257–2269, 2014.

*Chapter 7*

# Multiobjective optimization for smart grid system design

*Wei-Kang Hsieh*[1] *and Wei-Yu Chiu*[1]

This chapter proposes a framework for designing smart grid systems that considers multiple objectives. A grid is considered a combination of an electric grid and a network of transmission lines, substations, and transformers that delivers electricity from a power plant to consumers. A smart grid renders the grid efficient, provides a friendly environment for active grid participants, and improves the energy efficiency of the underlying power system. In a smart grid, it is desirable to optimize various objectives, such as minimizing the power consumption, maximizing the quality of service, and optimizing a stored energy level for emergency operations. To achieve these objectives, a multiobjective approach is investigated. First, objectives of the smart grid system are formulated as a multiobjective optimization problem (MOP). Multiobjective evolutionary algorithms are then employed to solve the MOP, yielding a set of approximate Pareto optimal solutions and an approximate Pareto front (APF). Based on the preference of a decision maker, the final solution is selected from among the obtained solutions according to the associated performance represented by the APF. A multiobjective approach to smart grid system designs is thus provided.

## 7.1 Introduction

With advances in science and technology, power grid technology has substantially progressed [1–3]. Smartness and robustness are essential requirements for developing modern power grids. A smart grid is a system that aims to protect the environment, improve energy efficiency, and reduce carbon emissions. Digital technology, which allows for two-way communication between the service providers and their customers, and sensing technologies along the transmission lines make the power grid smart. Smart grids enable many operations; for example, they can provide unprecedented opportunities to drive the energy industry toward a new era of reliability, availability,

[1]Department of Electrical Engineering and Innovation Center for Big Data and Digital Convergence, Yuan Ze University, Taoyuan 32003, Taiwan

and efficiency, thus contributing to the economy and environmental health. Led by the International Electrotechnical Commission [4], smart grids can be improved from various perspectives, such as reliability [5], security [6], safety, energy efficiency [7], environmental impact [8], and cost effectiveness [9]. A good smart grid design thus outperforms conventional grids in many aspects.

The design of a smart grid system is generally formulated as a single-objective optimization problem (SOP) [10,11]. The main goal of an SOP is to find an optimal solution that achieves the minimum or maximum value. However, with the increasing complexity in global smart grids, multiple objectives must often be realized [12,13]. The drawback of considering a problem as an SOP is that it cannot provide a set of alternative solutions that trade off different objectives. The need for practicability renders single-objective optimization under a set of constraints increasingly unfeasible, making smart grid system design involving a multiobjective optimization problem (MOP) imperative. A fair result that does not advantage one particular objective can be attained by solving an MOP, yielding a set of approximate Pareto optimal solutions and an approximate Pareto front (APF) [14]. A multiobjective evolutionary algorithm (MOEA) is used as a standard procedure for solving MOPs. During the solving process, dominated or infeasible points are gradually removed and nondominated points are retained. In this chapter, we introduce the main structures of MOEAs and apply them to solve associated MOPs. We investigate a specific MOEA, termed the multiobjective differential evolutionary (MODE) algorithm [15–17].

To illustrate the proposed methodology, we examine two examples of smart grid system designs. In the first example, we build a good relationship between the client and the service providers. The client desires an excellent level of quality of service (QoS). In general, QoS can be ensured by minimizing the probability of data loss [18]. Data loss results in the service provider being penalized, according to the service-level agreement (SLA); the SLA is a part of a service contract where a service is formally defined [19]. Using the SLA, a good relationship between the clients and service providers can be guaranteed. Service providers desire minimization of the electricity cost. Formulating the smart grid design problem as an MOP can enhance mutual trust and support between the client and service providers. The second example involves optimizing the microgrid performance, the profit of the power grid, and the stored energy levels for emergency operations. The development and advancement of smart grid systems are crucial for realizing energy efficiency and improving performance.

By using the proposed multiobjective (MO) approach for the smart grid system design, the designers can achieve Pareto optimality. In our design examples, we can improve the QoS and build a good relationship between the clients and service providers, and we provide a framework that increases the energy efficiency of the smart grid for emergency operations. In the following discussions, we first present our problem formulation and design examples; solution methods are then investigated; to validate the proposed methodology, we examine the numerical results for the MO design problems; and finally, the conclusions are provided.

## 7.2 Problem formulation

In this section, we illustrate the use of MO optimization for designing a smart grid system. Existing studies on grid system optimization have mostly considered a single-objective formulation. Practically, however, more than one objective, which may be mutually conflicting, must be achieved.

### *7.2.1 Model of MOP*

In mathematics and computer science, an optimization problem involves the search for best solutions that achieve the optimum value. A basic SOP can be formulated as

$$\begin{aligned} &\min\ f(\mathbf{x}) \\ &\text{s.t.}\ \ \mathbf{x} \in \mathrm{S} \end{aligned} \tag{7.1}$$

where $f(\mathbf{x})$ is the objective function to be minimized. Without loss of generality, minimization is considered. If maximization is desired, we can convert the function into a minimization problem by adding a minus sign to the objective function. In (7.1), $\mathbf{x}$ represents a vector of decision variables (i.e., quantities that a system designer can control, such as the number of active servers required for handling service requests from 7 a.m. to 10 a.m. in a data center). The feasible set S, generally described by equality and inequality constraints, is often expressed as

$$\mathrm{S} = \{\mathbf{x} \in R^M : h(\mathbf{x}) = 0, g(\mathbf{x}) \leq 0\} \tag{7.2}$$

Several methods can be employed to solve the SOPs, e.g., Newton's method, secant method, gradient descent, steepest descent method, conjugate gradient method, and generative method [20,21]. Newton's method is only guaranteed to converge if certain conditions are satisfied; if the starting point corresponds with a horizontal tangent line, then Newton's method fails. Newton's method can be inefficient because it requires the calculation of second-order derivatives. It can be too expensive to perform for some engineering design problems. We may consider the secant method as an alternative. The secant method is related to a root finding process and uses a series of secant lines to approximate an equation's root. The secant method can be considered a finite difference approximation of Newton's method. The gradient descent method uses the gradient of the given function, providing information about descent directions. The steepest descent method is basically a gradient descent method. It employs the step size that achieves the maximum amount of decrease of the objective function. The conjugate gradient method is a simple and effective modification of descent methods. It evaluates the search directions as the algorithm progresses. The generative method is a brute-force approach in which the design domain is divided into several equal intervals between its upper and lower bounds. The objective function is then evaluated at individual design levels. The generative method finds the global minimum in the design domain.

In contrast with SOPs, MOPs have multiple objectives that must be achieved simultaneously, thus yielding various challenges. The concept of MO optimization was first introduced by Vilfredo Pareto (1848–1923) and Francis Y. Edgeworth (1845–1926) [20]. This concept has since been rapidly developed. A typical MOP can be formulated as follows:

$$\begin{aligned} &\min \boldsymbol{f}(\mathbf{x}) \\ &\text{s.t. } \mathbf{x} \in \mathrm{S} \end{aligned} \tag{7.3}$$

with

$$\boldsymbol{f}(\mathbf{x}) = [f_1(\mathbf{x})\ f_2(\mathbf{x})\ \cdots\ f_N(\mathbf{x})]^T \tag{7.4}$$

where $\boldsymbol{f}$ is a vector-valued function consisting of objective functions $f_1, f_2, \ldots, f_N$. For MOPs, Pareto optimality, also called Pareto efficiency, is often adopted and is defined as follows [22].

**Definition 7.1** (Pareto Dominance). *A vector* $\mathbf{u} = [u_1\ u_2\ \cdots\ u_N]^T$ *dominates another vector* $\mathbf{v} = [v_1\ v_2\ \cdots\ v_N]^T$ *if the condition* $u_i \le v_i$ *holds true for all i and at least one inequality is strict. We denote* $\mathbf{u} \preceq \mathbf{v}$. *In the decision variable space, a point* $\mathbf{a}_1$ *dominates another point* $\mathbf{a}_2$ *with respect to* $\boldsymbol{f}$ *if* $\boldsymbol{f}(\mathbf{a}_1) \preceq \boldsymbol{f}(\mathbf{a}_2)$.

**Definition 7.2** (Pareto Optimal Solution). *In the decision variable space, if a point* $\mathbf{x}$ *is feasible and a feasible point that dominates it does not exist, then point* $\mathbf{x}$ *is a Pareto optimal solution.*

**Definition 7.3** (Pareto Optimal Set and Pareto Front). *The set of all Pareto optimal solutions is termed the Pareto optimal set. The image of the Pareto optimal set through the objective function* $\boldsymbol{f}$ *is termed the Pareto front.*

Figure 7.1 demonstrates Pareto optimality by connecting Pareto optimal solutions with the Pareto front [20]. Pareto optimal solutions $\mathbf{a}_1$–$\mathbf{a}_8$ in the solution space correspond to $\boldsymbol{f}(\mathbf{a}_1)$–$\boldsymbol{f}(\mathbf{a}_8)$ in the objective space. In the real world, most problems have more than one objective function, each with its unique individual optimal solution. Because the objective functions often mutually conflict, different optimal solutions

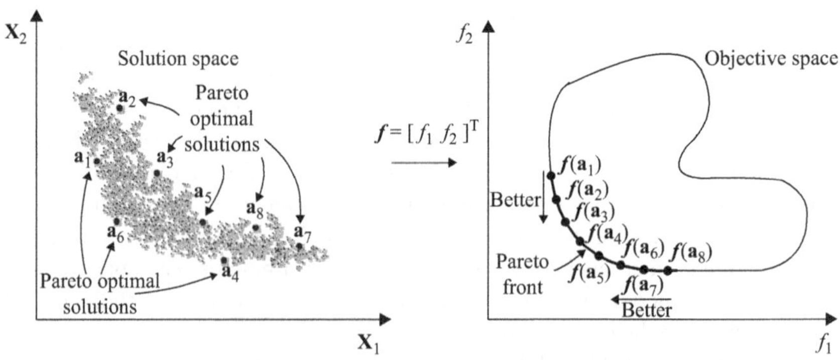


*Figure 7.1 Demonstration of Pareto optimality in the solution space and objective space*

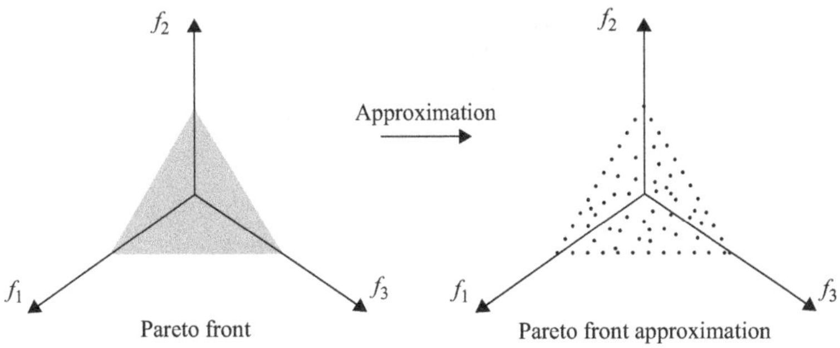


*Figure 7.2 Distribution example of Pareto front and Pareto front approximation modified from [23]*

correspond to the different objectives. A set of trade-off optimal solutions rather than one optimal solution is desired. In this case, no one solution can be considered better than the other with respect to all objective functions; we can only find a relatively better solution after trading off different objectives. Figure 7.2 shows an example of the Pareto front and Pareto front approximation [23]. MOEAs are typically employed to solve MOPs, and yield an APF. A true Pareto front has infinite points; therefore, we target finding a good approximation with finite points that is not too far from the true Pareto front. Pareto front approximation enables visualizing the effect of objectives on each other.

## *7.2.2 Design examples*

A smart grid system design problem usually involves more than one objective. In this section, we introduce two examples for designing a smart grid system through MO formulation.

### 7.2.2.1 Example 1: electricity cost vs. QoS

A data center is a facility used to house various systems such as computer systems, data storage systems, and telecommunications. The power consumption of data centers has increased rapidly over the last decade [24]. For service providers and clients, QoS is essential and often leads to a trade-off problem between power consumption and QoS. Higher QoS generally results in more power consumption and, hence, more electricity cost. In this example, we aim to balance power consumption and QoS. There are two objectives in our design: minimize the electricity cost of the data center and maximize the QoS for the clients. These two objectives are mutually conflicting. The following math models were obtained from References 25–28. Figure 7.3 illustrates the design process of this example, and Table 7.1 summarizes our symbols.

Let $ECost(t)$ denote the total electricity cost of the data center and is defined as [27]:

$$ECost(t) = m(t) * Po(t) * Pr(t) \tag{7.5}$$

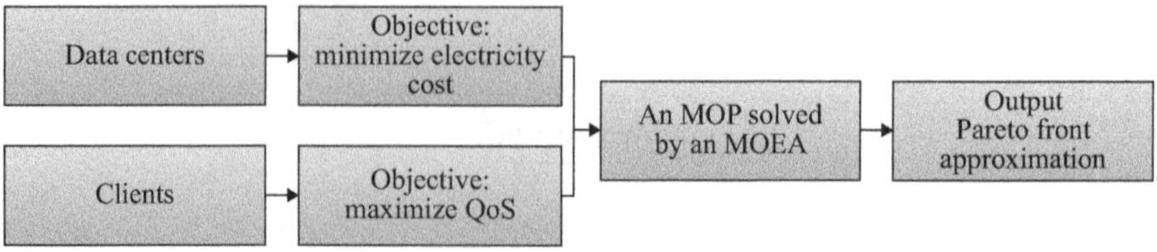


*Figure 7.3 Design process of Example 1*

*Table 7.1 Symbols used in Example 1*

| Symbol | Description |
|---|---|
| $m(t)$ | The number of active servers at time $t$ |
| $\mu(t)$ | The service rate |
| $\lambda(t)$ | The request rate |
| $\kappa$ | The service requests per second of a server |
| $Po(t)$ | The power consumption of each server |
| $Pr(t)$ | The electricity price at time $t$ |
| $Pl(\mu(t), \lambda(t))$ | The probability of data loss |

where $m(t)$ represents the number of active servers at time $t$, $Po(t)$ is the power consumption of each active server, and $Pr(t)$ is the electricity price at time $t$. The power consumption over time can be further expressed as [27]:

$$Po(t) = Af_r^{\delta(t)} + B \tag{7.6}$$

where $A$ and $B$ are some positive constants, $f_r$ represents the working frequency of a server, and $\delta(t) \in [2.5, 3]$ is a time-varying exponent. The first objective can be formulated as

$$\min_{m(t)} \; ECost(t) \tag{7.7}$$

We relate the QoS to the probability of data loss. The smaller the probability of data loss is, the better the QoS it provides. The QoS is thus modeled as

$$QoS(t) = -Pl(\mu(t), \lambda(t)) \tag{7.8}$$

where $Pl(\mu(t), \lambda(t))$ denotes the probability of data loss and is defined as [25,26]:

$$Pl(\mu(t), \lambda(t)) = \alpha(\mu(t), \lambda(t))e^{-\frac{1}{2}\min\limits_{n\geq 1} m_n(\mu(t),\lambda(t))} \tag{7.9}$$

with

$$\alpha(\mu(t), \lambda(t)) = \frac{1}{\lambda(t)\sqrt{2\pi}\sigma}e^{\frac{(\mu(t)-\lambda(t))^2}{2\sigma^2}} \int_{\mu(t)}^{\infty} (\gamma - \mu(t))e^{-\frac{(\gamma-\lambda(t))^2}{2\sigma^2}} d\gamma \tag{7.10}$$

and

$$m_n(\mu(t), \lambda(t)) = \frac{(D\mu(t) + n(\mu(t) - \lambda(t)))^2}{nC_{\lambda(t)}(0) + 2\sum_{l=1}^{n-1} C_{\lambda(t)}(l)(n-l)} \tag{7.11}$$

Here, $C_{\lambda(t)}(l)$ is the autocovariance function of the probability distribution of the service request rate (we denote $\sigma = C_{\lambda(t)}(0)$), $D$ is the deadline which depends on the SLA, and $\lambda(t)$ is the request rate [28].

The total service rate $\mu$ can be obtained using the following equation:

$$\mu(t) = \kappa * m(t) \tag{7.12}$$

where $\kappa$ denotes the service requests per second that can be handled by each server. The second objective can be formulated as follows:

$$\max_{m(t),\lambda(t)} \; QoS(t) \tag{7.13}$$

which can be further written as

$$\min_{m(t),\lambda(t)} \; Pl(\mu(t), \lambda(t)) \tag{7.14}$$

To simultaneously achieve the two objectives, we consider the following MOP:

$$\min_{m(t),\lambda(t)} \; \left[ECost(t) \;\; Pl(\mu(t), \lambda(t))\right]^T \tag{7.15}$$

#### 7.2.2.2 Example 2: overall utility vs. emergency operation

Dynamic pricing can help reshape or reduce power demands by varying the cost of power service overtime, and the power consumers who are sensitive to the energy price can reduce their power use [29]. In the second example, we consider an MO design for a microgrid system [30]. There are three objectives to be achieved: maximize the overall utility of the power grid, maximize the overall utility of microgrids, and maximize the stored energy level for emergency operation. The design process and symbols used in this example are presented in Figure 7.4 and Table 7.2, respectively.

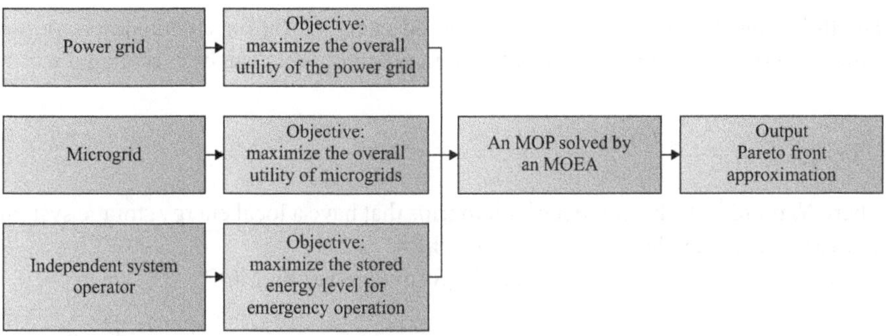


*Figure 7.4 Design process of Example 2*

*Table 7.2 Symbols used in Example 2*

| Symbol | Description |
|---|---|
| $U_d(\cdot)$ | The utility function of microgrids |
| $U_g(\cdot)$ | The utility function of the power grid |
| $p_{d_n}(t)$ | The power demand of microgrid $n$ |
| $p_g(t)$ | Total distributed power |
| $Pr(t)$ | The electricity price at time $t$ |

During grid operation, the power demand $p_{d_n}(t)$ of microgrid $n$ is modeled as

$$p_{d_n}(t) = f_{d_n}(Pr(t), b_n(t)) \tag{7.16}$$

where $b_n(t) > 0$ represents a nominal value of the base load, $Pr(t)$ represents the electricity price, and $f_{d_n}(\cdot)$ is a nonlinear function that relates the base load and electricity price to the power demand. The first objective can be formulated as

$$\max_{Pr(t)} \; U_d(p_{d_1}(t), \ldots, p_{d_N}(t), Pr(t)) \tag{7.17}$$

where $N$ represents the number of microgrids and $U_d(\cdot)$ is the overall utility function of microgrids. The second objective is to maximize the utility of the power grid. Let $p_{g_n}(t)$ be the power distributed between the $n$th microgrid and the power grid. The total distributed power $p_g(t)$ can be expressed as

$$p_g(t) = \sum_{n=1}^{N} p_{g_n}(t) \tag{7.18}$$

Let $U_g(p_g(t), Pr(t))$ denote the utility function of the power grid. The second objective can be formulated as

$$\max_{p_{g_n}(t), Pr(t)} \; U_g(p_g(t), Pr(t)) \tag{7.19}$$

The third objective is to maximize the stored energy level for emergency response, which is required by an independent system operator. This can be formulated as

$$\max_{p_{g_n}(t), Pr(t)} \; \sum_{n=1}^{N_s} s_n(t) \tag{7.20}$$

where $N_s$ represents the number of microgrids that have a local energy storage system, and $s_n(t)$ represents the stored energy level.

The microgrid system design can be formulated as an MOP:

$$\max_{p_{g_n}(t), Pr(t)} \left[ U_d(p_{d_1}(t), \ldots, p_{d_N}(t), Pr(t)) \;\; U_g(p_g(t), Pr(t)) \;\; \sum_{n=1}^{N_s} s_n(t) \right]^T \tag{7.21}$$

We can convert it into a minimization problem by adding a minus sign to the objective function:

$$\min_{p_{gn}(t),Pr(t)} \left[ -U_d(p_{d_1}(t),\ldots,p_{d_N}(t),Pr(t)) \quad - U_g(p_g(t),Pr(t)) \quad - \sum_{n=1}^{N_s} s_n(t) \right]^T \tag{7.22}$$

The reader can refer to Reference 30 for further discussions about MO designs for a multimicrogrid system.

## 7.3 Solution methods

In this section, we introduce the basic structures of an MOEA that can be applied to solve MOPs such as those in (7.15) and (7.22). Basically, an MOEA originates from a genetic algorithm (GA). GAs were developed to mimic some of the processes observed in natural evolution [31]. The general-purpose heuristic of the GA search algorithm mimics the natural selection process to find the optimal solutions. GAs differ from conventional deterministic search; they start with an initial set of random solutions, called the population. The best individuals are allowed to survive, mate, and produce offspring. Evolving solutions over time lead to better solutions.

An evolutionary algorithm (EA) is a variation of GAs and can often provide well-approximated solutions to all types of problems because it ideally does not make assumptions regarding the underlying fitness landscape. MOEAs perform additional operations to maintain multiple Pareto optimal solutions in the population. They often perform nondominated sorting, assign fitness to the population members, and preserve diversity among solutions of the same nondominated front. MOEAs have the following advantages: they can simultaneously process a set of possible solutions, enable finding several members of the Pareto optimal set in a single run of the algorithm, explore the entire search space, and be less susceptible to the shape or continuity of the Pareto front [20,32,33]. With the rapid development of effective techniques in MO optimization, many useful MOEAs have been developed and have been proposed for solving MOPs in many fields [34–39].

Differential evolution (DE) is a type of EAs originally used for solving optimization problems over a continuous domain, and an MODE algorithm is a useful MOEA for solving MOPs. It has been proven successful in searching for Pareto optimal solutions [15,17,40]. We briefly describe the MODE algorithm as follows. First, a population is generated randomly and the fitness functions are evaluated. Second, DE operations consisting of mutation, reflection, crossover, and selection are performed on the individuals of the population. The fitness functions of the trial vectors are evaluated. Then, an external archive of candidate solutions is updated by adding newly generated points, and removing dominated or infeasible points. Finally, the output of the Pareto front approximations and the selected design parameters can be attained. Figure 7.5 presents the algorithm flowchart. The $t_c$ and $t_{max}$ denote the algorithm counter and number of iterations, respectively. Population size $N_p$ is the number of

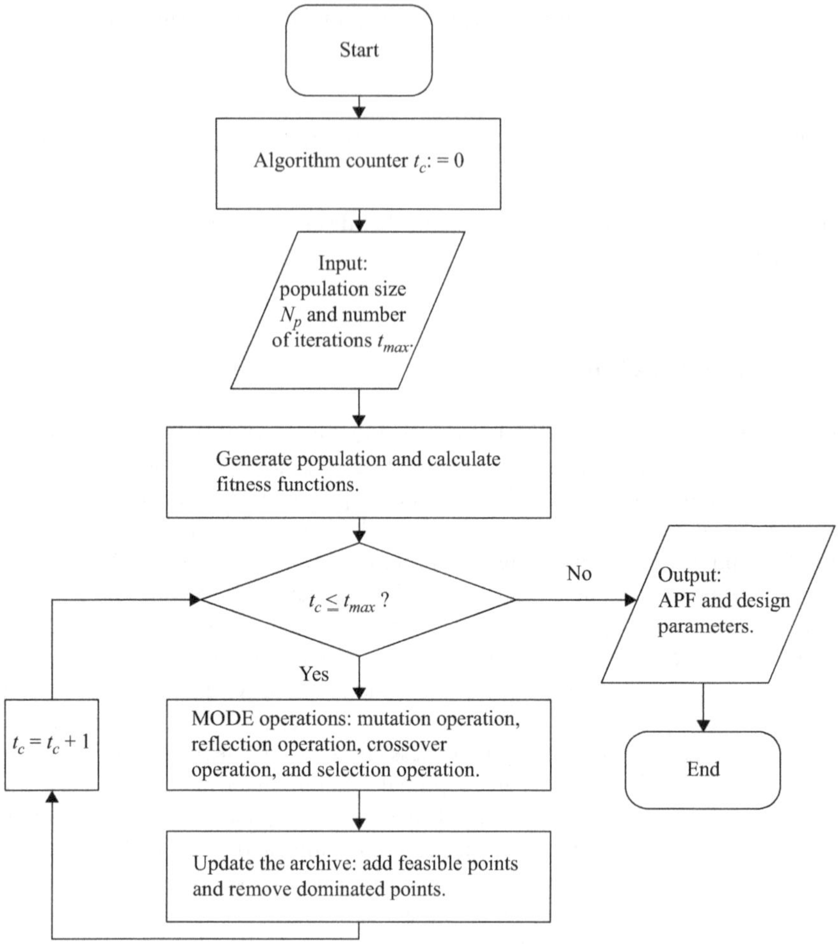


*Figure 7.5 Flowchart of the MODE algorithm modified from [41]*

points used for Pareto front approximations. In general, larger values of $t_{max}$ and $N_p$ yield better approximations to the true Pareto front.

## 7.4 Numerical results

In this section, numerical simulations were conducted to illustrate our proposed MO methodology. The MODE algorithm described in Figure 7.5 was applied to solve the design problem in (7.15) of Example 1. We have set the following parameters: $t_{max} = 500$ and $N_p = 50$ for the MODE algorithm; $A = 2.37$, $B = 70$, $\delta(t) = 3$, the

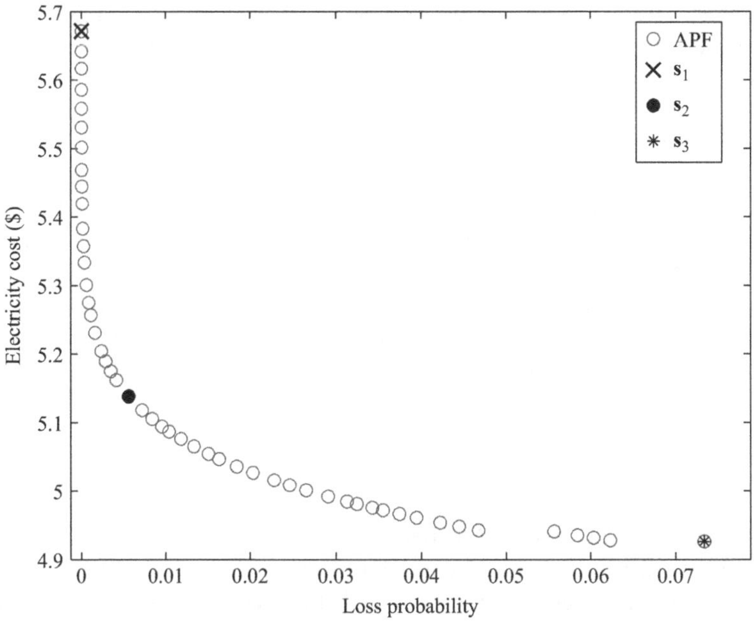


*Figure 7.6 Pareto front approximation of Example 1*

*Table 7.3 Performance evaluation of Example 1*

| | **λ(t)** | ***m(t)*** | **Electricity cost ($)** | **Loss probability (%)** |
|---|---|---|---|---|
| $\mathbf{s}_1$ | 250 | 3 104 | 5.6715 | 0.00072 |
| $\mathbf{s}_2$ | 250 | 2 813 | 5.1382 | 0.56 |
| $\mathbf{s}_3$ | 250 | 2 697 | 4.9263 | 7.33 |

CPU frequency $f_r = 3$ GHz, and thus $Po(t) = 91.33$ W in (7.6); $Pr(t) = 0.02$ \$/kWh in (7.5), $D = 300$ ms in (7.11), and $\kappa = 0.1$ in (7.12) [25]. In general, $\lambda(t)$ can serve as a decision variable, leading to a joint optimization problem considered by both the service provider and clients. For an illustrative purpose, suppose that the current request rate is $\lambda(t) = 250$ per second and $\sigma = C_{\lambda(t)}(0) = 258$ in (7.10). The service provider (or data center) aims to adjust the number of active servers so that the electricity cost of the data center can be minimized and the QoS for clients can be maximized.

Figure 7.6 shows the Pareto front approximation. Three nondominated vectors $\mathbf{s}_1, \mathbf{s}_2$, and $\mathbf{s}_3$ were chosen for further comparison. Table 7.3 summarizes the resulting performance. At $\mathbf{s}_1$, we can achieve the best QoS performance but have the largest electricity cost. At $\mathbf{s}_3$, we can have the least electricity cost but the worst QoS performance. At $\mathbf{s}_2$, we can achieve a fair outcome that does not advantage one particular objective, and both objectives can be attained to a certain extent at the same time. By having the APF, we are able to select a reasonable trade-off between the two objectives, illustrating the effectiveness of the proposed methodology.

In practice, the QoS depends on the SLA. A service provider receives the payment only if the service request is handled before the deadline, which is in accordance with the SLA. If the service provider cannot handle the service request before the deadline, it must be penalized by their clients. Furthermore, the QoS may strongly influence the mutual trust between clients and a service provider. It is desirable to use more power to provide a better QoS to the clients. This design example using the MO approach can provide an operator of the data center with a schedule for activating servers.

## 7.5 Conclusion

This chapter proposed a general MO framework for optimizing the smart grid system. The smart grid design is considered an MOP, following which the MOEAs are used to solve it. In one design example, we enabled higher QoS from service providers while reducing power consumption of a data center. In the other design example, we considered joint optimization among the power grid, microgrids, and independent system operator. These two examples validated our framework. For smart grid system designs, the proposed MO approach is expected to result in win–win situations, and be able to address other scenarios in which multiple objectives are involved.

## Acknowledgments

This work was supported in part by the Ministry of Science and Technology of Taiwan under Grant 102-2218-E-155-004-MY3, and in part by the Innovation Center for Big Data and Digital Convergence, Yuan Ze University, Taiwan.

## Bibliography

[1] M. Hashmi, S. Hanninen, and K. Maki, "Survey of smart grid concepts, architectures, and technological demonstrations worldwide," in *Proceedings of the IEEE PES Conference on Innovative Smart Grid Technologies*, Medellin, Colombia, Oct. 2011, pp. 1–7.

[2] J. Zhang, B.-C. Seet, T.-T. Lie, and C. F. Foh, "Opportunities for software-defined networking in smart grid," in *Proceedings of the International Conference on Information, Communications and Signal Processing*, Tainan, Taiwan, Dec. 2013, pp. 1–5.

[3] X. Miao, X. Chen, X.-M. Ma, G. Liu, H. Feng, and X. Song, "Comparing smart grid technology standards roadmap of the IEC, NIST and SGCC," in *Proceedings of the International Conference on Electricity Distribution*, Shanghai, China, Sep. 2012, pp. 1–4.

[4] W. R. Simpson and A. Greenspan, "The international electrotechnical commission and system level test," in *Proceedings of the IEEE International Conference on AUTOTESTCON*, Alexandria, VA, Sep. 1997, pp. 524–527.

[5] B. Falahati, Y. Fu, and L. Wu, "Reliability assessment of smart grid considering direct cyber-power interdependencies," *IEEE Transactions on Smart Grid*, vol. 3, no. 3, pp. 1515–1524, Sep. 2012.

[6] Y. Wang, D. Ruan, D. Gu, *et al.*, "Analysis of smart grid security standards," in *Proceedings of the IEEE International Conference on Computer Science and Automation Engineering*, Shanghai, China, Jun. 2011, pp. 697–701.

[7] D. Tsiamitros, D. Stimoniaris, N. Poulakis, *et al.*, "Advanced energy storage and demand-side management in smart grids using buildings energy efficiency technologies," in *Proceedings of the IEEE PES Innovative Smart Grid Technologies Conference Europe*, Istanbul, Turkey, Oct. 2014, pp. 1–6.

[8] E. A. Martinez-Cesena and J. Mutale, "Impact of wind speed uncertainty and variability on the planning and design of wind power projects in a smart grid environment," in *Proceedings of the IEEE PES International Conference on Innovative Smart Grid Technologies*, Manchester, UK, Dec. 2011, pp. 1–8.

[9] G. Ghatikar, J. Zuber, E. Koch, and R. Bienert, "Smart grid and customer transactions: the unrealized benefits of conformance," in *Proceedings of the IEEE Green Energy and Systems Conference*, Long Beach, CA, Nov. 2014, pp. 7–14.

[10] A. Andreotti, G. Carpinelli, F. Mottola, and D. Proto, "A review of single-objective optimization models for plug-in vehicles operation in smart grids – part I: theoretical aspects," in *Proceedings of the IEEE Power and Energy Society General Meeting*, San Diego, CA, Jul. 2012, pp. 1–8.

[11] A. Andreotti, G. Carpinelli, F. Mottola, and D. Proto, "A review of single-objective optimization models for plug-in vehicles operation in smart grids – part II: numerical applications to vehicles fleets," in *Proceedings of the IEEE Power and Energy Society General Meeting*, San Diego, CA, Jul. 2012, pp. 1–8.

[12] M. Caramia and P. Dell'Olmo, *Multi-objective Management in Freight Logistics*. London: Springer-Verlag, 2008.

[13] M. Ulbricht, "Single-objective vs. multi-objective scheduling algorithms for scheduling jobs in grid environment," in *Proceedings of the IEEE 10th International Symposium on Applied Machine Intelligence and Informatics*, Herl'any, Slovakia, Jan. 2012, pp. 411–414.

[14] S.-Y. Yuen and X. Zhang, "Multiobjective evolutionary algorithm portfolio: choosing suitable algorithm for multiobjective optimization problem," in *Proceedings of the IEEE Congress on Evolutionary Computation*, Beijing, China, Jul. 2014, pp. 1967–1973.

[15] B. V. Babu and A. M. Gujarathi, "Multi-objective differential evolution (MODE) for optimization of supply chain planning and management," in *Proceedings of the IEEE Congress on Evolutionary Computation*, Singapore, Sep. 2007, pp. 2732–2739.

[16] M. Zhao, R. Liu, W. Li, and H. Liu, "Multi-objective optimization based differential evolution constrained optimization algorithm," in *Proceedings of the Second WRI Global Congress on Intelligent Systems*, Wuhan, China, Dec. 2010, pp. 320–326.

[17] W.-Y. Chiu, "Multiobjective controller design by solving a multiobjective matrix inequality problem," *IET Control Theory & Application*, vol. 8, no. 16, pp. 1656–1665, Nov. 2014.

[18] N. Zaker, B. Kantarci, M. Erol-Kantarci, and H. T. Mouftah, "Quality-of-service-aware fiber wireless sensor network gateway design for the smart grid," in *Proceedings of the IEEE International Conference on Communication Workshops*, Budapest, Hungary, Jun. 2013, pp. 863–867.

[19] Y. Diao, L. Lam, L. Shwartz, and D. Northcutt, "Modeling the impact of service level agreements during service engagement," *IEEE Transactions on Network and Service Management*, vol. 11, no. 4, pp. 431–440, Dec. 2014.

[20] K.-H. Chang, *Design Theory and Methods Using CAD/CAE: The Computer Aided Engineering Design Series*. San Diego, CA: Elsevier Science, 2014.

[21] E. K. P. Chong and S. H. Zak, *An Introduction to Optimization*. Hoboken, NJ: Wiley-Interscience, 2013.

[22] A. Abraham and R. Goldberg, *Evolutionary Multiobjective Optimization: Theoretical Advances and Applications*. London: Springer-Verlag, 2006.

[23] M. Li, S. Yang, and X. Liu, "Diversity comparison of Pareto front approximations in many-objective optimization," *IEEE Transactions on Cybernetics*, vol. 44, no. 12, pp. 2568–2584, Dec. 2014.

[24] T. Wilde, A. Auweter, M. K. Patterson, *et al.*, "DWPE, a new data center energy-efficiency metric bridging the gap between infrastructure and workload," in *Proceedings of the International Conference on High Performance Computing and Simulation*, Bologna, Italy, Jul. 2014, pp. 893–901.

[25] M. Ghamkhari and H. Mohsenian-Rad, "Profit maximization and power management of green data centers supporting multiple SLAs," in *Proceedings of the International Conference on Computing, Networking and Communications*, San Diego, CA, Jan. 2013, pp. 465–469.

[26] M. Ghamkhari and H. Mohsenian-Rad, "Energy and performance management of green data centers: a profit maximization approach," *IEEE Transactions on Smart Grid*, vol. 4, no. 2, pp. 1017–1025, Jun. 2013.

[27] H. Shao, L. Rao, Z. Wang, X. Liu, Z. Wang, and K. Ren, "Optimal load balancing and energy cost management for internet data centers in deregulated electricity markets," *IEEE Transactions on Parallel and Distributed Systems*, vol. 25, no. 10, pp. 2659–2669, Oct. 2014.

[28] H. S. Kim and N. B. Shroff, "Loss probability calculations and asymptotic analysis for finite buffer multiplexers," *IEEE/ACM Transactions on Networking*, vol. 9, no. 6, pp. 755–768, Dec. 2001.

[29] W.-Y. Chiu, H. Sun, and H. V. Poor, "Demand-side energy storage system management in smart grid," in *Proceedings of the IEEE International Conference on Smart Grid Communications*, Tainan, Taiwan, Nov. 2012, pp. 73–78.

[30] W.-Y. Chiu, H. Sun, and H. V. Poor, "A multiobjective approach to multi-microgrid system design," *IEEE Transactions on Smart Grid*, vol. 6, no. 5, pp. 2263–2272, Sep. 2015.

[31] P. Guo, X. Wang, and Y. Han, "The enhanced genetic algorithms for the optimization design," in *Proceedings of the Third International Conference on Biomedical Engineering and Informatics*, Yantai, China, Oct. 2010, pp. 2990–2994.

[32] C. Bento, A. Cardoso, and G. Dias, *Progress in Artificial Intelligences*. Berlin: Springer-Verlag, 2005.

[33] K. Lucas and P. Roosen, *Emergence, Analysis and Evolution of Structures*. Berlin: Springer-Verlag, 2010.

[34] X. Ma, F. Liu, Y. Qi, X. Wang, L. Li, L. Jiao, M. Yin, and M. Gong, "A multiobjective evolutionary algorithm based on decision variable analyses for multiobjective optimization problems with large-scale variables," *IEEE Transactions on Evolutionary Computation*, vol. 20, no. 2, pp. 275–298, Apr. 2016.

[35] S. Greco, B. Matarazzo, and R. Słowiński, "Interactive evolutionary multiobjective optimization using dominance-based rough set approach," in *Proceedings of the IEEE Congress on Evolutionary Computation*, Barcelona, Spanish, Jul. 2010, pp. 1–8.

[36] P. A. N. Bosman, "On gradients and hybrid evolutionary algorithms for real-valued multiobjective optimization," *IEEE Transactions on Evolutionary Computation*, vol. 16, no. 1, pp. 51–69, Feb. 2011.

[37] H. Haleh, H. Iranmanesh, and H. Kor, "A new hybrid evolutionary algorithm for solving multi objective cell formation problem," in *Proceedings of the International Conference on Computers and Industrial Engineering*, Troyes, France, Jul. 2009, pp. 612–616.

[38] K. Deb, L. Zhu, and S. Kulkarni, "Multi-scenario, multi-objective optimization using evolutionary algorithms: initial results," in *Proceedings of the Congress on Evolutionary Computation*, Sendai, Japan, May 2015, pp. 1877–1884.

[39] R. Datta and A. Majumder, "Optimization of turning process parameters using multi-objective evolutionary algorithm," in *Proceedings of the IEEE Congress on Evolutionary Computation*, Barcelona, Spain, Jul. 2010, pp. 1–6.

[40] W.-Y. Chiu, "A multiobjective approach to resource management in smart grid," in *Proceedings of the International Conference on Control, Automation and Information Sciences*, Gwangju, South Korea, Dec. 2014, pp. 182–187.

[41] W.-Y. Chiu, "Pareto optimal controller designs in differential games," in *Proc. CACS International Automatic Control Conference*, Kaohsiung, Taiwan, Nov. 2014, pp. 179–184.

# Chapter 8

# Frequency regulation of smart grid via dynamic demand control and battery energy storage system

*Qi Zhu*[3], *Chuan-Ke Zhang*[1,3], *Wei Yao*[2,3] *and Lin Jiang*[3]

Balancing the active power between the generation side and the demand side to maintain the frequency is one of the main challenging problems of integrating the increased intermittent wind power to the smart grid. Although the energy storage system, such as battery energy storage system (BESS), has potential to solve this problem, the installation of the BESS with large capacity is limited by its high cost. This chapter investigates the frequency regulation of the smart grid working in the isolated mode with wind farms by introducing not only the BESS but also dynamic demand control (DDC) via controllable loads and the plug-in electric vehicles (PEVs) with vehicle-to-grid (V2G) service. First, modelling of a single-area load frequency control (LFC) system is obtained, which includes the wind farms equipped with variable-speed wind turbines, the simplified BESS, the air conditioner based DDC and the distributed PEVs. The LFC system contains traditional primary and supplementary control loops and three additional control loops of the BESS, the PEVs and the DDC, respectively. Then, state-space models of the closed-loop LFC scheme with/without communication delays in the control loops are derived, and the stability of the closed-loop system with time delays is investigated via the Lyapunov functional based method. Third, gains of proportional integral derivative (PID)-type controllers are tuned based on the $H_\infty$ performance analysis and the particle swarm optimization searching algorithm. Case studies are carried out for the single-area smart power grid through the MATLAB®/Simulink platform. Both the theoretical analysis and the simulation studies demonstrate the contribution of the DDC, the BESS, and the PEVs to frequency regulation, and the robustness of the designed PID-type LFC against the disturbances caused by the load changes and the intermittent wind power and the delays arising in the control loops via theoretical analysis and the simulation studies.

[1]School of Automation, China University of Geosciences, Wuhan 430074, China
[2]State Key Laboratory of Advanced Electromagnetic Engineering and Technology, Huazhong University of Science and Technology, Wuhan 430074, China
[3]Department of Electrical Engineering and Electronics, University of Liverpool, Liverpool L69 3GJ, UK

## 8.1 Introduction

The momentary imbalance between the power generation side and the demand side will lead to frequency deviation of a power system [1]. In order to ensure the frequency stability, frequency regulation through matching the supply and the demand is an important topic in the operating electricity markets [2]. In traditional regulated environment, the elimination of frequency deviation is achieved by adjusting the power output of generator units to track the demand changes [3]. In the current power grid, with the increasing development of the wind power generation system, wind power has become one of the main power supplies [4]. However, the wind power generation depends on the weather condition and the wind speed is time varying, stochastic and intermittent; and thus not easily controllable and dispatchable as the conventional generators and easily results in the imbalance of supply and demand. Moreover, for a power grid working in the isolated mode and with high penetration of wind power, the intermittent wind power injection would become an important factor of causing frequency deviation. Therefore, it is a challenging task to achieve the frequency regulation of the power grid with a high penetration level of wind power generations.

To integrate intermittent wind power into a power grid, spinning generation reserve with enough capacity are required to cover the period of no wind power outputs and thus will increase the operational cost [5]. Design of an advanced load frequency control (LFC) scheme is an alternative way to integrate more intermittent wind power, such as multi-stage fuzzy logic control based LFC scheme designed to guarantee the robustness against the disturbance caused by the intermittent wind power [6,7]. The grid-scale energy storage system, such as battery energy storage system (BESS), is an effective alternative of the backup generation capacity by charging or discharging based on frequency deviation [8,9], such as the BESS equipped with a controller based on area control error (ACE) [10]. However, the grid-scale of BESS requires some expensive auxiliary equipments and thus still not a feasible solution due to its high cost.

As the plug-in electric vehicles (PEVs) have drawn increasing attention in the transportation electrification in recent years, and considering most time PEVs are parked at home or work places, the battery of PEVs can be used as distributed energy storage to provide support of the grid operation, called vehicle-to-grid (V2G) service [11]. One function of the V2G services is to aggregate a large number of small batteries of individual PEV as an equivalent grid-scale BESS and then provide frequency regulation of smart grid via controlling the charging/discharging of these batteries [12–17]. Moreover, the PEVs have been adopted to suppress the frequency fluctuation in the power system with high penetration of renewable energy sources [18].

Demand-side response, such as dynamic demand control (DDC), can provide controllability from the load side to frequency regulation, rather than the conventional frequency regulation from the generator side, and is an effective way to reduce the spinning generation capacity [1]. The DDC method can self-adjust the usage of electricity based on the frequency deviation of the power system [19]. In Reference 20, the combination of the DDC and the BESS was proposed for frequency control, which has several advantages over other resources used for energy balancing and ancillary services, including relatively fast response time and high ramp rate, as well as low cost and high efficiency [21].

In order to implement the frequency regulation, inter-area power exchange and frequency derivation should be measured and transmitted to the control centre, and then the corresponding control signal calculated at the control centre should be transmitted to generation units taking part in the frequency regulation. In the traditional LFC scheme, the dedicated communication channel is usually used to transmit those signals, and the time delays arising are generally small and can be ignored due to the slow dynamic of frequency regulation [22]. However, in the modern smart power grid under deregulated environment, open communication networks are tended to be applied for transmitting those related signals, because of the feasibility, low cost and the bilateral contracts. With the introduction of the open communication channels, both constant delay and time-varying delay will be arisen in LFC problem [23]. Considering the characteristic of the communication channel, LFC scheme is a typical time-delay system. The stability analysis and controller design problems considering the communication delays have been investigated for traditional and/or deregulated power systems [22,24–27].

The usage of the DDC and the PEVs to support the frequency regulation usually requires to aggregate a large number of smaller units by a third-party aggregator via the communication networks. Open communication networks are preferred for communication between the aggregator and the large number of individual controllable loads or the PEVs, which will introduce time delays into the frequency control loops. In Reference 19, the time delays in the DDC loops are treated by using the different orders of Padé approximation. The field demonstration report shows that the delay between the aggregator and a PEV is less than 2 s via wireless communication [28]. Communication delays in the control loop have important impacts on the frequency regulation performance of the V2G service [14] and a large time delay may even destabilize the power grid with PEVs [15].

This chapter carries out the modelling, stability analysis and control design of frequency regulation of a smart power grid including wind farms, BESS, DDC and PEVs by considering the impact caused by time delays from communication networks in the control loops. First, the state-space model of the closed-loop LFC scheme with/without communication delays for the smart power grid is developed based on the simplified models of wind farms equipped with variable-speed wind turbines, the simple BESS, air conditioner based DDC, and the distributed PEV. Second, the Lyapunov stability theory based stability analysis method is given for the closed-loop LFC system with different embedded controllers. Third, a PID-type load frequency controller is tuned based on the $H_\infty$ performance analysis and the particle swarm optimization (PSO) searching algorithm. Case studies based on single-area smart power grid are carried out to investigate the contribution of the BESS, the PEVs, and the DDC to the frequency regulation and to verify the robustness of the designed PID controller against the power imbalance disturbances, the time delays, and the parameter uncertainties.

The rest of this chapter is organized as follows. The dynamic model of a single-area LFC scheme is presented in Section 8.2. Section 8.3 introduces the stability analysis methods for LFC system with time delays. Section 8.4 develops a tuning method to design a robust PID controller. In Section 8.5, case studies based on single-area smart power grid are presented. Conclusions are drawn in Section 8.6.

## 8.2 Dynamic model of smart grid for frequency regulation

This section describes the dynamic model of the smart grid including the wind farms, the BESS, the DDC, and the PEVs stations. The structure of the frequency regulation and the simplified models of the wind farm, the BESS, the DDC, and the PEVs are given at first. Then, the state-space models of the closed-loop LFC scheme equipped with a PID controller are constructed with and without considering the communication delays, respectively.

### *8.2.1 Structure of frequency regulation*

The smart power grid used in this chapter is shown in Figure 8.1, which can be operated in two alternative modes, i.e., grid-connect and island model mode [29]. When the smart grid is in grid-connected mode, the majority of the loads can be supplied by the connected main grid and the system could be controlled by distribution management system (DMS). When the smart power grid is in island operating mode, the system power flow is balanced by local generation, and the system is controlled by the local smart-grid dispatch system (SGDS). The signals of the system state are measured by distributed sensors and transmitted to the SGDS through communication channel. These signals are processed by the SGDS to generate control signals and sent back to each unit [29].

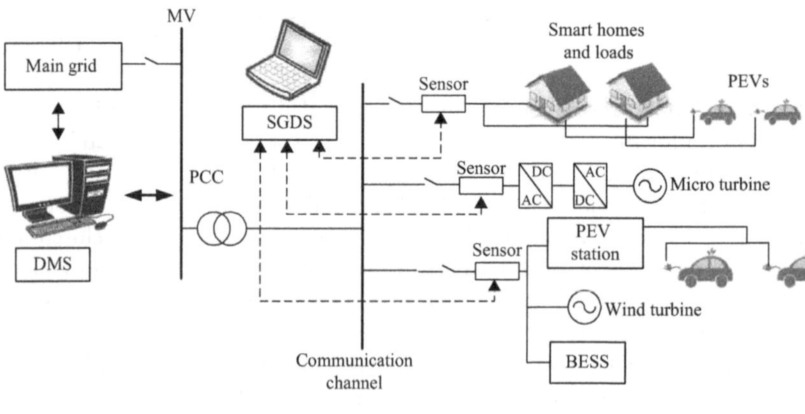


*Figure 8.1 Smart grid with wind turbines, the PEVs, the smart homes, and the BESS*

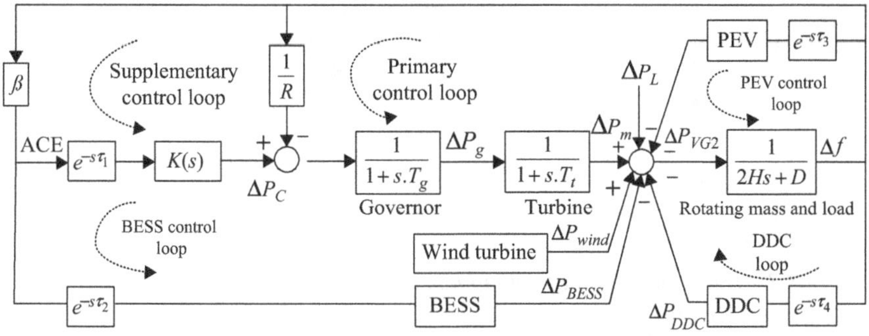


*Figure 8.2 The structure of the frequency regulation*

This chapter discusses the case where the smart power grid is in island operating mode, and the capacity of the micro traditional generator is not adequate to provide the local demands. The wind turbines can provide active power to local loads, while the wind power injection depends on the wind speed and is usually intermittent, which may often lead to the imbalance and result in the frequency deviation. The PEVs, the BESSs, and the controllable loads in the smart homes are used to compensate such unbalance. Then the structure of the frequency regulation for smart grid is given in Figure 8.2, in which the micro turbine is assumed to be a non-reheat turbine with the time constant, $T_t$.

From Figure 8.2, the relationship between the energy imbalance and the corresponding frequency deviation can be given as follows:

$$\begin{aligned}&\Delta P_m(s) + \Delta P_{wind}(s) - \Delta P_{BESS}(s) - \Delta P_L(s) - \Delta P_{DDC}(s) - \Delta P_{V2G}(s)\\ &\quad = 2Hs\Delta f(s) + D\Delta f(s)\end{aligned} \tag{8.1}$$

where $\Delta P_m$ is the generator mechanical output, $\Delta P_{wind}$ is the deviation of the wind power energy from the wind farm, $\Delta P_{BESS}$ is the power deviation from the BESS, $\Delta P_{DDC}$ is the power deviation from the DDC, $\Delta P_{V2G}$ is the power deviation from the PEVs, $\Delta P_L$ is the load change, $2H$ is the equivalent inertia constant, $D$ is the equivalent load-damping coefficient, and $\Delta f$ is the frequency deviation of smart grid.

As shown in Figure 8.2, there are five control loops taking part in the frequency regulation, including two generator-side loops (traditional primary and supplementary loops) and three demand-side loops (BESS, PEVs, and DDC loops). In primary control loop, the speed governor (with time constant, $T_g$) with the droop unit (speed droop, $R$) senses the frequency deviation and quickly decreases the deviation at steady state. The supplementary control loop (usually called LFC) is required to eliminate the steady-state error of the frequency deviation [19]. The load frequency controller, $K(s)$, commonly used in practice is the PID controller.

Three additional control loops are introduced for compensating the energy unbalance caused by the intermittent wind power injection. In the BESS loop, the BESS is controlled for charging or discharging based on the *ACE* signal, and the DDC and PEV loops are responding directly to the frequency deviation.

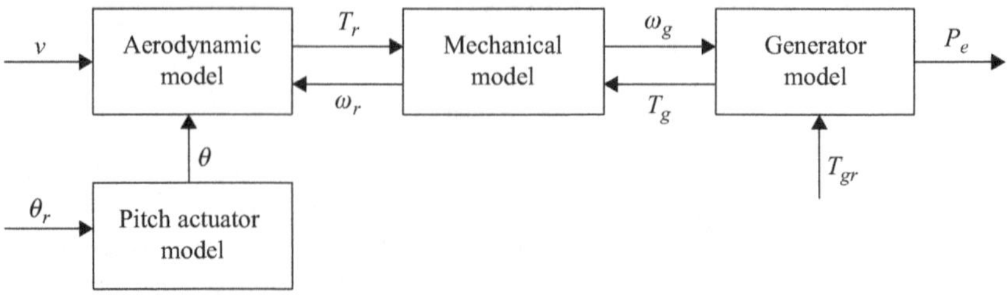


*Figure 8.3 Interconnection of submodels describing the characteristics of the wind turbine*

The measurements and control commands are transmitted through open communication channels, and the time delays may be introduced during the procedure of the transition. In Figure 8.2, the $\tau_1$, $\tau_2$, $\tau_3$, and $\tau_4$ show the communication time delays in LFC, the BESS, the PEV, and the DDC loops, respectively.

## *8.2.2 Wind farm with variable-speed wind turbines*

The wind power energy is developing very fast in recent years, and its intermittent characteristic may cause the energy imbalance of the smart grid. The part presents the dynamic model of a pitch-controlled variable-speed wind turbine [30,31]. The variable-speed turbine model is a complex non-linear system and consists of aerodynamics, turbine mechanics, generator dynamics, and pitch actuator dynamics parts. Figure 8.3 shows the relationships among them.

The aerodynamic blades on the rotor convert the kinetic energy of the wind into mechanical energy, effectively providing the torque, $T_r$, on the rotor:

$$T_r = \frac{P_r}{\omega_r} \tag{8.2}$$

where $\omega_r$ is the rotor speed, and the mechanical power absorbed from the wind $P_r$ is given as follows:

$$P_r = \frac{1}{2}\rho\pi R_{wind}^2 v^3 C_p \tag{8.3}$$

where $\rho$ is the air density, $R_{wind}$ is the wind radius, $v$ is the effective wind speed, and $C_p$ is the power coefficient which is a function of the blade pitch angle, $\theta$, and the tip speed ratio, $\lambda$, and the one used in this chapter is given as follows:

$$C_p(\lambda, \theta) = 0.22\left(\frac{116}{\lambda_t} - 0.6\theta - 5\right)e^{-\frac{12.5}{\lambda_t}} \tag{8.4}$$

where $\lambda_t$ is defined as

$$\frac{1}{\lambda_t} = \frac{1}{\lambda^{-1} + 0.129} - \frac{0.035}{(1.5\theta)^3 + 1} \tag{8.5}$$

with $\lambda = v/(R_{wind}\omega_r)$.

The blade pitch angle, $\theta$, is changed by a hydraulic/mechanical actuator. A simplified model of the blade dynamics is presented by the following first-order linear model:

$$\dot{\theta} = -\frac{1}{\tau_\theta}\theta + \frac{1}{\tau_\theta}\theta_r \tag{8.6}$$

where $\theta_r$ is the reference and the one control input for the wind turbine.

The mechanical model includes two parts, i.e., the rotor side and generator side. The dynamics on the two sides are described as follows:

$$\dot{\omega}_r J_r = T_r - T_{sr} \tag{8.7}$$

$$\dot{\omega}_g J_g = T_{sg} - T_g \tag{8.8}$$

where $J_r$ and $J_g$ are the inertia on the rotor side and generator side, respectively; $\omega_r$ and $\omega_g$ are the rotational speeds on the rotor side and generator side, respectively; $T_g$ and $T_r$ represent for rotor torque and generator torque, respectively; $T_{sg}$ and $T_{sr}$ are the torques on each side of the transmission, which are related by the gear ratio, $N_g$:

$$T_{sr} = T_{sg} N_g \tag{8.9}$$

and the torque at the rotor side of the transmission can be described by the twist of the flexible shaft:

$$T_{sr} = D_s\dot{\delta} + K_s\delta \tag{8.10}$$

where $D_s$ is the damping and $K_s$ is the spring constant, which can illustrate the dynamic nature of the shaft; and the twist of the flexible shaft, $\delta$, is determined by

$$\dot{\delta} = \omega_r - \frac{\omega_g}{N_g} \tag{8.11}$$

The generator power, $P_e$, is given by:

$$P_e = T_g \omega_g \tag{8.12}$$

where the generator torque, $T_g$, is controlled, however, it cannot be changed instantaneously. The dynamic response of the generator has therefore been modelled by a first-order linear model with time constant, $\tau_t$:

$$\dot{T}_g = -\frac{1}{\tau_t}T_g + \frac{1}{\tau_t}T_{gr} \tag{8.13}$$

where $T_{gr}$ is the reference and one of the control inputs for the wind turbine.

Some individual wind turbines are aggregated to represent a wind farm. It assumes that the wind speed for each wind turbine is same at the same time. Considering the clustering effect of wind turbines, the active power produced by a large-scale wind farm is expressed as follows:

$$\Delta P_{wind} = \sum_{j=1}^{N_{wind}} P_{e,j} - P_{wind,desired} \tag{8.14}$$

where $P_{wind,desired}$ is the desired active power output of the wind farm, $\Delta P_{wind}$ is the deviation of the actual power out from the desired value, $P_{e,j}$ is active power output of the $j$th wind turbines, and $N_{wind}$ is the total number of wind turbines in the wind farm [32].

### 8.2.3 Battery energy storage system

One alternative way for frequency regulation is to introduce the storage facilities, which provide energy stored in the low-load condition for the smart grid during peak load period. The BESS, an important storage facility, can provide fast active power compensation so as to improve the performance of the frequency control [10].

The main components of the BESS are an equivalent battery composed of parallel/series connected battery cells, a bridge converter, and the corresponding control scheme. The equivalent circuit of the BESS can be represented as a converter connected to an equivalent battery, as shown in Figure 8.4.

In the battery equivalent circuit, the terminal voltage of the battery, $E_{bt}$, is obtained as

$$E_{bt} = \frac{3\sqrt{6}}{\pi}E_t(\cos\alpha_1^o + \cos\alpha_2^o) - \frac{6}{\pi}X_{co}I_{bes} \tag{8.15}$$

where $E_t$ is the line to neutral root-mean-squared (RMS) voltage, $\alpha_1^o$ and $\alpha_2^o$ are the firing delay angle of converters 1 and 2, respectively, $X_{co}$ is the commutating reactance, and $I_{bes}$ stands for DC current flowing into battery, which can be given as follows:

$$I_{bes} = \frac{E_{bt} - E_{boc} - E_b}{r_{bt} + r_{bs}} \tag{8.16}$$

where $E_{boc}$ is battery open-circuit voltage, $E_b$ is the battery overvoltage, $r_{bt}$ and $r_{bs}$ are the connecting resistance and the internal resistance, respectively.

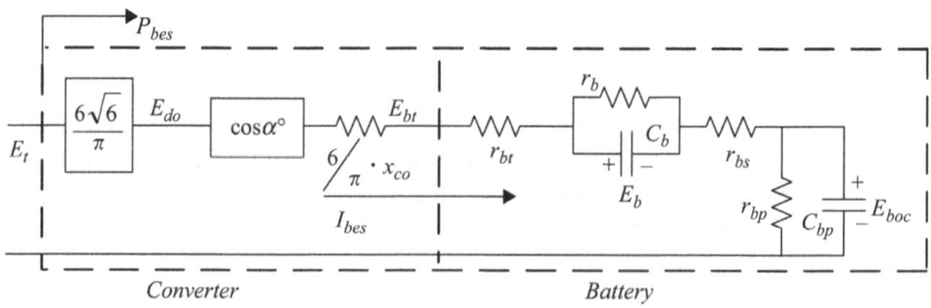


*Figure 8.4 Equivalent circuit of the BESS*

In the converter equivalent circuits, the active and reactive power absorbed by the BESS are given as follows:

$$P_{BESS} = \frac{3\sqrt{6}}{\pi} E_t I_{bes} (\cos \alpha_1^o + \cos \alpha_2^o) \tag{8.17}$$

$$Q_{BESS} = \frac{3\sqrt{6}}{\pi} E_t I_{bes} (\sin \alpha_1^o + \sin \alpha_2^o) \tag{8.18}$$

For the control scheme, two strategies, $P$–$Q$ modulation and $P$ modulation, are usually used to control the converter. The frequency regulation only considers the active power, thus the $P$ modulation ($\alpha^o := \alpha_1^o = -\alpha_2^o$) is considered in this chapter, namely

$$P_{BESS} = \frac{6\sqrt{6}}{\pi} E_t I_{bes} \cos \alpha^o = (E_{do} \cos \alpha^o) I_{bes} = E_{co} I_{bes} \tag{8.19}$$

$$Q_{BESS} = 0 \tag{8.20}$$

where $E_{co} = E_{do} \cos \alpha^o$ is the DC voltage without overlap.

Linearizing (8.19) and decomposing $\Delta E_{co}$ into two components, $\Delta E_p$ and $\Delta E_d$, yield

$$\Delta P_{BESS} = E_{co}^o \Delta I_{bes} + I_{bes}^o \Delta E_p + I_{bes}^o \Delta E_d \tag{8.21}$$

where the second term is to compensate the power deviation caused by the current deviation, i.e., $I_{bes}^o \Delta E_p = -E_{co}^o \Delta I_{bes}$. Then,

$$\Delta P_{BESS} = I_{bes}^o \Delta E_d \tag{8.22}$$

in which $\Delta E_d$ is used to respond the system disturbance, i.e., frequency regulation task. In this chapter, the $ACE$ signal is used as the feedback signal to control the BESS, i.e.,

$$\Delta E_d = \frac{K_{bes}}{1 + sT_{bes}} ACE \tag{8.23}$$

where $K_{bes}$ and $T_{bes}$ are the control loop gain and the measurement device time constant, respectively.

From (8.22) and (8.24), the simplified model of the BESS for frequency regulation can be given as follows:

$$\Delta P_{BESS} = \frac{I_{bes}^o K_{bes} \beta}{1 + sT_{bes}} \Delta f \tag{8.24}$$

### 8.2.4 Plug-in electric vehicles

The BESS with large capacity usually requires some expensive auxiliary equipments such that it is not a very economical way for frequency regulation. The PEVs have drawn increasing attention in recent years, and a number of the small batteries of those PEVs can be considered as an equivalent large-scale BESS. Therefore, large-scale PEVs have potential to provide frequency support to power grids. The battery of

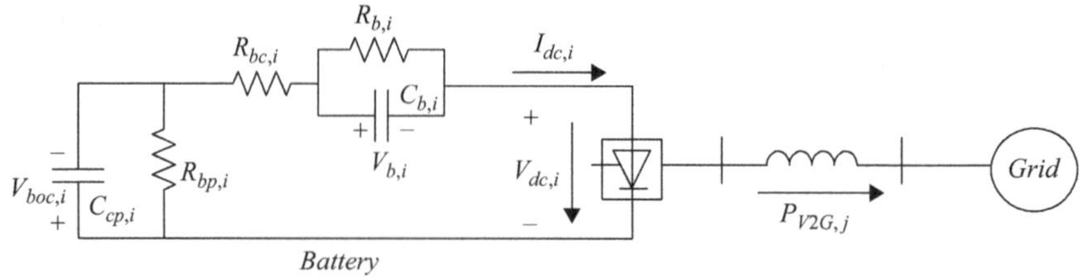


*Figure 8.5 Equivalent circuit of grid-connected PEV*

the PEV comprising parallel/series connected battery cells is connected to the distribution grid through the DC/AC inverter [33], and its equivalent circuit is shown in Figure 8.5.

The equivalent circuit reflects the discharging/charging characteristics of the battery (self-discharge, battery capacity storage, overvoltage, etc.) [14]. The dynamic equations of battery are given as follows:

$$-\frac{V_{boc,i} + V_{dc,i} + V_{b,i}}{R_{bc,i}} = I_{dc,i} \tag{8.25}$$

$$C_{cp,i}\frac{dV_{boc,i}}{dt} + \frac{V_{boc,i}}{R_{bp,i}} = I_{dc,i} \tag{8.26}$$

$$C_{b,i}\frac{dV_{b,i}}{dt} + \frac{V_{b,i}}{R_{b,i}} = I_{dc,i} \tag{8.27}$$

where $V_{boc,i}$ is the open-circuit voltage of battery, $V_{dc,i}$ is the battery's DC current, $V_{b,i}$ is the battery overvoltage, $R_{bc,i}$ is the resistance of the battery's terminals and inter-cell connections, $I_{dc,i}$ is the battery's DC current, $C_{cp,i}$ is the battery capacitance reflecting the main storage capacity, $R_{bp,i}$ is the self-discharge resistance, and $C_{b,i}$ and $R_{b,i}$ describe the transient overvoltage effects.

In this chapter, the active power losses in the inverter and the transformer are ignored, thus the active V2G power injected into the grid, $P_{V2G}$, can be given as follows:

$$P_{V2G,i} = I_{dc,i}V_{dc,i} \tag{8.28}$$

Linearizing (8.28) yields the incremental active power as follows:

$$\Delta P_{V2G,i} = V^{o}_{dc,i}\Delta I_{dc,i} + I^{o}_{dc,i}\Delta V_{dc,i}s \tag{8.29}$$

where $V^{o}_{dc,i}$ and $I^{o}_{dc,i}$ represent the DC voltage and current of the battery at the initial time, respectively. Decomposing the DC voltage, $\Delta V_{dc,i}$, into two components, $\Delta V_{r,i}$, $\Delta V_{s,r}$, yields

$$\Delta P_{V2G,i} = V^{o}_{dc,i}\Delta I_{dc,i} + I^{o}_{dc,i}(\Delta V_{r,i} + \Delta V_{s,i}) \tag{8.30}$$

Similar to the BESS, the partial adjustment of battery voltage is used to compensate the power change resulting from current deviation [14]. That is, the $I^{o}_{dc,i}\Delta V_{r,i}$ in (8.30) is to compensate power deviation resulting from $\Delta I_{dc,i}$ and the $I^{o}_{dc,i}\Delta V_{s,i}$

is the adjustment of V2G power responding to frequency regulation. So $i$th PEV output power is given as follows:

$$\Delta P_{V2G,i} = I^o_{dc,i} \Delta V_{s,i} \tag{8.31}$$

During taking part in the frequency regulation, the DC voltage, $\Delta V_{s,i}$, of the PEV is adjusted with the acquired frequency deviation signal, namely,

$$\Delta \dot{V}_{s,i} = \frac{1}{T_{b,i}}(-\Delta V_{s,i} - k_{v,i} \Delta f) \tag{8.32}$$

where $T_{b,i}$ represents time constant of battery power adjustment. Thus, the V2G power of $PEV_i$ responding to frequency regulation is

$$\Delta P_{V2G,i} = \frac{k_{b,i}}{1 + sT_{b,i}} \Delta f \tag{8.33}$$

where $k_{b,i} = -k_{v,i} I^o_{dc,i}$ is the battery gain, which represents the sensitivity of V2G power with respect to frequency deviation. It is determined by considering a trade-off between the effect of frequency regulation and the constrains of battery state of charge (SOC) [17].

Assume that there are $N$ PEVs in the smart grid discussed in this chapter can be used to provide frequency support, and consider that the $T_{b,i}$ in (8.33) is not very sensitive to frequency deviation, then the aggregated V2G power can be approximately presented as follows:

$$\Delta P_{V2G} = \sum_{i=1}^{N} \Delta P_{V2G,i} = \frac{k_{ev}}{1 + sT_{ev}} \Delta f \tag{8.34}$$

where the aggregated PEV gain $k_{ev} = \sum_{i=1}^{N} k_{b,i}$, and the aggregated time constant $T_{ev} = \sum_{i=1}^{N} \frac{T_{b,i}}{N}$.

### 8.2.5 Controllable air conditioner based DDC

Beside the controlling of the power support through the generators or storage facilities, the management of the power demand is an alternative way to solve the power imbalance problem of smart grids. Domestic electric appliances of the user side can be classified into five different groups based on their characteristics [34], in which the thermostatically controlled load (such as air conditioners, refrigerators, and water heaters) and domestic wet appliances with induction motors or heaters (e.g., washing machines and dryers) can be used as controllable load for frequency regulation, since they are relevant directly to the frequency derivation. In this chapter, a typical thermostatically controlled load, air conditioner, is considered as DDC loads, in which the controllable loads participate in the frequency regulation by adjusting their usage of electricity based on the frequency deviation [35].

The frequency-dependent characteristic of thermostatical load can be expressed as follows:

$$\Delta P_{DDC} = \Delta P_{LC} + D_{ac} \Delta \omega \tag{8.35}$$

where $D_{ac}$ is the reheat coefficient of the air conditioner, $\Delta\omega = 2\pi\Delta f$ is the deviation of the speed, and $\Delta P_{LC}$ is the controllable change in power consumed, which depends on the characteristic of the air conditioner and the set point of a smart thermostat, $\Delta T_{st}$, and can be represented as follows:

$$\Delta P_{LC} = \frac{mc_p}{EER}\Delta T_{st} \tag{8.36}$$

where $m$ is the mass of air flow, $c_p$ is the specific heat capacity of air, and $EER$ is the energy efficient ratio defined as the ratio of the capacity output to electricity input of an air conditioner [36].

The smart thermostat is usually controlled via an integral controller using the frequency deviation ($\Delta f$) from the control area as the input and the temperature set point as the output, namely,

$$\Delta T_{st} = \frac{\bar{K}}{s}\alpha\Delta f \tag{8.37}$$

where $\bar{K}$ is integral gain and $\alpha$ is a coefficient (given as 0.5 Rs/Hz in this chapter). The temperature set point is bounded and varies based on the weather condition and different time intervals in a day. In this chapter, these variations are ignored and the thermostat set-point is simply bounded as [24°C, 29°C].

From the aforementioned discussion, the load model with respect to frequency deviation of air conditioner is given as follows:

$$\Delta P_{ddc,i} = \left(\frac{0.5K_{ddc,i}}{s} + 2\pi D_{ac,i}\right)\Delta f \tag{8.38}$$

where $K_{ddc} = \frac{mc_p\bar{K}}{EER}$ is the combined integral gain.

Due to the small capacity of an individual air conditioner, it is necessary to aggregate a number of small domestic loads into a relative large and lumped DDC load to participate the LFC scheme. Assume that there are $M$ air conditioners in the smart homes of the grid, then the aggregated power of the DDC can be presented as follows:

$$\Delta P_{DDC} = \sum_{i=1}^{M}\Delta P_{ddc,i} = \left(\frac{0.5K_{DDC}}{s} + 2\pi D_{ac}\right)\Delta f \tag{8.39}$$

where $K_{DDC} = \sum_{i=1}^{M} K_{ddc,i}$ and $D_{ac} = \sum_{i=1}^{M} D_{ac,i}$.

### 8.2.6 State-space model of closed-loop LFC scheme

By taking into account the aforementioned models of the BESS, PEVs and DDC, and the model of single-area traditional LFC scheme discussed in Reference 22, the dynamic model of the LFC scheme shown in Figure 8.2 can be obtained as follows:

$$\begin{cases}\dot{x}(t) = Ax(t) + A_b x(t-\tau_2) + A_c x(t-\tau_3) + A_d x(t-\tau_4) + Bu(t-\tau_1) + Fv \\ y(t) = Cx(t)\end{cases} \tag{8.40}$$

where

$$x = \begin{bmatrix} \Delta f & \Delta P_{\mathrm{m}} & \Delta P_{\mathrm{g}} & \Delta P_{BESS} & \Delta P_{V2G} & \Delta P_{DDC} \end{bmatrix}^T$$

$$y = ACE$$

$$v = \Delta P_L - \Delta P_{wind}$$

$$A = \begin{bmatrix} -\frac{D}{2H} & \frac{1}{2H} & 0 & -\frac{1}{2H} & -\frac{1}{2H} & -\frac{1}{2H} \\ 0 & -\frac{1}{T_t} & \frac{1}{T_t} & 0 & 0 & 0 \\ -\frac{1}{RT_g} & 0 & -\frac{1}{T_g} & 0 & 0 & 0 \\ 0 & 0 & 0 & 0 & 0 & 0 \\ 0 & 0 & 0 & 0 & 0 & 0 \\ 0 & 0 & 0 & 0 & 0 & 0 \end{bmatrix}$$

$$A_b = \begin{bmatrix} 0 & 0 & 0 & 0 & 0 & 0 \\ 0 & 0 & 0 & 0 & 0 & 0 \\ 0 & 0 & 0 & 0 & 0 & 0 \\ \frac{I_{bes}^0 K_{bes} \beta}{T_{bes}} & 0 & 0 & -\frac{1}{T_{bes}} & 0 & 0 \\ 0 & 0 & 0 & 0 & 0 & 0 \\ 0 & 0 & 0 & 0 & 0 & 0 \end{bmatrix}$$

$$A_c = \begin{bmatrix} 0 & 0 & 0 & 0 & 0 & 0 \\ 0 & 0 & 0 & 0 & 0 & 0 \\ 0 & 0 & 0 & 0 & 0 & 0 \\ 0 & 0 & 0 & 0 & 0 & 0 \\ \frac{k_{ev}}{T_{ev}} & 0 & 0 & 0 & -\frac{1}{T_{ev}} & 0 \\ 0 & 0 & 0 & 0 & 0 & 0 \end{bmatrix}$$

$$A_d = \begin{bmatrix} 0 & 0 & 0 & 0 & 0 & 0 \\ 0 & 0 & 0 & 0 & 0 & 0 \\ 0 & 0 & 0 & 0 & 0 & 0 \\ 0 & 0 & 0 & 0 & 0 & 0 \\ 0 & 0 & 0 & 0 & 0 & 0 \\ 0.5K_{ddc} - \frac{2\pi DD_{ac}}{2H} & \frac{2\pi D_{ac}}{2H} & 0 & -\frac{2\pi D_{ac}}{2H} & -\frac{2\pi D_{ac}}{2H} & -\frac{2\pi D_{ac}}{2H} \end{bmatrix}$$

$$B = \begin{bmatrix} 0 & 0 & \frac{1}{T_g} & 0 & 0 & 0 \end{bmatrix}^T$$

$$F = \begin{bmatrix} -\frac{1}{2H} & 0 & 0 & 0 & 0 & -\frac{2\pi D_{ac}}{2H} \end{bmatrix}^T$$

$$C = \begin{bmatrix} \beta & 0 & 0 & 0 & 0 & 0 \end{bmatrix}$$

And the load frequency controller, $K(s)$, used in this chapter is the PID controller with the following form:

$$u(t) = -K_P ACE - K_I \int ACE\, dt - K_D \frac{d}{dt} ACE \tag{8.41}$$

where $K_P$, $K_I$, and $K_D$ stand for proportional, integral, and derivative gains, respectively; and $ACE$ is the area control error satisfying

$$ACE = \beta \Delta f \tag{8.42}$$

with $\beta$ being the frequency bias factor.

In order to simplify the analysis, the closed-loop system with a PID controller is transformed into a new system with a static output feedback controller. By defining the following virtual state vector and the output vector

$$\bar{x} = \begin{bmatrix} x^T & \int y^T dt \end{bmatrix}^T \tag{8.43}$$

$$\bar{y} = \begin{bmatrix} y^T & \int y^T dt & \frac{d}{dt} y^T \end{bmatrix}^T \tag{8.44}$$

the closed-loop system can be rewritten as follows:

$$\begin{cases} \dot{\bar{x}}(t) = \bar{A}\bar{x}(t) + \bar{A}_b \bar{x}(t - \tau_2) + \bar{A}_c \bar{x}(t - \tau_3) + \bar{A}_d \bar{x}(t - \tau_4) \\ \qquad - \bar{B}K\bar{C}\bar{x}(t - \tau_1) + (\bar{F} - \bar{B}K\bar{Q})v \\ \bar{y}(t) = \bar{C}\bar{x}(t) + \bar{Q}v \end{cases}$$

where

$$\bar{A} = \begin{bmatrix} A & 0 \\ C & 0 \end{bmatrix}, \quad \bar{A}_b = \begin{bmatrix} A_b & 0 \\ 0 & 0 \end{bmatrix}, \quad \bar{A}_c = \begin{bmatrix} A_c & 0 \\ 0 & 0 \end{bmatrix}, \quad \bar{A}_d = \begin{bmatrix} A_d & 0 \\ 0 & 0 \end{bmatrix}$$

$$\bar{B} = \begin{bmatrix} B \\ 0 \end{bmatrix}, \quad \bar{C} = \begin{bmatrix} C & 0 \\ 0 & 1 \\ CA & 0 \end{bmatrix}, \quad \bar{F} = \begin{bmatrix} F \\ 0 \end{bmatrix}, \quad \bar{Q} = \begin{bmatrix} 0 \\ 0 \\ CF \end{bmatrix}$$

$$K = [K_P \quad K_I \quad K_D] \tag{8.45}$$

By using the similar deriving procedure, the dynamic models for the special cases can be obtained. First, for the case that only primary and supplementary control loops (i.e., traditional LFC) are included, the closed-loop model is given as follows:

$$\begin{cases} \dot{\bar{x}}_1(t) = \bar{A}_1\bar{x}_1(t) - \bar{B}_1 K \bar{C}_1 \bar{x}_1(t - \tau_1) + (\bar{F}_1 - \bar{B}_1 K \bar{Q}_1)v \\ \bar{y}(t) = \bar{C}_1 \bar{x}_1(t) + \bar{Q}_1 v \end{cases}$$

where

$$\bar{x}_1 = \begin{bmatrix} \Delta f & \Delta P_{\mathrm{m}} & \Delta P_{\mathrm{g}} & \int y^T dt \end{bmatrix}^T$$

$$\bar{A}_1 = \begin{bmatrix} A_1 & 0 \\ C_1 & 0 \end{bmatrix}, \quad A_1 = \begin{bmatrix} -\frac{D}{2H} & \frac{1}{2H} & 0 \\ 0 & -\frac{1}{T_t} & \frac{1}{T_t} \\ -\frac{1}{RT_g} & 0 & -\frac{1}{T_g} \end{bmatrix}, \quad C_1 = \begin{bmatrix} \beta & 0 & 0 \end{bmatrix}$$

$$\bar{B}_1 = \begin{bmatrix} B_1^T & 0 \end{bmatrix}^T, \quad B_1 = \begin{bmatrix} 0 & 0 & \frac{1}{T_g} \end{bmatrix}^T, \quad \bar{C}_1 = \begin{bmatrix} C_1 & 0 \\ 0 & 1 \\ C_1A_1 & 0 \end{bmatrix}$$

$$\bar{F}_1 = \begin{bmatrix} F_1^T & 0 \end{bmatrix}^T, \quad F_1 = \begin{bmatrix} -\frac{1}{2H} & 0 & 0 \end{bmatrix}^T, \quad \bar{Q}_1 = [0 \quad 0 \quad C_1F_1]^T \tag{8.46}$$

Second, for the case of traditional LFC with the DDC part, the closed-loop model is given as follows:

$$\begin{cases} \dot{\bar{x}}_2(t) = \bar{A}_2\bar{x}_2(t) + \bar{A}_{d2}\bar{x}_2(t-\tau_4) - \bar{B}_2K\bar{C}_2\bar{x}_2(t-\tau_1) + (\bar{F}_2 - \bar{B}_2K\bar{Q}_2)v \\ \bar{y}(t) = \bar{C}_2\bar{x}_2(t) + \bar{Q}_2v \end{cases}$$

where

$$\bar{x}_2 = \begin{bmatrix} \Delta f & \Delta P_{\mathrm{m}} & \Delta P_{\mathrm{g}} & \Delta P_{DDC} & \int y^T dt \end{bmatrix}^T$$

$$\bar{A}_2 = \begin{bmatrix} A_2 & 0 \\ C_2 & 0 \end{bmatrix}, \quad A_2 = \begin{bmatrix} -\frac{D}{2H} & \frac{1}{2H} & 0 & -\frac{1}{2H} \\ 0 & -\frac{1}{T_t} & \frac{1}{T_t} & 0 \\ -\frac{1}{RT_g} & 0 & -\frac{1}{T_g} & 0 \\ 0 & 0 & 0 & 0 \end{bmatrix}$$

$$C_2 = \begin{bmatrix} C_1 & 0 \end{bmatrix}$$

$$\bar{A}_{d2} = \begin{bmatrix} 0 & 0 & 0 & 0 & 0 \\ 0 & 0 & 0 & 0 & 0 \\ 0 & 0 & 0 & 0 & 0 \\ 0.5K_{ddc} - \frac{2\pi DD_{ac}}{2H} & \frac{2\pi D_{ac}}{2H} & 0 & -\frac{2\pi D_{ac}}{2H} & 0 \\ 0 & 0 & 0 & 0 & 0 \end{bmatrix}$$

$$\bar{B}_2 = \begin{bmatrix} B_1^T & 0 & 0 \end{bmatrix}^T, \quad \bar{C}_2 = \begin{bmatrix} C_2 & 0 \\ 0 & 1 \\ C_2A_2 & 0 \end{bmatrix}$$

$$\bar{F}_2 = \begin{bmatrix} F_2^T & 0 \end{bmatrix}^T, \quad F_2 = \begin{bmatrix} F_1^T & -\frac{2\pi D_{ac}}{2H} \end{bmatrix}^T, \quad \bar{Q}_2 = [0 \quad 0 \quad C_2F_2]^T \tag{8.47}$$

Third, for the case where both the DDC and the BESS are introduced into the traditional LFC, the closed-loop model is given as follows:

$$\begin{cases}\dot{\bar{x}}_3(t) = \bar{A}_3\bar{x}_3(t) + \bar{A}_{b3}\bar{x}_3(t-\tau_2) + \bar{A}_{d3}\bar{x}_3(t-\tau_4) - \bar{B}_3K\bar{C}_3\bar{x}_3(t-\tau_1) \\ \qquad + (\bar{F}_3 - \bar{B}_3K\bar{Q}_3)v \\ \bar{y}(t) = \bar{C}_3\bar{x}_3(t) + \bar{Q}_3 v\end{cases}$$

where

$$\bar{x}_3 = \begin{bmatrix}\Delta f & \Delta P_{\mathrm{m}} & \Delta P_{\mathrm{g}} & \Delta P_{BESS} & \Delta P_{DDC} & \int y^T dt\end{bmatrix}^T$$

$$\bar{A}_3 = \begin{bmatrix}A_3 & 0 \\ C_3 & 0\end{bmatrix}, \quad A_3 = \begin{bmatrix}-\frac{D}{2H} & \frac{1}{2H} & 0 & -\frac{1}{2H} & -\frac{1}{2H} \\ 0 & -\frac{1}{T_t} & \frac{1}{T_t} & 0 & 0 \\ -\frac{1}{RT_g} & 0 & -\frac{1}{T_g} & 0 & 0 \\ 0 & 0 & 0 & 0 & 0 \\ 0 & 0 & 0 & 0 & 0\end{bmatrix}$$

$$C_3 = \begin{bmatrix}C_1 & 0 & 0\end{bmatrix}$$

$$\bar{A}_{b3} = \begin{bmatrix}0 & 0 & 0 & 0 & 0 & 0 \\ 0 & 0 & 0 & 0 & 0 & 0 \\ 0 & 0 & 0 & 0 & 0 & 0 \\ \frac{I_{bes}^0 K_{bes}\beta}{T_{bes}} & 0 & 0 & -\frac{1}{T_{bes}} & 0 & 0 \\ 0 & 0 & 0 & 0 & 0 & 0 \\ 0 & 0 & 0 & 0 & 0 & 0\end{bmatrix}$$

$$\bar{A}_{d3} = \begin{bmatrix}0 & 0 & 0 & 0 & 0 & 0 \\ 0 & 0 & 0 & 0 & 0 & 0 \\ 0 & 0 & 0 & 0 & 0 & 0 \\ 0 & 0 & 0 & 0 & 0 & 0 \\ 0.5K_{ddc} - \frac{2\pi DD_{ac}}{2H} & \frac{2\pi D_{ac}}{2H} & 0 & -\frac{2\pi D_{ac}}{2H} & -\frac{2\pi D_{ac}}{2H} & 0 \\ 0 & 0 & 0 & 0 & 0 & 0\end{bmatrix}$$

$$\bar{B}_3 = \begin{bmatrix}B_1^T & 0 & 0 & 0\end{bmatrix}^T, \quad \bar{C}_3 = \begin{bmatrix}C_3 & 0 \\ 0 & 1 \\ C_3A_3 & 0\end{bmatrix}$$

$$F_3 = \begin{bmatrix}F_1^T & 0 & -W\end{bmatrix}^T, \quad \bar{F}_3 = \begin{bmatrix}F_3^T & 0\end{bmatrix}^T, \quad \bar{Q}_3 = [0 \quad 0 \quad C_3F_3]^T \quad (8.48)$$

## 8.3 Delay-dependent stability analysis

The system stability is the basic requirement of the smart power grid, and the communication time delays arising from model (8.45) will degrade the system performance even cause the instability. In this section, a delay-dependent stability analysis method is developed based on the Lyapunov functional method and the linear matrix inequality (LMI), including the proof of the stability criterion and the procedure of the calculation of the delay margin.

In model (8.45), the time delays of different control loops are different, which makes the analysis complex. Hence, some modifications are made to simplify the analysis. That is, the time delays in LFC and BESS control loops caused from the measurement and the transmission of the ACE signal are assumed to be identical, and the time delays in PEV and DDC loops arising due to the measurement of the frequency deviation are also assumed to be identical, i.e., $\bar{\tau}_1 = \tau_1 = \tau_2$ and $\bar{\tau}_2 = \tau_3 = \tau_4$. This chapter investigates the initial stability, thus the disturbance is not taken into account. Then, the closed-loop system shown in (8.45) can be rewritten as follows:

$$\dot{\bar{x}}(t) = \bar{A}\bar{x}(t) + (\bar{A}_b - \bar{B}K\bar{C})\bar{x}(t - \bar{\tau}_1) + (\bar{A}_c + \bar{A}_d)\bar{x}(t - \bar{\tau}_2) \tag{8.49}$$

## 8.3.1 Delay-dependent stability criterion

The stability criterion is important to judge the system stability. By using the Lyapunov stability theory and the Wirtinger inequality and following the similar procedure of Reference 25, the following delay-dependent stability criterion can be obtained.

**Theorem 8.1.** *Consider the following time-delay system:*

$$\dot{x}(t) = A_0x(t) + A_1x(t - \tau_1) + A_2x(t - \tau_2) \tag{8.50}$$

*For given scalars $\tau_i$ satisfying $0 = \tau_0 \leq \tau_1 \leq \tau_2$, the above system is asymptotically stable if there exist symmetric positive-definite matrices $P_1 > 0$, $Q_1 > 0$, $Q_2 > 0$, $R_1 \geq 0$, and $R_2 \geq 0$ such that the following LMI holds:*

$$\Xi = \Phi + \Phi_1 + \Phi_2 < 0$$

*where*

$$\Phi = \begin{bmatrix} e_1 \\ \tau_1 e_4 \\ (\tau_2 - \tau_1)e_5 \end{bmatrix}^T P_1 \begin{bmatrix} e_0 \\ e_1 - e_2 \\ e_2 - e_3 \end{bmatrix} + \begin{bmatrix} e_0 \\ e_1 - e_2 \\ e_2 - e_3 \end{bmatrix}^T P_1 \begin{bmatrix} e_1 \\ \tau_1 e_4 \\ (\tau_2 - \tau_1)e_5 \end{bmatrix}$$

$$\Phi_1 = e_1^T Q_1 e_1 - e_2^T Q_1 e_2 + e_2^T Q_2 e_2 - e_3^T Q_2 e_3 + \tau_1^2 e_0^T R_1 e_0 + (\tau_2 - \tau_1)^2 e_0^T R_2 e_0$$

$$\Phi_2 = -\begin{bmatrix} e_1 - e_2 \\ e_1 + e_2 - 2e_4 \end{bmatrix}^T \begin{bmatrix} R_1 & 0 \\ 0 & 3R_1 \end{bmatrix} \begin{bmatrix} e_1 - e_2 \\ e_1 + e_2 - 2e_4 \end{bmatrix} - \begin{bmatrix} e_2 - e_3 \\ e_2 + e_3 - 2e_5 \end{bmatrix}^T \begin{bmatrix} R_2 & 0 \\ 0 & 3R_2 \end{bmatrix} \begin{bmatrix} e_2 - e_3 \\ e_2 + e_3 - 2e_5 \end{bmatrix}$$

$$
\begin{aligned}
e_0 &= \begin{bmatrix} A_0, & A_1, & A_2, & 0, & 0 \end{bmatrix} \\
e_1 &= \begin{bmatrix} I, & 0, & 0, & 0, & 0 \end{bmatrix} \\
e_2 &= \begin{bmatrix} 0, & I, & 0, & 0, & 0 \end{bmatrix} \\
e_3 &= \begin{bmatrix} 0, & 0, & I, & 0, & 0 \end{bmatrix} \\
e_4 &= \begin{bmatrix} 0, & 0, & 0, & I, & 0 \end{bmatrix} \\
e_5 &= \begin{bmatrix} 0, & 0, & 0, & 0, & I \end{bmatrix}
\end{aligned} \tag{8.51}
$$

*Proof.* Construct a candidate Lyapunov function as

$$
\begin{aligned}
V(t) = {} & \begin{bmatrix} x(t) \\ \int_{t-\tau_1}^{t} x(s)ds \\ \int_{t-\tau_2}^{t-\tau_1} x(s)ds \end{bmatrix}^T P_1 \begin{bmatrix} x(t) \\ \int_{t-\tau_1}^{t} x(s)ds \\ \int_{t-\tau_2}^{t-\tau_1} x(s)ds \end{bmatrix} + \int_{t-\tau_1}^{t} x^T(s)Q_1x(s)ds \\
& + \tau_1 \int_{-\tau_1}^{0} \int_{t+\theta}^{t} \dot{x}^T(s)R_1\dot{x}(s)\, ds\, d\theta + \int_{t-\tau_2}^{t-\tau_1} x^T(s)Q_2x(s)ds \\
& + (\tau_2 - \tau_1) \int_{-\tau_2}^{-\tau_1} \int_{t+\theta}^{t} \dot{x}^T(s)R_2\dot{x}(s)\, ds\, d\theta
\end{aligned}
$$

where $P_1$, $Q_1$, $Q_2$, $R_1$, and $R_2$ are positive-definite symmetric matrices, which means the positive of the function, i.e., $V(t) \geq \varepsilon_1 ||x(t)||^2$ with $\varepsilon_1 > 0$.

Calculating the derivative of (8.52) yields

$$
\begin{aligned}
\dot{V}(t) = {} & 2 \begin{bmatrix} x(t) \\ \int_{t-\tau_1}^{t} x(s)ds \\ \int_{t-\tau_2}^{t-\tau_1} x(s)ds \end{bmatrix}^T P_1 \begin{bmatrix} \dot{x}(t) \\ x(t) - x(t-\tau_1) \\ x(t-\tau_1) - x(t-\tau_2) \end{bmatrix} \\
& + x^T(t)Q_1x(t) - x^T(t-\tau_1)Q_1x(t-\tau_1) \\
& + x^T(t-\tau_1)Q_2x(t-\tau_1) - x^T(t-\tau_2)Q_2x(t-\tau_2) \\
& + \tau_1^2\dot{x}^T(t)R_1\dot{x}(t) + (\tau_2-\tau_1)^2\dot{x}^T(t)R_2\dot{x}(t) \\
& - \tau_1 \int_{t-\tau_1}^{t} \dot{x}^T(s)R_1\dot{x}(s)ds - (\tau_2-\tau_1) \int_{t-\tau_2}^{t-\tau_1} \dot{x}^T(s)R_2\dot{x}(s)ds
\end{aligned} \tag{8.52}
$$

It follows from Wirtinger-based integral inequality [37] that

$$\tau_1 \int_{t-\tau_1}^{t} \dot{x}^T(s) R_1 \dot{x}(s) ds + (\tau_2 - \tau_1) \int_{t-\tau_2}^{t-\tau_1} \dot{x}^T(s) R_2 \dot{x}(s) ds \tag{8.53}$$

$$\geq \begin{bmatrix} x(t) - x(t-\tau_1) \\ x(t) + x(t-\tau_1) - 2\int_{t-\tau_1}^{t} \frac{x(s)}{\tau_1} ds \end{bmatrix}^T \begin{bmatrix} R_1 & 0 \\ 0 & 3R_1 \end{bmatrix} \begin{bmatrix} x(t) - x(t-\tau_1) \\ x(t) + x(t-\tau_1) - 2\int_{t-\tau_1}^{t} \frac{x(s)}{\tau_1} ds \end{bmatrix}$$

$$+ \begin{bmatrix} x(t-\tau_1) - x(t-\tau_2) \\ x(t-\tau_1) + x(t-\tau_2) - 2\int_{t-\tau_2}^{t-\tau_1} \frac{x(s)}{\tau_2 - \tau_1} ds \end{bmatrix}^T \begin{bmatrix} R_2 & 0 \\ 0 & 3R_2 \end{bmatrix}$$

$$\times \begin{bmatrix} x(t-\tau_1) - x(t-\tau_2) \\ x(t-\tau_1) + x(t-\tau_2) - 2\int_{t-\tau_2}^{t-\tau_1} \frac{x(s)}{\tau_2 - \tau_1} ds \end{bmatrix}$$

then applying (8.51) and (8.53) yields

$$\dot{V}(t) \leq \bar{\xi}^T(t) \Xi \bar{\xi}(t) \leq 0$$

where

$$\bar{\xi}(t) = \left[ x^T(t), \quad x^T(t-\tau_1), \quad x^T(t-\tau_2), \quad \int_{t-\tau_1}^{t} \frac{x^T(s)}{\tau_1} ds, \quad \int_{t-\tau_2}^{t-\tau_1} \frac{x^T(s)}{\tau_2 - \tau_1} ds \right]^T \tag{8.54}$$

Therefore, the system is asymptotically stable if $P_1 > 0, Q_1 \geq 0, Q_2 \geq 0, R_1 \geq 0$, and $R_2 \geq 0$ and (8.51) holds. This completes the proof. □

### *8.3.2 Delay margin calculation*

The aforementioned stability criterion (Theorem 8.1) shows the relationship between the stability and the value of the time delays. Based on this criterion, one can judge the stability of the system with the given time delays, and also find the admissible maximal delay value, so-called delay margin, that the system starts to become instability. As reported in Reference 25, the calculation of the delay margin of the system is an important issue during the delay-dependent stability analysis.

There are two time delays in system (8.49). One can find the delay margin of the one of the delays, based on Theorem 8.1, by fixing the other time delay. That is, the delay margin of $\bar{\tau}_1$ (or $\bar{\tau}_2$) should be a function of the delay $\bar{\tau}_2$ (or $\bar{\tau}_1$), i.e.,

$$\bar{\tau}_{\max i} = f_s(\bar{\tau}_j), \quad i = 1, j = 2; \; i = 2, j = 1 \tag{8.55}$$

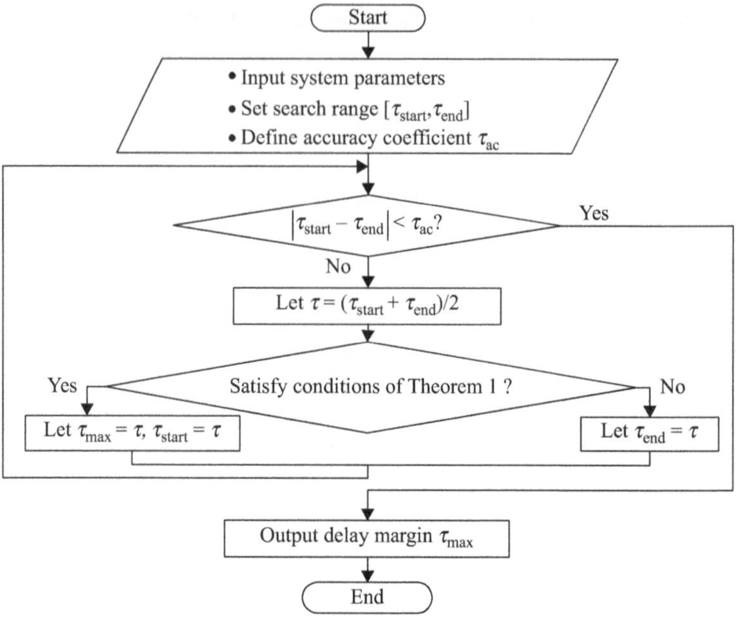


*Figure 8.6 The simplified procedure of delay margin calculation*

As mentioned in Reference 38, the method combining the *feasp* solver in MATLAB/LMI toolbox and the binary search algorithm can be applied to calculate the delay margin. The simplified procedure is described in Figure 8.6.

## 8.4 Delay-dependent robust controller design

In this section, a PID controller tuning method is developed based on the delay-dependent $H_\infty$ performance analysis and the PSO searching algorithm. The delay-dependent $H_\infty$ performance analysis is carried out to derive a criterion, which constructs the relationships among the delay bound, the robust performance index, and the control gains. Then, based on the criterion, the tuning of the controller gains is transformed into an optimization problem solved by standard PSO algorithm.

In model (8.45), four delays exist in the different control loops. To simplify the analysis, those delays are assumed to be identical, i.e., $h = \tau_1 = \tau_2 = \tau_3 = \tau_4$. Then, the closed-loop system shown in (8.45) can be rewritten as follows:

$$\begin{cases} \dot{x}(t) = Ax(t) + (A_{bc} - \bar{B}KC_y)x(t-h) + (B_\omega - \bar{B}K\bar{Q})\omega(t) \\ z(t) = C_z x(t) \end{cases} \tag{8.56}$$

where $x(t) = \bar{x}(t)$, $A = \bar{A}$, $A_{bc} = \bar{A}_b + \bar{A}_c + \bar{A}_d$, $\omega(t) = v(t)$, $B_\omega = \bar{F}$, $C_y = \bar{C}$, and $C_z = [1 \quad 0 \quad 0 \quad 0 \quad 0 \quad 0 \quad 0]$.

### 8.4.1 Delay-dependent $H_\infty$ performance analysis

For the above system, by using the Lyapunov–Krasovskii functional method, the relationships among the delay bound, the robust performance index, and the control gains can be described by the following criterion.

**Theorem 8.2.** *Consider the closed-loop system (8.56), for the delay bound h, the $H_\infty$ performance index $\gamma$, and the controller gains $K = [K_P \; K_I \; K_D]$, if there exist symmetric matrices P, Q, and R, such that the following LMIs hold:*

$$P > 0, \quad Q > 0, \quad R > 0 \tag{8.57}$$

$$\begin{aligned}\Phi = Sym & \left\{ \begin{bmatrix} e_1 \\ he_3 \end{bmatrix}^T P \begin{bmatrix} e_s \\ e_1 - e_2 \end{bmatrix} \right\} \\ & - \begin{bmatrix} e_1 - e_2 \\ e_1 + e_2 - 2e_3 \end{bmatrix}^T \begin{bmatrix} R & 0 \\ 0 & 3R \end{bmatrix} \begin{bmatrix} e_1 - e_2 \\ e_1 + e_2 - 2e_3 \end{bmatrix} \\ & + e_1^T Q e_1 - e_2^T Q e_2 + h^2 e_s^T R e_s + e_1^T C_z^T C_z e_1 - \gamma^2 e_4^T e_4 \\ < 0 & \end{aligned}$$

*where $e_s = [A, A_d - \bar{B}KC_y, 0, B_w - \bar{B}K\bar{Q}]$, $e_1 = [I, 0, 0, 0]$, $e_2 = [0, I, 0, 0]$, $e_3 = [0, 0, I, 0]$, $e_4 = [0, 0, 0, I]$, then the system is stable and has $H_\infty$ performance index, $\gamma$, against a non-zero disturbance for any delays smaller than h.*

*Proof.* Choose an LKF candidate as follows:

$$\begin{aligned} V(x_t) = & \begin{bmatrix} x(t) \\ \int_{t-h}^{t} x(s)ds \end{bmatrix}^T P \begin{bmatrix} x(t) \\ \int_{t-h}^{t} x(s)ds \end{bmatrix} + \int_{t-h}^{t} x^T(s) Q x(s) ds \\ & + h \int_{-h}^{0} \int_{t+\theta}^{t} \dot{x}^T(s) R \dot{x}(s) ds \end{aligned} \tag{8.58}$$

where $P$, $Q$, and $R$ are symmetrical matrices. It can be find that the positive of the LKF, i.e. ($V(t) > 0$), can be ensured if LMI (8.57) holds.

Calculating the derivative of LKF and using Wirtinger-based integral inequality [37] to estimate the yields

$$\begin{aligned}
&\dot{V}(x_t)\\
&= 2\begin{bmatrix} x(t) \\ \int_{t-h}^{t} x(s)ds \end{bmatrix}^T P \begin{bmatrix} \dot{x}(t) \\ x(t) - x(t-h) \end{bmatrix}\\
&\quad + x^T(t)Qx(t) - x^T(t-h)Qx(t-h) + h^2\dot{x}^T(t)R\dot{x}(t) - h\int_{t-h}^{t} \dot{x}^T(s)R\dot{x}(s)ds \quad (8.59)\\
&\le 2\begin{bmatrix} x(t) \\ \int_{t-h}^{t} x(s)ds \end{bmatrix}^T P \begin{bmatrix} \dot{x}(t) \\ x(t) - x(t-h) \end{bmatrix}\\
&\quad + x^T(t)Qx(t) - x^T(t-h)Qx(t-h) + h^2\dot{x}^T(t)R\dot{x}(t) \quad (8.60)\\
&\quad - \begin{bmatrix} x(t) - x(t-h) \\ x(t) + x(t-h) - 2\int_{t-h}^{t} \frac{x(s)}{h}ds \end{bmatrix}^T \begin{bmatrix} R & 0 \\ 0 & 3R \end{bmatrix} \begin{bmatrix} x(t) - x(t-h) \\ x(t) + x(t-h) - 2\int_{t-h}^{t} \frac{x(s)}{h}ds \end{bmatrix}
\end{aligned}$$

By defining the following notations:

$$\begin{aligned}
\zeta(t) &= \left[x(t),\ x(t-h),\ \int_{t-h}^{t} \frac{x(s)}{h}ds, \omega(t)\right]\\
\dot{x}(t) &= e_s\zeta(t), \quad e_s = [A, A_d - \bar{B}KC_y, 0, B_w - \bar{B}K\bar{Q}]\\
x(t) &= e_1\zeta(t), \quad e_1 = [I, 0, 0, 0]\\
x(t-h) &= e_2\zeta(t), \quad e_2 = [0, I, 0, 0]\\
\int_{t-h}^{t} \frac{x(s)}{h}ds &= e_3\zeta(t), \quad e_3 = [0, 0, I, 0]\\
\omega(t) &= e_4\zeta(t), \quad e_4 = [0, 0, 0, I]
\end{aligned}$$

it follows (8.60) and (8.58) that

$$\dot{V}(x_t) + z^T(t)z(t) - \gamma^2\omega^T(t)\omega(t) < \zeta^T(t)\Phi\zeta(t) < 0 \quad (8.61)$$

where $\Phi$ is defined in (8.58). Thus, based on (8.61) and $V(x_0) = 0$, $V(x_\infty) \ge 0$, the following holds:

$$\int_0^\infty [z^T(s)z(s) - \gamma^2\omega^T(s)\omega(s)]ds < V(x_0) - V(x_\infty) < 0 \quad (8.62)$$

Therefore,

$$\sqrt{\frac{\int_0^\infty z^T(s)z(s)ds}{\int_0^\infty \omega^T(s)\omega(s)ds}} \le \gamma \quad (8.63)$$

which means the system is stable and has a $H_\infty$ performance index, $\gamma$.

Based on the above discussion, it can conclude that if LMIs (8.57) and (8.58) hold, the system is stable when without disturbance and has a $H_\infty$ performance index, $\gamma$, against a non-zero disturbance, for any delays smaller than $h$. This completes the proof. □

**Remark 8.1.** *Theorem 8.2 gives the relationships among the delay bound $h$, the $H_\infty$ performance index $\gamma$, and the controller gains $K$. As discussed in Reference 25, based on Theorem 8.2, for the fixed delay bound $h$ and the controller gains $K$, one can find the minimal value of the performance index $\gamma_{\min}$, i.e., delay-dependent $H_\infty$ performance analysis. That is, the minimal value $\gamma_{\min}$ is a function of the delay bound $h$ and the controller gains $K$, as described as follows:*

$$\gamma_{\min} = f(h, K) = f(h, K_P, K_I, K_D) \tag{8.64}$$

*How to calculate the minimal value, $\gamma_{\min}$, for fixed delay bound $h$ and controller gains $K$ can be found in Reference 25 and is omitted here.*

## 8.4.2 Controller gain tuning based on the PSO algorithm

For a built communication channel, the time-delay upper bound can be estimated from the transmitted data with time stamp. Then, for such delay bound, different controller gains lead to different performance index calculated via (8.64). Thus, to provide the optimal robust performance for a preset time delay, the tuning of control gain can be transformed to the following optimization problem:

$$\begin{aligned} \textit{Minimize}: \quad & \gamma_{\min} = f(h, K_P, K_I, K_D) \\ \textit{subject to}: \quad & K_{Pmin} \le K_P \le K_{Pmax} \\ & K_{Imin} \le K_I \le K_{Imax} \\ & K_{Dmin} \le K_D \le K_{Dmax} \end{aligned} \tag{8.65}$$

The above optimization problem can be solved by different optimization algorithms. This chapter chooses the PSO searching algorithm as it is a meta-heuristic algorithm and has been widely used due to its decent performance in numerical optimization [39,40]. The simplified flow chart of the PID gains tuning via the $H_\infty$ performance index and the PSO algorithm is shown in Figure 8.7.

In Figure 8.7, for the initialization step, the following parameters should be given or calculated:

- Set the time-delay upper bound, $h$.
- Set position bounds, $X_{\min}$ and $X_{\max}$, velocity bounds, $V_{\min}$ and $V_{\max}$ and the size of particles, $N$; and obtain random populations of the positions $X_0$ within $[X_{\min}, X_{\max}]$ (i.e., $N$ sets of gains $K = [K_P\ K_I\ K_D]$) and the velocities $V_0$ within $[V_{\min}, V_{\max}]$.
- Set the maximal iteration times, $k_{\max}$, and the initial iteration times $i = 0$.
- Evaluate the fitness (i.e., $H_\infty$ performance index, $\gamma_{\min}$ shown in (8.64)) for $N$ particles, $f(h, X_{0,j}), j = 1, 2, \ldots, N$, via Theorem 8.2 and Remark 8.1.

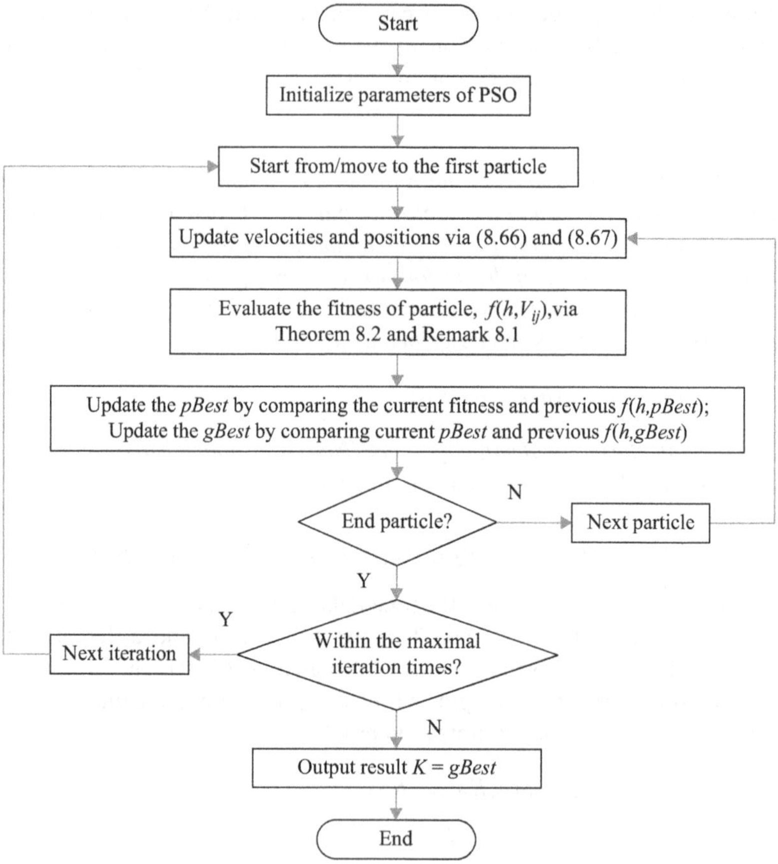


*Figure 8.7 Simplified flow chart of the PID tuning*

- Find the best position for each particle, *pBest*, and the best position within all particles, *gBest*, i.e., set $f(h, pBest_j) = f(h, X_{0,j}), j = 1, 2, \ldots, N$ and $pBest = X_0$, and set $gBest = X_{o,k}, k \in \{1, 2, \ldots, N\}$ such that $f(h, gBest) = \min f(h, X_{0,j}), j = 1, 2, \ldots, N$.

For the step of updating the velocities and the positions, the following renewing conditions are applied:

$$V_{i+1} = \begin{cases} V_{\min}, & \text{if } V_{i+1} < V_{\min} \\ wV_i + c_1 \cdot \text{rand} \cdot (pBest - X_i) + c_2 \cdot \text{rand} \cdot (gBest - X_i), & \text{if } V_{i+1} \in [V_{\min}, V_{\max}] \\ V_{\max}, & \text{if } V_{i+1} > V_{\max} \end{cases} \tag{8.66}$$

where $V_i$ and $V_{i+1}$ are, respectively, the velocities in $i$th and $(i+1)$th iteration, $w$ is the inertia weight, $c_1$ and $c_2$ are the acceleration constants, and rand is a value randomly generated between 0 and 1; and

$$X_{i+1} = \begin{cases} X_{\min}, & \text{if } X_{i+1} < X_{\min} \\ X_i + V_{i+1}, & \text{if } X_{i+1} \in [X_{\min}, X_{\max}] \\ X_{\max}, & \text{if } X_{i+1} > X_{\max} \end{cases} \tag{8.67}$$

where $X_i$ and $X_{i+1}$ are, respectively, the positions in $i$th and $(i+1)$th iteration.

## 8.5 Case studies

Case studies are carried out based on the smart power grid operating in the island mode, shown in Figure 8.1, which consists of traditional micro generation, some smart homes with controllable loads, a PEV station, and a wind farm. The micro generation equips with a non-heat turbine with the capacity of 800 MW, and its generation rate constrain (GRC) is assumed to be ±0.1 pu/min [3]. In the smart homes, there are 200 air conditioners that can provide the DDC service. Assume that there are total 60 PEVs in the considered area, and the ones parking in the PEV station or at smart homes can participate the frequency support service. The wind farm consists 65 variable-speed wind turbines and every turbine's capacity is around 3.04 MW. The total wind farm output power is 197 MW (about 20% of total output power). Assume that all wind turbines are identical and the cut-in wind speed, the rated wind speed, and the furling speed are 2, 18, and 25 m/s, respectively [1]. The related parameters are listed in Table 8.1 and recalled from the literature [10,14,41]. All corresponding simulations are carried out by using MATLAB 7.10.0 (R2011a) running on a PC.

*Table 8.1 The parameters of test smart power grid*

| Parameter | Value | Parameter | Value | Parameter | Value |
|---|---|---|---|---|---|
| $T_g$ (s) | 0.08 | $D_{ac}$ | 0.03 | $r_{bs}$ (Ω) | 0.013 |
| $T_t$ (s) | 0.38 | $EER$ | 3.75 | $J_g$ (kg m$^2$) | 10 |
| $R$ (Hz pu) | 2.5 | $k_b$ (kW/Hz) | 11.75 | $J_r$ (kg m$^2$) | 90 000 |
| $2H$ (pu s) | 0.1667 | $T_b$ (ms) | 50 | $k_s$ (N/m) | $8 \times 10^6$ |
| $D$ (pu/Hz) | 0.0083 | $I^0_{bes}$ (kA) | 4.426 | $D_s$ (s$^{-1}$) | $8 \times 10^4$ |
| $\beta$ (pu/Hz) | 0.55 | $K_{bes}$ (kV/pu MW) | 100 | $N_g$ | 24.6 |
| $c_p$ (J/K) | 1.01 | $T_{bes}(s)$ | 0.026 | $R_{wind}$ (m) | 14.5 |
| $m$ (pu/s) | 0.35 | $\alpha$ (°) | 15 | $\tau_\theta$ (s) | 0.15 |
| $K_{DDC}$ | 10 | $r_{bt}$ (Ω) | 0.0167 | $\tau_T$ (s) | 0.1 |

### 8.5.1 Robust controller design

At first, the robust LFC is designed based on the simplified state-space model developed in Section 8.2 and the robust PID controller design method presented in Section 8.4.

In order to simplify the design procedure, the time delays in different loops in the model are assumed to be identical and here it is preset to be 0.2 s. And, the related initial parameters of the PSO mentioned in Section 8.4 are given as follows: position bounds, $X_{max} = 1$, $X_{min} = -1$ (i.e., $\{K_P, K_I, K_D\} \in [-1, 1]$); velocity bounds, $V_{max} = 1$, $V_{min} = -1$; population size $N = 50$; maximum iteration times $k_{max} = 50$; inertia weight, $w = 0.4$; acceleration constants, $c_1 = c_2 = 2$. By following the controller gains tuning method shown in Figure 8.7, the following PID controller can be obtained:

$$K = [-0.34438 \quad -0.07727 \quad -0.31065] \tag{8.68}$$

### 8.5.2 Contribution of the DDC, BESS, and PEV to frequency regulation

How the additionally connected DDC, BESS, and PEV contribute to improve the frequency control performance is shown in this part.

For the case where the DDC, BESS, and PEV do not participate in the frequency regulation, the responds of the frequency deviation for the system-connected wind turbines with different output power (0.0912 pu and 0.1216 pu) are given in Figure 8.8. From the results, the frequency deviation can recover to small fluctuation range when the low penetration of wind power injects in the power system (0.0912 pu). While the wind farm output power becomes to 0.1216 pu, the frequency regulation is no longer in steady state. It can be concluded that the traditional LFC scheme can only maintain the power system in steady state for the case of small intermittent wind power injection.

When the connected wind power is 0.1611 pu, the frequency deviations for two cases (with or without DDC, BESS, and PEV) are shown in Figure 8.9. It is easily found that, compared with the case with only LFC, the case with DDC, BESS, and PEV can provide better dynamic performances, shorter settling time, and smaller overshoot and effectively solve problem caused by the intermittent wind power energy, which shows that the introduction of the DDC, BESS, and PEVs enhances the transient response of frequency regulation and improves the frequency control performance. Moreover, although the time delays of 0.2 s are included in the control loops during the simulation, the frequency deviation is quickly convergent to schedule values, which shows the effectiveness of the proposed PID controller.

The LFC scheme with different capacities of DDC, BESS, and PEV are also tested. Figures 8.10–8.12 show the responds of the frequency deviation for those cases. From the results, it can be found that the DDC/BESS with bigger capacity or more PEVs participating in frequency regulation can recover the frequency deviation in steady-state quicker and fluctuation smaller.

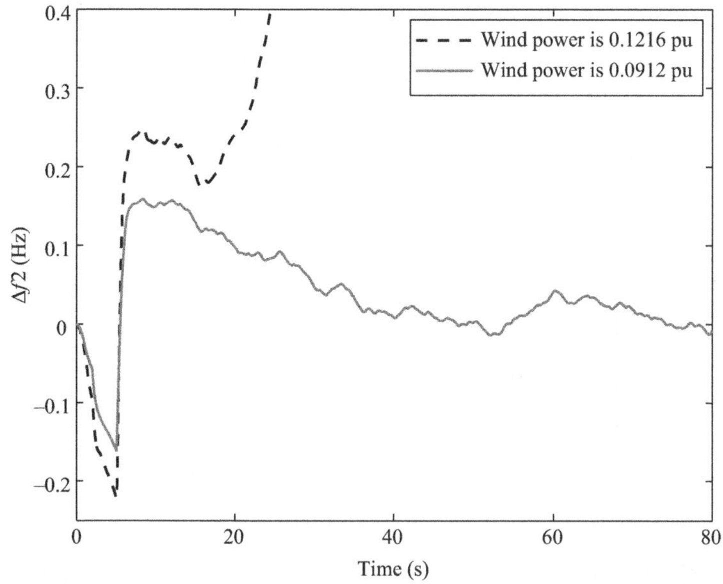


*Figure 8.8 Frequency deviation in the LFC model when different value of wind power energy in the model*

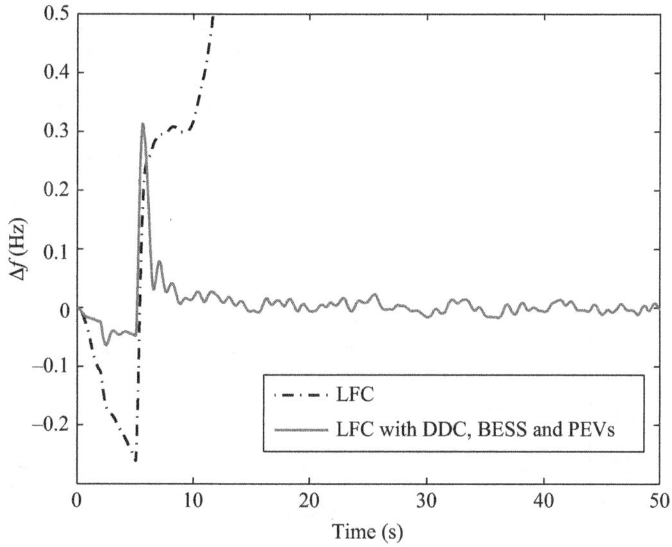


*Figure 8.9 Frequency deviation in LFC and LFC with DDC, BESS, and PEVs model*

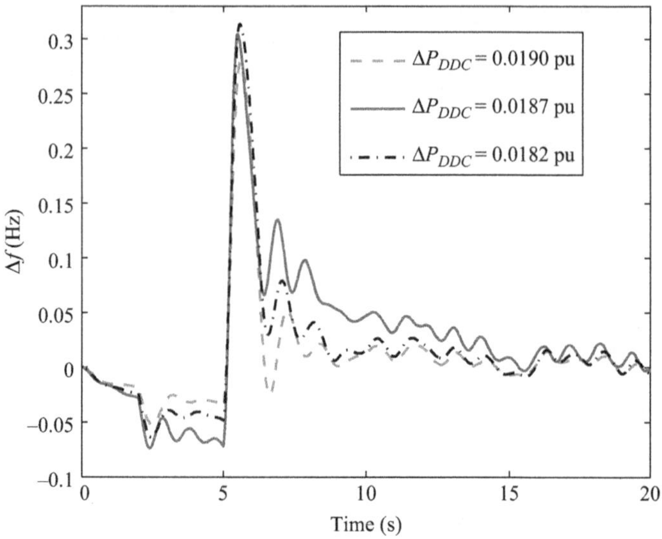


*Figure 8.10 Frequency deviation from LFC with BESS, PEVs, and different sizes of DDC model*

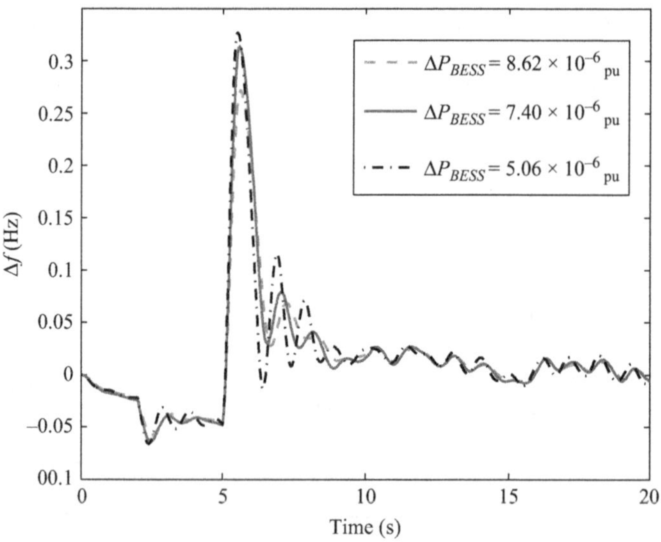


*Figure 8.11 Frequency deviation from LFC with DDC, PEVs, and different sizes of BESS model*

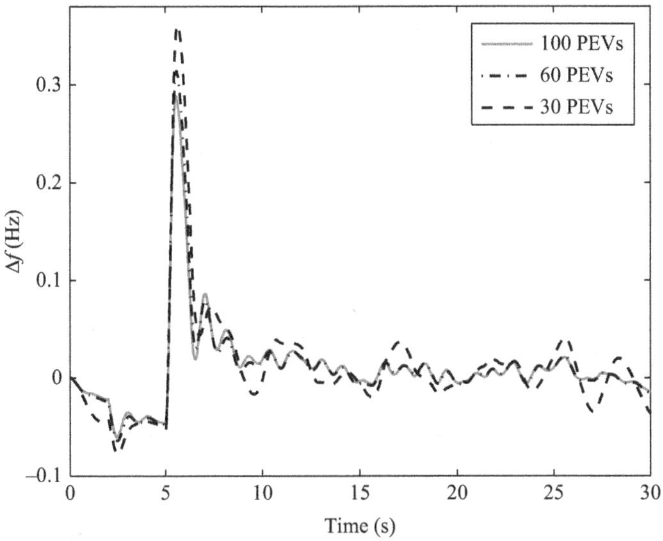


*Figure 8.12 Frequency deviation from LFC with DDC, BESS, and different number of PEVs model*

*Table 8.2 The $\gamma_{min}$ for different time delays*

| Case A | | Case B | | Case C | |
|---|---|---|---|---|---|
| **$\tau$ (s)** | **$\gamma_{min}$** | **$\tau$ (s)** | **$\gamma_{min}$** | **$\tau$ (s)** | **$\gamma_{min}$** |
| 0.001 | 1.4820 | 0.001 | 1.4820 | 0.001 | 1.4820 |
| 0.05 | 1.3470 | 0.05 | 1.3470 | 0.05 | 1.3470 |
| 0.10 | 1.3250 | 0.10 | 1.3240 | 0.10 | 1.3220 |
| 0.15 | 1.6240 | 0.15 | 1.6240 | 0.15 | 1.6240 |
| 0.20 | 2.4850 | 0.20 | 2.4810 | 0.20 | 2.4800 |
| 0.25 | 5.8870 | 0.25 | 5.6390 | 0.25 | 5.6380 |

### *8.5.3 Robustness against to load disturbances*

The important aim of the frequency regulation is to minimize the frequency derivation of the system after appearing the load disturbances. And the time delays, if considered, also effect the frequency performance. Such disturbance rejection capability for different delays is described by the delay-dependent $H_\infty$ performance index defined in Section 8.4.

Assume that the time delays in different control loops are identical. By using the method given in Section 8.4, the performance index, $\gamma_{min}$, of closed-loop LFC scheme are calculated and some typical values are listed in Table 8.2, where Cases A, B, and C, respectively, indicate the LFC with only DDC, with DDC and BESS,

*Table 8.3 Dynamic performance indices (ITAE and ITSE)*

| Parameter change (%) | Case A | | Case B | | Case C | |
|---|---|---|---|---|---|---|
| | **ITAE** | **ITSE** | **ITAE** | **ITSE** | **ITAE** | **ITSE** |
| +20 | 51.66 | 0.9890 | 35.43 | 0.5627 | 23.10 | 0.3215 |
| +15 | 49.49 | 0.9085 | 34.72 | 0.5292 | 22.04 | 0.2933 |
| +10 | 47.29 | 0.8325 | 33.86 | 0.4949 | 21.00 | 0.2667 |
| +5 | 45.02 | 0.7591 | 32.89 | 0.4600 | 19.97 | 0.2415 |
| 0 | 42.74 | 0.6886 | 31.86 | 0.4256 | 18.96 | 0.2178 |
| −5 | 40.54 | 0.6223 | 30.75 | 0.3917 | 17.99 | 0.1956 |
| −10 | 38.40 | 0.5595 | 29.57 | 0.3585 | 17.05 | 0.1750 |
| −15 | 36.31 | 0.4998 | 27.32 | 0.3261 | 16.12 | 0.1561 |
| −20 | 34.26 | 0.4434 | 27.01 | 0.2947 | 15.21 | 0.1391 |

with DDC, BESS, and PEV. The results show that the value of the performance index increases with the increasing of the time delay. The smaller performance index, $\gamma_{\min}$, indicates that the closed-loop LFC scheme has better disturbance rejection capability, which means that the smart power grid can accept more load disturbances, such as intermittent wind power energy. Thus, the injection of the DDC, BESS, and PEV indeed improves the capability of introducing intermittent wind power energy.

### *8.5.4 Robustness against to parameters uncertainties*

The PID controller gains aforementioned are tuned for the nominal system parameters. However, in reality, there exist uncertainties in the system parameters due to measurement errors, operation condition change, etc., as well in the controller gains during the implementation procedure. Therefore, the robustness against to those parameter uncertainties is also tested. To indicate the dynamic performances, the following two indices, the integral of the time multiplied absolute value of the error (ITAE) and the integral of the time multiplied square of the error (ITSE), with respect to ACE are defined:

$$ITAE = \int_0^t t|ACE|dt \tag{8.69}$$

$$ITSE = \int_0^t tACE^2 dt \tag{8.70}$$

For three cases defined in the previous subsection, uncertainties in system parameters (within ±20%) are simulated, and some typical values of the ITAE and the ITSE are given in Table 8.3. And Figure 8.13 shows the frequency deviation of LFC with DDC, BESS, and PEV under system parameters in normal value, increasing, and decreasing 20%, respectively.

From the results, it can be found that the designed PID controller stabilizes the closed-loop LFC system even the uncertainties exist, which shows the robustness of the controller against to parameter uncertainties. Meanwhile, it can be found that

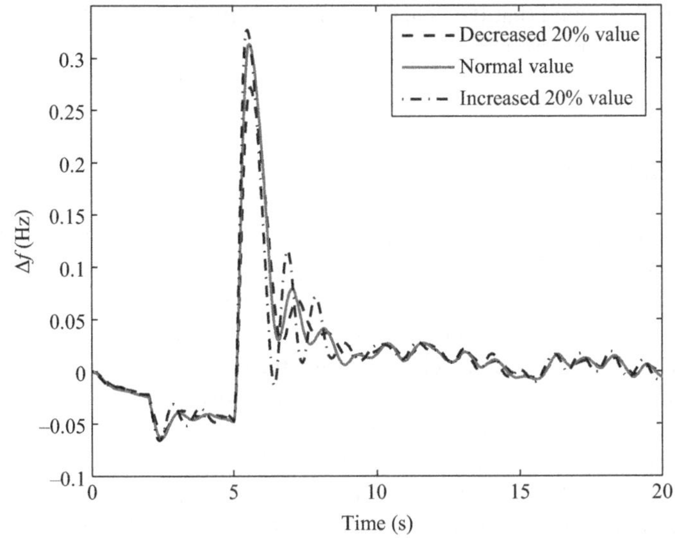


*Figure 8.13* *Frequency deviation of LFC with DDC, BESS, and PEV under system parameters variation*

*Table 8.4* *The delay margins in LFC with DDC, BESS, and PEV model*

| $\bar{\tau}_1$ (s) | Calculated $\bar{\tau}_2$ (s) | Simulated $\bar{\tau}_2$ (s) |
|---|---|---|
| 0 | 1.20 | 1.39 |
| 0.05 | 1.13 | 1.40 |
| 0.10 | 1.14 | 1.42 |
| 0.15 | 1.11 | 1.44 |
| 0.20 | 1.31 | 1.45 |
| 0.25 | 1.30 | 1.46 |

the introduction of DDC, BESS, and PEV can stabilize the system more quickly with smaller fluctuation and provide better performance indices.

## *8.5.5 Robustness against to time delays*

The PID controller gains given in Section 8.5.1 are tuned by setting all delays to be identical. However, in reality, there usually exist different delays for different loops. Assume that the time delays in LFC and BESS loops are same, and the ones in DDC and PEV loops are also same, i.e., $\bar{\tau}_1 = \tau_1 = \tau_2$ and $\bar{\tau}_2 = \tau_3 = \tau_4$. This part calculates the delay margins by using the method given in Section 8.3.

The delay margins for the case of the LFC with DDC, BESS, and PEV are listed in Table 8.4, in which the delay margins obtained by using simulation method are also given. It is found that the calculated delay margins are smaller than simulated delay margins, but these two kinds of results are very similar, which shows that the method in Section 8.3 is feasible to calculate multiple delay margins.

## 8.6 Conclusion

In this chapter, the contribution of the DDC, the BESS, and the PEV for frequency regulation has been investigated for a smart power grid operating in isolated mode with intermittent wind power injections. First, the smart power grid, including the BESS, the DDC, the PEVs, and the wind farm, is modelled as a single-area LFC system, and the state-space model considering the communication time delays in the control loops has been constructed. Then, the delay-dependent stability analysis and PID controller design methods have been developed by using the Lyapunov theory and the LMI technique. Case studies based on the single-area power system with the DDC, the BESS, and the PEV have been carried out. Analysis and simulation results have demonstrated that the LFC with the DDC, the BESS, and the PEV can provide better performance compared with other cases, which means the combined DDC, BESS, and PEVs can effectively alleviate the power unbalance caused by intermittent wind power. Moreover, simulation results have also demonstrated the effectiveness of the proposed PID controller, including the robustness against to the load disturbances, the parameter uncertainties, and multiple time delays. In addition, delay margins calculated based on the developed stability analysis method and the ones obtained via trial-and-error simulation method are very closed to each other, which shows that the proposed delay-dependent stability analysis method is effective and can be used to theoretically investigate the robustness of the controller to multiple time delays.

## Bibliography

[1] Short J. A., Infield D. G., Freris L. L.: 'Stabilization of grid frequency through dynamic demand control', *IEEE Trans. Power Syst.*, 2007, **22**, (3), pp. 1284–1293.

[2] Shashi K. P., Soumya R. M., Nand K.: 'A literature survey on load-frequency control for conventional and distribution generation power systems', *Renew. Sust. Energ. Rev.*, 2013, **25**, pp. 318–334.

[3] Bevrani H.: *Robust Power System Frequency Control* (Springer, Berlin, 2009).

[4] Yao W., Lee K. Y., Texas W.: 'A wind farm configuration for load-following control and its application to primary frequency support', in *Proceedings of the North American Power Symposium*, Boston, MA, Aug. 2011, pp. 1–6.

[5] Kyoho R., Goya T., Mengyan W., *et al.*: 'Thermal units communication with demand response to optimize battery storage capacity', in *Proceedings of the IEEE 10th PEDS International Conference*, Kitakyushu, Japan, Apr. 2013, pp. 1207–1212.

[6] Dhanalakshmi R., Palaniswami S.: 'Application of multi-stage fuzzy logic control for load frequency control of an isolated wind diesel hybrid power system', in *Proceedings of the International Conference on GTEC*, Chennai, India, Dec. 2011, pp. 309–315.

[7] Bevrani H., Daneshmand P. R.: 'Fuzzy logic based load frequency control concerning high penetration of wind turbines', *IEEE Syst. J.*, 2012, **6**, (1), pp. 173–180.

[8] Mohod S. W., Aware M. V.: 'Micro wind power generator with battery energy storage for critical load', *IEEE Syst. J.*, 2012, **6**, (1), pp. 118–125.

[9] Yao D. L., Choi S. S., Tseng K. L., Lie T. T.: 'Determination of short-term power dispatch schedule for a wind farm incorporated with dual-battery energy storage scheme', *IEEE Trans. Sust. Energy*, 2012, **3**, (1), pp. 74–84.

[10] Aditya S. K., Das D.: 'Battery energy storage for load frequency control of an interconnected power system', *Electric Power Syst. Res.*, 2001, **58**, (3), pp. 179–185.

[11] Ma Y., Houghton T., Cruden A., Infield D.: 'Modeling the benefits of vehicle-to-grid technology to a power system', *IEEE Trans. Power Syst.*, 2012, **27**, (2), pp. 1012–1020.

[12] Shimizu K., Masuta T., Ota Y., Yokoyama A.: 'Load frequency control in power system using vehicle-to-grid system considering the customer convenience of electric vehicles', in *Proceedings of the 2010 International Conference on POWERCON*, Hangzhou, China, Oct. 2010, pp. 1–8.

[13] Pillai J. R., Bak-Jensen B.: 'Integration of vehicle-to-grid in the western Danish power system', *IEEE Trans. Sust. Energy*, 2011, **2**, (1), pp. 12–19.

[14] Yang H., Chung C. Y., Zhao J.: 'Application of plug-in electric vehicles to frequency regulation based on distributed signal acquisition via limited communication', *IEEE Trans. Power Syst.*, 2013, **28**, (2), pp. 1017–1026.

[15] Pham T. N., Trinh H., Hien L. V.: 'Load frequency control of power systems with electric vehicle and diverse transmission links using distributed functional observers', *IEEE Trans. Smart Grid*, 2016, **7**, (1), pp. 238–252.

[16] Datta M., Senjyu T.: 'Fuzzy control of distributed PV inverters/energy storage systems/electric vehicles for frequency regulation in a large power system', *IEEE Trans. Smart Grid*, 2013, **4**, (1), pp. 479–488.

[17] Ota Y., Taniguchi H., Nakajima T, Liyanage K. M., Baba J., Yokoyama A.: 'Autonomous distributed V2G (vehicle-to-grid) satisfying scheduled charging', *IEEE Trans. Smart Grid*, 2012, **3**, (1), pp. 559–564.

[18] Luo X., Xia S., Chan K. W.: 'A decentralized charging control strategy for plug-in electric vehicles to mitigate wind farm intermittency and enhance frequency regulation', *J. Power Sources*, 2014, **248**, pp. 604–614.

[19] Pourmousavi S. A., Nehrir M. H.: 'Introducing dynamic demand response in the LFC model', *IEEE Trans. Power Syst.*, 2014, **29**, (4), pp. 1562–1572.

[20] Cheng Y., Tabrizi M., Sahni M., Povedano A., Nichols D.: 'Dynamic available AGC based approach for enhancing utility scale energy storage performance', *IEEE Trans. Smart Grid*, 2014, **5**, (2), pp. 1070–1078.

[21] Oldewurtel F., Borsche T., Bucher M., *et al.*: 'A framework for and assessment of demand response and energy storage in power systems', in *Proceedings of the 2013 IREP Symposium Conference*, Rethymno, Greece, Aug. 2013, pp. 1–24.

[22] Jiang L., Yao W., Wu Q. H., Wen J. Y., Cheng S. J.: 'Delay-dependent stability for load frequency control with constant and time-varying delays', *IEEE Trans. Power Syst.*, 2012, **27**, (2), pp. 932–941.

[23] Bhowmik S., Tomsovic K., Bose A.: 'Communication models for third party load frequency control', *IEEE Trans. Power Syst.*, 2004, **19**, (1), pp. 543–548.

[24] Zhang C. K., Jiang L., Wu Q. H., He Y., Wu M.: 'Delay-dependent robust load frequency control for time delay power systems', *IEEE Trans. Power Syst.*, 2013, **28**, (3), pp. 2192–2201.

[25] Zhang C. K., Jiang L., Wu Q. H., He Y., Wu M.: 'Further results on delay-dependent stability of multi-area load frequency control', *IEEE Trans. Power Syst.*, 2013, **28**, (4), pp. 4465–4474.

[26] Bevrani H., Hiyama T.: 'Robust decentralised PI based LFC design for time delay power systems', *Energy Convers. Manage.*, 2008, **49**, (2), pp. 193–204.

[27] Bevrani H., Hiyama T.: 'On load-frequency regulation with time delays: design and real-time implementation', *IEEE Trans. Energy Convers.*, 2009, **24**, (1), pp. 292–300.

[28] Brooks A. N.: *Vehicle-to-Grid Demonstration Project: Grid Regulation Ancillary Service with a Battery Electric Vehicle* (California Environmental Protection Agency, Air Resources Board, Research Division, 2002).

[29] Tang Y., Yang J., Yan J., He H.: 'Intelligent load frequency controller using GrADP for island smart grid with electrical vehicles and renewable resources', *Neurocomputing*, 2015, **170**, pp. 406–416.

[30] Thomsen S. C., Poulsen N. K.: 'A disturbance decoupling nonlinear control law for variable speed wind turbines', in *Proceedings of the 15th Mediterranean Conference*, Athens, Greece, Jun. 2007, pp. 1–6.

[31] Boukhezzar B., Siguerdidjane H.: 'Nonlinear control of a variable-speed wind turbine using a two mass model', *IEEE Trans. Energy Convers.*, 2011, **26**, (1), pp. 149–162.

[32] Chang-Chien L., Sun C., Yeh Y.: 'Modeling of wind farm participation in AGC', *IEEE Trans. Power Syst.*, 2013, **29**, (3), pp. 1204–1211.

[33] Hajizadeh A., Golkar M. A.: 'Intelligent power management strategy of hybrid distributed generation system', *Proc. Int. J. Elect. Power Energy Syst.*, 2007, **29**, (10), pp. 783–795.

[34] Samarakoon K., Ekanayake J., Jenkins N.: 'Investigation of domestic load control to provide primary frequency response using smart meters', *IEEE Trans. Smart Grid*, 2012, **3**, (1), pp. 282–292.

[35] Zhu Q., Yao W., Jiang L., Luo C., Wu Q. H.: 'Load frequency control with dynamic demand control for deregulated power system', in *Proceedings of the 2014 IEEE PES General Meeting*, Washington, DC, Jul. 2014, pp. 1–5.

[36] Lu N., Chassin D. P., Widergren S. E.: 'Modeling uncertainties in aggregated thermostatically controlled loads using a state queueing model', *IEEE Trans. Power Syst.*, 2005, **20**, (2), pp. 725–733.
[37] Seuret A., Gouaisbaut F.: 'Wirtinger-based integral inequality: application to time-delay systems', *Automatica*, 2013, **49**, (9), pp. 2860–2866.
[38] Balas G., Chiang R., Packard A., Safonov, M.: *Robust Control Toolbox User's Guide* (MathWorks, Natick, MA, 2010).
[39] Ting T., Rao M., Loo C.: 'A novel approach for unit commitment problem via an effective hybrid particle swarm optimization', *IEEE Trans. Power Syst.*, 2006, **21**, (1), pp. 411–418.
[40] Zhao J., Wen F., Dong Z. Y., Xue Y., Wong K. P.: 'Optimal dispatch of electric vehicle and wind power using exchange particle swarm optimization', *IEEE Trans. Ind. Inf.*, 2012, **8**, (4), pp. 889–899.
[41] Jay D., Swarup K. S.: 'Frequency restoration using dynamic demand control under smart grid environment', in *Proceedings of the IEEE PES ISGT*, Kollam, India, Dec. 2011, pp. 311–315.

*Chapter 9*

# Distributed frequency control and demand-side management

*E. Devane*[1], *A. Kasis*[2], *C. Spanias*[3], *M. Antoniou*[2] *and I. Lestas*[2,3]

## 9.1 Introduction

Reliable and efficient control of bus frequencies is crucial to the operation of any power grid. Classically, frequency control is implemented on the generation side via primary and secondary control schemes, with issues of optimality addressed at a much slower timescale via the optimal power flow (OPF) problem. In recent years, power systems have undergone significant changes, such as the liberalisation of the electricity markets, the introduction of new generation technologies, and the increased penetration of renewable energy sources. This expansion of power systems along with the stochastic nature of renewable energy inevitably lead to a need for faster, more efficient, and more reliable frequency control mechanisms. Furthermore, in a smart grid paradigm, frequency control schemes can be highly distributed due to the participation of the demand-side. Such approaches, incorporating control on both generators and loads, have the potential to reduce operational costs, improve system security, and increase the overall economic efficiency of the network's operation. In this chapter, we will discuss various approaches to distributed frequency control, paying particular attention to the incorporation of demand-side management and to the economic optimality of the schemes. We begin by describing the key concepts that will be considered.

### *9.1.1 Frequency control in the power grid*

The foremost objective in any power system is to reliably balance generation and demand. There are three main control schemes implemented in the grid in practice in order to achieve this, each corresponding to both a different timescale and a different control aim. The timescales for these control phases are illustrated in Figure 9.1. Initially, the operating set points of the grid's generators are set to balance the operator's

[1]Cambridge Centre for Analysis, University of Cambridge, Cambridge, UK
[2]Department of Engineering, University of Cambridge, Cambridge, UK
[3]Department of Electrical Engineering, Computer Engineering and Informatics, Cyprus University of Technology, Limassol, Cyprus.

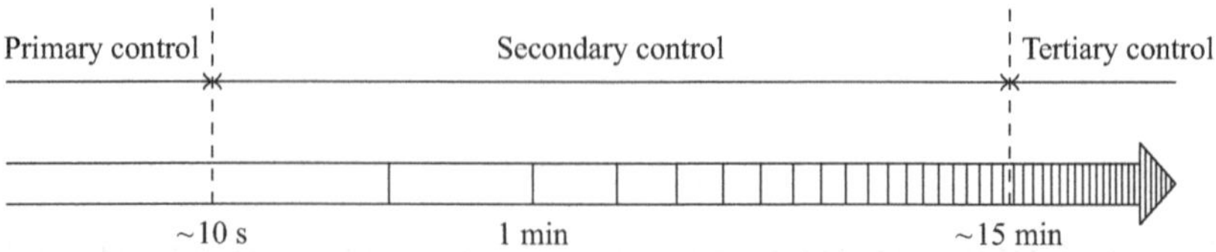


*Figure 9.1 Typical timescales of the main control schemes in the power grid. Typical values taken from Reference 1. Note that time is indicated on a logarithmic scale*

best projections (based upon demand forecasting[4]) of demand over the next time period. This describes the tertiary phase of control, which operates on a timescale of ~15 minutes to hours. Balance between supply and demand and security of system operation are then maintained in real time through the faster schemes of secondary and primary control. Both of these control stages are based on control mechanisms at the generating units that aim to satisfy the instantaneous demand and regulate the grid frequency to its nominal value; as such they are termed frequency control schemes. We now describe more precisely the goals of each of these separate schemes. A more in-depth description of each of these control stages and how they are implemented in practice can be found in various textbooks, such as References 2–4.

The main objective of primary control is to ensure that unexpected changes in generation/demand can be met on short timescales. Once a disturbance in the system[5] occurs, primary control schemes kick in within seconds to automatically adjust the speed governors on generator turbines so as to alter the generation profile in such a way that the extra power required is provided. An adverse consequence of changing the rotation rates of synchronous generators is that the grid frequency, given at each bus as the rate of change of the voltage phase angle, will deviate from its nominal value[6]. The frequency will drift downwards if there is an excess of demand over supply, and vice versa. In order to prevent these deviations growing too large, which could damage grid infrastructure and appliances connected to the grid, primary control schemes are designed in a way so as to regulate the frequency towards a steady-state value lying within permitted tolerances of nominal. Furthermore, such schemes are usually distributed in nature, with each turbine governor acting based upon its own local frequency measurement, in order to enable the control actuations to be enacted at the requisite fast timescales. A typical example of such a scheme is droop control, which is usually a proportional feedback scheme that gets activated when the frequency deviation is sufficiently large.

[4]Although the usage of individual consumers is highly variable, individual users' consumption is uncorrelated on short timescales, meaning that aggregate system demand can be predicted with relatively high accuracy even up to the day-ahead level.

[5]Such a disturbance can be, for example, an unexpected additional load connecting to the grid or a sudden drop in generation capability.

[6]The nominal frequency is a constant, uniform frequency that represents the target frequency at which we wish the system to operate. In Europe, this would typically be 50 Hz, while in North America it is 60 Hz.

Primary control can usually meet the excess demand; however, it typically will result in nonzero deviations of grid frequency away from nominal. Therefore, secondary control is required to adjust the generation profiles such that nominal frequency is restored, while also maintaining balance between generation and demand. This is achieved by means of a centralised control action that permits communication with turbine controllers and also activates necessary spinning reserve in order to restore the bus frequencies to the common nominal value. Due to this additional communication overhead and the need to activate spinning reserve, which is typically available at a timescale of tens of seconds to minutes, secondary control is generally active in the time frame from $\sim$10 s to $\sim$15 min after a disturbance. In addition to frequency restoration, secondary control is also responsible for ensuring that tie-line power flows between neighbouring areas are close to their scheduled values. Automatic generation control (AGC), e.g. References 5–8, corresponds to a class of typical implementations of secondary control.

**Remark 9.1.** *In the most common case of small disturbances, the above control schemes are usually sufficient to maintain power balance and avoid frequency collapse. However, in the event of larger disturbances, more aggressive control approaches can become necessary. One example of these is load shedding, whereby certain loads are disconnected from the grid. Commercial opportunities exist (see Reference 9 for example) for large consumers to offer a load reduction service to the grid operator for use in such situations. In emergencies, such as if multiple generators fail, other loads may also have to be disconnected in order to avoid frequency collapse and the resulting infrastructure damage. If this is needed in the UK, the National Grid issues a Demand Control Imminent notification, requesting distributors to reduce their total loads [10]. A recent example of this was on 27 May 2008 (see Reference 11), when concurrent failures of two power plants resulted in a loss of* 1 510 MW *of generation capacity and required significant load shedding to stabilise the grid frequency. The consequent blackouts affected an estimated* 500 000 *customers.*

### *9.1.2 Optimality in frequency control*

In principle, there could be many possible set points that can be chosen at the tertiary control stage such that the equations governing the power flows in the network admit a feasible solution under the projected active and reactive power demands. Economic efficiency, however, motivates the idea of trying to select the choice that will minimise the projected operating costs while satisfying the physical and operational constraints of the network. This leads to the idea of economic dispatch, which is defined in Reference 12 as 'the operation of generation facilities to produce energy at the lowest cost to reliably serve consumers, recognising any operational limits of generation and transmission facilities'. The corresponding optimisation problem that aims to determine a cost-minimal generator dispatch that yields a feasible generation and controllable demand schedule (i.e. can meet the projected demands) and also satisfies all reasonable physical limitations on system operation (generator limits, bus voltage limits, line flow constraints, etc.) is known as the OPF problem. The complexity of this problem, however, makes a direct implementation of OPF in real time generally infeasible. Consequently, typical implementations of primary and secondary control often rebalance the network without consideration of the optimality of their actions. Therefore, in order to further increase the economic efficiency of the grid's operation, it is desirable to investigate whether it is possible to incorporate some form of optimality within the primary and secondary control loops. In particular, control schemes should be considered that minimise some measure of the cost incurred due to the changes in the generation and load profiles, while still achieving the objectives of primary and secondary control described in Section 9.1.1.

### *9.1.3 Demand-side management*

The rising focus on renewable sources of energy is expected to increase their penetration in modern power systems. However, due to the fluctuations associated with renewable power generation, imbalances between generation and demand are expected to become more frequent. This will inevitably result in a loss of efficiency with severe consequences to the grid, as spinning reserves with fast response times are in general a costly investment. Demand-side management can potentially contribute significantly in addressing this issue due to the ability of controllable loads to provide fast response to power imbalances until backup conventional generation can be brought online to provide the necessary additional power. The changes in demand could be incurred by household appliances like air conditioners, refrigerators, and water or space heaters. For the considered timescale, ranging from a few seconds to a few minutes, the altered demand for these appliances is not expected to significantly affect consumers' convenience and usage.

The idea of using controllable devices for power balancing and frequency restoration dates back to the 1970s [13], while the use of frequency as a control signal for demand response was patented [14] in the US in 1979. The patent stated that consumers that are not critically dependent on energy could make use of this concept, which was named frequency adaptive power-energy re-scheduler (FAPER). More recently, the British firm ResponsiveLoad Limited has patented a frequency-dependent switching controller using various frequency limits to change the mode of appliance operation [15]. However, only in recent years has significant research attention focussed on the use of demand management [16] both for primary [17,18] and secondary control [19,20]. Simulations in all these studies have consistently shown significant improvements in system performance and reliability.

The practical benefits of using controllable devices have furthermore been demonstrated by several field tests. For instance, demand response schemes, together with various other smart grid technologies, have been successfully trialled in Ontario, Canada, by the company Hydro One [21]. Additionally, a demand response scheme was successfully implemented by the Pacific Northwest National Laboratory in 2006–2007, with 112 homes changing their electric consumption for water and space heating using price signals [22].

Based upon the promising results from field tests such as these, several recent initiatives and smart grid related projects encourage the use of smart appliances and demand-side management. For example, the IntelliGrid project devised by the Electric Power Research Institute (EPRI) in the US provides recommendations to maximise utilisation using existing infrastructure, and demand response is included as a tool to achieve this objective. Several aspects of its proposed architecture have been implemented by utility companies and there are plans for further demonstration projects which include the incorporation of demand response schemes based upon price signals [23]. Moreover, Grid2030 [24] is a vision statement for the US electrical system, proposed by various important stakeholders, describing a pathway for the future evolution of the power grid in terms of generation, transmission, distribution, and storage with demand response included. The GRID4EU project [25], mainly funded by the European Commission with the cooperation of six distribution companies, aims for the development and study of smart grid technologies. It specified targets on the use of more renewable energy sources, connected to distribution networks, to encourage customer participation in electricity markets and to develop means of demand-side management.

In order to allow the large-scale integration of controllable devices within the network, it is necessary to be able to derive stability guarantees that ensure that their inevitably highly distributed nature and their interaction with the electromechanical behaviour of the power system and existing frequency control schemes will not lead to damaging instabilities. Recent studies have tried to address the stability issue by providing stability guarantees for several classes of generation/controllable demand dynamics on arbitrary networks. It is an open research problem to determine the least conservative generation and controllable demand dynamics that allow for such stability guarantees on general network topologies.

A further issue raised is that of fairness in power allocation among controllable loads. It is crucial to be able to justify that any enforced deviations in user device consumption levels are equitably distributed. Recent attempts to address this issue have generally considered the use of control schemes that guarantee that all equilibria of the power network system solve some appropriately constructed optimisation problem that ensures the desired fairness in allocation. Several such schemes have been proposed in the literature for both primary and secondary control, with the common feature being the presence of some synchronising variable in order to ensure agreement of marginal costs. The main difference between these two classes of problems is that frequency can be used as a synchronising variable for primary control, allowing fairness to be achieved without any exchange of additional information, whereas for secondary control an exchange signal needs to be communicated in order for fair allocations to be guaranteed.

## 9.2 Swing equation dynamics

In order to investigate various classes of primary and secondary control schemes, consider a network model consisting of a collection of buses $N$ and a set of flow lines $E$ joining them. The set of buses $N$ is comprised of buses $G$, which have nonzero inertia, and load buses $L = N \backslash G$ with no inertia. The flow lines $E$ are assumed to be lossless and each line $(i,k) \in E$ is characterised by its susceptance $B_{ik} = B_{ki} > 0$. Furthermore, we assume that the graph $(N,E)$ is directed with arbitrary direction[7] so that if $(i,k) \in E$ then $(k,i) \notin E$, and we denote by $k: k \to i$ and $k: i \to k$, respectively, the sets of predecessor and successor buses of any $i \in N$. The bus frequencies then satisfy the following swing equation dynamics (e.g. References 26 and Section 5.1 of Reference 4):

$$\frac{d\eta_{ik}}{dt} = \omega_i - \omega_k, \quad (i,k) \in E, \tag{9.1a}$$

$$M_i \frac{d\omega_i}{dt} = -p_i^L + p_i^M - (d_i^c + d_i^u) - \sum_{k:i\to k} p_{ik} + \sum_{k:k\to i} p_{ki}, \quad i \in G, \tag{9.1b}$$

$$0 = -p_i^L - (d_i^c + d_i^u) - \sum_{k:i\to k} p_{ik} + \sum_{k:k\to i} p_{ki}, \quad i \in L, \tag{9.1c}$$

$$p_{ik} = B_{ik} V_i V_k \sin \eta_{ik} - p_{ik}^{nom}, \quad (i,k) \in E. \tag{9.1d}$$

In the system (9.1), the terms are defined as follows:

- $\eta_{ik}$ – the phase difference between buses $i, k \in N$; equal to $\theta_i - \theta_k$ in terms of the individual voltage phases,
- $V_i$ – the voltage magnitude at a bus $i \in N$,

and the following variables represent deviations of the respective quantities from their values $\omega_i^{nom}, p_{ik}^{nom}, p_i^{M,nom}, d_i^{c,nom}, d_i^{u,nom}$ at the system's nominal operating point:

- $\omega_i$ – the frequency at a bus $i \in N$; equal to $\frac{d\theta_i}{dt}$ in terms of the individual voltage phases,
- $p_{ik}$ – the active power flow along a line $(i,k) \in E$,
- $p_i^M$ – the active power injected by the generation at a bus $i \in G$,
- $d_i^c$ – the controllable demand at a bus $i \in N$,
- $d_i^u$ – the uncontrollable, frequency-dependent demand and generation damping at a bus $i \in N$.

The constant $M_i$ represents the generation inertia at a bus $i \in G$, while the value $p_i^L$ is a given step-change in uncontrollable, frequency-independent load at a bus $i \in N$. The aim of the control schemes that will be considered is to appropriately stabilise the system (9.1) by fairly and optimally allocating this change in $p_i^L$ to available generating units and controllable loads.

**Remark 9.2.** *It can be seen that the dynamics in* (9.1) *depend directly on the bus voltage magnitudes* $V_i$ *through* (9.1d). *Thus, the evolution of the variables in* (9.1) *is coupled with the dynamics of these bus voltages. Several models exist describing how these voltage variables evolve with time, many of which are described in e.g. References 3,4,26,27. However, since the timescales for primary and secondary control are relatively short, and the time constant of voltage evolution in a typical*

[7] Note that the analysis of the system that follows is unaltered by any change of graph ordering.

*synchronous generator model is substantial, it is often reasonable to assume that the voltage magnitudes are fixed. By normalising the variables appropriately, we can just assume that these magnitudes take the value* 1. *This assumption is often made so as to simplify the analysis in general network topologies and also in studies where issues of optimality are addressed.*

## 9.3 Primary frequency control

As discussed in Section 9.1.1, the aim of primary control is to couple the system (9.1) with some control dynamics for the generation injections $p_i^M$ and the controllable loads $d_i^c$ such that the solutions of the interconnected system balance generation and load and stabilise frequency. Global power balance can be ensured by summing (9.1b) and (9.1c) over all buses whenever the frequency is constant. Thus, to achieve the goals of primary control, it is sufficient to ensure convergence of the frequency variables in (9.1) to appropriate constant values, independent of time. It should be noted that, as is clear from the form of (9.1a), all grid frequencies synchronise to a single unique value at equilibrium.

### *9.3.1 Historical development*

The first paper to directly address the question of frequency stability in power systems from an analytical perspective was [28], in which Steinmetz considered how problematic oscillatory behaviour can be avoided in a model of two coupled synchronous generators. Due to the unavailability of computing power, many of the investigations that followed focussed either on simple models of this form or on practical experiments or field tests. Several major instability issues in the North American grid in the 1970s (e.g. Toronto, January 1974 and Missouri/Illinois, February 1978) led to a proliferation in frequency stability investigations. For example, the paper [29] proved stability for two classes of distributed frequency control schemes using a linearized swing equation model, while Kundur *et al.* [30] considered the impact of such controllers on the generation system as a whole. The substantial textbook [3] presented a thorough discussion of the types of control scheme that can be implemented, detailing the mathematical frameworks within which they should be analysed. For illustration, we recount the most common implementation of primary control, which is governor droop control whereby the turbine governor increases the mechanical torque of the generator in direct proportion with its measurement of the frequency's local deviation from nominal. The performance of these droop control schemes within the European grid between 1976 and 1996 was reviewed in Reference 31, while also various studies based on more detailed models can be found for example in References 3,4,32,33.

### *9.3.2 Passivity conditions for stability analysis*

A key structural property that has often been used in the analysis of large-scale systems is the notion of passivity. Examples of its application within power system studies date back to [34], which considered the swing dynamics (9.1) with a first-order model for the voltage dynamics and passivity-based techniques were used to deduce stability. More recently, passivity studies in power systems have typically used the framework of port-Hamiltonian systems (described in Reference 35). Examples of this approach include References 36–39. Two additional interesting analyses, which demonstrated that passivity-based approaches can also be applied to study the behaviour of (9.1) under dynamic control schemes for $p_i^M$, can be found in References 40,41. Both of these papers considered zero controllable demand and linear damping, i.e. $d_i^c = 0$ and $d_i^u = D_i\omega_i$. The former made use of passivity arguments to prove stability when the linearized form of (9.1) is coupled with the controller:

$$p_i^M = -K_i(s)\omega_i \tag{9.2}$$

for any positive real transfer function $K_i(s)$. This corresponds to generalising classical droop control to permit arbitrary passive generation controllers. The latter paper included also first-order voltage dynamics and used similar passivity arguments to demonstrate asymptotic stability of the frequency dynamics in (9.1) under the class of droop controllers:

$$p_i^M = -K_i^\omega \omega_i - K_i^V(V_i - V_i^{nom}) \tag{9.3}$$

with constants $K_i^\omega, K_i^V > 0$, in models where frequency regulation resources are shared between several AC subgrids interconnected across DC links. It will further be seen below that passivity approaches offer excellent potential for guaranteeing convergence in (9.1) under broad classes of primary frequency control dynamics. Furthermore, when used in conjunction with appropriate input–output conditions on the system dynamics they can lead to an optimality interpretation for the equilibrium point reached.

### 9.3.3 Economic optimality and fairness in primary control

We discussed in Section 9.1.2 the idea that the dispatch of all available generation resources is typically optimised in the tertiary control step in order to schedule set points that minimise some measure of cost. However, on the shorter timescales of primary and secondary control, traditionally generation outputs are altered in response to disturbances with the sole intention of rebalancing of the supply and avoiding damaging instabilities. As such, the new generation levels will in general fail to be optimal and furthermore certain generators may be unduly burdened with a disproportionate share of the total allocation. This presents opportunities for significant economic efficiency improvements, while the fairness issue becomes particularly profound if user appliances are to be used to provide controllable load for frequency control in the manner described in Section 9.1.3. These ideas motivate attempting to design the controllers used in primary/secondary control in such a way that their steady states either achieve optimality for an appropriate cost minimisation or provide some form of fairness of the allocation among ancillary service providers.

It is evident that, in order to achieve either optimality or fairness, a nonzero synchronising variable is necessary in order to enable all buses to adapt their generation and controllable demand levels so as to attain equal marginal costs. In primary control, the synchronisation of frequencies allows the frequency deviation from nominal to be used for this purpose, allowing distributed control to be achieved. An example of this was described in Reference 42, where the authors demonstrated that if the units' droop coefficients are selected proportionally, then the power changes resulting from the droop control are also shared proportionally among them. Furthermore, an interesting result shown in Reference 43 implies that, in principle, given any DC-OPF formulation, it is possible to choose appropriate droop coefficients such that the steady states are optimal.

An alternative approach was taken in Reference 44, in which the authors considered the linearized dynamics in (9.1) with constant voltages, damping $d_i^u = D_i\omega_i$, fixed $p_i^M$, and controllable demand $d_i^c$. They then formulated the optimisation problem:

$$\begin{aligned}
\underset{\omega, d^c}{\text{minimise}} \quad & \sum_{i\in N} \left(C_{di}(d_i^c) + \tfrac{1}{2}D_i\omega_i^2\right) \\
\text{subject to} \quad & \sum_{i\in G} p_i^M = \sum_{i\in N} (d_i^c + D_i\omega_i + p_i^L), \\
& d_i^{c,min} \le d_i^c \le d_i^{c,max}, \quad \forall i \in N,
\end{aligned} \tag{9.4}$$

which explicitly penalises changes in controllable demand from nominal via an arbitrary strictly convex disutility function $C_{di}$ and changes in frequency from nominal via the quadratic term, subject to requiring power balance and boundedness of the demand values. It was then proved, under appropriate assumptions, that the distributed static control dynamics:

$$d_i^c = \left[(C'_{di})^{-1}(\omega_i)\right]_{d_i^{c,min}}^{d_i^{c,max}} \tag{9.5}$$

ensure convergence of the variables in (9.1) to equilibrium points which are optimal for (9.4). This analysis demonstrated that it is possible to design control dynamics such that the load-side participation in primary control can be assured of achieving a prescribed measure of economic optimality.

A similar approach was used (without linearizations) in Reference 45 to show that convergence to optimality for an analogous optimisation problem (with an added cost term $\sum_{i\in G} C_i(p_i^M)$ penalising deviations in generation from nominal) is also attained with the nontrivial second-order generation dynamics:

$$\left.\begin{aligned}
\frac{d\alpha_i}{dt} &= -\frac{1}{\tau_{g,i}}\alpha_i + \frac{1}{\tau_{g,i}}p_i^c, \\
\frac{dp_i^M}{dt} &= -\frac{1}{\tau_{b,i}}p_i^M + \frac{1}{\tau_{b,i}}\alpha_i,
\end{aligned}\right\} \quad i \in G \tag{9.6}$$

for $p_i^M$, provided that the command signal is $p_i^c = [(C_i')^{-1}(-\omega_i)]_{p_i^{c,min}}^{p_i^{c,max}}$ and this satisfies the gain constraint:

$$|p_i^c(\omega_i) - p_i^c(\omega_i^*)| \leq L_i|\omega_i - \omega_i^*| \tag{9.7}$$

for some constant $L_i < D_i$ in some neighbourhood of any equilibrium frequency $\omega_i^*$. Therefore, we see that control schemes for both generation and controllable load can be coupled to (9.1) so as to achieve the goals of primary control and also achieve minimality for an appropriately constructed optimisation problem.

Similar approaches to these have been employed in numerous recent papers within the literature. Examples include [46], which used the linearized framework to describe a distributed dynamic load control scheme that ensures convergence to equilibria minimising (9.4), and [47], which directly interpreted the interconnection of (9.1) with static generation and load controllers including deadband regions as an optimisation scheme for a cost minimisation problem.

### 9.3.4 Supply passivity framework for demand-side integration

We now discuss a general framework, in which generation and load dynamics are combined under a single condition of combined supply passivity, that allows stability of the swing dynamics (9.1) to be guaranteed and permits the integration of broad classes of distributed controllable demand dynamics. Furthermore, by combining this condition with appropriate input–output properties of the dynamics, we can obtain an optimality interpretation of such schemes that ensures fairness in the resulting power allocations. This general framework was first discussed in Reference 48, and the full details can be found in Reference 49.

The general framework will be defined in terms of dynamical systems with input $u(t) \in \mathbb{R}$, state $x(t) \in \mathbb{R}^m$, and output $y(t) \in \mathbb{R}$ with a state-space realisation of the following form:

$$\begin{aligned} \frac{dx}{dt} &= f(x,u), \\ y &= g(x,u), \end{aligned} \tag{9.8}$$

where $f : \mathbb{R}^m \times \mathbb{R} \to \mathbb{R}^m$ is locally Lipschitz and $g : \mathbb{R}^m \times \mathbb{R} \to \mathbb{R}$ is continuous. We will consider systems having the property that given any constant input $u(t) \equiv \bar{u} \in \mathbb{R}$, there exists a unique[8] locally asymptotically stable equilibrium point $\bar{x} \in \mathbb{R}^m$, i.e. $f(\bar{x}, \bar{u}) = 0$. We can then define the static input-state characteristic map $k_x : \mathbb{R} \to \mathbb{R}^m$ by $k_x(\bar{u}) := \bar{x}$. Based on this, we can also define the static input–output characteristic map $k_y : \mathbb{R} \to \mathbb{R}$,

$$k_y(\bar{u}) := g(k_x(\bar{u}), \bar{u}). \tag{9.9}$$

[8]This uniqueness could be relaxed to require only isolated equilibrium points; however, we assume it here to simplify the presentation.

We then close the loop in (9.1) by supposing that each of the variables $p_i^M$, $d_i^c$, and $d_i^u$ are determined as outputs from systems of the form of (9.8) with inputs given by the local bus frequency measurements $\omega_i$, viz.:

$$\left.\begin{aligned} \frac{dx^{M,i}}{dt} &= f^{M,i}(x^{M,i}, -\omega_i), \\ p_i^M &= g^{M,i}(x^{M,i}, -\omega_i), \end{aligned}\right\} \quad i \in G, \tag{9.10a}$$

$$\left.\begin{aligned} \frac{dx^{c,i}}{dt} &= f^{c,i}(x^{c,i}, -\omega_i), \\ d_i^c &= g^{c,i}(x^{c,i}, -\omega_i), \end{aligned}\right\} \quad i \in N, \tag{9.10b}$$

$$\left.\begin{aligned} \frac{dx^{u,i}}{dt} &= f^{u,i}(x^{u,i}, -\omega_i), \\ d_i^u &= g^{u,i}(x^{u,i}, -\omega_i), \end{aligned}\right\} \quad i \in N. \tag{9.10c}$$

The systems (9.10a) and (9.10b) represent the primary control dynamics associated with the generation at bus $i \in G$ and with the controllable demand at bus $i \in N$. Since the variables $p_i^M$ and $d_i^c$ can be controlled, we have freedom in our analysis to design certain properties of the dynamics in (9.10a) and (9.10b). By contrast, $d_i^u$ represents uncontrollable load and the dynamics in (9.10c) are thus fixed.

We suppose that the interconnected systems (9.1)–(9.10) admit some equilibrium $(\eta^*, \omega^*, x^{M,*}, x^{c,*}, x^{u,*}, p^*, p^{M,*}, d^{c,*}, d^{u,*})$ for which the phase differences satisfy the security constraint $|\eta_{ik}^*| < \frac{\pi}{2}$ for all $(i,k) \in E$. As discussed previously, our aim is to ensure that these equilibria are stable and are also solutions to an appropriately constructed network optimisation problem.

The key idea is now to combine the dynamics in (9.10) to consider the behaviour of the net supply deviation variables:

$$s_i^G = p_i^M - (d_i^c + d_i^u), \quad i \in G, \tag{9.11a}$$

$$s_i^L = -(d_i^c + d_i^u), \quad i \in L. \tag{9.11b}$$

As the variables in (9.11) evolve according to the dynamics in (9.10), $s_i^G$ and $s_i^L$ can be viewed as outputs from these combined dynamical systems with inputs $-\omega_i$. The broad class of dynamics in (9.10) that will be considered is then defined by the following passivity condition on these supply dynamics.

**Definition 9.3.** A system of the form (9.8) is said to be locally input-strictly passive about the constant input values $\bar{u}$ and the constant state values $\bar{x}$ if there exist open neighbourhoods $U$ of $\bar{u}$ and $X$ of $\bar{x}$ and a continuously differentiable, positive semidefinite function $V(x)$ (the storage function) such that, for all $u \in U$ and all $x \in X$,

$$\dot{V}(x) \leq (u - \bar{u})^T (y - \bar{y}) - \phi(u - \bar{u}),$$

where $\phi$ is a positive definite function and $\bar{y} = k_y(\bar{u})$.

**Assumption 9.4.** *The combined system dynamics in* (9.10) *with inputs* $-\omega_i$ *and outputs given by* (9.11a) *for* $i \in G$ *and* (9.11b) *for* $i \in L$ *are each locally input-strictly passive about equilibrium. Moreover, the storage functions have strict local minima at equilibrium.*

The fact that we assume only a passivity property for the supply dynamics without specifying the precise form of the systems (9.10) allows the framework to include a broad class of generation and load dynamics. In particular, this permits a large amount of flexibility in the design of the controllable load dynamics (9.10b). Furthermore, it should be noted that the presence of damping in the dynamics (9.10c) often naturally contributes a degree of strict passivity to the supply dynamics, so even non-passive behaviour from the controllable loads can be tolerated.

For linear systems, it is straightforward to verify the conditions in Assumption 9.4 either by checking strict positive realness of the transfer function or numerically using the Kalman-Yakubovich-Popov (KYP) lemma. Examples of classes of dynamics for which Assumption 9.4 are satisfied include sign-preserving static nonlinearities, first-order linear dynamics, and higher-order dynamics provided sufficiently large damping is present.

The passivity condition in Assumption 9.4 is sufficient to lead to the desired stability result. We now combine this approach with design conditions on the input–output behaviour of the control dynamics in (9.10a) and (9.10b) that allow an optimisation interpretation of the limiting steady states to be inferred. To do this, we consider the following simple optimisation problem:

$$\underline{\text{OSLC:}}\quad \begin{cases} \underset{p^M, d^c, d^u}{\text{minimise}} & \sum_{i\in G} C_i(p_i^M) + \sum_{i\in N}\left(C_{di}(d_i^c) + \int_0^{d_i^u} h_i^{-1}(z)\,dz\right) \\ \text{subject to} & \sum_{i\in G} p_i^M = \sum_{i\in N}(d_i^c + d_i^u + p_i^L), \\ & p_i^{M,min} \le p_i^M \le p_i^{M,max}, \quad \forall i \in G, \\ & d_i^{c,min} \le d_i^c \le d_i^{c,max}, \quad \forall i \in N. \end{cases} \tag{9.12}$$

In the Optimal Supply and Load Control (OSLC) problem (9.12), $p_i^{M,min}$, $p_i^{M,max}$, $d_i^{c,min}$, and $d_i^{c,max}$ are the bounds for generation and controllable demand deviations, respectively, at bus $i$. The equality constraint specifies balance between frequency-independent load and the total generation plus all the frequency-dependent loads. Some cost is naturally incurred due to any change in frequency which alters uncontrollable demand. This is represented by an integral cost in terms of the function $h_i$, which we determine from the dynamics in (9.10c) as

$$h_i(\bar{u}) := k_{d_i^u}(-\bar{u}) \quad \text{for all} \quad \bar{u} \in \mathbb{R}. \tag{9.13}$$

In addition, we impose the following natural conditions in order to penalise deviations from nominal values.

**Assumption 9.5.** *The cost functions* $C_i$ *and* $C_{di}$ *are continuously differentiable and strictly convex. Additionally, the first derivative of* $h_i^{-1}(z)$ *is nonnegative for all* $z \in \mathbb{R}$.

The differentiability and strict convexity conditions in Assumption 9.5 are reasonable for most common cost functions and are sufficient to allow the use of the Karush–Kuhn–Tucker (KKT) conditions [50] within the optimality proof. To include dynamics incorporating deadbands, it is possible to permit discontinuities in the derivatives by using the subgradient KKT conditions [51]. The assumption on $h_i$ corresponds to the static input–output characteristic of the uncontrollable frequency-independent load dynamics (9.10c) being increasing, which is reasonable since we expect the contribution of such loads to increase with frequency.

Under the stated assumptions,[9] we then obtain the following result, which guarantees stabilisation of the frequencies by the dynamics considered, and also the optimality of the resulting steady state for the OSLC problem (9.12).

**Theorem 9.6.** *If Assumptions 9.4 and 9.5 are satisfied and the control dynamics in* (9.10a) *and* (9.10b) *are chosen such that:*

$$k_{p_i^M}(-\bar{u}) = [(C_i')^{-1}(-\bar{u})]_{p_i^{M,min}}^{p_i^{M,max}} \quad \text{and} \quad k_{d_i^c}(-\bar{u}) = [(C_{di}')^{-1}(\bar{u})]_{d_i^{c,min}}^{d_i^{c,max}} \tag{9.14}$$

*hold for all $\bar{u} \in \mathbb{R}$, then for any initial conditions sufficiently close to the equilibrium set, the solutions of the systems* (9.1)–(9.10) *are guaranteed to converge to an equilibrium point that is globally minimal for the OSLC problem* (9.12).

**Remark 9.7.** *The proof of Theorem 9.6 proceeds in two stages. First, the local asymptotic stability of the equilibrium set is proved by invoking Assumption 9.4 to generate a Lyapunov function to use LaSalle's Invariance Principle. Second, the optimality of these equilibrium points is deduced by showing that the equilibrium conditions for the systems* (9.1), (9.10) *satisfy the KKT conditions under the specifications* (9.14). *The full proof can be found in Reference 49.*

The conclusions of Theorem 9.6 thus demonstrate that, under small enough perturbations in the system loadings $p_i^L$, the primary control dynamics in (9.10a) and (9.10b) are guaranteed to stabilise the grid frequencies in (9.1) and furthermore to ensure convergence of the generation and controllable load allocations to optimality for the cost-minimisation problem (9.12). The local nature of this stability result arises naturally due to the nonlinearity of the dynamics considered. An estimate of the region of attraction can be obtained by means of the Lyapunov function used in the proof of the theorem.

It can be seen that the optimisation problem (9.12) recovers (9.4) as a special case. Furthermore, the control dynamics (9.5) and (9.6) correspond to subsystems (9.10a) and (9.10b) such that Assumption 9.4 is satisfied when $d_i^u = D_i\omega_i$. Hence, it follows that studies such as References 44,45 fit within the general supply passivity

[9]A simple technical condition stipulating that, at any time instant, the values of $\omega_i$ at buses $i \in L$ should be locally uniquely defined by the other state variables in (9.1) is also required. This condition can easily be verified in a distributed manner and is expected to hold in all realistic system models. See Reference 49 for a discussion of this assumption.

framework described here. Moreover, it can be shown that Assumption 9.4 is preserved under weaker conditions than the gain constraint in (9.7), meaning that stability and optimality can be deduced in less restrictive parameter regimes. Additionally, various other interesting control schemes fit within this framework, for instance the distributed dynamic load control:

$$\frac{dd_i^c}{dt} = -(C_i'(d_i^c) - \omega_i), \quad i \in N. \tag{9.15}$$

Thus, by considering this general framework, stability and optimality can be verified for broad classes of interesting primary control dynamics for both generation and controllable load.

## 9.4 Secondary frequency control

Once primary control has stabilised the grid frequency, secondary control is employed, as described in Section 9.1.1, to regulate this frequency back towards its nominal value. The control schemes for generation $p_i^M$ and controllable load $d_i^c$ thus need to be designed so as to drive the solutions of (9.1) to an equilibrium with frequency deviation equal to zero. In contrast to primary control, the fact that the frequency deviations return to zero means that a different variable needs to be used for synchronisation, if an optimal allocation is desired. Therefore, the control schemes involved will typically require some communicated variable. It should be noted that this communication requirement makes it important to carefully consider which devices to include within secondary control as the participation of large numbers of small loads could imply a significant communication overhead. It is therefore likely that provision of secondary control resources might best be restricted to generators and larger controllable loads. In that case, the remaining loads (which may have been controllable and hence represented within the $d_i^c$ terms for the purpose of primary control) would be modelled within the uncontrollable terms $d_i^u$.

### 9.4.1 *Historical development*

Studies concerned with frequency regulation through AGC date back to the 1950s, with References 52,53 mainly focussing on tie-line based control techniques. Over the following decades, many more studies were performed on the topic, mainly considering linearized models of two/multi-area systems [54–56]. An *n*-area system was investigated in Reference 54, which considered noninteraction between frequency and tie-line controllers, while in Reference 55 the authors studied a multi-area power network model and gave recommendations to improve stability margins, comparing them with the existing regulations of the North American Power Systems Committee. Linear analysis is only justifiable in the presence of small perturbations. This has been noted in studies considering system nonlinearities such as governor deadband [57] and nonlinear tie-line bias control [58]. Artificial intelligence techniques have been employed to permit the study of models that change according to the operating conditions, thus allowing a more realistic representation of power systems. For instance, neural network approaches have offered many advantages in the study of systems operating in nonlinear regimes. This approach was applied on a four-area system with nonlinear turbine dynamics in Reference 59 and on a single-area and a two-area system in Reference 60. In both cases, there were significant performance improvements compared to integral control action. It should be noted that further to the neural network approach, fuzzy logic [61] and genetic algorithm techniques [62,63] have also been applied to this problem, all demonstrating satisfactory performance characteristics.

In the early days, the AGC problem was dealt with through centralised control strategies [54,55,64]. This approach has the important limitation of requiring communication, computation, and storage infrastructures. Decentralised control techniques appeared later in an effort to deal with these complexities [65–67]. For further discussion and a more thorough review on AGC, refer Reference [68].

### *9.4.2 Economic optimality and fairness in secondary control*

As we have previously discussed in Section 9.1.1, the aim of secondary frequency control is to restore frequency to its nominal value. Traditionally, this task is carried out by generators who adapt their power production by using frequency as a control signal, usually via some integral control action. As mentioned in Section 9.1.2, this results in new generation levels that fail in general to be economically optimal. Furthermore, if controllable devices' participation in secondary control is desired, then it is important to be able to ensure fairness in the power allocation between them. Thus, analogously to the discussion for primary control in Section 9.3.3, these topics present research opportunities to derive control schemes that would ensure fairness and economic optimality in secondary frequency control.

As previously discussed, in order to ensure equality of the users' marginal costs so as to achieve optimality, a nonzero synchronising variable is required. This makes the frequency deviations employed for this purpose in Section 9.3.3 unsuitable to use here, since in secondary control the frequency returns to its nominal value. Therefore, a different variable needs to be synchronised, which presents the need for a communication network to provide the necessary information exchange for that synchronisation to happen.

Several recent studies have attempted to devise control schemes such that the steady-state conditions coincide with the solutions of an appropriately constructed optimisation problem which ensures economic optimality and/or fairness in power allocation between loads. An example of such a study is Reference 69, which considered a linearization of the swing equations (9.1), damping terms $d^u = D_j\omega_j$ with $D_j > 0$, first-order dynamics for generation, and constant demand. The authors then posed the optimisation problem:

$$\begin{aligned} \underset{p^M,p}{\text{minimise}} \quad & \sum_{i\in N} C_i(p_i^M) \\ \text{subject to} \quad & \sum_{i\in G} p_i^M = \sum_{i\in G}\left(p_i^L + \sum_{k:i\to k} p_{ik} - \sum_{k:k\to i} p_{ki}\right), \end{aligned} \tag{9.16}$$

which penalises deviations from nominal value in power generation via a strictly convex function $C_i$. If the equality constraint above is satisfied at equilibrium in the dynamics described by (9.1), then $\sum_{i\in N} d_i^u = \sum_{i\in N} D_i\omega_i = \omega \sum_{i\in N} D_i = 0$ holds at steady state, which immediately implies that frequency does indeed return to its nominal value. The authors then introduced auxiliary variables to represent the exchange of information between buses and demonstrated that the closed-loop system dynamics resemble a dual form of the optimisation problem (9.16). This equivalence allowed the desired steady-state optimisation property to be guaranteed. A similar approach has also been followed in a number of other studies, all including a constraint within the optimisation problem to guarantee that frequency would return to its nominal value at steady state. In Reference 70, for example, a similar system including controllable demand and constant generation was studied. The following optimisation problem, called frequency-preserving optimal load control (FP-OLC) was constructed:

$$
\begin{aligned}
\underset{d^c,d^u,p,R}{\text{minimise}} \quad & \sum_{i\in N}\left(C_i(d_i^c) + \frac{(d_i^u)^2}{2D_i}\right) \\
\text{subject to} \quad & p_i^M - (d_i^c + d_i^u) = \sum_{k:i\to k} p_{ik} - \sum_{k:k\to i} p_{ki}, \qquad i \in N, \\
& p_i^M - d_i^c = \sum_{k:i\to k} R_{ik} - \sum_{k:k\to i} R_{ki}, \qquad i \in N,
\end{aligned}
\tag{9.17}
$$

penalising the deviation in controllable demand from its nominal value, again via a strictly convex function $C_j$. The formulation of this optimisation problem ensured that any optimal solution has zero steady-state deviation in frequency. In addition, the inclusion within the cost function of a term that is proportional to the frequency deviation, which does not change the solution of the problem, allowed the authors to show via the KKT conditions that for optimality, the frequency, and the Lagrange multiplier associated with the first constraint needed to be equivalent. Furthermore, the Lagrange variable associated with the second constraint can be conveniently thought of as representing a power command signal. This approach motivated the following intuitive dynamics for this power command signals, which could be used as the requisite exchange variables:

$$\dot{p}_i^c = p_i^M - d_i^c - \sum_{k:i\to k} R_{ik} + \sum_{k:k\to i} R_{ik}, \quad i \in G, \tag{9.18a}$$

$$\dot{p}_i^c = -d_i^c - \sum_{k:i\to k} R_{ik} + \sum_{k:k\to i} R_{ki}, \quad i \in L, \tag{9.18b}$$

$$\dot{R}_{ik} = p_i^c - p_k^c, \quad (i,k) \in E. \tag{9.18c}$$

It should be noted that the communication graph implicit within (9.18) does not necessarily need to be the same as the graph representing the power network. The controllable demand values are then specified by

$$d_j^c = C_j'^{-1}(\omega_j + p_j^c). \tag{9.19}$$

The dynamics in (9.18c) ensure synchronisation of the power command variables, and therefore enable the closed-loop dynamics (9.1), (9.18), (9.19) to achieve optimal solution of the FP-OLC problem (9.17). This analysis shows that it is possible to design control schemes for controllable loads that will ensure fairness in the power allocation between them.

A similar approach was adopted in Reference 19, using the same dynamics as in Reference 70 but adding constraints on power transfers. In Reference 71, the nonlinear swing equations (9.1) were used together with the second-order generation dynamics described in (9.6) above and static controllable loads, and distributed control schemes were proposed for the generation/controllable demand such that the equilibrium points considered were optimal for a prescribed optimisation problem. Stability of the system was guaranteed through the imposition of a gain condition, similar to the one described in (9.7). A further study which utilised the nonlinear swing equations and also included voltage dynamics within its analysis is Reference 20. Furthermore, in Reference 72 the authors imposed steady-state conditions that ensure that the power injection alterations in micro-grids are proportional to the users' droop coefficients, guaranteeing fair power allocation. Finally, Zhang and Papachristodoulou [73] demonstrated the solution of an OPF optimisation problem with constraints on transmission and load power consumption, through the use of gradient-based distributed control laws within a nonlinear swing equation model with first-order generation dynamics.

### *9.4.3 Stability guarantees via a dissipativity framework*

It can be shown that the framework introduced in Section 9.3.4 can be extended to deduce stability and optimality for broad classes of systems relevant for secondary control. In particular, consider any bus that participates in secondary frequency control, and note that it can be regarded as a subsystem with inputs given by the power transfers into the bus and the information signals transmitted to it from other buses, and outputs given by the local frequency and its own information signal which will be transmitted to other buses. It can easily be shown that the interconnection matrix associated with these bus dynamics is passive, which is a property that can be exploited to deduce local conditions through which network stability and optimality can be deduced (see e.g. the early stability results in Reference 74, and also more recent studies such as References 75–77). In analogy to Section 9.3.4, the problem therefore reduces to one of imposing appropriate input–output conditions on the local bus dynamics so that the equilibrium points of interest are stable and solve a desired network optimisation problem. It can be shown [78] that a dissipativity condition on the aggregate generation dynamics at each bus can be sufficient to deduce stability in a general network. It should also be noted that power command dynamics, as in (9.18), are an important part of these mechanisms, as they lead to a synchronising variable through which network optimality can be deduced, while their structure also ensures that the frequency returns to its nominal value.

## 9.5 Future challenges

The results and studies described in this chapter illustrate that appropriate design can potentially enable distributed demand-side management schemes to be incorporated in modern power systems, despite their often highly distributed nature. Nevertheless, there are still many challenges that need to be addressed from both a theoretical and implementation perspective.

As mentioned in Remark 9.2, most of the current works on distributed frequency control in which issues of stability and optimality are jointly addressed have adopted dynamics similar to (9.1) so as to facilitate the analysis. Although the proposed schemes behave well when tested on more complex and detailed network models, it would be interesting to expand the current models used for theoretical analysis on general network topologies so as to also take into account additional features such as the dependence on bus voltages and the presence of various mechanisms that assist network operation, such as exciters and power system stabilisers (PSSs). It is, therefore, a challenge for researchers to adopt more advanced network models and embed voltage dynamics and reactive power flows into their analysis when issues of stability and optimality are addressed within a common framework.

As discussed in Section 9.1.3, demand-side management by means of controllable loads could improve power system operation and efficiency. However, the presence of a large number of loads contributing to frequency control could lead to undesirable situations. As noted in References 79–81, load-side distributed frequency control schemes can potentially result in problematic situations due to the inherent limit cycle behaviour of many types of loads, such as thermostatically controlled units. The latter could lead to undesirable synchronisation phenomena that could cause large transients and oscillatory behaviour in the frequency of the overall system. The need to avoid the synchronisation of loads that participate in frequency control was addressed in Reference 82, where the authors suggested an approach that is based on the introduction of randomisation in the control policies. The analysis of such schemes is therefore important and it would be interesting for these to be further investigated at the network level.

The technological advances in power electronics during the last decades have also facilitated the development of electronic equipment that offer the ability to handle large amounts of power. This has led to an increasing use of such technologies in existing power systems. These electronic devices, called flexible AC transmission system (FACTS), are based on electronic power converters and provide the ability to make quick adjustments that can enhance the ability to control a power system [83–85]. The incorporation of FACTS devices in distributed frequency control mechanisms could therefore be an additional feature that can be exploited since they can provide extensive control of both the voltage and the power flow [86–88]. It would be rather interesting to use the framework described in Section 9.3.4 in order to investigate distributed control schemes through which more advanced formulations of network optimisation problems can be solved, with voltage constraints and reactive power also taken into account. This will allow to improve the efficiency of the grid, by allowing versions of the OPF problem to be solved at faster timescales, without compromising the stability of the network. Furthermore, the ability of FACTS devices to control the flow of active and reactive power, along with their fast-acting nature, makes them easily applicable to contribute to both primary and secondary frequency and voltage control.

## Bibliography


[1] M. Beck and M. Scherer, *Overview of Ancillary Services*. SwissGrid, Berlin, 2010.

[2] H. Bevrani, *Robust Power System Frequency Control*. Springer, Berlin, 2009.

[3] P. Kundur, *Power System Stability and Control*. McGraw-Hill, New York, NY, 1994.

[4] J. Machowski, J. Bialek, and J. Bumby, *Power System Dynamics: Stability and Control*. John Wiley & Sons, New York, NY, 2011.

[5] IEEE Technical Committee, "Standard definitions of terms for automatic generation control on electric power systems," *IEEE Transactions on Power Apparatus and Systems*, vol. 89, no. 6, pp. 1356–1364, 1970.

[6] I. Ibraheem, P. Kumar, and D. P. Kothari, "Recent philosophies of automatic generation control strategies in power systems," *IEEE Transactions on Power Systems*, vol. 20, no. 1, pp. 346–357, 2005.

[7] N. Jaleeli, L. S. VanSlyck, D. N. Ewart, L. H. Fink, and A. G. Hoffmann, "Understanding automatic generation control," *IEEE Transactions on Power Systems*, vol. 7, no. 3, pp. 1106–1122, 1992.

[8] J. Nanda and B. Kaul, "Automatic generation control of an interconnected power system," *Proceedings of the Institution of Electrical Engineers*, vol. 125, no. 5, pp. 385–390, 1978.

[9] M. Duffield, "Commercial opportunities for back-up generation (including diesel generators) and load reduction via National Grid, the National Electricity Transmission System Operator (NETSO) for England, Scotland, Wales and Offshore," *Claverton Energy Research Group*, 2009.

[10] N. Williams, "How we manage risk," *National Grid Connecting*, 2015.

[11] M. Milner and G. Wearden, "Q&A: Blackout Britain," *The Guardian*, 2008.

[12] Congress of the United States of America, "Energy policy act," *Public Law*, vol. 109, p. 54, 2005.

[13] F. C. Schweppe, R. D. Tabors, J. L. Kirtley, H. R. Outhred, F. H. Pickel, and A. J. Cox, "Homeostatic utility control," *IEEE Transactions on Power Apparatus and Systems*, vol. 99, no. 3, pp. 1151–1163, 1980.

[14] F. C. Schweppe, "US patent 4317049," *Massachusetts Institute of Technology*, 1979.

[15] D. Hirst, "UK patent gb2361118," *ResponsiveLoad Ltd.*, 2001.

[16] J. A. Short, D. G. Infield, and L. L. Freris, "Stabilization of grid frequency through dynamic demand control," *IEEE Transactions on Power Systems*, vol. 22, no. 3, pp. 1284–1293, 2007.

[17] D. Trudnowski, M. Donnelly, and E. Lightner, "Power-system frequency and stability control using decentralized intelligent loads," in *Proceedings of the IEEE Power and Energy Society General Meeting*, Dallas, TX, US, pp. 1453–1459, 2006.
[18] A. Molina-Garciá, F. Bouffard, and D. S. Kirschen, "Decentralized demand-side contribution to primary frequency control," *IEEE Transactions on Power Systems*, vol. 26, no. 1, pp. 411–419, 2010.
[19] E. Mallada, C. Zhao, and S. Low, "Optimal load-side control for frequency regulation in smart grids," in *Proceedings of the 52nd Annual Allerton Conference on Communication, Control, and Computing (Allerton)*, Monticello, IL, US, pp. 731–738, 2014.
[20] S. Trip, M. Burger, and C. D. Persis, "An internal model approach to frequency regulation in inverter-based microgrids with time-varying voltages," in *Proceedings of the 53rd IEEE Conference on Decision and Control (CDC)*, Los Angeles, CA, US, pp. 223–228, 2014.
[21] A. Bettencourt and J. Malenfant, "Hydro One approach to smart grid," *Presentations to the OEB's Smart Grid Advisory Committee*, 2013.
[22] J. B. D. Hammerstrom, D. Chassin, G. Horst, *et al.*, "Pacific Northwest GridWise testbed demonstration projects, part II: grid friendly appliance project," *Pacific Northwest National Laboratory*, 2007.
[23] Electric Power Research Institute, "Enabling a smart grid by applying information and communication technologies," *IntelliGrid Program 2013 Annual Overview*, 2014.
[24] Office of Electric Transmission and Distribution, "Grid 2030, a national vision for electricity's second 100 years," *United States Department of Energy*, 2003.
[25] European Commission, "Grid4EU." Available from http://www.grid4eu.eu/. Accessed on: 24-07-2016.
[26] A. R. Bergen and V. Vittal, *Power Systems Analysis*. Prentice Hall, Englewood Cliffs, NJ, 1999.
[27] M. Ilic and J. Zaborszky, *Dynamics and Control of Large Electric Power Systems*. Wiley–IEEE Press, Hoboken, NJ, US, 2001.
[28] C. P. Steinmetz, "Power control and stability of electric generating stations," *Transactions of the American Institute of Electrical Engineers*, vol. 2, no. 39, pp. 1215–1287, 1920.
[29] E. J. Davison and N. Tripathi, "The optimal decentralized control of a large power system: load and frequency control," *IEEE Transactions on Automatic Control*, vol. 23, no. 2, pp. 312–325, 1978.
[30] P. Kundur, D. C. Lee, J. P. Bayne, and P. L. Dandeno, "Impact of turbine generator overspeed controls on unit performance under system disturbance conditions," *IEEE Transactions on Power Apparatus and Systems*, vol. 104, no. 6, pp. 1262–1269, 1985.
[31] H. W. Weber, H. P. Asal, and E. Grebe, "Characteristic numbers of primary control in the UCPTE power system and future requirement," in *Proceedings of the ETG Summer Meeting*, Berlin, Germany, pp. 395–405, 1997.

[32] P. M. Anderson and A. A. Fouad, *Power System Control and Stability*. Wiley–IEEE Press, Hoboken, NJ, US, 2002.

[33] L. Grigsby, *Power System Stability and Control*. CRC Press, Boca Raton, FL, 2012.

[34] H. Miyagi and A. R. Bergen, "Stability studies of multimachine power systems with the effects of automatic voltage regulators," *IEEE Transactions on Automatic Control*, vol. 31, no. 3, pp. 210–215, 1986.

[35] A. J. van der Schaft and B. M. Maschke, "Port-Hamiltonian systems on graphs," *SIAM Journal on Control and Optimization*, vol. 51, no. 2, pp. 906–937, 2013.

[36] B. Maschke, R. Ortega, and A. J. van der Schaft, "Energy-based Lyapunov functions for forced Hamiltonian systems with dissipation," *IEEE Transactions on Automatic Control*, vol. 45, no. 8, pp. 1498–1502, 2000.

[37] Y. Wang, D. Cheng, C. Li, and Y. Ge, "Dissipative Hamiltonian realization and energy-based $L_2$-disturbance attenuation control of multimachine power systems," *IEEE Transactions on Automatic Control*, vol. 48, no. 8, pp. 1428–1433, 2003.

[38] S. Fiaz, D. Zonetti, R. Ortega, J. M. A. Scherpen, and A. J. van der Schaft, "A port-Hamiltonian approach to power network modelling and analysis," *European Journal of Control*, vol. 19, no. 6, pp. 477–485, 2013.

[39] S. Y. Caliskan and P. Tabuada, "Compositional transient stability analysis of multimachine power networks," *IEEE Transactions on Control of Network Systems*, vol. 1, no. 1, pp. 4–14, 2014.

[40] J. Liu, B. H. Krogh, and B. E. Ydstie, "Decentralized robust frequency control for power systems subject to wind power variability," in *Proceedings of the IEEE Power and Energy Society General Meeting*, Detroit, MI, US, pp. 1–8, 2011.

[41] M. Andreasson, R. Wiget, D. V. Dimarogonas, K. H. Johansson, and G. Andersson, "Distributed primary frequency control through multi-terminal HVDC transmission systems," in *Proceedings of the American Control Conference (ACC)*, Chicago, IL, US, pp. 5029–5034, 2015.

[42] J. W. Simpson-Porco, F. Dörfler, and F. Bullo, "Synchronization and power sharing for droop-controlled inverters in islanded microgrids," *Automatica*, vol. 49, no. 9, pp. 2603–2611, 2013.

[43] F. Dörfler, J. Simpson-Porco, and F. Bullo, "Breaking the hierarchy: distributed control & economic optimality in microgrids," *IEEE Transactions on Control of Network Systems*, 2016. To appear.

[44] C. Zhao, U. Topcu, N. Li, and S. H. Low, "Design and stability of load-side primary frequency control in power systems," *IEEE Transactions on Automatic Control*, vol. 59, no. 5, pp. 1177–1189, 2014.

[45] C. Zhao and S. H. Low, "Optimal decentralized primary frequency control in power networks," in *Proceedings of the 53rd IEEE Conference on Decision and Control (CDC)*, Los Angeles, CA, US, pp. 2467–2473, 2014.

[46] E. Mallada, C. Zhao, and S. H. Low, "Optimal load-side control for frequency regulation in smart grids," in *Proceedings of the 52nd Annual*

*Allerton Conference on Communication, Control, and Computing (Allerton)*, Monticello, IL, US, pp. 731–738, 2014.

[47] S. You and L. Chen, "Reverse and forward engineering of frequency control in power networks," in *Proceedings of the 53rd IEEE Conference on Decision and Control (CDC)*, Los Angeles, CA, US, pp. 191–198, 2014.

[48] A. Kasis, E. Devane, and I. Lestas, "On the stability and optimality of primary frequency regulation with load-side participation," in *Proceedings of the 54th IEEE Conference on Decision and Control (CDC)*, Osaka, Japan, 2015.

[49] A. Kasis, E. Devane, and I. Lestas, "Primary frequency regulation with load-side participation: stability and optimality," *arXiv:1602.02800*, Feb. 2016.

[50] S. Boyd and L. Vandenberghe, *Convex Optimization*. Cambridge University Press, Cambridge, 2004.

[51] R. T. Rockafellar, *Convex Analysis*. Princeton University Press, Princeton, NJ, 1970.

[52] N. Cohn, "Some aspects of tie-line bias control on interconnected power systems," *IEEE Transactions on Power Apparatus and Systems*, vol. 75, no. 3, pp. 1415–1436, 1956.

[53] C. Concordia and L. K. Kirchmayer, "Tie-line power and frequency control of electric power systems," *Transactions of the American Institute of Electrical Engineers*, vol. 72, no. 2, pp. 562–572, 1953.

[54] G. Quazza, "Noninteracting controls of interconnected electric power systems," *IEEE Transactions on Power Apparatus and Systems*, vol. 7, no. PAS-85, pp. 727–741, 1966.

[55] O. I. Elgerd and C. E. Fosha, "Optimum megawatt-frequency control of multiarea electric energy systems," *IEEE Transactions on Power Apparatus and Systems*, vol. 4, no. PAS-89, pp. 556–564, 1970.

[56] E. C. Tacker, T. W. Reddoch, O. T. Pan, and T. D. Linton, "Automatic generation control of electric energy systems – a simulation study," *IEEE Transactions on Systems, Man and Cybernetics*, vol. SMC-3, no. 4, pp. 403–405, 1973.

[57] F. F. Wu and V. S. Dea, "Describing-function analysis of automatic generation control system with governor deadband," *Electric Power Systems Research*, vol. 1, no. 2, p. 113–116, 1978.

[58] B. Oni, H. Graham, and L. Walker, "Investigation of non-linear tie line bias control of interconnected power systems," *IEEE Transactions on Power Apparatus and Systems*, vol. PAS-100, no. 5, pp. 2350–2356, 1981.

[59] H. L. Zeynelgil, A. Demiroren, and N. S. Sengor, "The application of ANN technique to automatic generation control for multi-area power system," *International Journal of Electrical Power & Energy Systems*, vol. 24, no. 5, pp. 345–354, 2002.

[60] D. K. Chaturvedi, P. S. Satsangi, and P. K. Kalra, "Load frequency control: a generalised neural network approach," *International Journal of Electrical Power & Energy Systems*, vol. 21, no. 6, pp. 405–415, 1999.

[61] C. S. Indulkar and B. Raj, "Application of fuzzy controller to automatic generation control," *Electric Machines and Power Systems*, vol. 23, no. 2, pp. 209–220, 1995.

[62] Z. M. Al-Hamouz and H. N. Al-Duwaish, "A new load frequency variable structure controller using genetic algorithms," *Electric Power Systems Research*, vol. 55, no. 1, pp. 1–6, 2000.

[63] D. Rerkpreedapong, A. Hasanovic, and A. Feliachi, "Robust load frequency control using genetic algorithms and linear matrix inequalities," *IEEE Transactions on Power Systems*, vol. 18, no. 2, pp. 855–861, 2003.

[64] C. E. F. O. I. Elgerd, "The megawatt-frequency control problem: a new approach via optimal control theory," *IEEE Transactions on Power Apparatus and Systems*, vol. 4, no. PAS-89, pp. 563–577, 1970.

[65] H. Kawabata and M. Kido, "A decentralized scheme of load frequency control of power system," *Electrical Engineering in Japan*, vol. 102, no. 4, pp. 100–106, 1982.

[66] M. S. Calovic, "Automatic generation control: decentralized area-wise optimal solution," *Electric Power Systems Research*, vol. 7, no. 2, pp. 115–139, 1984.

[67] M. Aldeen, "Interaction modelling approach to distributed control with application to power systems," *International Journal of Control*, vol. 53, no. 5, pp. 1035–1054, 1991.

[68] P. Kumar and D. P. Kothari, "Recent philosophies of automatic generation control strategies in power systems," *IEEE Transactions on Power Systems*, vol. 20, no. 1, pp. 346–357, 2005.

[69] N. Li, L. Chen, C. Zhao, and S. H. Low, "Economic automatic generation control," *Technical Report*, 2013.

[70] E. Mallada and S. H. Low, "Distributed frequency-preserving optimal load control," in *Proceedings of the International Federation of Automatic Control World Congress (IFAC)*, Cape Town, South Africa, 2014.

[71] C. Zhao, E. Mallada, and S. H. Low, "Distributed generator and load-side secondary frequency control in power networks," in *Proceedings of the 49th Annual Conference on Information Sciences and Systems (CISS)*, Baltimore, MD, US, pp. 1–6, 2015.

[72] H. Bouattour, J. W. Simpson-Porco, F. Dörfler, and F. Bullo, "Further results on distributed secondary control in microgrids," in *Proceedings of the 52nd IEEE Conference on Decision and Control (CDC)*, Firenze, Italy, pp. 1514–1519, 2013.

[73] X. Zhang and A. Papachristodoulou, "A real-time control framework for smart power networks with star topology," in *Proceedings of the American Control Conference (ACC)*, Washington, DC, United States, pp. 5062–5067, 2013.

[74] P. Moylan and D. Hill, "Stability criteria for large-scale systems," *IEEE Transactions on Automatic Control*, vol. 23, no. 2, pp. 143–149, 1978.

[75] J. T. Wen and M. Arcak, "A unifying passivity framework for network flow control," *IEEE Transactions on Automatic Control*, vol. 49, no. 2, pp. 162–174, 2004.

[76] I. Lestas, "Large scale heterogeneous networks, the Davis–Wielandt shell and graph separation," *SIAM Journal on Control and Optimization*, vol. 50, no. 4, pp. 1753–1774, 2012.

[77] I. Lestas, "On network stability, graph separation, interconnection structure and convex shells," in *Proceedings of the 50th IEEE Conference on Decision and Control and the European Control Conference (CDC/ECC)*, Orlando, FL, US, pp. 4257–4263, 2011.

[78] A. Kasis, E. Devane, and I. Lestas, "Stability and optimality of distributed schemes for secondary frequency regulation in power networks," In *Proceedings of the 55th IEEE Conference on Decision and Control (CDC), (Las Vegas, Nevada, US), 2016*. To appear.

[79] C. Hinrichs, U. Vogel, and M. Sonnenschein, "Modelling and evaluation of desynchronization strategies for controllable cooling devices," in *Proceedings of the Sixth Vienna International Conference on Mathematical Modelling (Mathmod)*, Vienna, Austria, pp. 220–225, 2009.

[80] M. Stadler, W. Krause, M. Sonnenschein, and U. Vogel, "The adaptive fridge – comparing different control schemes for enhancing load shifting of electricity demand," in *Proceedings of the 21st Conference on Informatics for Environmental Protection (Enviroinfo)*, Warsaw, Poland, pp. 199–206, 2007.

[81] J. Short, D. G. Infield, and L. L. Freris, "Stabilization of grid frequency through dynamic demand control," *IEEE Transactions on Power Systems*, vol. 22, no. 3, pp. 1284–1293, 2007.

[82] P. Angeli and P.-A. Kountouriotis, "A stochastic approach to 'dynamic-demand' refrigerator control," *IEEE Transactions on Control Systems Technology*, vol. 20, no. 3, pp. 581–592, 2012.

[83] Y.-H. Song and A. Johns, *Flexible AC Transmission Systems (FACTS)*. Institution of Engineering and Technology, London, 1999.

[84] N. G. Hingorani and L. Gyugyi, *Understanding FACTS*. John Wiley & Sons, New York, NY, 2000.

[85] M. Eslami, H. Shareef, A. Mohamed, and M. Khajehzadeh, "A survey on flexible AC transmission systems (FACTS)," *Przeglad Elektrotechniczny*, vol. 88, no. 1A, pp. 1–11, 2012.

[86] M. Abido, "Power system stability enhancement using FACTS controllers: a review," *The Arabian Journal for Science and Engineering*, vol. 34, no. 1B, pp. 153–172, 2009.

[87] G. Ramana, B. S. Ram, and P. Engg, "Power system stability improvement using FACTS with expert systems," *International Journal of Advances in Engineering & Technology*, vol. 1, no. 4, pp. 395–404, 2011.

[88] B. Singh, "Applications of FACTS controllers in power systems for enhance the power system stability: a state-of-the-art," *International Journal of Reviews in Computing*, vol. 6, pp. 40–69, 2011.

## Chapter 10

# Game theory approaches for demand side management in the smart grid

*Georgia Asimakopoulou*[1] *and Nikos Hatziargyriou*[1]

## 10.1 Introduction

During the last decades, the electricity systems have undergone significant changes in their structure and operation mainly driven by growing environmental awareness and the ever increasing application of information and communication technologies (ICTs). The installation of units producing energy from renewable sources both in the medium voltage (MV) and low voltage (LV) networks and the new challenges with regard to their integration posed by the intermittent nature of their generation; the rollout of smart meters that establish bidirectional communication between interacting entities; the advent of flexible demand and new types of loads, such as electric vehicles, create new conditions and a complex environment, in which a plethora of entities influencing each other with their decisions (e.g., Energy Services Providers, retailers, Distribution System Operator, Transmission System Operator, prosumers, Regulatory Authority) are called upon to operate.

In the distribution network, in particular, the changes are even more significant as the integration of an increased number of individual resources, increases the complexity of its structure and operation. The aggregation and coordination of distributed resources to appear as a controlled entity to the upstream network is particularly attractive. To this end, the concept of the microgrid (MG) has been introduced during the past years, as a structure that allows application of new control methods, in both interconnected, but also isolated mode of operation from the main grid, in order to maximize the benefits obtained from multiple individual resources connected to the distribution network [1].

Small local resources (as compared to the electricity system in total), when organized together and aggregated by a representing entity, inevitably affect the operation of the electricity market. Such an entity, responsible for managing the local resources, also interacts with the wholesale market in order to acquire at the prevailing market price, e.g., marginal price, any energy deficit that might arise. Thus, the optimal

[1]School of Electrical and Computer Engineering, National Technical University of Athens, Athens, Greece

decision regarding the energy production of aggregated local resources that minimizes the cost, although made at a local level, has an impact on the total system load. This, in turn, affects the optimal dispatch of conventional production units and, thus, the clearing price of the wholesale electricity market, with which the aggregating entity performs transactions.

From the customers' side, the form of the pricing scheme is of great interest, since, in combination with several other parameters regarding demographic characteristics (income, education, number of inhabitants per household) affects the decisions regarding energy consumption. On the other hand, the managing entity's profitability is affected by the customers' decisions regarding their consumption, by the form of the pricing scheme and by the prevailing conditions in the wholesale market with which energy transactions take place.

Evidently, in order to operate in such an environment, the various actors of the electricity marketplace will need to develop new services that call for new tools, methods, and algorithms, as traditional approaches might prove to be insufficient to handle the novel operating conditions and incapable of taking full advantage of the increased volume of available information. To this end, game theory is identified as particularly appropriate for simulating competitive or cooperative relationships and interactions between entities whose decision-making at a local level affects one another.

Situations such as these, with interdependent decision-making entities interacting in a hierarchical framework, have been identified as soon as 1934 when von Stackelberg first described the leader-follower strategy [2]. Following a parallel course, von Neumann and Morgenstern [3] established the min-max problem in game theory. It took, however, several decades, for the intervention of control theory and mathematical programming principles to develop techniques for solving such complex problems, now generally termed "multilevel programming problems."

Bilevel programming problems, the simplest and most widely used and studied from multilevel programming problems, constitute a mathematical representation of a Stackelberg game. Such a game is characterized by two levels of hierarchy; the upper level (leader) and the lower level (follower) coupled through the decision variables of each level (Figure 10.1). The decision-making process on each level encompasses the optimization of an objective function representing the interests of each entity with respect to a variable under the control of the specific entity. Even though the interacting entities cannot immediately prescribe the decision of the other, they nevertheless, affect indirectly the process, as each entity's decision variables participate in the objective function of the counterparty.

More specifically, the so-called leader has control over variable $x$ in order to minimize the objective function $F(x, y)$, satisfying the inequality $G(x, y)$ and equality $H(x, y)$ constraints. Given variable $x$ selected by the leader and observed by the follower, the latter optimally selects variable $y$ minimizing $f(x, y)$ subject to the respective inequality $g(x, y)$ and equality $h(x, y)$ constraints.

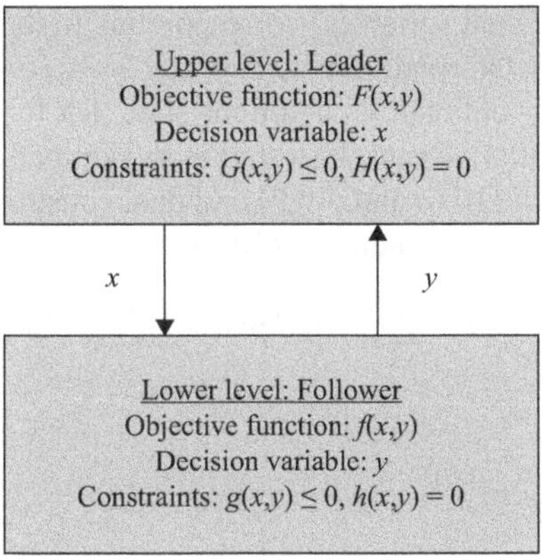


*Figure 10.1 Schematic diagram of the decision-making process in a bilevel framework*

What we obtain is merely an optimization problem with a second optimization problem in its constraints, that is described mathematically by (10.1)–(10.3).

$$\min_{x} F(\boldsymbol{x},\boldsymbol{y}) \tag{10.1}$$

subject to

$$\mathbf{G}(\boldsymbol{x},\boldsymbol{y}) \leq 0, \mathbf{H}(\boldsymbol{x},\boldsymbol{y}) = 0 \tag{10.2}$$

where

$$\boldsymbol{y} \in arg\left\{\min_{y} f(\boldsymbol{x},\boldsymbol{y}) \text{ s.t. } \mathbf{g}(\boldsymbol{x},\boldsymbol{y}) \leq \mathbf{0}, \mathbf{h}(\boldsymbol{x},\boldsymbol{y}) = 0\right\} \tag{10.3}$$

The constraints of the two levels appear here in vector format. This means that if, for example, $\mathbf{G}(\boldsymbol{x},\boldsymbol{y}) \leq 0$ comprises $m$ inequalities, these can be equivalently written as $G_m(\boldsymbol{x},\boldsymbol{y}) \leq 0, \forall m$.

In general, solving such problems is inherently difficult. Under convexity conditions, however, certain transformation techniques can be applied for reformulating the bilevel programming problem to an equivalent problem that can be solved using established algorithms. One classical approach involves replacing the lower-level problem (10.3) with the Karush–Kuhn–Tucker (KKT) necessary optimality conditions in order to obtain a single-level problem.

The Lagrangian of problem (10.3) is

$$\mathcal{L}(\boldsymbol{x},\boldsymbol{y},\mu_i,\lambda_j) = f(\boldsymbol{x},\boldsymbol{y}) + \sum_{i} \mu_i g_i(\boldsymbol{x},\boldsymbol{y}) + \sum_{j} \lambda_j h_j(\boldsymbol{x},\boldsymbol{y}) \tag{10.4}$$

where $\mu_i$ and $\lambda_j$ are the dual variables[2] corresponding to the inequality and equality constraints, respectively. On condition that $f$, $g_i$ and $h_j$ are continuously differentiable, the first order necessary optimality conditions state that for $\boldsymbol{y}^*$, $\boldsymbol{\lambda}^*$, $\boldsymbol{\mu}^*$ to be a local minimum of problem (10.3), it must be a stationary point of $\mathcal{L}$, i.e., (10.5) must hold, it must satisfy the primal (10.6) and (10.7) and dual constraint feasibility (10.8) and the complementary slackness conditions (10.9).

$$\nabla_y \mathcal{L}(\boldsymbol{x},\boldsymbol{y}) = \nabla_y f(\boldsymbol{x},\boldsymbol{y}) + \sum_i \mu_i \nabla_y g_i(\boldsymbol{x},\boldsymbol{y}) + \sum_j \lambda_j \nabla_y h_j(\boldsymbol{x},\boldsymbol{y}) = 0 \tag{10.5}$$

$$\mathbf{h}(\boldsymbol{x},\boldsymbol{y}) = 0 \tag{10.6}$$

$$g_i(\boldsymbol{x},\boldsymbol{y}) \leq 0, \quad \forall i \tag{10.7}$$

$$\mu_i \geq 0, \quad \forall i \tag{10.8}$$

$$\mu_i g_i(\boldsymbol{x},\boldsymbol{y}) = 0, \quad \forall i \tag{10.9}$$

Thus, the single-level problem consists of the upper-level objective function constrained by (10.11)–(10.14).[3]

$$\min_x F(\boldsymbol{x},\boldsymbol{y}) \tag{10.10}$$

$$\mathbf{G}(\boldsymbol{x},\boldsymbol{y}) \leq 0, \mathbf{H}(\boldsymbol{x},\boldsymbol{y}) = 0 \tag{10.11}$$

$$\nabla_y f(\boldsymbol{x},\boldsymbol{y}) + \sum_i \mu_i \nabla_y g_i(\boldsymbol{x},\boldsymbol{y}) + \sum_j \lambda_j \nabla_y h_j(\boldsymbol{x},\boldsymbol{y}) = 0 \tag{10.12}$$

$$0 \leq -g_i(\boldsymbol{x},\boldsymbol{y}) \perp \mu_i \geq 0, \quad \forall i \tag{10.13}$$

$$\mathbf{h}(\boldsymbol{x},\boldsymbol{y}) = 0 \tag{10.14}$$

Nonlinearity and nonconvexity of the single-level problem pose two significant challenges in finding the optimal solution. The first one can be tackled by applying linearization techniques, such as the Strong Duality Theorem (SDT) [4] and the big-M formulation [5] to transform the single-level problem to a mixed-integer linear programming (MILP) problem. This reformulation allows tackling the second challenge as well, since MILP problems have been extensively studied and existing methods can be used for solving such problems (e.g., enumeration algorithms, heuristic methods, cutting plane methods). Even then, however, one should be careful in interpreting the

[2]Dual variables, also known as Lagrange multipliers in equality constrained problems or KKT multipliers in generally constrained optimization problems (with equality and inequality constraints), form the basis of the classical method of mathematical optimization for finding extrema of a general multivariate function subject to a variety of equality and inequality constraints. Apart from enabling solving such problems using iterative methods, these multipliers also have an interesting physical interpretation especially when the problem under study is a cost minimization problem. In that case, they are also named "shadow prices" and express the effect on the objective function when the respective constraint is marginally relaxed.
[3]Equations (10.7)–(10.9) are written in the more compact but equivalent format of (10.13).

results, as the solution of the derived MILP problem corresponds to the solution of the original bilevel programming problem only in the optimistic case, i.e., when it is assumed that the leader is able to influence the decision of the follower to the one that suits him best [6].

### 10.1.1 Related bibliography

Notwithstanding the difficulties and constrictions mentioned in the previous paragraph, game theoretic frameworks have been used by several researchers for modeling the implications of demand response and demand-side management on parties involved in such operations, where interdependent decision-making processes implicate discrete competing entities with conflicting goals. In what follows, some selected relevant applications are presented. For a more extensive review of applications of game theory concepts in the smart grid, the interested reader could refer to Reference 7, while References 8,9 offer a more general view over bilevel programming.

In the context of the smart grid, in Reference 10, energy management of the resources is formulated as an energy consumption game between the users of the grid. A distributed algorithm is developed for determining the pricing scheme and the resulting electricity consumption schedule of household appliances.

Maharjan *et al.* [11] present a model for describing the interactions of several utility companies with various consumers, each one maximizing its revenues and its payoff, respectively. The utilities solve an optimal price-setting problem, while the consumers solve an optimal power consumption problem. The two types of problems are coupled through the unit prices set by the utilities and the power consumption selected by the consumers based on the unit prices. The Stackelberg game is solved using a distributed algorithm.

In Reference 12, the behavior of customers, described by a utility function, is affected by the pricing strategy of a company that delivers electricity to retail customers of different types (residential, commercial, and small industrial). The company, interested in maximizing the profit, optimally selects the time-of-use pricing scheme anticipating the optimal response of demand customers to the prices announced. Here, the customers' lower-level problem is replaced by the optimal response of the users to electricity prices calculated analytically and the resulting problem is solved using fmincon function of MATLAB® Optimization Toolbox.

A two-level framework is developed in Reference 13 for modeling the interaction between multiple utility companies and the residential consumers. Competition among utilities is also simulated and a distributed algorithm is used for solving the problem. On the customer side, residential users are characterized by an evolutionary behavior where each one belonging to the population observes and replicates the strategies of other users in the same population.

In Reference 14, an energy management scheme is proposed for a smart community comprising local resources (residential consumers with distributed generation (DG) units and an entity managing shared facilities, such as water pumps, lighting, energy storage), where transactions with the main grid also take place. A Stackelberg game is formulated in order to model the energy trading process among consumers with DG units that decide the amount of the excess energy and, based on the announced prices, whether this amount of energy will be sold to the managing entity or the main grid.

In Reference 15, a management scheme for a demand response Aggregator is proposed. Within the framework of the interaction with the electricity market, the Aggregator selects the optimal level of demand reduction that maximizes the profit. While interacting with consumers, the Aggregator seeks to minimize the aggregate inconvenience of consumers due to load curtailment by selecting the price function for the offered curtailment according to which payments to consumers will be made.

In Reference 16 the authors present the use of Prospect Theory (PT) to enhance game theoretic approaches to demand-side management. More specifically, PT is used for simulating the decision-making process regarding energy management of storage devices and demand resources. This course of action is selected by the authors, since PT provides the mathematical tools for achieving a realistic simulation of the -often difficult to predict- behavior of individual energy consumers taking into account deviations from the prescripts of classical utility theory and conventional game theoretic approaches.

### *10.1.2 Overview*

In the following paragraphs, two applications of the bilevel decision-making framework relevant to smart grids are presented. In the first application, the interaction of an entity responsible for managing several distributed energy resources (DERs), such as DG units and flexible loads, with the wholesale market is modeled. In the second application, the bilevel framework is applied for modeling the interdependence of the decision-making process of a DER Aggregator with the local resources that he represents and manages. The solution methodology used in both cases follows the common practice of transforming the bilevel programming problem to an equivalent one-level problem that can be solved using commercially available software. Several scenarios are tested and relevant results are presented. It should be noted that these applications are inspired by or built upon the models presented in References 17 and 18, respectively.

## 10.2 Bilevel decision framework for optimal energy procurement of DERs

As discussed in Section 10.1, local resources, such as DG units and flexible loads, collectively described by the term DERs, each one insignificantly small in terms of capacity compared to the entire electricity system, cannot individually participate in market transactions. The energy produced, consumed, or curtailed locally needs to be managed by an entity, such as a DER Aggregator, responsible for representing these resources in the various market operations and capable of participating in them. It should be noted that local resources do not necessarily belong to the same part of the electricity grid, i.e., they can be connected to various feeders. Thus, the DER Aggregator is considered as an entity that holds a portfolio of several DERs and local load. When favorable, priority is given to the local consumption of the produced energy. Any part of the load that cannot be satisfied by the local production is served by energy bought from the wholesale market. Thus, the DER Aggregator acts as an intermediate between the retail customers and the wholesale market. In order to participate in the wholesale market, however, the DER Aggregator needs to define the day-ahead operation and, by extension, decide the optimal schedule for the local resources and their aggregated electricity profile. When the total capacity of the local resources is significant, changes in the aggregated profile affect the total system load, the operation of the wholesale market, and the prevailing market prices. Since the DER Aggregator keeps the energy balance of his portfolio, any eventual energy deficit must be covered by purchasing energy from the wholesale market at the system marginal price (SMP). Evidently, while deciding upon the optimal dispatch of the local resources minimizing the total procurement cost, the entity responsible for managing the DERs at a local level has to take into account the impact of this decision (i.e., the energy purchased from the wholesale market) on the SMP.

This interdependence between the DER energy management problem and the (wholesale) market clearing process is modeled as a bilevel programming problem. Focus is given to the DER energy procurement problem, which is placed at the upper level. At the lower level, the wholesale market operation is simulated. The two problems are coupled through the decision variables of each level; at the upper level these are the aggregated net load and at the lower level the SMP. More specifically, in the day-ahead horizon, the DER Aggregator decides upon the optimal dispatch of the local resources keeping the energy balance of the portfolio. Any demand not satisfied locally is aggregated and forms the load represented by the DER Aggregator in the wholesale market. Given this load and the remaining system load not managed by the DER Aggregator under study, the market clearing takes place. Note that the SMP is the dual variable of the system energy balance constraint and that it is assumed that the local DG production is insufficient for covering the entire load of the local energy consuming resources. The structure of the bilevel problem is presented in Figure 10.2.

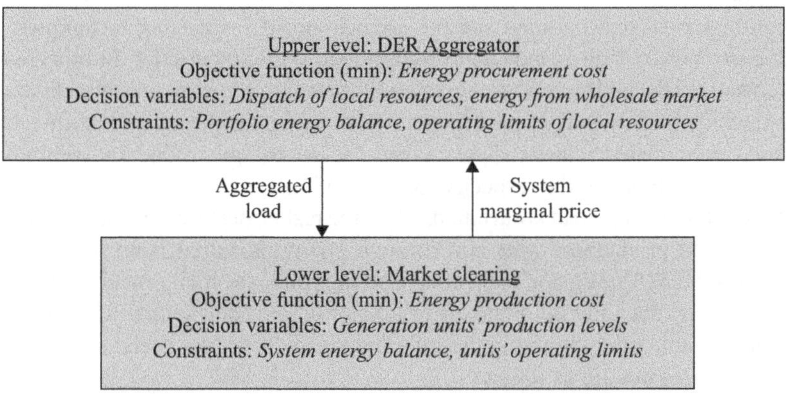


*Figure 10.2 Structure of the interaction of a DER Aggregator with the wholesale electricity market*

### 10.2.1 Nomenclature

*Indices and sets*

$t \in \mathcal{T}$ Set of dispatch intervals
$i \in \mathcal{I}$ Set of DG units
$k \in \mathcal{K}$ Set of customers with load curtailment
$u \in \mathcal{U}$ Set of conventional units

*Parameters*

$\alpha_i$ Unit cost of local production of DG unit $i$ (euros/MWh)
$\gamma_k$ Unit cost of load curtailment of consumer $k$ (euros/MWh)
$c_u$ Unit cost of conventional unit $u$ (euros/MWh)
$\bar{p}_i$ Operating limit of DG unit $i$ (MW)
$\bar{d}_k$ Maximum available load curtailment of consumer $k$ (MW)
$\bar{g}_u$ Operating limit of conventional unit $u$ (MW)
$l_t$ Forecasted inflexible load served by the DER Aggregator during hour $t$ (MW)
$L_t$ Forecasted system load during hour $t$ (MW)

*Variables*

$e_t$ Energy acquired from the wholesale market at time $t$ or Aggregator's net load (MW)
$p_{i,t}$ Energy production of DG unit $i$ at time $t$ (MW)
$d_{k,t}$ Load curtailment of consumer $k$ at time $t$ (MW)
$g_{u,t}$ Energy production of conventional unit $u$ at time $t$ (MW)
$\lambda_t$ System marginal price (euros/MWh)

### 10.2.2 Model

The upper-level problem comprises the decision-making process of a DER Aggregator managing local resources. These include DG units and consumers with flexible and inflexible loads. The inflexible part of the load is a priori known to the DER Aggregator through historical observations and the application of forecasting techniques. The flexible part of the load is modeled through load curtailment bids. In other words, each consumer declares per hour the amount of available energy curtailment and the price at which he is willing to be remunerated for implementing the curtailment. The DG units are modeled analogously, i.e., they declare energy production volumes and offer prices for injecting the respective amount of energy.

Part of the flexible load is curtailed, while the rest is met through energy produced from the local production units and through energy acquired from the wholesale market. The DER Aggregator's objective is to minimize the total procurement cost of his portfolio by deciding upon the energy quantity that each source will contribute, while at the same time satisfying the inflexible load. Thus, the decision variables of the DER Aggregator are as follows:

1. $e_t, t = 1, \ldots, T$: energy acquired from the wholesale market at time $t$
2. $p_{i,t}, i = 1, \ldots, I, t = 1, \ldots, T$: energy production of the $i$th DG unit at time $t$
3. $d_{k,t}, k = 1, \ldots, K, t = 1, \ldots, T$: energy curtailment of $k$th consumer at time $t$

The price at which each DG unit $i \in I$ is remunerated for the amount of energy produced $p_{i,t}$ is $\alpha_i$, while the respective price for consumer $k \in K$ for curtailing the amount $d_{k,t}$ is $\gamma_k$. Information regarding these unit prices is considered to be available to the DER Aggregator either through contractual agreements or through consumption/production declarations. The energy from the wholesale market $e_t$ is bought at the SMP $\lambda_t$, a variable that is acquired by solving the Market Clearing Problem.

At the lower level, the Market Clearing Problem is solved. This encompasses the decisions regarding the optimal dispatch of conventional units that minimizes the total production cost, while respecting the energy balance and the operation limits of the units.

Given the above, the problem of the DER Aggregator is formulated as follows. Note that $\varphi_{u,t}^{min}$ and $\varphi_{u,t}^{max}$ are the dual variables of unit $u$'s operational constraints (minimum and maximum production, respectively) and $\lambda_t$ is the dual variable of the energy balance constraint. According to the interpretation of dual variable $\lambda_t$ in all economic dispatch problems, this variable signifies the effect on the system cost of a marginal change in the energy balance constraint.

$$\min_{e_t, p_{i,t}, d_{k,t}} \sum_{t \in \mathcal{T}} \left( \lambda_t \cdot e_t + \sum_{i \in \mathcal{I}} \alpha_i \cdot p_{i,t} + \sum_{k \in \mathcal{K}} \gamma_k \cdot d_{k,t} \right) \tag{10.15}$$

subject to

$$l_t - \sum_{k \in \mathcal{K}} d_{k,t} = e_t + \sum_{i \in \mathcal{I}} p_{i,t}, \quad \forall t \tag{10.16}$$

$$0 \leq d_{k,t} \leq \bar{d}_k, \quad \forall k, t \tag{10.17}$$

$$0 \leq p_{i,t} \leq \bar{p}_i, \quad \forall i, t \tag{10.18}$$

where $\lambda_t = arg\Big\{$

$$\min_{\substack{g_{u,t}, \lambda_t, \\ \varphi_{u,t}^{\min}, \varphi_{u,t}^{\max}}} \sum_{t \in \mathcal{T}} \sum_{u \in \mathcal{U}} c_u \cdot g_{u,t} \tag{10.19}$$

$$0 \leq g_{u,t} \leq \bar{g}_u : \varphi_{u,t}^{min}, \varphi_{u,t}^{max}, \quad \forall u, t \tag{10.20}$$

$$L_t + e_t = \sum_{u \in \mathcal{U}} g_{u,t} : \lambda_t, \forall t \tag{10.21}$$

$\Big\} \forall t.$

The energy procurement cost is given by (10.15). It incorporates three cost components: for buying energy from the wholesale market $(\lambda_t \cdot e_t)$, from the local DG units $\sum_{i \in I} \alpha_i \cdot p_{i,t}$, and for remunerating the offered load curtailment $\sum_{k \in K} \gamma_k \cdot d_{k,t}$. Minimization of the cost function is performed while respecting the energy balance (10.16) (i.e., the remaining load after subtracting load curtailment $\sum_{k \in K} d_{k,t}$ from the total inflexible load $l_t$ is satisfied by acquiring energy from the wholesale market $e_t$ and by the local production units $\sum_{i \in I} p_{i,t}$) and the operational limits of the distributed resources (10.17) and (10.18). At the lower level, the production of the conventional units is optimally selected in order to minimize the total production cost (10.19) while observing the operating limits (10.20) and the energy balance constraint (10.21). The last one ensures that the total system load (i.e., the forecasted load $L_t$ and the load that balances the Aggregator's portfolio $e_t$) is satisfied by energy produced from conventional units. Note that the SMP $\lambda_t$ is the dual variable of the energy balance constraint of the Market Clearing Problem.

Drawing a parallel to the general formulation presented in Section 10.1, for the upper-level optimization problem: variables $\boldsymbol{x}$ are $e_t, \forall t, p_{i,t}, \forall i, t$, and $d_{k,t}, \forall k, t$; the objective function $F(\boldsymbol{x}, \boldsymbol{y})$ is (10.15); by reorganizing the terms in (10.17) and (10.18) the upper-level inequality constraints can take the form $\mathbf{G}(\boldsymbol{x}, \boldsymbol{y}) \leq 0$, while there are no equality constraints. Regarding the lower-level problem: the optimization variables $\boldsymbol{y}$ are $g_{u,t}, \forall u, t$ the objective function $f(\boldsymbol{x}, \boldsymbol{y})$ is (10.19); by reorganizing the terms in (10.20), we obtain inequalities of the form $g(\boldsymbol{x}, \boldsymbol{y}) \leq 0$; the equality constraints $\mathbf{h}(\boldsymbol{x}, \boldsymbol{y})$ are given by (10.21).

### 10.2.3 Solution methodology

The two-level problems (10.15)–(10.21) is equivalently transformed into an one-level mathematical programming problem with equilibrium constraints (MPEC) by replacing the lower-level problem with the KKT conditions as presented in (10.12)–(10.14). Note that this transformation is admissible since the lower-level problem is continuous and linear over a convex set, and, thus, convex. More specifically, the Lagrangian of the lower-level problem is the following:

$$\begin{aligned}
&\mathcal{L}(g_{u,t}, \lambda_t, \varphi_{u,t}^{min}, \varphi_{u,t}^{max}) \\
&\quad = \sum_{t \in \mathcal{T}} \sum_{u \in \mathcal{U}} c_u \cdot g_{u,t} + \sum_{t \in \mathcal{T}} \lambda_t \cdot \left( L_t + e_t - \sum_{u \in \mathcal{U}} g_{u,t} \right) \\
&\quad + \sum_{t \in \mathcal{T}} \sum_{u \in \mathcal{U}} ((-\varphi_{u,t}^{min}) \cdot g_{u,t} + \varphi_{u,t}^{max} \cdot (g_{u,t} - \bar{g}_u)), \quad \forall u, t
\end{aligned} \tag{10.22}$$

and the necessary optimality conditions are

$$\nabla_{g_{u,t}} \mathcal{L} = c_u - \lambda_t - \varphi_{u,t}^{min} + \varphi_{u,t}^{max} = 0, \quad \forall u, t \tag{10.23}$$

$$0 \leq -g_{u,t} + \bar{g}_u \perp \varphi_{u,t}^{max} \geq 0, \quad \forall u, t \tag{10.24}$$

$$0 \leq g_{u,t} \perp \varphi_{u,t}^{min} \geq 0, \quad \forall u, t \tag{10.25}$$

$$L_t + e_t = \sum_{u=1}^{U} g_{u,t}, \quad \forall t \tag{10.26}$$

Note that (10.23) can be incorporated in (10.25) by replacing $\varphi_{u,t}^{min}$ with $c_u - \lambda_t + \varphi_{u,t}^{max}$.

However, the derived problem is a nonlinear one due to the bilinear product $\lambda_t \cdot e_t$ in (10.15) and the complementarity conditions (10.24) and (10.25). In order to use commercially available solvers appropriate for linear problems, commonly used linearization techniques are applied. According to the SDT [4], the value of the objective function of the primal and of the dual problem at the optimal solution is identical. More specifically, the objective function of the dual problem is (10.22), while (10.23) holds. By multiplying (10.23) with $g_{u,t}$ and applying the SDT, the objective function of the dual problem takes the form presented on the right-hand side of (10.27). Thus, it is possible to substitute the bilinear term $\lambda_t \cdot e_t$ in (10.15) by a linear expression by rearranging the terms in (10.27).

$$\sum_{t=1}^{T}\sum_{u=1}^{U} c_u \cdot g_{u,t} = \sum_{t=1}^{T} \lambda_t \cdot (L_t + e_t) - \sum_{t=1}^{T}\sum_{u=1}^{U} \varphi_{u,t}^{max} \cdot \bar{g}_u \tag{10.27}$$

Furthermore, the big-M formulation [5] is appropriate for replacing complementary slackness conditions of the form $0 \leq \lambda \perp s \geq 0$ with the linear expressions $0 \leq \lambda \leq M \cdot \delta$ and $0 \leq s \leq M \cdot (1 - \delta)$, using an auxiliary binary variable $\delta$ and a sufficiently large positive constant $M$. Thus, the following equivalent expressions for (10.24) and (10.25) are obtained:

$$0 \leq -g_{u,t} + \bar{g}_u \leq M \cdot \delta_{u,t}^2, \quad \forall u, t \tag{10.28}$$

$$0 \leq \varphi_{u,t}^{max} \leq M \cdot (1 - \delta_{u,t}^2), \quad \forall u, t \tag{10.29}$$

$$0 \leq c_u - \lambda_t + \varphi_{u,t}^{max} \leq M \cdot \delta_{u,t}^1, \quad \forall u, t \tag{10.30}$$

$$0 \leq g_{u,t} \leq M \cdot (1 - \delta_{u,t}^1), \quad \forall u, t \tag{10.31}$$

After the aforementioned linearizations are applied, the resulting MILP problem is as follows:

$$\min_{\substack{e_t, p_{i,t}, d_{k,t}, g_{u,t}, \\ \varphi_{u,t}^{max}, \lambda_t, \delta_{u,t}^1, \delta_{u,t}^2}} \sum_{t=1}^{T} \left( \sum_{u=1}^{U} (c_u \cdot g_{u,t} + \varphi_{u,t}^{max} \cdot \bar{g}_u) - \lambda_t \cdot L_t + \sum_{i=1}^{I} \alpha_i \cdot p_{i,t} + \sum_{k=1}^{K} \gamma_k \cdot d_{k,t} \right) \tag{10.32}$$

subject to constraints (10.16)–(10.18), (10.26), and (10.28)–(10.31). Optimization is performed with respect to the upper-level variables $e_t$, $p_{i,t}$, $d_{k,t}$, the lower-level variables $g_{u,t}$, the dual variables of the lower-level constraints $\varphi_{u,t}^{max}$, $\lambda_t$ and the auxiliary binary variables $\delta_{u,t}^1$ and $\delta_{u,t}^2$.

## *10.2.4 Implementation*

### 10.2.4.1 Input data—scenarios

The time horizon of the optimization is 24 h. Input parameters of the problem are the characteristics of the local resources and of the conventional units (cost parameters $\alpha_i$, $\gamma_k$, $c_u$, operation limits $\bar{p}_i$, $\bar{d}_k$, $\bar{g}_u$) as well as the forecasted inflexible load, part of which is represented by the Aggregator ($l_t$), while the rest of it constitutes the system load ($L_t$) that must be satisfied through production of energy by the conventional units. This set of data comprises the basic scenario, with 22% of the total forecasted inflexible load represented by the Aggregator, i.e., $L_t$ is 3.5 times higher than $l_t$.

Table 10.1 Technical and economic characteristics of conventional units

| Conventional unit index, $u$ | Upper production limit, $\bar{g}_u$ (MW) | Cost, $c_u$ (euros/MWh) |
|---|---|---|
| 1 | 15 | 40 |
| 2 | 20 | 45 |
| 3 | 25 | 50 |
| 4 | 30 | 55 |
| 5 | 15 | 60 |
| 6 | 25 | 65 |
| 7 | 15 | 70 |
| 8 | 20 | 75 |
| 9 | 25 | 80 |
| 10 | 30 | 85 |

Table 10.2 Technical and economic characteristics of local resources (DG units, consumers with curtailable loads)

| DG unit index, $i$ | Production limit, $\bar{p}_i$ (MW) | Cost, $\alpha_i$ (euros/MWh) | Consumer index, $k$ | Curtailment limit, $\bar{d}_k$ (MW) | Cost, $\gamma_k$ (euros/MWh) |
|---|---|---|---|---|---|
| 1 | 7 | 81.8 | 1 | 4.0 | 77.3 |
| 2 | 4 | 83.0 | 2 | 2.5 | 79.6 |
| 3 | 6 | 83.6 | 3 | 4.5 | 82.5 |
| 4 | 9 | 83.8 | 4 | 5.0 | 85.4 |
| 5 | 9 | 84.3 | 5 | 1.0 | 88.1 |
| 6 | 6 | 86.4 | 6 | 1.5 | 89.5 |
| 7 | 1 | 87.3 | 7 | 3.5 | 89.9 |
| 8 | 8 | 89.0 | 8 | 2.5 | 95.9 |
| 9 | 5 | 91.6 | 9 | 2.0 | 96.7 |
| 10 | 3 | 96.2 | 10 | 3.0 | 97.7 |
| 11 | 2 | 97.0 | 11 | 1.0 | 98.9 |
| 12 | 3 | 97.3 | 12 | 3.5 | 104.6 |
| 13 | 10 | 97.3 | 13 | 4.5 | 105.4 |
| 14 | 1 | 101.0 | 14 | 0.5 | 120.2 |
| 15 | 4 | 109.1 | 15 | 4.0 | 122.7 |
| 16 | 7 | 109.3 | 16 | 5.0 | 128.3 |
| 17 | 5 | 115.1 | 17 | 3.0 | 128.5 |
| 18 | 10 | 116.7 | 18 | 0.5 | 134.0 |
| 19 | 8 | 117.3 | 19 | 1.5 | 136.9 |
| 20 | 2 | 118.6 | 20 | 2.0 | 138.9 |

Apart from the basic scenario, a sensitivity analysis on the forecasted and system load is performed, in order to examine the impact of the relevant size of the Aggregator's inflexible load in comparison to the system load. One case is the high-impact scenario, where the Aggregator represents 44% of the total forecasted inflexible load,

*Table 10.3 Forecasted system load and inflexible load served by the Aggregator per scenario*

| | System load, $L_t$ (MW) | | | Inflexible load, $l_t$ (MW) | | |
|---|---|---|---|---|---|---|
| | **Basic** | **Low impact** | **High impact** | **Basic** | **Low impact** | **High impact** |
| t1 | 129.7 | 146.0 | 87.9 | 44.0 | 27.7 | 85.8 |
| t2 | 117.6 | 132.4 | 79.7 | 40.0 | 25.2 | 77.9 |
| t3 | 117.2 | 131.8 | 79.6 | 39.5 | 24.9 | 77.0 |
| t4 | 113.1 | 127.0 | 77.5 | 37.5 | 23.6 | 73.2 |
| t5 | 110.0 | 123.3 | 75.9 | 35.9 | 22.6 | 70.0 |
| t6 | 112.1 | 125.2 | 78.3 | 35.6 | 22.4 | 69.3 |
| t7 | 124.3 | 137.9 | 89.4 | 36.7 | 23.1 | 71.6 |
| t8 | 145.8 | 159.7 | 110.1 | 37.5 | 23.6 | 73.2 |
| t9 | 155.3 | 169.8 | 118.2 | 39.1 | 24.6 | 76.2 |
| t10 | 163.7 | 178.7 | 125.2 | 40.5 | 25.5 | 79.0 |
| t11 | 161.0 | 176.2 | 121.8 | 41.2 | 26.0 | 80.4 |
| t12 | 159.0 | 174.9 | 118.1 | 43.0 | 27.1 | 83.9 |
| t13 | 162.3 | 178.5 | 120.6 | 43.8 | 27.6 | 85.5 |
| t14 | 162.4 | 178.0 | 122.2 | 42.3 | 26.6 | 82.5 |
| t15 | 157.4 | 171.9 | 120.2 | 39.2 | 24.7 | 76.3 |
| t16 | 164.8 | 179.9 | 125.9 | 41.0 | 25.8 | 79.9 |
| t17 | 169.2 | 185.2 | 128.2 | 43.1 | 27.2 | 84.1 |
| t18 | 176.0 | 193.2 | 131.8 | 46.5 | 29.3 | 90.7 |
| t19 | 195.6 | 216.1 | 142.9 | 55.4 | 34.9 | 108.1 |
| t20 | 197.5 | 219.3 | 141.5 | 59.0 | 37.1 | 115.0 |
| t21 | 195.4 | 217.3 | 139.1 | 59.2 | 37.3 | 115.5 |
| t22 | 179.4 | 200.6 | 124.8 | 57.4 | 36.2 | 112.0 |
| t23 | 162.5 | 182.2 | 112.2 | 53.0 | 33.4 | 103.4 |
| t24 | 150.0 | 168.4 | 102.8 | 49.7 | 31.3 | 96.9 |

i.e., the system's load is only 1.3 times higher than the Aggregator's inflexible load. The second case is a low-impact scenario, where the Aggregator represents 14% of the total forecasted inflexible load. This means that the system load is significantly higher than the inflexible load served by the Aggregator; in fact, $L_t$ is 6.1 times higher than $l_t$. It should be noted that in all scenarios the total forecasted load (i.e., the sum of system and Aggregator's load) during the 24-h horizon is the same, but the part of the load represented in the market transactions by the Aggregator varies from scenario to scenario, thus reflecting different degrees of load aggregation in the low-impact scenario the Aggregator represents 14% of the total forecasted inflexible load, in the basic scenario 22% and in the high-impact scenario 44%.

Ten conventional units with varying economical and technical characteristics (Table 10.1) are considered. Regarding the DER units and the consumers with curtailable loads, the input parameters are presented in Table 10.2. Table 10.3 presents the forecasted system load and inflexible load per scenario for the 24-h period considered.

#### 10.2.4.2 Computational issues

The equivalent MILP problem consist of 3,217 equations, 2,473 real variables, and 744 binary variables and is solved in less than 1 s using CPLEX 12.5 [19] (a branch-and-cut method for searching the enumeration tree) under the General Algebraic Modeling System (GAMS) [20] running on an Intel®Core™2 Duo with two processors at 2.99 GHz and 1.96 GM of RAM. The relative termination tolerance is set to $10^{-8}$.

### *10.2.5 Results*

Figure 10.3(a) presents the energy supply mix of the DER Aggregator portfolio per hour for the basic scenario. The total load as forecasted is covered by energy acquired

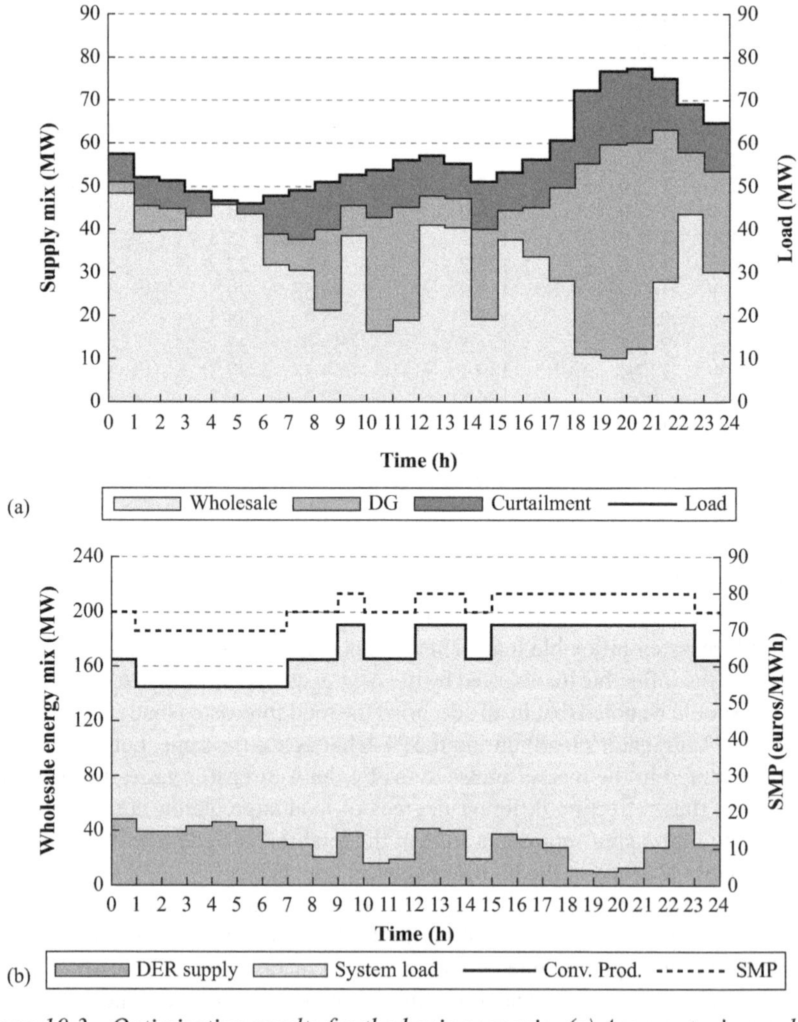


*Figure 10.3 Optimization results for the basic scenario. (a) Aggregator's supply mix and (b) wholesale market energy mix*

from the wholesale market and the local production units. Furthermore, part of the load is curtailed. Figure 10.3(b) presents the energy mix in the wholesale market for the same scenario, i.e., the composition of the total load covered by production from conventional units. Figures 10.4 and 10.5 present the relevant optimization results for the low- and high-impact scenarios, respectively.

Regarding the wholesale energy mix (conventional production), the energy supplied to the DER Aggregator evidently increases when moving from the low-impact scenario (277 MWh) to the high-impact scenario (1,076 MWh). The basic scenario lies in between, with 611 MWh of the Aggregator's load supplied by energy acquired from the wholesale market (Table 10.4). This is attributed to the specific characteristics of the scenarios, with the high-impact scenario corresponding to the highest load

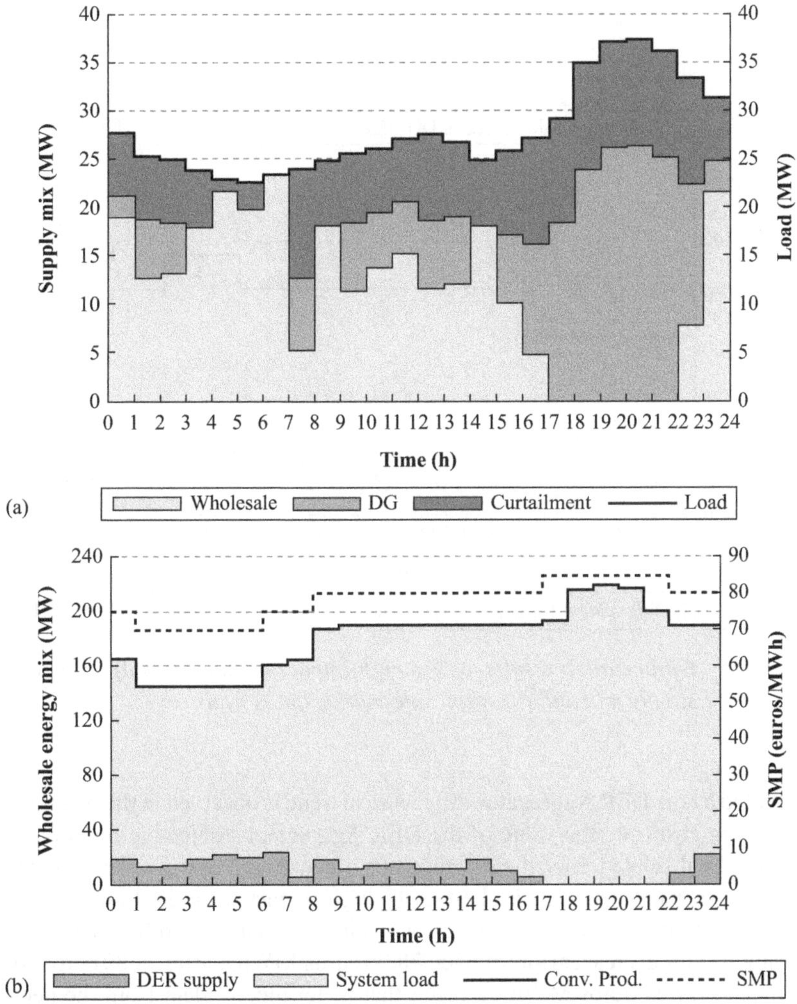


*Figure 10.4 Optimization results for the low-impact scenario. (a) Aggregator's supply mix and (b) wholesale market energy mix*

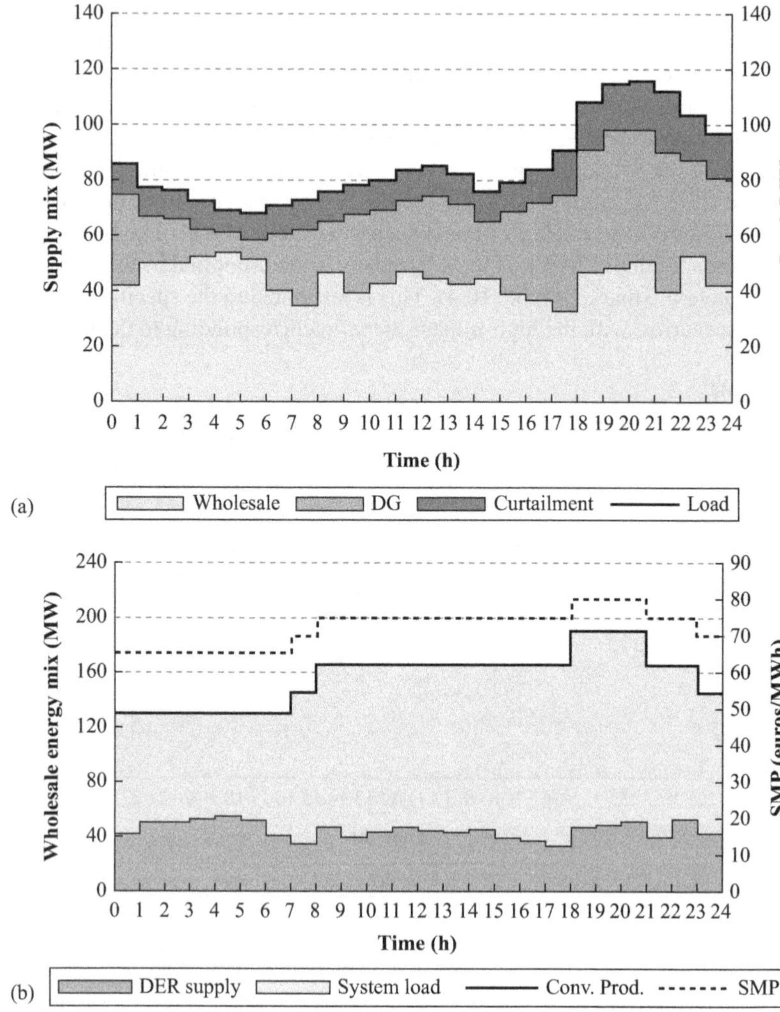


*Figure 10.5 Optimization results for the high-impact scenario. (a) Aggregator's supply mix and (b) wholesale market energy mix*

represented by the DER Aggregator. An identical trend is observed in the Aggregator's procurement cost (i.e., the value of the DER Aggregator's objective function at the optimum). In absolute terms, the Aggregator's procurement cost is at its lowest in the low-impact scenario and at its highest in the high-impact scenario. In order to avoid misinterpreting this result, the Aggregator's unit cost is calculated for each scenario. Interestingly, this decreases (low impact: 78.96 euros/MWh, basic: 78.80 euros/MWh, high impact: 77.21 euros/MWh), a result that is not independent from the decrease observed in the SMP. The higher the forecasted load represented by the DER Aggregator, the lower the SMP (low impact: 78.97 euros/MWh, basic: 76.48 euros/MWh,

*Table 10.4 Aggregated results for the three scenarios*

| | Low impact | Basic | High impact |
|---|---|---|---|
| **Aggregator's load (MWh)** | **668** | **1,060** | **2,067** |
| Energy from wholesale market (MWh) ([% of aggregated load]) | 277 (41%) | 611 (58%) | 1,076 (52%) |
| DG units' production (MWh) ([% of aggregated load]) | 209 (31%) | 257 (24%) | 685 (33%) |
| Load curtailment (MWh) ([% of aggregated load]) | 182 (27%) | 192 (18%) | 306 (15%) |
| **Conventional production (MWh)** | **4,350** | **4,292** | **3,750** |
| System load (MWh) ([% of conventional production]) | 4,073 (94%) | 3,681 (86%) | 2,674 (71%) |
| Aggregator's net load (MWh) ([% of conventional production]) | 277 (6%) | 611 (14%) | 1,076 (29%) |
| Aggregator's procurement cost (euros) | 52,735 | 83,539 | 159,620 |
| System cost (euros) | 343,557 | 336,125 | 273,550 |
| Aggregator's unit cost (euros/MWh) | 78.96 | 78.80 | 77.21 |
| Average SMP (euros/MWh) | 78.97 | 78.32 | 72.95 |

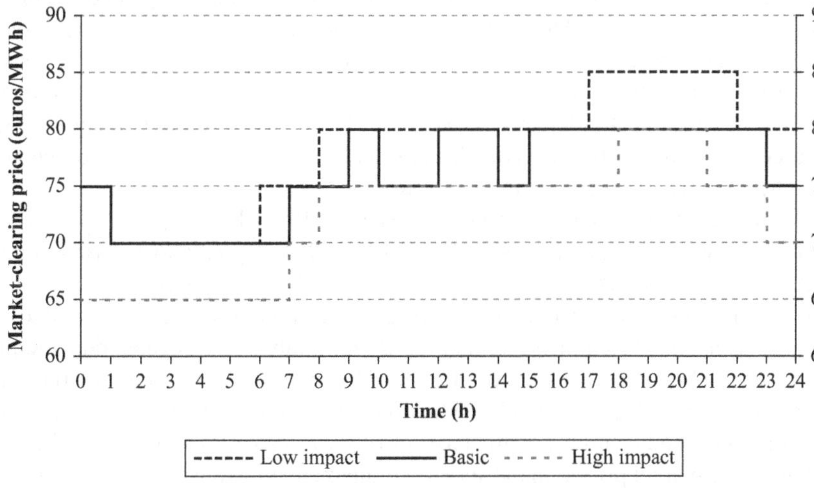


*Figure 10.6 Wholesale market-clearing prices per scenario*

high impact: 72.95 euros/MWh). This effect can also be observed in the SMP time series (Figure 10.6).

These results indicate that the higher degree of aggregation of local resources, their coordination by means of a DER Aggregator and the integration of the inflexible load in the portfolio of such an Aggregator, rather than as part of the system load, is beneficiary to the entity managing the DERs and contributes in lowering the system cost both in absolute and in relative terms.

Some final notes regarding the model presented in this paragraph follow. In the previously presented optimization problem, the network topology has not been taken into account. This could be done by also considering the fact that the DERs managed by the Aggregator could belong to different feeders, and adding the DC power flow equations to the lower-level problem, as presented in Reference 21. This way, it is possible to study several scenarios encompassing congestion issues as well. Furthermore, even though the model simulates a period of 24 h, it can be easily upscaled to incorporate a more extensive optimization horizon or a greater variety of local resources (e.g., energy storage).

## 10.3 Bilevel decision framework for optimal energy management of DERs

Apart from acting as an intermediate between the retail customers and the wholesale market, an interaction studied in Section 10.2, the DER Aggregator's activities also encompass managing the local resources. This can be seen as an extension of the role of the supplier in a more complex environment, as is the future smart grid, where local markets operate and advanced metering infrastructure and price signaling is available.

The DER Aggregator is responsible for composing, managing, and supplying a portfolio of a variety of DERs. Distributed producers, curtailable/shiftable loads, etc., that cannot participate in market procedures, are equipped with all necessary paraphernalia facilitating bidirectional exchange of information and are contracted with the DER Aggregator. It is necessary for the DER Aggregator to manage the local resources in a way that balances the declaration he submits to the market procedures (day-ahead, intraday, balancing markets, etc.).

In contrast to the model presented in Section 10.2, the perspective taken here is directed towards the operations performed by a DER Aggregator in managing the portfolio of entities. As such, the model includes the DER Aggregator's revenues maximization problem, the operating management of local sources and the entities' response decisions. From the system point of view, only an aggregated demand profile is seen. This profile is the result of the decision-making process of the DER Aggregator optimizing his portfolio valuation, while taking into account on one side the economical and technical characteristics of local resources and on the other side the wholesale market prices. In the specific implementation, the DERs that constitute the Aggregator's portfolio are a subset of the total electricity distribution network users and may be dispersed over several feeders of the network. Thus, the described model does not account for energy flows and losses.

Managing local entities is achieved through price incentives (not direct control signals) to which, each local entity responds by deciding upon the respective energy volume. This decision affects the profitability of the DER Aggregator and, for that reason, it is important to take into account the parameters of local demand and production entities. More specifically, the DER Aggregator seeks to maximize the revenues, by selecting the prices offered for the energy bought from/sold to the local

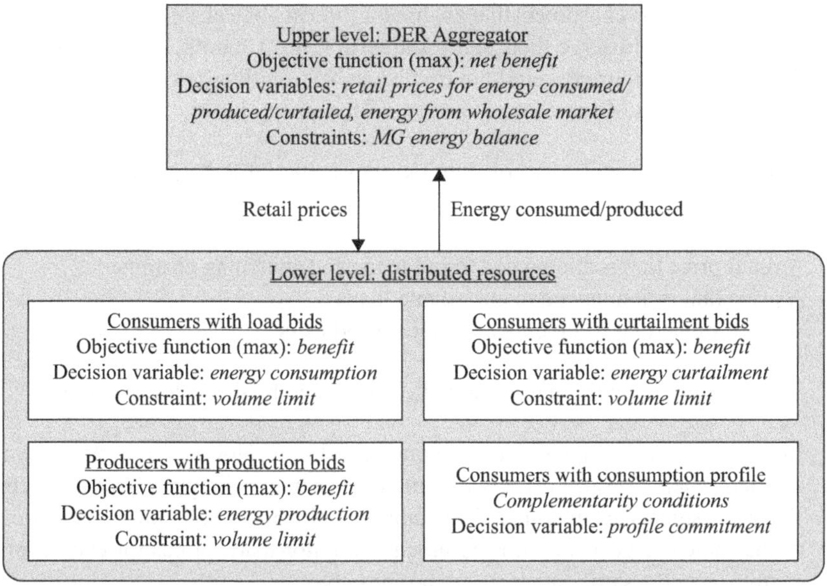


*Figure 10.7 Structure of the interaction of a DER Aggregator with his customers*

resources, while keeping the energy balance (any energy surplus or deficit is traded with the wholesale market). In turn, local resources, taking into account the price signals of the Aggregator, select the desired energy volumes (offered to/demanded from the Aggregator) in order to maximize their individual welfare. Thus, interdependence between the DER Aggregator and the distributed resources is established.

The interdependence is modeled as a bilevel optimization problem with the DER Aggregator acting as the leader by selecting the price incentives and the DERs acting as the followers by responding to the prices announced by the DER Aggregator (Figure 10.7). In order for the leader to determine the offered prices that apply during a midterm horizon, it is important to acquire the behavioral characteristics of the DERs. In this model, the response of these entities is modeled through volume-willing price curves, representing a notional welfare for each entity. This choice is justified by the fact that parameters of such functions can be based on information practically available to the Aggregator, e.g., by observation of historical data, contractual statements, or consumers'/producers' declarations. In the same logic, parameters of fundamental technical constraints, constitute also practically available information, and can be embedded, as well, in the DERs decision model.

A variety of DERs can be considered as followers of the interaction:

- Curtailable loads: consumers that accept demand shedding, which is reimbursed at a specific bid/willing price.
- Bid loads/price elastic loads: consumers that adjust their hourly demand depending on the retail price.

- Switching loads: consumers that are medium-term price elastic, that would select the Aggregator to serve their entire midterm (e.g., 1 month) demand profile, if the average retail price would not exceed an offer/willing price threshold.
- Production units: dispatchable sources, with capacity limits.

Considering the above set of entities' types, the DER Aggregator's decisions include:

- A retail price that is announced to bid loads and switching customers.
- A price that is announced to curtailable loads.
- A price signal that is announced to local producers.

The decision upon these price signals is governed by the following rules:

- In general, the price signals have an hourly resolution. However, time zones can be defined, where the same price is applied for each hour belonging to the same zone.
- The price signals are announced and applied uniformly to all potential customers of the same category, i.e., it is assumed that tailor-made prices for each explicit customer, in the same category, are not possible.

The scope of the presented decision-making model is primarily a midterm portfolio composition and management. Thus forecasted wholesale market prices are used. The horizon could be narrowed down to a day-ahead window, thus adapted to respective market procedures. This application could be suitable, only with respect to prices offered to flexible and dispatchable entities and to the extent that control and signaling technology provides these entities with the ability to respond within a reasonable time interval to price signals. In any case, it is assumed that perfect competition is achieved in the wholesale market, rendering the Aggregator a price-taker, i.e., incapable of influencing the price of the wholesale market. Thus, the wholesale price is considered as an input parameter of the problem, and can be obtained by application of forecasting techniques on historical data.

### 10.3.1 Nomenclature

*Indices and sets*

| | |
|---|---|
| $dr \in \mathcal{DR}$ | Set of distributed resources, where *dr* includes $db \in \mathcal{DB}$ for consumers with load bids, $dc \in \mathcal{DC}$ for consumers with curtailable loads, and $pb \in \mathcal{PB}$ for producers with production bids |
| $dp \in \mathcal{DP}$ | Set of consumers with demand profiles |
| $s \in \mathcal{S}$ | Set of steps of the submitted bids of the distributed resources |
| $p \in \mathcal{P}$ | Set of characteristic periods |
| $z \in \mathcal{Z}$ | Set of dispatch time zones |
| $t \in \mathcal{T}$ | Set of dispatch intervals |
| $t_z \in \mathcal{T}_z$ | Subset of dispatch intervals in time zone $z$ of period $p$ |

*Parameters*

| | |
|---|---|
| $Nz$ | Number of dispatch time zones |
| $w(p)$ | Scenario weight of each characteristic period |
| $P_{dr}^{p,t,s}$, $Q_{dr}^{p,t,s}$ | Price-quantity pair at period $p$, dispatch interval $t$, and block $s$ for the hourly priced bids submitted by distributed resource $dr$ (euros/MWh and MWh, respectively) |
| $P_{dp}$, $Q_{dp}^{p,t}$ | Price-quantity pair of profile consumer $dp$ at period $p$ and dispatch interval $t$ (euros/MWh and MWh, respectively); the quantity $Q_{dp}^{p,t}$ for a given profile consumer $dp$ may be different at each period $p$ and dispatch interval $t$ |
| $\overline{RP}$ | Unit charge for the baseline load of consumers with curtailable loads (euros/MWh) |
| $\overline{Q}_{dc}^{p,t}$ | Demand volume without curtailment (baseline load) at period $p$ and dispatch interval $t$ of consumer $dc$ (MWh) |
| $SMP^{p,t}$ | Forecasted wholesale price (SMP) at period $p$ and dispatch interval $t$ (euros/MWh) |

*Variables*

| | |
|---|---|
| $RP^{p,z}$ | Retail price for consumers with load bids and demand profiles in time zone $z$ of period $p$ (euros/MWh) |
| $CP^{p,z}$ | Curtailment price for consumers with curtailable loads, in time zone $z$ of period $p$ (euros/MWh) |
| $PP^{p,z}$ | Production price for producers with production bids, in time zone $z$ of period $p$ (euros/MWh) |
| $e^{p,t}$ | Amount of energy from the wholesale market at period $p$ and dispatch interval $t$ (MWh) |
| $x_{dr}^{p,t,s}$ | Cleared quantity of distributed resource $dr$ at period $p$, dispatch interval $t$, and block (MWh) |
| $x_{dp}$ | Commitment of profile consumer $dp$ |

## *10.3.2 Model*

### 10.3.2.1 Upper-level problem

The upper-level problem of the bilevel model encompasses the Aggregator's decisions regarding the optimal pricing scheme for the energy bought from/sold to the local resources (variables $RP^{p,z}$, $CP^{p,z}$, $PP^{p,z}$), as well as regarding the energy volume from the wholesale market ($e^{p,t}$) during a midterm horizon. During that horizon, the parameters of the problem can vary significantly. For the Aggregator, it is important to be able to plan ahead of time his actions incorporating in his decision-making process different scenarios regarding the various parameters of the problem. This is achieved by defining different characteristic periods $p$, each one with a different occurrence frequency $w(p)$.

In total, the Aggregator minimizes the net cost for supplying the loads with energy during all characteristic periods. The objective function (10.33) represents that cost, i.e., the expenses for buying energy $e^{p,t}$ from the wholesale market at price $SMP^{p,t}$, for buying energy $x_{pb}^{p,t,s}$ from the local production units at price $PP^{p,z}$ and for remunerating the curtailed load minus the revenues from selling energy $Q_{dp}^{p,t}$ and $x_{db}^{p,t,s}$ at price $RP^{p,z}$ to the customers with load profiles and load bids, respectively.

$$\min_{e^{p,t},RP^{p,z},PP^{p,z},CP^{p,z}} \sum_{p,z}\sum_{t\in\mathcal{T}_z} w(p)\Big[SMP^{p,t}e^{p,t} + \sum_{pb,s} PP^{p,z}x_{pb}^{p,t,s} + \sum_{dc,s}(CP^{p,z}x_{dc}^{p,t,s} - \overline{RP}\,\overline{Q}_{dc}^{p,t})$$

$$-\sum_{dp} RP^{p,z}x_{dp}Q_{dp}^{p,t} - \sum_{db,s} RP^{p,z}x_{dp}^{p,t,s}\Big] \tag{10.33}$$

The cost minimization problem is constrained by the energy balance constraint (10.34):

$$\sum_{dc,s}(\bar{Q}_{dc}^{p,t} - x_{dc}^{p,t,s}) + \sum_{dp} x_{dp}Q_{dp}^{p,t} + \sum_{db,s} x_{db}^{p,t,s} = e^{p,t} + \sum_{pb,s} x_{pb}^{p,t,s}, \quad \forall p,t \tag{10.34}$$

This last equation ensures that the total load belonging to the Aggregator's portfolio is satisfied by energy acquired from the wholesale market and from DG units.

The term $\sum_{dc,s}(CP^{p,z}x_{dc}^{p,t,s} - \overline{RP}\bar{Q}_{dc}^{p,t})$ in the objective function of the DER Aggregator corresponds to the minus net benefit obtained from consumers with curtailable loads. It derives as the difference between the expenses for remunerating the curtailed load $\sum_s x_{dc}^{p,t,s}$ at price $CP^{p,z} - \overline{RP}$ (first term of (10.35)) and the revenues from supplying the residual load of the consumers with curtailment offers $(\bar{Q}_{dc}^{p,t} - \sum_s x_{dc}^{p,t,s})$ at the predefined price $\overline{RP}$ (second term of (10.35)). $\bar{Q}_{dc}^{p,t}$ is the baseline consumption that has been agreed when signing the supply contract and is used for calculating the offered demand reduction. Evidently, the term $\overline{RP}\bar{Q}_{dc}^{p,t}$ does not affect the optimization results; it is only given here for completeness and is subsequently ignored.

$$(CP^{p,z} - \overline{RP})\sum_s x_{dc}^{p,t,s} - \overline{RP}\left(\bar{Q}_{dc}^{p,t} - \sum_s x_{dc}^{p,t,s}\right) \tag{10.35}$$

The energy volumes of the various local resources are not known by the Aggregator a priori, at least in a direct way, but are obtained by solving the lower-level problems. Because the Aggregator cannot control them directly, they are considered as parameters in the upper-level problem.

When comparing the above problem with the general formulation of Section 10.1, it is evident that variables $\boldsymbol{x}$ are now $e^{p,t}, \forall p,t, RP^{p,z}, \forall p,z, PP^{p,z}, \forall p,z$, and $CP^{p,z}, \forall p,z$; the objective function $F(\boldsymbol{x},\boldsymbol{y})$ is (10.33); (10.34) is the equality constraint $\mathbf{H}(\boldsymbol{x},\boldsymbol{y}) = 0$, while there are no inequality constraints.

#### 10.3.2.2 Lower-level problems

Consumers with load bids, consumers with curtailable loads and producers with production bids, are modeled as entities optimizing the notional benefit, i.e., the consumer's or producer's surplus, as expressed by (10.36)–(10.37), (10.38)–(10.39), (10.40)–(10.41), respectively (dual variables appear next to each constraint separated

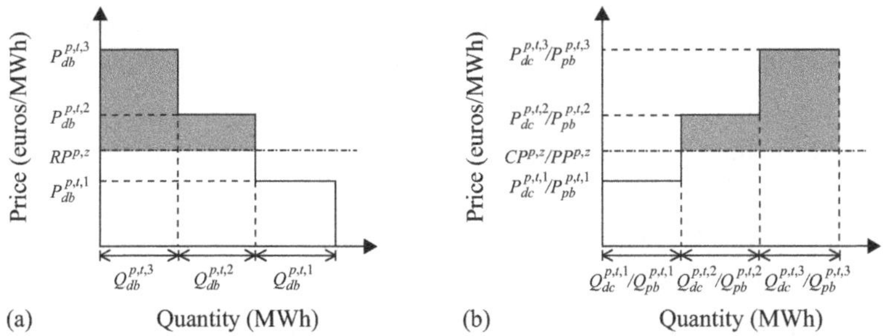


*Figure 10.8 Price-quantity pairs for: (a) a consumer with load bids db and (b) for a producer with production bids pb or a consumer with curtailment bids dc at period p and dispatch interval t*

by a colon). Their behavior is described by a set of price-quantity pairs that are different for each entity and considered as input data known to the Aggregator in advance. These pairs are highly dependent on individual preferences and characteristics of the various lower-level entities.

In Figure 10.8(a), a set of price-quantity pairs for consumers with load bids is presented. The overall set practically describes a utility function that expresses the welfare valuation of the specific good (electricity) measured in monetary units as perceived by the customer. As is common, the amount of energy demanded decreases as the price increases, i.e., the envelope of the stepwise function has a negative slope. When the price announced is $RP^{p,z}$, then, according to the utility function, the customer selects the energy volume that maximizes the consumer's surplus, i.e., the shaded area between the inverse demand curve and the horizontal line at the prevailing retail price. This surplus maximization problem is cast mathematically as in (10.36) and (10.37) ($\varphi_{db}^{p,t,s}$ is the dual variable of the volume limit constraint).

$$\forall db \begin{cases} \min\limits_{x_{db}^{p,t,s}} \sum\limits_{s} (RP^{p,z} - P_{db}^{p,t,s}) x_{db}^{p,t,s} & (10.36) \\ s.t. x_{db}^{p,t,s} \leq Q_{db}^{p,t,s} : \varphi_{db}^{p,t,s}, \quad \forall p, t, s & (10.37) \end{cases}$$

For consumers with curtailment bids and producers with production bids, the utility function has a positive slope (as presented in Figure 10.8(b)) and the entities maximize the producer's surplus based on the announced production $PP^{p,z}$/curtailment $CP^{p,z}$ price, respectively.

Thus, the producers' problem is cast mathematically as in (10.38) and (10.39) ($\gamma_{pb}^{p,t,s}$ is the dual complement variable of the volume limit constraint).

$$\forall pb \begin{cases} \min\limits_{x_{pb}^{p,t,s}} \sum\limits_{s} (P_{pb}^{p,t,s} - PP^{p,z}) x_{pb}^{p,t,s} & (10.38) \\ s.t. \ x_{pb}^{p,t,s} \leq Q_{pb}^{p,t,s} : \gamma_{pb}^{p,t,s}, \quad \forall p, t, s & (10.39) \end{cases}$$

As already mentioned, the problem of the consumers with curtailable loads encompasses maximization of the notional benefit that has two components:

1. The first one is associated with the remuneration received for the energy volume of the load curtailed, at price $(CP^{p,z} - \overline{RP})$, and is the net surplus $\sum_s ((CP^{p,z} - \overline{RP}) - P_{dc}^{p,t,s})x_{dc}^{p,t,s}$.
2. The second one is obtained by the reduced total cost for buying electricity, since, after the curtailment bid $x_{dc}^{p,t,s}$ is accepted, a lower load $\overline{Q}_{dc}^{p,t} - \sum_s x_{dc}^{p,t,s}$ remains to be served at price $\overline{RP}$. The perceived benefit is, thus, $\sum_s \overline{RP}x_{dc}^{p,t,s}$.

Consequently, the objective function of this type of loads is the surplus derived by the curtailment *and* the lowered costs for electricity ($\xi_{dc}^{p,t,s}$ is the dual complement variable of the volume limit constraint).

$$\forall dc \begin{cases} \min\limits_{x_{dc}^{p,t,s}} \sum\limits_s (P_{dc}^{p,t,s} - CP^{p,z})x_{dc}^{p,t,s} & (10.40) \\ s.t.\ x_{dc}^{p,t,s} \leq Q_{dc}^{p,t,s} : \xi_{dc}^{p,t,s}, \quad \forall p,t,s & (10.41) \end{cases}$$

Finally, the behavior of the consumers with demand profiles incorporates a decision regarding their entire midterm profile. The problem is formulated directly by a set of complementarity conditions (10.42) and (10.43) ($\tau_{dp}$ is the dual complement variable of the limit constraint), as proposed in Reference 22:

$$\forall dp \begin{cases} x_{dp} \leq 1 \perp \tau_{dp} \geq 0 & (10.42) \\ -P_{dp} + \frac{\sum_{p,z} RP^{p,z}}{Nz} + \tau_{dp} \geq 0 \perp x_{dp} \geq 0 & (10.43) \end{cases}$$

Variable $x_{dp}$ expresses the status of demand profile *dp*; either the entire profile $Q_{dp}^{p,t}$ for all periods and time intervals is dispatched ($x_{dp} = 1$), or it is not dispatched at all ($x_{dp} = 0$), depending on the sign of the difference $-P_{dp} + \sum_{p,z} RP^{p,z}/Nz$. This way, it is possible to model the behavior of consumers that are medium-term price elastic. In case the average retail price announced by the DER Aggregator exceeds a specific threshold, this type of consumers has high proclivity of selecting a different supplier, thus modeling the customer switching effect observed in a retail market which favors competition.

In each of the previously defined lower-level problems, the price levels $RP^{p,z}$, $CP^{p,z}$, and $PP^{p,z}$ are parameters and cannot be influenced, in a direct way, by the local entities, but are derived from the optimization at the upper level of the problem.

Regarding the lower-level problem: the optimization variables $\boldsymbol{y}$ are $x_{db}^{p,t,s}, \forall db, p, t, s, x_{dc}^{p,t,s}, \forall dc, p, t, s, x_{pb}^{p,t,s}, \forall pb, p, t, s, x_{dp}, \forall dp$; there is more than one objective function $f(\boldsymbol{x}, \boldsymbol{y})$, i.e., (10.36), (10.38), and (10.40); the inequality constraints are (10.37), (10.39), and (10.41)–(10.43), while there are no equality constraints.

### 10.3.3 Solution methodology

The two-level interaction is formulated mathematically as a one-level optimization model, by adding the KKT optimality conditions of each lower-level problem.

Note that these transformations are admissible since the lower-level problems are continuous and linear (and, thus, convex).

More specifically, the Lagrangian functions of the lower-level problems are as follows:

$$\mathcal{L}_{db}(x_{db}^{p,t,s}, \varphi_{db}^{p,t,s}) = \sum_{p,z,s} \sum_{t \in \mathcal{T}_z} (RP^{p,z} - P_{db}^{p,t,s}) x_{db}^{p,t,s} + \sum_{p,t,s} \varphi_{db}^{p,t,s} \cdot (x_{db}^{p,t,s} - Q_{db}^{p,t,s}), \quad \forall db \tag{10.44}$$

$$\mathcal{L}_{pb}(x_{pb}^{p,t,s}, \gamma_{pb}^{p,t,s}) = \sum_{p,z,s} \sum_{t \in \mathcal{T}_z} (P_{pb}^{p,t,s} - PP^{p,z}) x_{pb}^{p,t,s} + \sum_{p,t,s} \gamma_{pb}^{p,t,s} \cdot (x_{pb}^{p,t,s} - Q_{pb}^{p,t,s}), \quad \forall pb \tag{10.45}$$

$$\mathcal{L}_{dc}(x_{pb}^{p,t,s}, \xi_{dc}^{p,t,s}) = \sum_{p,z,s} \sum_{t \in \mathcal{T}_z} (P_{dc}^{p,t,s} - CP^{p,z}) x_{dc}^{p,t,s} + \sum_{p,t,s} \xi_{dc}^{p,t,s} \cdot (x_{dc}^{p,t,s} - Q_{dc}^{p,t,s}), \quad \forall dc \tag{10.46}$$

and the necessary optimality conditions (10.47)–(10.52)

$$\nabla_{x_{db}^{p,t,s}} \mathcal{L}_{db} = RP^{p,z} - P_{db}^{p,t,s} + \varphi_{db}^{p,t,s} = 0, \quad \forall db, p, t, s \tag{10.47}$$

$$x_{db}^{p,t,s} \leq Q_{db}^{p,t,s} \perp \varphi_{db}^{p,t,s} \geq 0, \quad \forall db, p, t, s \tag{10.48}$$

$$\nabla_{x_{pb}^{p,t,s}} \mathcal{L}_{pb} = P_{pb}^{p,t,s} - PP^{p,z} + \gamma_{pb}^{p,t,s} = 0, \quad \forall pb, p, t, s \tag{10.49}$$

$$x_{pb}^{p,t,s} \leq Q_{pb}^{p,t,s} \perp \gamma_{pb}^{p,t,s} \geq 0, \quad \forall pb, p, t, s \tag{10.50}$$

$$\nabla_{x_{dc}^{p,t,s}} \mathcal{L}_{dc} = P_{dc}^{p,t,s} - CP^{p,z} + \xi_{dc}^{p,t,s} = 0, \quad \forall dc, p, t, s \tag{10.51}$$

$$x_{dc}^{p,t,s} \leq Q_{dc}^{p,t,s} \perp \xi_{dc}^{p,t,s} \geq 0, \quad \forall dc, p, t, s \tag{10.52}$$

are appended to the upper-level problems (10.33) and (10.34) as complementarity constraints (10.53)–(10.58).

- Consumers with consumption bids

$$0 \leq RP^{p,z} - P_{db}^{p,t,s} + \varphi_{db}^{p,t,s} \perp x_{db}^{p,t,s} \geq 0, \quad \forall db, p, t, s \tag{10.53}$$

$$0 \leq Q_{db}^{p,t,s} - x_{db}^{p,t,s} \perp \varphi_{db}^{p,t,s} \geq 0, \quad \forall db, p, t, s \tag{10.54}$$

- Producers with production bids

$$0 \leq P_{pb}^{p,t,s} - PP^{p,z} + \gamma_{pb}^{p,t,s} \perp x_{pb}^{p,t,s} \geq 0, \quad \forall pb, p, t, s \tag{10.55}$$

$$0 \leq Q_{pb}^{p,t,s} - x_{pb}^{p,t,s} \perp \gamma_{pb}^{p,t,s} \geq 0, \quad \forall pb, p, t, s \tag{10.56}$$

- Consumers with curtailable loads

$$0 \leq P_{dc}^{p,t,s} - CP^{p,z} + \xi_{dc}^{p,t,s} \perp x_{dc}^{p,t,s} \geq 0, \quad \forall dc, p, t, s \tag{10.57}$$

$$0 \leq Q_{dc}^{p,t,s} - x_{dc}^{p,t,s} \perp \xi_{dc}^{p,t,s} \geq 0, \quad \forall dc, p, t, s \tag{10.58}$$

The resulting formulation is a mathematical programming problem with complementarity constraints (MPCC) with nonlinearities due to the complementary slackness conditions and due to the existence of bilinear products $RP^{p,z}x_{db}^{p,t,s}$, $CP^{p,z}x_{dc}^{p,t,s}$, $PP^{p,z}x_{pb}^{p,t,s}$, and $RP^{p,z}x_{dp}$ in the objective function (10.33). These types of nonlinearities are tackled by applying the SDT ((10.59)–(10.62)) and the big-M formulation ((10.63)–(10.82)).

$$CP^{p,z}x_{dc}^{p,t,s} = P_{dc}^{p,t,s}x_{dc}^{p,t,s} + Q_{dc}^{p,t,s}\xi_{dc}^{p,t,s}, \quad \forall dc, p, t, s \tag{10.59}$$

$$RP^{p,z}x_{db}^{p,t,s} = P_{db}^{p,t,s}x_{db}^{p,t,s} - Q_{db}^{p,t,s}\varphi_{db}^{p,t,s}, \quad \forall db, p, t, s \tag{10.60}$$

$$PP^{p,z}x_{pb}^{p,t,s} = P_{pb}^{p,t,s}x_{pb}^{p,t,s} + Q_{pb}^{p,t,s}\gamma_{pb}^{p,t,s}, \quad \forall pb, p, t, s \tag{10.61}$$

$$\left(\sum_{p,z} RP^{p,z}/Nz - P_{dp}\right)x_{dp} = -1 \cdot \tau_{dp}, \quad \forall dp \tag{10.62}$$

$$Q_{db}^{p,t,s} - x_{db}^{p,t,s} \leq (1 - \delta_{db,1}^{p,t,s})M, \quad \forall db, p, t, s \tag{10.63}$$

$$\varphi_{db}^{p,t,s} \leq \delta_{db,1}^{p,t,s}M, \quad \forall db, p, t, s \tag{10.64}$$

$$0 \leq RP^{p,z} - P_{db}^{p,t,s} + \varphi_{db}^{p,t,s}, \quad \forall db, p, t, s \tag{10.65}$$

$$RP^{p,z} - P_{db}^{p,t,s} + \varphi_{db}^{p,t,s} \leq (1 - \delta_{db,2}^{p,t,s})M, \quad \forall db, p, t, s \tag{10.66}$$

$$x_{db}^{p,t,s} \leq \delta_{db,2}^{p,t,s}M, \quad \forall db, p, t, s \tag{10.67}$$

$$Q_{pb}^{p,t,s} - x_{pb}^{p,t,s} \leq (1 - \delta_{pb,1}^{p,t,s})M, \quad \forall pb, p, t, s \tag{10.68}$$

$$\gamma_{pb}^{p,t,s} \leq \delta_{pb,1}^{p,t,s}M, \quad \forall pb, p, t, s \tag{10.69}$$

$$0 \leq P_{pb}^{p,t,s} - PP^{p,z} + \gamma_{pb}^{p,t,s}, \quad \forall pb, p, t, s \tag{10.70}$$

$$P_{pb}^{p,t,s} - PP^{p,z} + \gamma_{pb}^{p,t,s} \leq (1 - \delta_{pb,2}^{p,t,s})M, \quad \forall pb, p, t, s \tag{10.71}$$

$$x_{pb}^{p,t,s} \leq \delta_{pb,2}^{p,t,s}M, \quad \forall pb, p, t, s \tag{10.72}$$

$$Q_{dc}^{p,t,s} - x_{dc}^{p,t,s} \leq (1 - \delta_{dc,1}^{p,t,s})M, \quad \forall dc, p, t, s \tag{10.73}$$

$$\xi_{dc}^{p,t,s} \leq \delta_{dc,1}^{p,t,s} M, \quad \forall dc, p, t, s \tag{10.74}$$

$$0 \leq P_{dc}^{p,t,s} - CP^{p,z} + \xi_{dc}^{p,t,s}, \quad \forall dc, p, t, s \tag{10.75}$$

$$P_{dc}^{p,t,s} - CP^{p,z} + \xi_{dc}^{p,t,s} \leq (1 - \delta_{dc,2}^{p,t,s}) M, \quad \forall dc, p, t, s \tag{10.76}$$

$$x_{dc}^{p,t,s} \leq \delta_{dc,2}^{p,t,s} M, \quad \forall dc, p, t, s \tag{10.77}$$

$$1 - x_{dp} \leq (1 - \delta_{dp,1}) M, \quad \forall dp \tag{10.78}$$

$$\tau_{dp} \leq \delta_{dp,1} M, \quad \forall dp \tag{10.79}$$

$$0 \leq \sum_{p,z} RP^{p,z}/Nz - P_{dp} + \tau_{dp}, \quad \forall dp \tag{10.80}$$

$$\sum_{p,z} RP^{p,z}/Nz - P_{dp} + \tau_{dp} \leq (1 - \delta_{dp,2}) M, \quad \forall dp \tag{10.81}$$

$$x_{dp} \leq \delta_{dp,2} M, \quad \forall dp \tag{10.82}$$

The resulting MILP problem is as follows:

$$\begin{aligned} \min \sum_{p} w(p) \Big[ & \sum_{t} SMP^{p,t} e^{p,t} + \sum_{t,pb,s} (P_{pb}^{p,t,s} x_{pb}^{p,t,s} + Q_{pb}^{p,t,s} \gamma_{pb}^{p,t,s}) \\ & + \sum_{t,dc,s} (P_{dc}^{p,t,s} x_{dc}^{p,t,s} + Q_{dc}^{p,t,s} \xi_{dc}^{p,t,s}) - \sum_{t,dp} (P_{dp} x_{dp} - \tau_{dp}) Q_{dp}^{p,t} \\ & - \sum_{t,db,s} (P_{db}^{p,t,s} x_{db}^{p,t,s} - Q_{db}^{p,t,s} \varphi_{db}^{p,t,s}) \Big] \end{aligned} \tag{10.83}$$

subject to constraints (10.34) and (10.59)–(10.82). Optimization is performed with respect to the upper-level variables $e^{p,t}$, $RP^{p,z}$, $CP^{p,z}$, $PP^{p,z}$, the lower-level variables $x_{db}^{p,t,s}$, $x_{dc}^{p,t,s}$, $x_{pb}^{p,t,s}$, $x_{dp}$, the dual variables of the lower-level constraints $\varphi_{db}^{p,t,s}$, $\xi_{dc}^{p,t,s}$, $\gamma_{pb}^{p,t,s}$, $\tau_{dp}$, and the auxiliary binary variables $\delta_{dr,1}^{p,t,s}$, $\delta_{dr,2}^{p,t,s}$, $\delta_{dp,1}$, $\delta_{dp,2}$.

The results of the optimization provide the optimal prices for the DER Aggregator, applied to the entire horizon, which maximize his profit, as well as the quantities corresponding to these prices that optimize the objective functions of the lower-level entities and lead the interaction to equilibrium.

### *10.3.4 Implementation*

#### 10.3.4.1 Input data—scenarios

The simulation concerns a time horizon of 24 h (one characteristic period $\mathcal{P} = \{p_1\}$; scenario weight $w(p_1) = 1$; 24 hourly dispatch intervals $\mathcal{T} = \{t_1, \ldots, t_{24}\}$). Each customer category comprises 20 entities with different price-quantity bids: consumers with load bids (27–47 euros/MWh, 8–40 MWh), consumers with curtailable loads (19–27.5 euros/MWh, 2.78–6.75 MWh, baseline: 8.3–20.5 MWh), producers with production bids (10–15 euros/MWh, 14.9–59.3 MWh), consumers with load profiles (15–42 euros/MWh, peak load: 18–40 MWh). Wholesale market prices vary between 14.3 and 30 euros/MWh.

In order to test different forms of pricing on the side of the retailer and their impact on the decisions of the local resources, two scenarios regarding the number of time zones, into which the 24-h period is divided, are examined: (A) one zone of 24 h each and (B) two zones of 12 h each. Grouping of the hours of the day to the various zones is predefined and is performed by grouping together the hours with similar wholesale market prices.

#### 10.3.4.2 Computational issues

The equivalent MILP problem comprises 8,785 equations, 5,868 real variables, 2,920 binary variables, and is solved in less than 1 min using CPLEX 12.5 [19] under GAMS [20] running on an Intel®Core™2 Duo with two processors at 2.99 GHz and 1.96 GM of RAM. The relative termination tolerance is set to $10^{-8}$.

### *10.3.5 Results*

The optimal price levels for Case A (one price zone) for each type of entity are as follows: retail price (*RP*) 28 euros/MWh, curtailment price 21.2 euros/MWh, and production price 15 euros/MWh. For these price levels selected by the DER Aggregator, the optimal response of the local entities, in terms of volumes, is presented in Figure 10.9.

The supply mix on Figure 10.9(a) corresponds to the various producing entities (either local or not), i.e., DG units, load curtailment, and energy bought from wholesale market. The total energy production, according to the energy balance, equals the total demand of the Aggregator's portfolio. Thus, the envelope of the area of the two graphs in Figure 10.9 is identical. Figure 10.9(b) presents the composition of the Aggregator's load, i.e., the baseline load of consumers with curtailment bids, the aggregated demand profiles that are "dispatched" and the aggregated load bids that are accepted.

Evidently, with the prices announced to the local resources invariable from hour to hour (Case A), the entities have no other motive for determining their behavior, apart from any inherent restrictions. In fact, the total demand is independent from the prices prevailing in the wholesale market and the energy acquired from the wholesale market is simply used to keep the energy supply–demand balance of the Aggregator's portfolio.

In order to test the impact of zonal pricing, Case B is examined, where two price zones are defined. The respective results are presented in Figure 10.10, while the optimal price levels are presented in Figure 10.11. Interestingly, though not surprisingly, when the Aggregator has more degrees of freedom in determining price levels, these are selected to be in line with the variations in the wholesale electricity market, i.e., the retail prices are higher during the hours when the wholesale prices are also

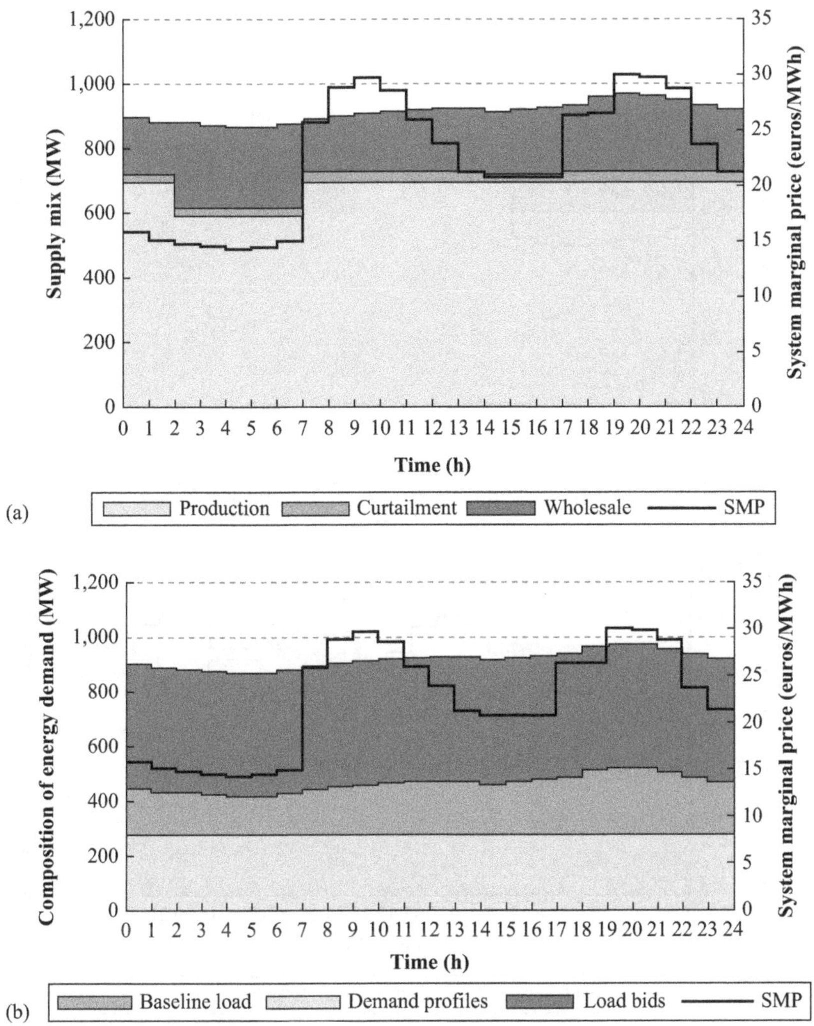


*Figure 10.9 Optimization results per dispatch interval for the case with one price zone along with the wholesale market prices (SMP): (a) supply mix and (b) composition of energy demand*

high. As a result, the local entities adjust their response to these prices by selecting higher production and curtailment levels and lower demand during hours of high price levels. Furthermore, the energy acquired from the wholesale market and the total energy demand, represented by the Aggregator, are now negatively correlated with the SMPs.

These changes are also reflected in the resulting revenues of the Aggregator. In Case A the total revenues (i.e., the negative of the objective function (10.83)) are 61,199 euros, while in Case B the total revenues are by 14.3% higher and equal to

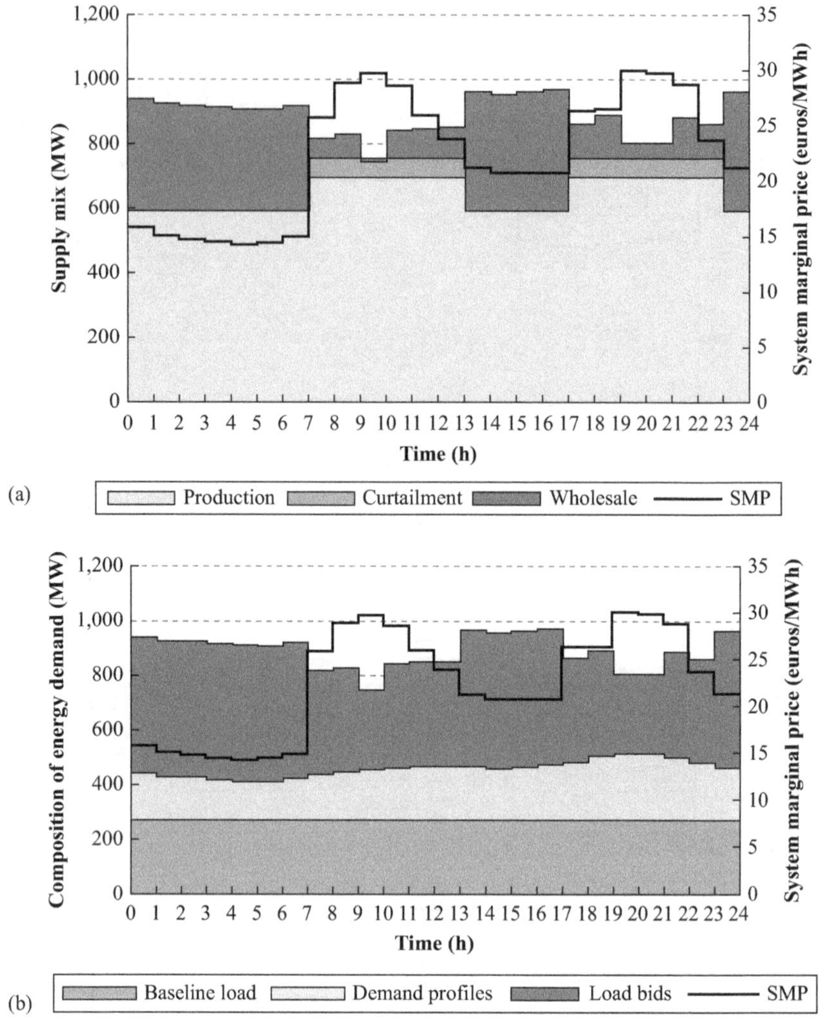


*Figure 10.10 Optimization results per dispatch interval for the case with two price zones along with the wholesale market prices (SMP): (a) supply mix and (b) composition of energy demand*

69,946 euros. It is then evident that, as expected, the definition of price zones for the energy produced, consumed, or curtailed locally by entities managed and represented in the market transactions by an aggregating entity is a task that affects the profitability of the entity.

In the specific application, only one characteristic period is simulated. However, the DER Aggregator model is general in the sense that it allows simulating a variety of different time periods and dispatch intervals, as desired. For example, a month could be modeled either as a set of 30 periods (days) of 24 intervals (hours) or as one

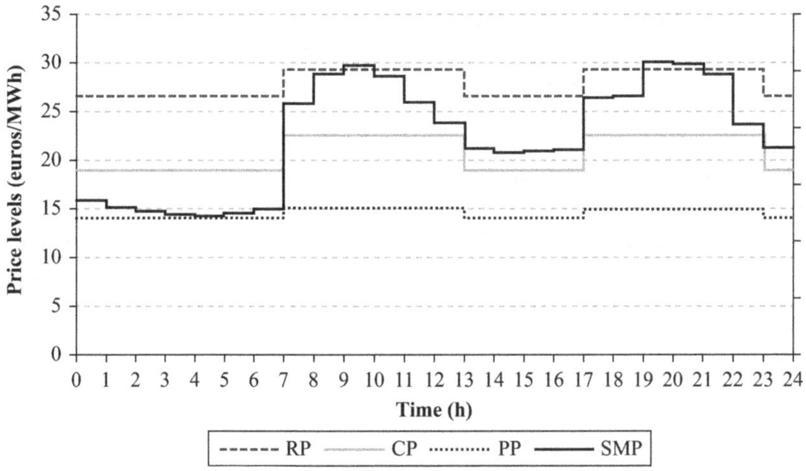


*Figure 10.11 Optimal price levels for Case B for each type of local entity (RP: retail price for demand customers; CP: price for load curtailment; PP: price for local generation; and SMP: wholesale market price)*

period of 720 intervals (hours). Depending on the point of focus, this model could prove useful for any retail market stakeholder deciding upon the form of the pricing scheme to offer to the local resources under his control, taking into account not only external factors such as the wholesale market prices but also the customer side and its characteristics.

## 10.4 Conclusions

For every entity not indifferent to or unaffected by changes in the environment in which it operates, any decision taken locally inevitably reflects the inherent characteristics as well as the influence of extrinsic parameters. In turn, the coexistence of various entities, possibly with conflicting goals, forms an environment in which each individual is called upon to act. The interrelation between the decision-making process of interacting entities formulates a complex environment, where the behavior of each entity influences and is influenced by the behavior of the rest. In this respect, game theory is considered particularly useful for describing situations such as these, as it offers the appropriate normative framework for giving form to problems that would have been otherwise difficult to cast in a mathematical manner. One category includes hierarchical decision-making, that corresponds to a Stackelberg (leader-follower) game, which can be described mathematically by a bilevel programming problem.

Such problems have been extensively studied and, although not lacking in limitations or complexity issues, have been put to use for modeling situations where several interrelated entities affect with their decisions one another. Such models are suitable to solve complex distributed decision problems also in the context of smart grid, as shown in the two practical examples presented in this chapter. These examples simulate the interactions (both with the wholesale market and with his customers) of an Aggregator responsible for managing local resources that form key components of the smart grid, i.e., flexible and inflexible loads and DG units.

The equivalent transformation to a MILP problem by employing appropriate techniques facilitates the solution, as commercially available software can be used. Furthermore, once defined, upscaling and extending the problems to incorporate either a more extended clientele or different types of resources and costs (e.g., storage facilities, capacity charges) can be easily done, albeit at the expense of computational time. This is, however, of secondary importance in planning problems solved prior to the real-time operation, such as those presented here.

## Bibliography

[1] Hatziargyriou, N. (eds.). *Microgrids: Architectures and Control*. London: John Wiley & Sons Ltd; 2014.

[2] von Stackelberg, H. *Market Structure and Equilibrium*. Trans. D. Bazin, L. Urch, R. Hill. Berlin: Springer-Verlag; 2011.

[3] von Neumann, J., Morgenstern, O. *Theory of Games and Economic Behavior*. Princeton, NJ: Princeton University Press; 1944.

[4] Luenberger D.G., Ye Y. *Linear and Nonlinear Programming*. 3rd edn. New York, NY: Springer; 2008.

[5] Fortuny-Amat, J., McCarl, B. "A representation and economic interpretation of a two-level programming problem". *Journal of the Operational Research Society*. 1981;32(9):783–792.

[6] Dempe, S. *Foundations of Bilevel Programming: Nonconvex Optimization and Its Applications*, vol. 61. Dordrecht: Kluwer Academic Publishers; 2002.

[7] Saad, W., Han, Z., Poor, H.V., Basar, T. "Game-theoretic methods for the smart grid: An overview of microgrid systems, demand-side management, and smart grid communications". *IEEE Signal Processing Magazine*. 2012;29(5): 86–105.

[8] Dempe, S. "Annotated bibliography on bilevel programming and mathematical programs with equilibrium constraints". *Optimization*. 2003;52(3):333–359.

[9] Colson, B., Marcotte, P., Savard G. "An overview of bilevel optimization". *Annals of Operations Research*. 2007;153(1):235–256.

[10] Mohsenian-Rad, A., Wong, V.W.S., Jatskevich, J., Schober, R., Leon-Garcia, A. "Autonomous demand-side management based on game theoretic energy consumption scheduling for the future smart grid". *IEEE Transactions on Smart Grid*. 2010;1(3):320–331.

[11] Maharjan, S., Zhu, Q., Zhang, Y., Gjessing, S., Basar, T. "Dependable demand response management in the smart grid: A Stackelberg game approach". *IEEE Transactions on Smart Grid*. 2013;4(1):120–132.

[12] Yang, P., Tang, G., Nehorai, A. "A game-theoretic approach for optimal time-of-use electricity pricing". *IEEE Transactions on Power Systems*. 2013;28(2):884–892.
[13] Chai, B., Chen, J., Yang, Z., Zhang, Y. "Demand response management with multiple utility companies: A two-level game approach". *IEEE Transactions on Smart Grid*. 2014;5(2):722–731.
[14] Tushar, W., Chai, B., Yuen, C., *et al*. "Three-party energy management with distributed energy resources in smart grid". *IEEE Transactions on Industrial Electronics*. 2015;62(4):2487–2498.
[15] Nekouei, E., Alpcan, T., Chattopadhyay, D. "Game-theoretic frameworks for demand response in electricity markets". *IEEE Transactions on Smart Grid*. 2015;6(2):748–758.
[16] Saad, W., Glass, A.L., Mandayam, N.B., Poor, H.V. "Toward a consumer-centric grid: A behavioral perspective". *Proceedings of the IEEE*. 2016; 104(4):865–882.
[17] Asimakopoulou, G.E., Dimeas, A.L., Hatziargyriou, N.D. "Leader-follower strategies for energy management of multi-microgrids". *IEEE Transactions on Smart Grid*. 2013;4(4):1909–1916.
[18] Asimakopoulou, G.E., Vlachos, A.G., Hatziargyriou, N.D. "Hierarchical decision making for aggregated energy management of distributed resources". *IEEE Transactions on Power Systems*. 2015;30(6):3255–3264.
[19] GAMS Development Corporation. GAMS – The Solver Manuals. Washington, DC, USA. [Online]. Available at: http://www.gams.com/help/topic/gams.doc/solvers/allsolvers.pdf.
[20] GAMS Development Corporation. General Algebraic Modeling System (GAMS) Release 23.8.2. Washington, DC, USA, 2012.
[21] Baringo, L., Conejo, A.J. "Wind power investment: A Benders decomposition approach". *IEEE Transactions on Power Systems*. 2012;27(1):433–441.
[22] Vlachos, A.G., Biskas P.N. "Adjustable profile blocks with spatial relations in the day-ahead electricity market". *IEEE Transactions on Power Systems*. 2013;28(4):4578–4587.

*Part III*

# Smart grid communications and networking

*Chapter 11*

# Cyber security of smart grid state estimation: attacks and defense mechanisms

*Jinping Hao[1], Robert J. Piechocki[1], Dritan Kaleshi[1], Woon Hau Chin[1] and Zhong Fan[1]*

Compared to traditional power grids, a smart grid tends to be much more reliable, efficient, and intelligent due to the remarkable advancements in sensing, monitoring, control technologies, and also the tight integration with the cyber infrastructure and advanced computing and communications technologies. Nevertheless, this integration may lead to new vulnerabilities to cyber-attacks on the power grid systems. Cyber-attacks are reported as one of the main potential threats to the reliable operation of the power systems [1]. In this chapter, the problem of malicious false data injection attacks (FDIAs) against the supervisory control and data acquisition (SCADA) system is investigated.

In smart grids, SCADA systems obtain power system information such as power flows and bus voltages through remote terminal units (RTUs). These measurements are then often used for state estimation in the energy management system (EMS), which builds real-time electricity network models and perform various control and planning tasks. The complex network connections, as well as the Internet, make SCADA systems susceptible to potential FDIAs, in which adversaries aim to contaminate the measurements collected from RTUs and bias the state estimation in order to mislead the operation of the power system. Therefore, it is crucially important to understand the behavior of the attackers, so that appropriate countermeasures can be designed to either protect the system from attacks beforehand or identify the malicious data injections in the measurements.

Adversaries can launch attacks through hacking RTUs such as sensors in substations. In consideration of the accessibility of sensors and also hacking cost, attackers always tend to control only a few meters to implement a successful attack [2]. Thus an attack, which can be modeled as a vector, is designed as sparse as possible. Existing sparse attack strategies are mainly designed to bias certain targeted system states [3]. In this chapter, a general scenario is considered, in which adversaries can

[1]Department of Electrical and Electronic Engineering (MVB), School of Engineering, University of Bristol, Bristol, UK

access arbitrary measurements to change arbitrary system states in state estimations. To the best of our knowledge, there is no feasible algorithm that can efficiently construct highly sparse undetectable attacks in this case. An efficient and effective attack vector construction algorithm is proposed, which can quickly generate highly sparse attack vectors in this scenario. Sparse stealth attacks in a specific scenario are also considered. An optimization-based algorithm is introduced to generate sparse-targeted attack vectors to bias specified state variables with the consideration that a subset of measurements is protected.

There are two approaches to defend against the malicious data attacks. The first is to protect the system beforehand from being attacked by adversaries. This can be achieved by either protecting a number of measurements from being modified to prevent stealth attacks, or monitoring state information directly by the deployment of advanced measuring devices such as phasor measurement units (PMUs). It is not economical or necessary to secure all measurements to prevent attacks due to high cost. Instead, stealth malicious attacks can be prevented by protecting a carefully selected subset of measurements. The challenge of this approach is to quickly search the effective small measurement subset to make them immune to attacks. The second approach to deal with malicious attacks is to identify the injected false data in measurements and then either abandon the contaminated data or correct them. In this chapter, both approaches of defending against malicious attacks are discussed.

In this chapter, a fast greedy search method is proposed to quickly find a subset of measurements to be protected to defend against stealth attacks. This fast method can find a subset with the same size as that from using brute-force search in nearly all cases. Then a detection algorithm is discussed with consideration of the case that only partial noise contaminated measurements are collected. The detection algorithm takes advantages of the sparsity of attacks and can detect both random errors and carefully designed stealth false data in the measurements. Additionally, this algorithm can also recover the true measurements in addition to detecting errors in measurements. The proposed algorithms are tested based on the IEEE test systems [4] with software MATPOWER [5].

## 11.1 Power system state estimation and FDIAs

A power grid system is a sophisticated network which connects a number of electric power generators to various consumers through extensive transmission lines. It is extremely important to monitor the states of this complex system such that various control and planning tasks can be performed and the reliable operation of the power system is guaranteed. In power system, state estimation [6] is used to estimate system states through a number of sensor measurements and is a useful and necessary function in EMSs.

State estimation aims to acquire the states of the system, which refer to voltage magnitudes and phasor angles at all buses. Directly measuring the state information would be vulnerable to measurement errors or telemetry failures. Therefore, state estimation makes use of redundant power measurements from RTUs, including active and reactive branch power flows, bus power injections, and voltage magnitudes. Measurements may be subject to errors due to various reasons such as errors from measuring devices, errors in communication links, and also errors from the changes in the system. The redundancy of the measurements allows state estimation under the removal of bad measurements. Additionally, bad data detection (BDD) is an important function in state estimator, which can detect those measurements with errors to achieve robust state estimation. However, coordinated FDIAs can be hardly identified by traditional BDD methods.

If adversaries possess the knowledge of power grid network topology and power line impedance information, they can inject false data into measurements which can bypass the existing techniques of BDD and are undetectable [2]. Recent research has also proposed that even when power grid information is unavailable, undetectable attacks can still be accomplished [7]. Existing techniques for BDD are based on the residue testing. Large errors in measurements can result in large residues. However, a carefully designed attack can bypass BDD by injecting errors into measurements without causing residue variations.

### *11.1.1 State estimation*

We consider a power transmission grid, which consists of $N+1$ buses and $L$ transmission lines. The state of the system includes $N+1$ bus voltage magnitudes and $N+1$ voltage phasor angles. One of the buses is chosen as reference bus and the voltage phasor is set to 0. Therefore, the state vector $\mathbf{x} \in \mathbb{R}^{2N+1}$ includes $N$ bus voltage phasor angles and $N+1$ bus voltage magnitudes. The measurements that are used to estimate the system state consist of branch power flows, bus injections, and bus voltage magnitudes. In full AC power flow state estimation, these measurements and the sate vector have the following non-linear relationship:

$$\mathbf{z} = \mathbf{h}(\mathbf{x}) + \mathbf{e} \tag{11.1}$$

where $\mathbf{z} \in \mathbb{R}^M$ is the vector of non-synchronized power measurements and $M$ is the number of measurements. $\mathbf{h}(\mathbf{x})$ represents a set of non-linear functions that relate measurements to the state variables. $\mathbf{e}$ denotes the measurement errors which can be modeled as zero-mean Gaussian distributed random variables.

Transitional approach to estimate state vector utilized weighted least squares (WLS) method [8], which is to solve the following optimization problem:

$$\min\ F(\mathbf{x}) = \frac{1}{2}(\mathbf{z} - \mathbf{h}(\mathbf{x}))^T \mathbf{W}^{-1}(\mathbf{z} - \mathbf{h}(\mathbf{x})) \tag{11.2}$$

where $\mathbf{W}$ denotes the weighting matrix whose diagonal elements equal to the noise covariance.

In DC power flow model, bus voltage magnitudes are already known and equal to one unit. Neglecting all shunt susceptances and series branch resistances, the above AC power flow state estimation model is approximated by a linearized form. Therefore, state variables that are estimated consist of only $N$ voltage phasor angles. All reactive power are neglected and the power measurements include only active power flows and active power injections. In DC power flow state estimation, power measurements have the following linear relationships with state variables:

$$\mathbf{z} = \mathbf{H}\theta + \mathbf{e} \tag{11.3}$$

where $\mathbf{H}$ is the measurement matrix. The measurements $\mathbf{z}$ in this case comprise only $L$ active power flows and $N + 1$ power injections. The state variables are voltage phasors, which are denoted by $\theta$. $\mathbf{e}$ represents the zero-mean Gaussian measurement noise.

The linearized state estimation is much easier to solve. Commonly used methods include maximum likelihood and WLS algorithms [6]. Specifically, WLS has the following objective function in this case:

$$F(\theta) = \frac{1}{2}(\mathbf{z} - \mathbf{H}\theta)^T\mathbf{W}^{-1}(\mathbf{z} - \mathbf{H}\theta) \tag{11.4}$$

The solution can be easily obtained as follows:

$$\hat{\theta} = (\mathbf{H}^T\mathbf{W}^{-1}\mathbf{H})^{-1}\mathbf{H}^T\mathbf{W}^{-1}\mathbf{z} \tag{11.5}$$

Due to the limited accuracy of the measuring devices and also the telecommunication medium, random errors may exist in the measurements. If the measurements contain bad data that are beyond a realistic limit, significant errors may be caused in determining the state of the system. Therefore, BDD is a crucial function in the state estimator. Existing BDD methods are mainly based on the measurement residue, which is the differences between the obtained measurements and that derived from the estimated state variables [9]:

$$\mathbf{r} = \mathbf{z} - \mathbf{H}\hat{\theta} \tag{11.6}$$

A typical traditional BDD method is the largest normalized residue (LNR) test [10], in which the measurement is identified as bad data if the corresponding LNR is greater than a specified identification threshold $\tau$, namely,

$$\max_i \frac{|r_i|}{\sigma_i} \geq \tau \tag{11.7}$$

where $\sigma_i$ is the standard deviation of the $i$th residue $r_i$. Bad data are then filtered out and the redundant measurements can still ensure the accuracy of the state estimation solution.

### *11.1.2 Malicious FDIAs*

When malicious data are injected into measurements, (11.3) in DC power flow model becomes:

$$\mathbf{z_a} = \mathbf{H}\theta + \mathbf{a} + \mathbf{e} \tag{11.8}$$

where vector $\mathbf{a} \in \mathbb{R}^M$ denotes the malicious attack data injected by an attacker. The $i$th non-zero element in this attack vector $a_i$ indicates that the $i$th measurement is compromised by the attacker. It is notable that attack vector $\mathbf{a}$ tends to be sparse as the attackers always intend to launch attacks with least effort, namely, to compromise as few measurements as possible.

Additive errors in the measurements can be detected and eliminated in the state estimator. Thus, a randomly injected attack can easily be identified and can hardly threaten the system. However, a carefully designed coordinated malicious attack can bypass the existing residue-based BDD methods [2]. If adversaries have the information of power grid network topologies and power line impedance information, they can inject stealth attacks into measurements. Specifically, an attack $\mathbf{a}$ is undetectable if it is a linear combination of the column vectors of measurement matrix $\mathbf{H}$, i.e.,

$$\mathbf{a} = \mathbf{Hc} \tag{11.9}$$

where $\mathbf{c} \in \mathbb{R}^N$ is any arbitrary vector.

Then the measurements with attacks $\mathbf{z_a}$ can be written as the following form:

$$\mathbf{z_a} = \mathbf{H}\theta + \mathbf{a} + \mathbf{e} = \mathbf{H}(\theta + \mathbf{c}) + \mathbf{e} \tag{11.10}$$

In this case, the system would regard the manipulated state vector $(\theta + \mathbf{c})$ as the real value in the state estimator. The vector $\mathbf{c}$ indicates the errors that are added on the state variables introduced by attack vector $\mathbf{a}$. Thus, measurements $\mathbf{z_a}$ can bypass the BDD. This can be further verified as the residue in this case is the same as that when attack $\mathbf{a}$ does not exist. Denoting $\mathbf{r_a}$ as the residue and $\hat{\theta}_\mathbf{a}$ as the estimated state vector when attack vector $\mathbf{a}$ is presented in measurements, we have:

$$\mathbf{r_a} = \mathbf{z_a} - \mathbf{H}\hat{\theta}_\mathbf{a} = \mathbf{z} + \mathbf{a} - \mathbf{H}(\hat{\theta} + \mathbf{c}) = \mathbf{z} - \mathbf{H}\hat{\theta} + \mathbf{a} - \mathbf{Hc} = \mathbf{z} - \mathbf{H}\hat{\theta} \tag{11.11}$$

which is the same as that in (11.6) when there is no attack. Therefore, carefully designed stealth attacks would not cause residue variations and thus can bypass BDD in the state estimator.

## 11.2 Stealth attack strategies

Stealth FDIAs were first discussed in Reference 2. Based on this work, a lot of efforts have been made to design effective sparse attack construction algorithms and the corresponding countermeasures. Authors in Reference 11 developed a general optimization framework for constructing sparse attack vectors when a subset of measurements is protected from being modified by attackers. In Reference 3, the sparse attack construction model was extended to a distributed framework. Authors in Reference 12 considered sparse attacks with injections into critical measurements, which are essential for the observability of the power grids and sensitive to attacks. In Reference 13, methods of finding both strong stealth attacks and also optimal weak malicious data attacks (non-stealth attacks when power grid network topology is unknown) with the aim of reducing the number of compromised measurements were discussed.

As mentioned previously, attackers would always tend to launch attacks with least effort. Thus, attack vectors have to be designed as sparse as possible. The second criteria of designing attack vectors are to ensure they are undetectable, namely, an attack vector $\mathbf{a}$ has to satisfy the condition (11.9). In some specific scenarios, there may exit other restrictions for designing attack vectors. For example, some of the measurements are protected from being modified. This can be achieved by either applying more advanced measuring devices, or implementing protection mechanisms [14]. In this case, stealth attacks need to be designed without contaminating the protected measurements.

In this section, methods of constructing sparse stealth attacks are proposed for two typical scenarios. The first scenario considered in this section is a general case in which it assumes that adversaries can compromise arbitrary measurements in order to bias arbitrary state variables in state estimation. To the best of our knowledge, there is no feasible algorithm that can efficiently construct highly sparse undetectable attack vectors in this case. It is observed that undetectable attack vectors to compromise the minimum number of measurements in this scenario can be found using brute-force search. However, this is not practical due to the high complexity. The other scenario considers a specific case that adversaries aim to launch stealth attacks in order to modify specific state variables. Existing attack strategies are mainly considered for this scenario. In this section, an optimization-based framework is introduced to construct sparse attack vector to bias-specified state variables with the consideration that a subset of measurements is protected.

In the literature, there are few algorithms that can efficiently construct highly sparse undetectable attack vectors in a general scenario. In Reference 2, it is discussed that the optimal undetectable attack vector to compromise the minimum number of measurements can be found using brute-force search method. However, this is not applicable in practice due to the high complexity. An efficient and effective attack vector construction algorithm is proposed in this section. This method can quickly generate highly sparse attack vectors in a general scenario. Authors in Reference 2 have demonstrated that stealth attack vectors always exist when the number of measurements that can be contaminated exceeds a certain value. In this section, it is shown that the proposed algorithm can inject stealth attacks by manipulating only a much smaller number of measurements with high probability.

### *11.2.1 Random attacks*

In a general scenario, there are no restrictions on the accessibility to power measurements for adversaries. Compromising different measurements are assumed to have the same costs and difficulties. Also, no particular state variables are specified for the attacks. Adversaries can hack arbitrary measurements to bias any state variables in an attack. Therefore, the main consideration for an attacker to launch stealth attacks is to compromise as few measurements as possible. These attacks are denoted as random attacks.

A random attack vector $\mathbf{a}$ therefore has to satisfy condition (11.9) while be designed as sparse as possible. Since $\mathbf{a}$ is a linear combination of the columns of measurement matrix $\mathbf{H}$, it is possible to generate sparse attack vector from column transformations of $\mathbf{H}$ as in this way the number of non-zero elements in the transformed column vector can be reduced. However, the sparsity cannot be guaranteed using this method to construct attack vector $\mathbf{a}$. A brute-force approach can make it possible to find the optimal attack vector, which has the minimum number of non-zero elements. But brute-force method is complex and time-consuming, which makes it not feasible in practice.

Authors in Reference 2 also proposed that using a projection matrix **P** can lead to an equivalent criteria to generate attack vector **a** satisfying stealth condition (11.9). Let projection matrix $\mathbf{P} = \mathbf{H}(\mathbf{H}^T\mathbf{H})^{-1}\mathbf{H}^T$, then for any vector **a**, we have $\mathbf{Pa} = \mathbf{a}$ if and only if **a** is a linear combination of column vectors of **H**, i.e., $\mathbf{a} = \mathbf{Hc}$. This relationship can be expressed as follows:

$$\mathbf{Pa} = \mathbf{H}(\mathbf{H}^{\mathrm{T}}\mathbf{H})^{-1}\mathbf{H}^{\mathrm{T}}\mathbf{Hc} = \mathbf{Hc} = \mathbf{a} \tag{11.12}$$

Let $\mathbf{G} = \mathbf{P} - \mathbf{I}$, from the above equation, it is easily known that $(\mathbf{P} - \mathbf{I})\mathbf{a} = 0$. Then this relationship can be written as the following form:

$$\mathbf{Ga} = 0 \tag{11.13}$$

which is an equivalent criteria to the condition $\mathbf{a} = \mathbf{Hc}$. Therefore, a vector **a** satisfies $\mathbf{a} = \mathbf{Hc}$ if and only if $\mathbf{Ga} = 0$, with $\mathbf{G} = \mathbf{H}(\mathbf{H}^{\mathrm{T}}\mathbf{H})^{-1}\mathbf{H}^{\mathrm{T}} - \mathbf{I}$.

This criterion can be used to generate attack vectors in certain scenarios such as when attackers tend to compromise a known subset of measurements, which are not protected from attacks [2]. The sparsities of the attack vectors are not addressed in this scenario. For random attacks, a straightforward way to find sparse attack vectors using criteria (11.13) can be formulated using the following optimization problem:

$$\min \|\mathbf{a}\|_0 \quad \text{s.t.} \quad \mathbf{Ga} = 0,\ \mathbf{a} \neq 0 \tag{11.14}$$

where $\|\cdot\|_0$ denotes the $l_0$-norm, which counts the number of non-zero elements in a vector. The above is a non-convex problem and finding the solution to this problem is highly complex. Therefore, it is not feasible in practice to construct sparse attack vectors by solving this problem.

It is known from (11.13) that undetectable attack vectors must be in the null space of the matrix **G**. A null space of matrix is defined as $Null(\mathbf{G}) = \{\mathbf{v} \in \mathbb{R}^M\, |\mathbf{Gv} = 0\}$. Similar to the column transformations of matrix **H**, performing column transformations to the vectors in null space of matrix **G** can generate a vector with reduced number of non-zero elements. This method cannot guarantee the sparsity of the vector and thus is not applicable. Thus, a more effective and efficient method to construct highly sparse random attack vectors in a general scenario is desired.

#### 11.2.1.1 Sparse attack construction algorithm

Rather than solving the complex problem in (11.14), or performing column transformations, we propose an algorithm taking advantage of null space of matrix **G** to construct sparse attack vectors. The aim is to quickly construct highly sparse attack vectors. The algorithm proposed in this section can obtain attack vectors with different sparsities with high probabilities depending on the noise level in measurements.

Measurements are obtained from RTUs such as sensors at the substations. These data are then transmitted through communication networks to the control center, which is far away from field sensors. Therefore, measurements are inevitably subject to random errors, or noise. These noises may be from the measuring devices and process, or communication links. Although measurements with big errors are filtered out in state estimation, noise within the range below the threshold still commonly exists. The noise can be modeled following Gaussian distribution $\mathcal{N}(0, \sigma^2)$ and the noise variance vector is $\Sigma_e$.

With the modern state estimation technologies, the system is designed to be tolerant to measurement deviations within a certain level. Thus in control center, state estimation can cope with measurement noise within this level. This fact provides the possibility for attackers to construct attack vectors designed based on the column vectors in $Null(\mathbf{G})$. Those small elements in the vector in $Null(\mathbf{G})$ can be dealt with as noise if their average energy, denoted as $\Sigma_B$, is within the range of the variance of the noise $\Sigma_e$, namely, $\Sigma_B \leq \Sigma_e$. Therefore, attackers can launch stealth attacks by only injecting errors into the measurements corresponding to the large elements in the chosen vector in $Null(\mathbf{G})$. In this way, the number of compromised measurements can be greatly reduced.

Since BDD identifies bad measurements using a threshold of the residues, it also indicates a threshold in the measurements noise. The errors beyond this value in the measurements would incur residue variations exceeding the threshold used for bad data identifications. Otherwise, the measurement errors are tolerable in the state estimations. Since the vectors in $Null(\mathbf{G})$ contain only a small number of relatively large elements, we can always find vectors from $Null(\mathbf{G})$ that having most of their elements within the range, which is tolerable to the system if they are added on measurements as errors. Then the proposed algorithm constructs the attack vectors by choosing these vectors and keeping only relatively large elements and forcing the small elements below a certain threshold to 0.

The vectors from $Null(\mathbf{G})$ with only relatively large elements are detectable as they are not in the null space of matrix $\mathbf{G}$ and thus do not satisfy stealth condition (11.9). However, with consideration of the noise in measurements, these vectors are undetectable if they are injected into the measurements as attacks. This is because the random errors in measurements can make up the missing small elements in the attack vectors and make them undetectable.

Define a shrinkage operation $\mathcal{S}_t$ as follows:

$$\mathcal{S}_t(x) := \begin{cases} \frac{x}{|x|-t} \max(|x| - t,\ 0) & |x| \neq t \\ x & |x| = t \end{cases} \tag{11.15}$$

where $t$ denotes the threshold. The variable $x$ is shrunk to 0 if it is smaller than the threshold $t$.

The attack vector construction procedure can then be designed as follows: given measurement matrix $\mathbf{H}$, compute matrix $\mathbf{G}$ as well as a basis matrix $\mathbf{U}$ of its null space $Null(\mathbf{G})$. Then the column vector $\mathbf{u}$ with the largest variance from all column vectors is selected:

$$\mathbf{u} = \arg\max_i (var(\mathbf{u}_i)) \tag{11.16}$$

---

**Algorithm 1** Sparse stealth attacks construction

---

Input: $\mathbf{H} \in \mathbb{R}^{M \times N}$, $C > 0, t > 0$.
Procedure:

1. Compute $\mathbf{G} = \mathbf{H}(\mathbf{H}^T\mathbf{H})^{-1}\mathbf{H}^T - \mathbf{I}$.
2. Get the standard basis matrix $\mathbf{U}$ of $Null(\mathbf{G})$ so that $i$th column $\mathbf{u}_i$: $\mathbf{G}\mathbf{u}_i = 0$.
3. Find column vector $\mathbf{u}$ in $\mathbf{U}$: $\mathbf{u} = \arg\max_i (var(\mathbf{u}_i))$.
4. Scale up/down vector $\mathbf{u}$ by $\varepsilon$: $\mathbf{u}' = \varepsilon\mathbf{u}$ and $\varepsilon = \frac{C}{\max(\mathbf{u})}$.
5. Shrink the vector using the threshold $t$ to obtain the sparse attack vector $\mathbf{a}$: $\mathbf{a} = \mathscr{S}_t(\mathbf{u}')$.

Output: $\mathbf{a}$.

---

where $\mathbf{u}_i$ denotes the $i$th column vector in $\mathbf{U}$. Then the chosen vector is scaled up, or down, until the maximal element reaches a pre-designed attack value $C$. The last step is to force the small element below the threshold $t$ to 0:

$$\mathbf{a} = \mathscr{S}_t(\varepsilon\mathbf{u}) \tag{11.17}$$

where $\varepsilon$ is the scaling parameter: $\varepsilon = \frac{C}{\max(\mathbf{u})}$. The whole procedure is concluded in Algorithm 1.

It is notable that threshold $t$ should be carefully chosen with consideration of both sparsity and evading BDD. A higher threshold $t$ can not only generate a sparser attack vector $\mathbf{a}$ but also increase the possibility of being detected. Since the measurement noise follows $\mathscr{N}(0, \sigma^2)$, it is assumed that all noise variables are within the range of $[-3\sigma, 3\sigma]$ (otherwise it will be identified as bad data).

### *11.2.2 Numerical results*

The performance of Algorithm 1 which generates highly sparse undetectable attack vectors is tested in different scenarios based on the IEEE test systems [4]. The MATLAB® package MATPOWER [5] is used to simulate the power system.

Figures 11.1 and 11.2 display the probabilities of successfully generating undetectable attack vectors with different levels of sparsity and different attack level ratios (ALRs), respectively. An attack is regarded as successful when the maximum value in residue vector does not exceed that without attacks. Sparsity ratio (SR) is defined as $k/M$, where $k$ is the number of non-zero elements in $\mathbf{a}$ and $M$ is the size of $\mathbf{a}$. The ALR is defined as the maximum attack value $C$ to the mean value of the state variables: $\frac{C}{mean(\theta)}$. Generally, these figures reveal that there are high probabilities for Algorithm 1 to successfully generate highly sparse undetectable attacks.

The noise in the simulation is modeled as Gaussian distributed with zero mean. The signal-to-noise ratio (SNR) indicates the noise level compared with true measurements in the simulation. The noise may due to measuring devices and process,

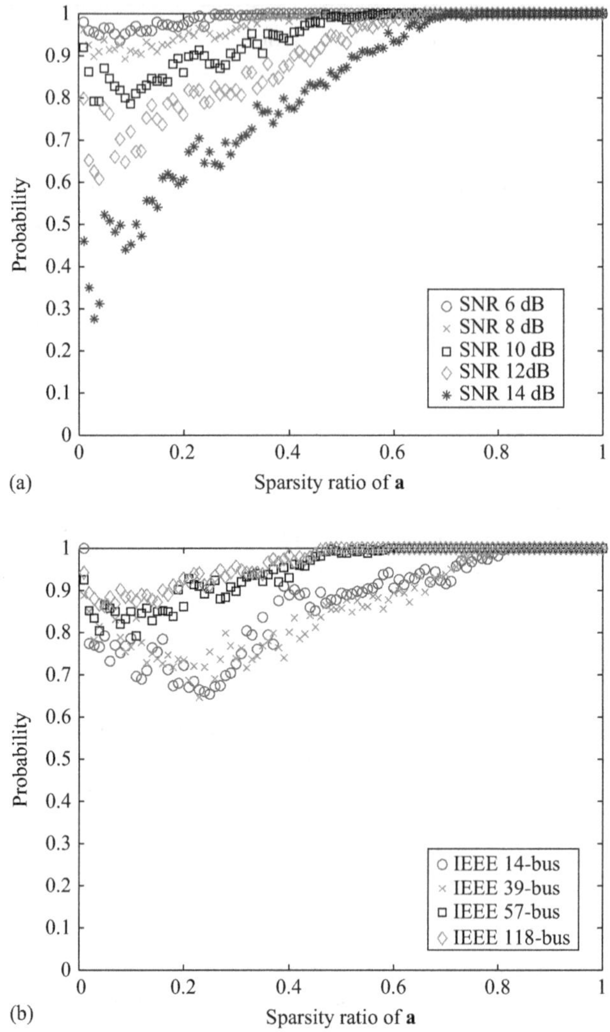


*Figure 11.1 Probabilities of successful attack injections (a) under different SNRs for IEEE 57-bus system, ALR = 0.5; (b) for different bus systems, ALR = 0.5 and SNR = 10 dB*

or due to the communication channel noise. Figure 11.1 depicts the probabilities of successfully generating stealth attacks with different SRs. The attacks with different SRs are achieved by choosing appropriate shrinkage threshold in Algorithm 1. The ALR is set to 0.5 in Figure 11.1. Figure 11.2 reveals the probabilities of successful stealth attack injections with different ALRs and the attack vectors are generated with SRs equaling 0.4.

Figures 11.1(a) and 11.2(a) compare the performances of Algorithm 1 under different SNRs in IEEE 57-bus test system. It is clear that in a relatively noisy case,

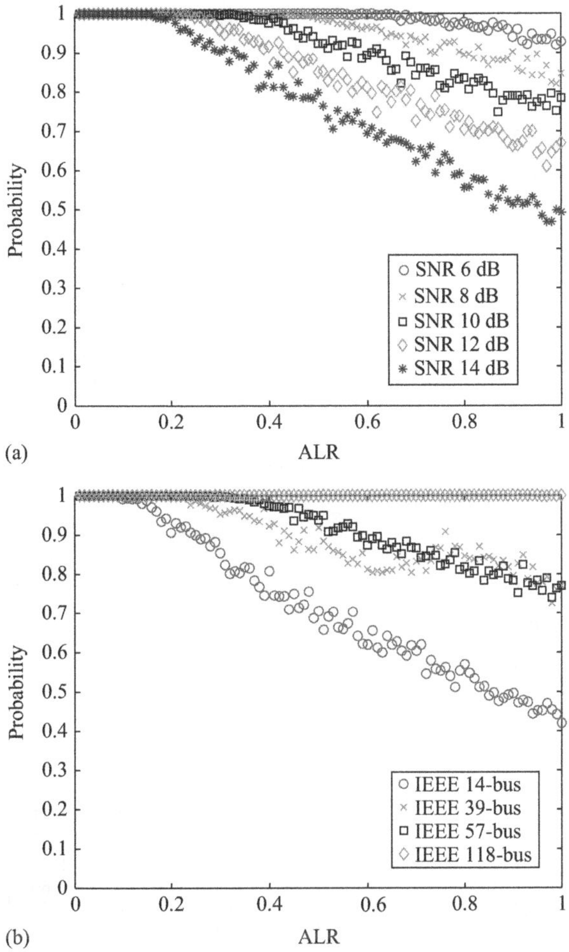


*Figure 11.2 Probabilities of successful attack injections (a) under different SNRs for IEEE 57-bus system, SR = 0.4; (b) for different bus systems, SR = 0.4 and SNR = 10 dB*

the probability of a successful attack is extremely high (close to 1). In the low noise case, there is also a high probability of injecting a successful highly sparse undetectable malicious attack, e.g., the success probability is still around 50% to inject an attack with SR = 0.1 when SNR is 14 dB.

The algorithm is also assessed using different power grid system models, which is shown in Figures 11.1(b) and 11.2(b). It is notable that in a larger bus system, Algorithm 1 can provide a better performance even for extremely sparse attacks, or high ALRs. For example, the success ratio is around 90% for IEEE 118-bus system to generate stealth attacks with SR lower than 0.1, compared with 75% for IEEE 14-bus system shown in Figure 11.1(b). This probability is 100% for IEEE 118-bus system

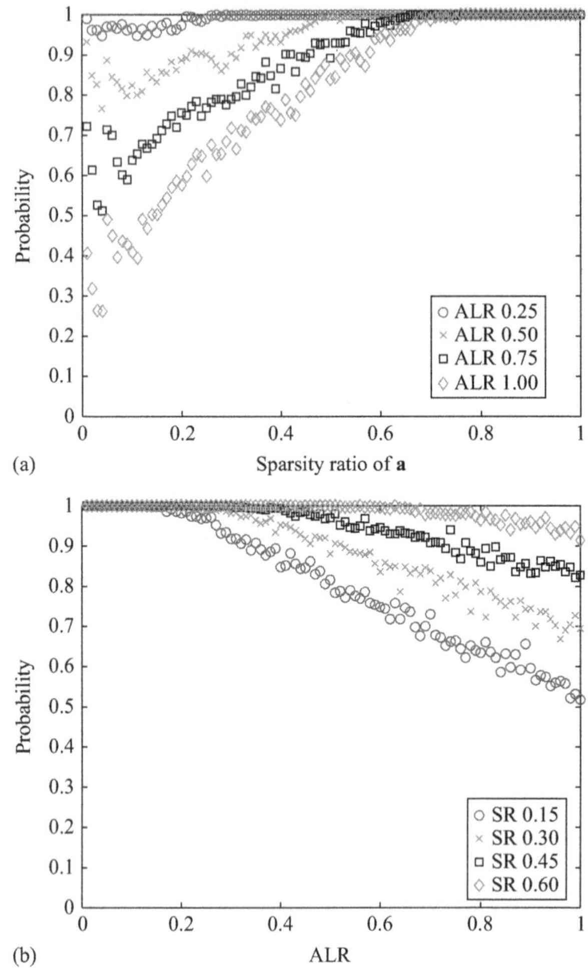


*Figure 11.3 Probabilities of successful attack injections (a) for different attack level ratios for IEEE 57-bus system, SNR = 10 dB; (b) different sparsity ratios in IEEE 57-bus system, SNR = 10 dB. Panel (b) utilizes random columns in Null(**G**) rather than that with largest variance*

to generate attacks with ALR = 1, compared to 80% for IEEE 57-bus system and 42% for IEEE 14-bus system shown in Figure 11.2(b). Therefore, it can be anticipated that the algorithm would have a better performance in a real power system, which is much larger than the tested systems.

Additionally, it can be seen from Figure 11.1 that it is always harder to inject sparser attacks while Figure 11.2 reveals that attacks with higher values would be more likely to be detected. The simulation in Figure 11.3 is conducted in IEEE 57-bus system with SNR = 10 dB. The two figures display the success probabilities for injecting attacks with different ALRs and SRs. It is notable that in Figure 11.3(b) the algorithm utilizes randomly selected column vectors in the basis matrix of *Null*(**G**), instead of that with the largest variance. The results imply that using randomly chosen column vectors in basis matrix **U** can also successfully inject undetectable attacks with high probabilities.

It is known that stealth attacks having $m - n$ non-zero entries can always be found [2]. In IEEE 57-bus system, this value is 80 non-zero elements, for which the SR is about 59%. However, by using Algorithm 1 there is still a high probability that attackers can inject undetectable attacks with SRs lower than 59%. Even for an attack with SR lower than 0.05, the success rate is still around 80% when the SNR is 10 dB and ALR is 0.5, which is shown in Figure 11.1(a).

### *11.2.3 Target attacks*

In practice, adversaries may intend to modify specific state variables. Stealth attacks in this scenario are denoted as targeted attacks. In this case, the amounts of the elements in the targeted subset in the vector **c** are fixed. Adversaries need to inject coordinated data into SCADA measurements to bias specific state variables. Sparse attack vector construction methods for targeted attacks have already been widely studied in the literature, such as References 3,11.

Additionally, certain measurements may also be protected, in which case adversaries would not be able to compromise these secured measurements. This can be achieved by either utilizing more advanced measuring devices or implementing protection mechanisms. The number of protected measurements is usually small as it is not economical, or necessary to protect a majority or all measurements. Therefore, sparse attack vectors need to be carefully designed to contaminate specific state variables without compromising those protected measurements.

The criteria for designing attack vectors in this scenario can thus be concluded as follows: first, an attack vector **a** has to satisfy stealth condition (11.9) to ensure they are stealthy to BDD. Then **a** should be sparse so that adversaries can compromise as few measurements as possible. Third, when a number of measurements are protected, attackers cannot contaminate those measurements. The corresponding elements in **a** have to be 0. Additionally, there may exist other requirements for an attack, such as to compromise certain particular measurements [12]. In this case, the elements in **a** corresponding to those particular measurements are pre-designed.

Let $\mathscr{I}$ denote the set containing indices of state variables that are specifically targeted. $\bar{\mathscr{I}}$ is the complementary set of $\mathscr{I}$ and denotes the indices of state variables that can be arbitrarily chosen in order to launch coordinated stealth attacks. Measurement matrix **H** is a set of column vectors: $[\mathbf{h}_1, \mathbf{h}_2, \ldots, \mathbf{h}_N]$, where $\mathbf{h}_i$ denotes the $i$th column vector of **H**. A stealth attack vector **a** can then be written as follows:

$$\mathbf{a} = \mathbf{Hc} = \sum_{i \in \mathscr{I}} \mathbf{h}_i c_i + \sum_{j \in \bar{\mathscr{I}}} \mathbf{h}_j c_j \tag{11.18}$$

In a targeted attack, the value of $c_i, i \in \mathscr{I}$ is fixed and pre-designed to be added onto the corresponding state variable $\theta_i$. Let $\mathbf{b} = \sum_{i \in \mathscr{I}} \mathbf{h}_i c_i$, which is pre-designed by attackers and its value is fixed. The attack vector **a** is then designed based on the pre-designed vector **b**. From the above equation, it is known that

$$\mathbf{a} - \mathbf{b} = \sum_{j \in \bar{\mathscr{I}}} \mathbf{h}_j c_j = \mathbf{H}_{\bar{\mathscr{I}}} \mathbf{c}_{\bar{\mathscr{I}}} \tag{11.19}$$

where $\mathbf{H}_{\bar{\mathscr{I}}}$ denotes the submatrix of **H** containing columns corresponding to the indices in $\bar{\mathscr{I}}$, i.e., $\mathbf{H}_{\bar{\mathscr{I}}} = [\mathbf{h}_{j_1}, \mathbf{h}_{j_2}, \ldots, \mathbf{h}_{j_{|\bar{\mathscr{I}}|}}]$, where $j_i \in \bar{\mathscr{I}}$ for $1 \leq i \leq |\bar{\mathscr{I}}|$.

Defining the projection matrix $\mathbf{P} = \mathbf{H}(\mathbf{H}^T\mathbf{H})^{-1}\mathbf{H}^T$, (11.18) can be transformed to the following form by left-multiplying both sides of (11.19) by $\mathbf{P}_{\bar{\mathscr{I}}}$:

$$\mathbf{P}_{\bar{\mathscr{I}}}(\mathbf{a} - \mathbf{b}) = \mathbf{P}_{\bar{\mathscr{I}}}\mathbf{H}_{\bar{\mathscr{I}}}\mathbf{c}_{\bar{\mathscr{I}}} = \mathbf{H}_{\bar{\mathscr{I}}}(\mathbf{H}_{\bar{\mathscr{I}}}^T\mathbf{H}_{\bar{\mathscr{I}}})^{-1}\mathbf{H}_{\bar{\mathscr{I}}}^T\mathbf{H}_{\bar{\mathscr{I}}}\mathbf{c}_{\bar{\mathscr{I}}} = \mathbf{H}_{\bar{\mathscr{I}}}\mathbf{c}_{\bar{\mathscr{I}}} = \mathbf{a} - \mathbf{b} \tag{11.20}$$

where $\mathbf{P}_{\bar{\mathscr{I}}} = \mathbf{H}_{\bar{\mathscr{I}}}(\mathbf{H}_{\bar{\mathscr{I}}}^T\mathbf{H}_{\bar{\mathscr{I}}})^{-1}\mathbf{H}_{\bar{\mathscr{I}}}^T$. From this equation, we can then easily obtain the following relationship between the attack vector **a** and the pre-designed vector **b**:

$$(\mathbf{P}_{\bar{\mathscr{I}}} - \mathbf{I})\mathbf{a} = (\mathbf{P}_{\bar{\mathscr{I}}} - \mathbf{I})\mathbf{b} \tag{11.21}$$

Let matrix $\mathbf{G} = \mathbf{P} - \mathbf{I}$, the above equation can be further simplified as follows:

$$\mathbf{G}_{\bar{\mathscr{I}}}\mathbf{a} = \mathbf{G}_{\bar{\mathscr{I}}}\mathbf{b} \tag{11.22}$$

where $\mathbf{G}_{\bar{\mathscr{I}}} = \mathbf{P}_{\bar{\mathscr{I}}} - \mathbf{I}$. This equation presents the relationship between **a** and **b** while it also ensures **a** is stealthy, which satisfies condition $\mathbf{a} = \mathbf{Hc}$. Therefore, targeted attack vectors can be designed based on this relationship.

Since a subset of measurements are protected, those elements in attack vector **a** should be restricted to 0. Assume that the set $\mathscr{P}$ consists of the indices of those protected measurements, then **a** has to be designed satisfying $\mathbf{a}_{\mathscr{P}} = 0$. Let $\mathbf{y} = \mathbf{B}_{\bar{\mathscr{I}}}\mathbf{b}$, to find the sparsest attack vector **a**, which has the smallest number of non-zero elements, adversaries can solve the following optimization problem:

$$\min_{\mathbf{c}} \|\mathbf{Hc}\|_0 \quad \text{s.t. } \mathbf{G}_{\bar{\mathscr{I}}}\mathbf{Hc} = \mathbf{y} \ \mathbf{H}^{\mathscr{P}}\mathbf{c} = 0 \tag{11.23}$$

Solving the above problem will be difficult since it is non-convex. Therefore, we apply the $l_1$-relaxation to the above problem, where we optimize $l_1$-norm rather than the $l_0$-norm of the attack vector. The sparse attack vector **a** can then be obtained by solving the following optimization problem:

$$\min_{\mathbf{c}} \|\mathbf{Hc}\|_1 \quad \text{s.t. } \mathbf{G}_{\bar{\mathscr{I}}}\mathbf{Hc} = \mathbf{y} \ \mathbf{H}^{\mathscr{P}}\mathbf{c} = 0 \tag{11.24}$$

where $\mathbf{H}^{\mathscr{P}}$ denotes the submatrix of **H** containing only rows with indices in set $\mathscr{P}$. Minimizing $l_1$-norm of a vector, where $\|\mathbf{v}\|_1 = \sqrt{\sum_i |v_i|}$, can promote it to be sparse. This problem is convex and is well discussed in the field of compressive sensing [15] and can be quickly solved.

When other requirements exist, for example, certain measurement is of particular interest, then additional constraints can be added to the optimization problem (11.24). A typical example is that the *i*th measurement is a critical measurement. Adversaries may ensure the critical measurement is compromised in order to make the attacks more easily accomplished, or more effective, by designing attacks from the following optimization problem:

$$\min_{\mathbf{c}} \|\mathbf{Hc}\|_1 \quad \text{s.t. } \mathbf{G}_{\bar{\mathscr{J}}}\mathbf{Hc} = \mathbf{y} \; \mathbf{H}^{\mathscr{P}}\mathbf{c} = 0 |\mathbf{H}^{i}\mathbf{c}| \geq \tau \tag{11.25}$$

where $\tau > 0$, which denotes the designed smallest value that is injected into the power measurement.

## *11.2.4 Numerical results*

Targeted attack construction method in (11.24) is assessed under different attack conditions in which different percentage of total state variables are assumed to be modified. The convex optimization toolbox CVX [16] is used for solving convex optimization problems discussed above. The targeted state variables are randomly selected. It is assumed that 2.5% of the total measurements are protected and they are randomly chosen in the simulation. It can be observed from Figure 11.4 that, in order to precisely alter specified state variables, the coordinated attack vectors cannot be highly sparse. Thus, attackers need to compromise a number of measurements to launch targeted stealth attacks. Highly sparse attacks can only be achieved when the percentage of targeted state variables is extremely low for certain test systems. For example, SR can be less than 0.1 for IEEE 39-bus system when a small number of state variables are targeted. The figure also shows that in some cases, SR of attacks is 0. They correspond to the cases that: for a certain targeted set of state variables, no feasible attack vectors exist when these measurements are protected. Therefore, it implies that when certain carefully selected measurements are protected, attackers may not be able to inject targeted stealth attacks.

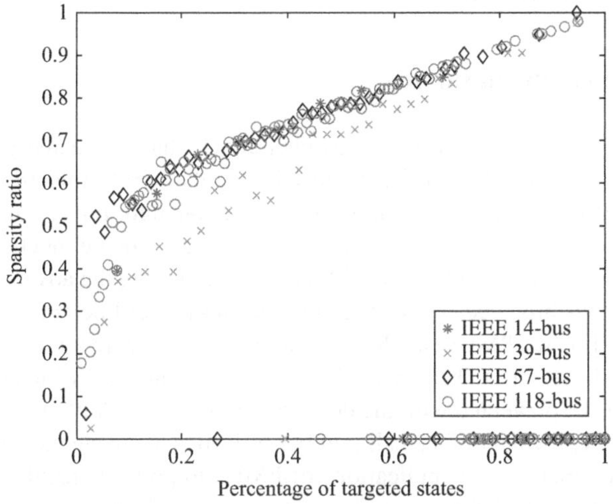


*Figure 11.4 Sparsity of* **a** *under different attack conditions*

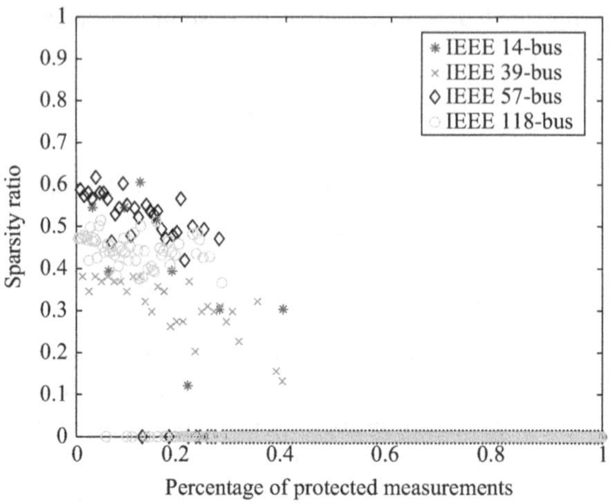


*Figure 11.5 Sparsity of* **a** *under different attack conditions*

Figure 11.5 shows the SR of constructed attack vector **a** when a number of measurements are protected from being compromised by attackers. In the simulation, 10% of the total state variables are targeted. The locations of the protected measurements are randomly chosen. The figure reveals that, as the number of protected measurements is increased, the case that no feasible attack vectors exist becomes more and more often. In the simulation, no feasible attack vectors can be found, which indicates the whole system is protected, when more than about 40% of the total measurements are protected. It is notable that this figure can smaller if the protection set is carefully chosen rather than randomly selected. The protection subset searching methods will be discussed in the next section.

## 11.3 Defense mechanisms

The defense schemes against malicious data attacks have attracted great attention. Defending against attacks can be achieved by either protecting the system from being attacked, or utilizing robust false data detection algorithms. Sophisticated and advanced measurement devices have emerged and can provide much more accurate measurements compared to traditional RTUs. For example, the advanced PMUs can provide time synchronized bus voltage phasor and power line current phasor measurements [17]. Thus, PMUs can directly provide system state information and more robust state estimation can be achieved. The malicious attacks, which bias state variables, can thus be prevented with the deployment of PMUs. Authors in Reference 18 discussed the deployment of a small number of PMUs to enable the detection of bad data in measurements. The applications of PMUs to prevent stealth malicious data attacks [11,19], and also for robust state estimations [17,20] have been discussed in the literature.

Another protection strategy is to protect the traditional SCADA measurements from being modified by attackers. Due to the budgetary constraints and also the legacy of the measurement devices and the communications, it is not feasible to replace the whole existing installed system so as to secure all measurements. In fact, protecting a number of carefully selected measurements can prevent stealth attacks from bypassing BDD in state estimator [21,22]. Thus, it is not necessary to protect all measurements to defend against stealth attacks. In Reference 22 the protection scheme was considered under the constraint of budget. Author in Reference 21 explored the minimal measurement subset that is required to be protected to defend against attacks using brute-force search. This method is time-consuming and only feasible for small-size networks. The vulnerabilities with respect to FDIAs in AC state estimation problem are assessed in [vulnerability], which provides guidance on which measurements are vulnerable and need increased protection.

The second defense strategy is to implement robust false data detection algorithms. Traditional false data detection methods are based on residue test, such as Reference 23. However, these residue-based methods cannot protect state estimation from carefully designed stealth attacks, which was first proposed in Reference 2. Therefore, more robust false data detection algorithms are required in order to identify malicious injected false data and achieve the robust state estimations.

With the advent of smart grid, new detection methods have been proposed. A survey of the detection methods was provided in Reference 24. In Reference 13, generalized likelihood ratio test is introduced to detect weak FDIAs. The adaptive CUSUM test-based detection mechanism introduced in References 25,26 is also designed for non-stealth attacks. Authors in Reference 27 proposed stealth false data detection methods using machine learning. In Reference 28, an effective method capable of detecting false data as well as recovering the real state information was proposed.

In this section, both measurement protection strategy and stealth false data attack detection algorithm are discussed. We first look into the strategy of protecting a minimum number of measurements to defend against the stealth attacks. In Reference 21, the minimum number of protecting measurements are derived from brute-force search method. We propose a more efficient algorithm that can quickly obtain a measurement protection subset to prevent stealth attacks. If a vector $\mathbf{p}$ is used to indicate the measurement protection condition, i.e., $p_i = 1$ if the $i$th measurement is protected and $p_i = 0$ otherwise, it is obvious that the vector $\mathbf{p}$ is sparse. Then, the problem is to search the sparsest $\mathbf{p}$ in order to protect the system from stealth attacks by protecting the least number of measurement sensors.

The detection algorithm is also discussed in this section. A manifest characteristic of the maliciously injected false data in measurements is that only a small part of measurements tend to be compromised, i.e., the attack vectors are sparse. This characteristic can be applied in the designing of stealth attack detection algorithms [28]. In this section, we address the measurement noise in the stealth attack detection algorithm. Additionally, it may happen that certain measurements are lost due to various reasons, such as the measurement device failures, lost communication links, or those measurements are distinctly abnormal and abandoned. Therefore, the stealth attack detection algorithm is discussed in the noise case with the consideration that only partial measurements are obtained.

### 11.3.1 Strategic protection

Increasing the number of protected measurements can make the stealth attacks more difficult to be accomplished. It is obvious that malicious data attacks can be completely prevented by securing all measurements. However, it is not economical or necessary to protect all measurements from malicious attacks by replacing the existing system. A feasible and efficient algorithm is proposed to quickly find small protecting measurement subsets, which have the same size as that from brute-force method [21] in nearly all cases.

Denote the set $\mathscr{P} \subset \{1, 2, \ldots, M\}$ as the measurement subset in which measurements are secured and the complementary set $\bar{\mathscr{P}}$ denotes the index of those measurements that can be contaminated by attackers. Let $\mathscr{I}$ denotes the set containing indices of state variables that are specifically targeted. $\bar{\mathscr{I}}$ is the complementary set of $\mathscr{I}$ and denotes the indices of state variables that can be arbitrarily chosen in order to launch coordinated stealth attacks. Similar to (11.24), adversaries can construct the sparse-targeted attack vector **a** by solving:

$$\min_{\mathbf{c}} \|\mathbf{H}^{\bar{\mathscr{P}}}\mathbf{c}\|_1 \quad \text{s.t. } \mathbf{G}_{\bar{\mathscr{I}}}\mathbf{H}\mathbf{c} = \mathbf{y} \; \mathbf{H}^{\mathscr{P}}\mathbf{c} = 0 \tag{11.26}$$

where $\mathbf{G}_{\bar{\mathscr{I}}} = \mathbf{H}_{\bar{\mathscr{I}}}(\mathbf{H}^T_{\bar{\mathscr{I}}}\mathbf{H}_{\bar{\mathscr{I}}})^{-1}\mathbf{H}^T_{\bar{\mathscr{I}}} - \mathbf{I}$, and $\mathbf{H}_{\bar{\mathscr{I}}}$ denotes the submatrix of $\mathbf{H}$ containing columns corresponding to the indices in $\bar{\mathscr{I}}$. $\mathbf{H}^{\mathscr{P}}$ and $\mathbf{H}^{\bar{\mathscr{P}}}$ denote the submatrix of $\mathbf{H}$ containing only rows with indices in set $\mathscr{P}$ and $\bar{\mathscr{P}}$, respectively. Vector $\mathbf{y}$ is predesigned by the attacker and relates to the designed errors on targeted state variables. Specifically, $\mathbf{y} = \mathbf{G}_{\bar{\mathscr{I}}}\mathbf{H}_{\mathscr{I}}\mathbf{c}_{\mathscr{I}}$ where $\mathbf{c}_{\mathscr{I}}$ is the designed errors on state variables. The $l_1$-norm of a vector $\mathbf{v}$ is defined as $\|\mathbf{v}\|_1 = \left(\sum_i |v_i|\right)$. Solving the above problem can generate sparse attack vectors which manipulate specific state variables when certain measurements are protected.

If the protection set $\mathscr{P}$ is properly chosen, problem (11.26) would have no solutions. In this case, targeted attack vectors aiming at specific state variables would not exist with measurements in set $\mathscr{P}$ being protected. Giving specified vector $\mathbf{c}_{\mathscr{I}}$, which is the targeted subset vector of $\mathbf{c}$, the straightforward method is to protect all measurements in the set corresponding to all non-zero elements in **a** that $\mathbf{a} = \mathbf{H}_{\mathscr{I}}\mathbf{c}_{\mathscr{I}}$. In this way, it probably requires a large number of measurements to be protected since **a** may not be desirably sparse. The challenge is to find the smallest protection set that can prevent targeted attacks. The brute-force search method, which is discussed in Reference 21, can guarantee the smallest possible measurement protection sets are obtained. However, this method is extremely complex as it requires to exhaust every possible measurements combination. Thus, it is not feasible in practice to search the measurement subset using brute-force approach.

When a certain measurement is secured, attackers may need to compromise more measurements or inject extra errors into the rest of the measurements in order to launch targeted attacks without being detected. From (11.18) we have:

$$\mathbf{a} = \mathbf{Hc} = \mathbf{b} + \mathbf{H}_{\bar{\mathscr{I}}}\mathbf{c}_{\bar{\mathscr{I}}} \tag{11.27}$$

where $\mathbf{b}$ represents pre-desired injections. When a certain measurement is protected, e.g., $a_i = 0$, and targeted injection $\mathbf{b}$ is fixed, adversaries have to inject proper errors on state variables in untargeted set $\bar{\mathscr{I}}$ to obtain sparse and undetectable attack $\mathbf{a}$ without compromising measurement $a_i$.

It can be seen that in order to protect specific state variables, protecting certain measurements can always be more effective than others. For example, it is more important to secure the measurements corresponding to the non-zero elements in $\mathbf{b}$ than those with indices corresponding to the zero elements in $\mathbf{b}$. If a subset $\mathscr{P}$ of the total measurements are protected, we have:

$$\mathbf{a}_{\mathscr{P}} = \mathbf{b}_{\mathscr{P}} + \mathbf{H}_{\bar{\mathscr{I}}}^{\mathscr{P}}\mathbf{c}_{\bar{\mathscr{I}}} = 0 \tag{11.28}$$

$$-\mathbf{b}_{\mathscr{P}} = \mathbf{H}_{\bar{\mathscr{I}}}^{\mathscr{P}}\mathbf{c}_{\bar{\mathscr{I}}} \tag{11.29}$$

Therefore, when measurements in set $\mathscr{P}$ are protected, adversaries have to find proper solution of $\mathbf{c}_{\bar{\mathscr{I}}}$ to inject stealth attack with fixed targeted errors $\mathbf{b}$. If the rank of $\mathbf{H}_{\bar{\mathscr{I}}}^{\mathscr{P}}$ is smaller than protection size $|\mathscr{P}|$ and the augmented matrix with vector $\mathbf{b}_{\mathscr{P}}$ can increase the rank, namely, $rank([\mathbf{H}_{\bar{\mathscr{I}}}^{\mathscr{P}}|\mathbf{b}_{\mathscr{P}}]) = rank(\mathbf{H}_{\bar{\mathscr{I}}}^{\mathscr{P}}) + 1$, then vector $\mathbf{b}_{\mathscr{P}}$ is independent with the column vectors in $\mathbf{H}_{\bar{\mathscr{I}}}^{\mathscr{P}}$. Therefore, in this case, $\mathbf{c}_{\bar{\mathscr{I}}}$ satisfying (11.29) does not exist, indicating that the system is successfully protected from targeted attacks with $\mathbf{b}$. Otherwise, when vector $\mathbf{b}_{\mathscr{P}}$ does not increase the rank of matrix $\mathbf{H}_{\bar{\mathscr{I}}}^{\mathscr{P}}$, i.e., $rank([\mathbf{H}_{\bar{\mathscr{I}}}^{\mathscr{P}}|\mathbf{b}_{\mathscr{P}}]) = rank(\mathbf{H}_{\bar{\mathscr{I}}}^{\mathscr{P}})$, then vector $\mathbf{b}_{\mathscr{P}}$ is in the column space of matrix $\mathbf{H}_{\bar{\mathscr{I}}}^{\mathscr{P}}$. In this case, there exist solutions of $\mathbf{c}_{\bar{\mathscr{I}}}$, which means adversaries can still find attack vectors to launch targeted attacks. The problem is then to find the best solution to obtain highly sparse $\mathbf{a}$.

With specifically targeted injections $\mathbf{b}$, attackers always find the sparsest attack vector $\mathbf{a}$. From (11.27), it is known that making a certain subset $\mathscr{P}$ of the measurements immune to attacks can result in another sparse attack vector $\mathbf{a}$ which would contaminate more measurements. This makes the attacks more difficult to be accomplished. Therefore, it can be deduced that protecting certain measurement would result in a larger $\|\mathbf{a}\|_1$ value than that of protecting another measurement. Protecting these measurements would be more effective than others and these measurements can be regarded as critical measurements to targeted attacks. Based on this idea, given a specified targeted state bias vector $\mathbf{c}_{\mathscr{I}}$, a greedy method can be designed to find those more effective protection indices until the stealth attack vectors do not exist anymore. This algorithm can be concluded as follows:

Algorithm 2 presents the greedy search method to find a small protection subset of measurements with the knowledge of existing protection set $\mathscr{P}$ and targeted error

---

**Algorithm 2** Greedy subset searching

---

Input: $\mathbf{H}$, $\mathscr{I}$, $\mathbf{c}_{\mathscr{I}}$, $\mathscr{P}$.
Initialize: $\mathbf{G}_{\bar{\mathscr{I}}} = \mathbf{H}_{\bar{\mathscr{I}}}(\mathbf{H}^T_{\bar{\mathscr{I}}}\mathbf{H}_{\bar{\mathscr{I}}})^{-1}\mathbf{H}^T_{\bar{\mathscr{I}}} - \mathbf{I}$, $\mathbf{y} = \mathbf{G}_{\bar{\mathscr{I}}}\mathbf{H}_{\mathscr{I}}\mathbf{c}_{\mathscr{I}}$, $\mathscr{P}' = \mathscr{P}$, $k = 1$, $\mathscr{P}_k = \mathscr{P}'$.
Iteration: At the $k$th iteration:
Compute the complementary set $\bar{\mathscr{P}}$ of $\mathscr{P}'$.
**For** $i = 1 : |\bar{\mathscr{P}}|$
  Put the $i$th entry in $\bar{\mathscr{P}}$ into $\mathscr{P}_k$: $\mathscr{P}_k = \mathscr{P}' \cup \bar{\mathscr{P}}_i$.
  Checking the feasibility of finding $\mathbf{c}$ from (11.26).
  **If** feasible
    Compute $\chi_i = \|\mathbf{Hc}\|_1$.
  **else**
    $\mathscr{P}' = \mathscr{P}_k$; Quit the iteration.
  **end**
**end**
Find index $i$ such that $\chi_i$ has the largest value.
Update set $\mathscr{P}' = \mathscr{P}' \cup \bar{\mathscr{P}}_i$.
Output: $\mathscr{P}'$.

---

vector $\mathbf{c}_{\mathscr{I}}$. At each iteration, the algorithm tries every measurement in the unprotected indices set $\bar{\mathscr{P}}$ and assume it is protected. The feasibility of constructing stealth attack vector $\mathbf{a}$ is checked when every measurement is assumed protected. If the stealth attack vector does not exist when $z_i$, $i \in \bar{\mathscr{P}}$ is protected, then the protection set is found as $\mathscr{P} \cup \{i\}$. Otherwise, if the stealth attack vector exists when every measurement in $\bar{\mathscr{P}}$ is tested, the algorithm increases the protection set $\mathscr{P}$ by selecting the most important measurement index $i$, which leads to the largest value of $\chi_i$ when $z_i$ is protected. The selection process continues until the stealth-targeted attack vector does not exist. It is notable that the obtained subset using Algorithm 1 depends on the solutions of $\|\mathbf{Hc}\|_1$ during the indices selection process. Different indices may be chosen if solutions were not optimally computed.

For a large power grid system, it is not feasible to find the smallest protection subset to prevent any of undetectable attacks that satisfy $\mathbf{a} = \mathbf{Hc}$ by brute-force search. Instead, we can protect the union set of those subsets selected for protecting every single-state variable $c_i$, $i \in \{1, 2, \ldots, N\}$. Since the proposed algorithm can quickly find a small subset to protect any state variable $c_i$, $i \in \{1, 2, \ldots, N\}$, in this way the protection subset can be easily found to protect the whole system from any stealth attacks satisfying (11.9). The search procedure can be concluded in Algorithm 3.

This method cannot guarantee that the smallest subset can be found, but it provides at least a quasi-optimal subset that contains a slightly larger number of elements than the smallest subset. Most importantly, this method is fast and feasible in practice.

**Algorithm 3** Minimal subset selection

Input: $\mathbf{H}$.
Initialize: $\mathscr{P} = \emptyset$.
**For** $i = 1 : N$
  Let $\mathscr{I} = \{i\}$.
  Find $\mathscr{P}_i$ using Algorithm 1.
**end**
$\mathscr{P} = \mathscr{P}_1 \cup \mathscr{P}_2 \cup \cdots \cup \mathscr{P}_n$.
Output: $\mathscr{P}$.

*Table 11.1 The number of measurements in protection sets from two methods*

| Test systems | Algorithm 3 | Brute-force search |
|---|---|---|
| IEEE 9-bus | 9 | 8 |
| IEEE 14-bus | 15 | 13 |
| IEEE 39-bus | 59 | – |
| IEEE 57-bus | 84 | – |
| IEEE 118-bus | 187 | – |

In the worst case, to find a protection set with $k$ elements, the algorithm needs to test the feasibility $\mathscr{K}_a$ times:

$$\mathscr{K}_a = Mk - \frac{k(k-1)}{2} \tag{11.30}$$

This figure is much smaller than that using brute-force method, where it needs to test $\sum_{i=1}^{k-1} \binom{M}{i} + 1$ combinations in the best case to find the protection set with $k$ elements. Although our proposed algorithm may not find the global optimum solution, it provides some flexibility. For example, when a small protection subset $\mathscr{P}$ is obtained but it is not possible to protect a certain selected measurement $z_i$, $i \in \mathscr{P}$ in practice, the algorithm can then find a sub-optimal subset instead.

## 11.3.2 Numerical results

To compare the sizes of protection subsets generated by the proposed algorithm and brute-force method, which was discussed in Reference 21, we apply different power test systems. Table 11.1 shows the number of measurements in protection subset found by two methods. The IEEE 9-bus system contains 17 total measurements, and IEEE 14-bus system has 33 total measurements.

It can be seen that for the first two test systems the smallest protection sets generated from the proposed algorithm contain only slightly more measurements than that from brute-force method. In IEEE 14-bus system, the difference of this number

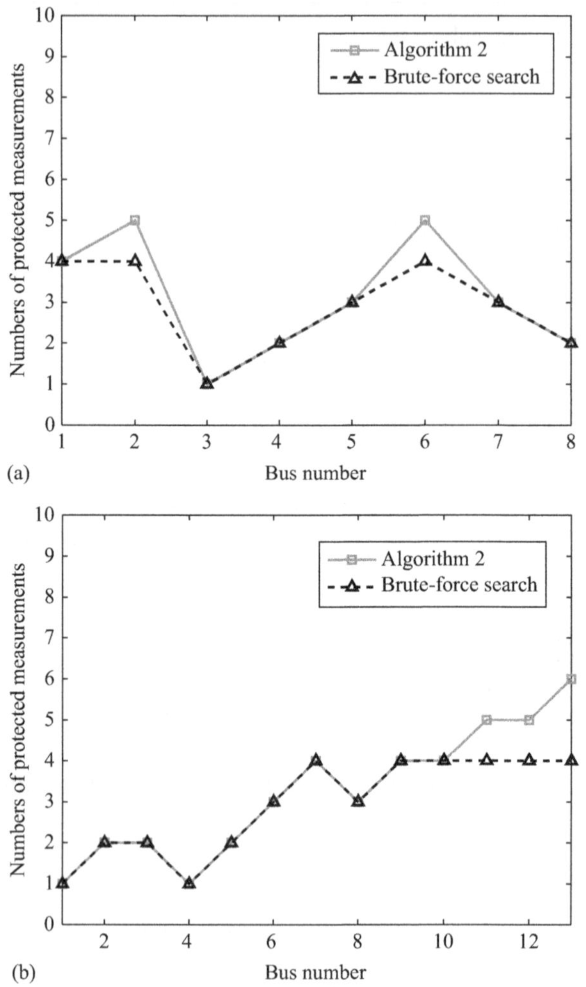


*Figure 11.6 Number of protected measurements to protect every single-state variable from being targeted for (a) IEEE 9-bus system and (b) IEEE 14-bus system*

for two algorithms is quite small compared to the total number of 33 measurements. Thus, Algorithm 3 can find protection subsets with a similar number of elements. The results from the proposed algorithm for other larger test systems are also provided. However, it is not feasible to obtain the results from brute-force method for these larger power systems since it would take extremely long time.

In order to further compare the performance of the proposed algorithm with that of the brute-force method, which can provide optimal measurement subset, we look into the protection subsets for defending against stealth attacks targeted on a single-state variable. Figure 11.6 displays the number of elements in the smallest protection

*Table 11.2 The number of testing times for two algorithms to find protection subsets*

| Test systems | Bus number | 1 | 2 | 3 | 4 | 5 | 6 | 7 |
|---|---|---|---|---|---|---|---|---|
| IEEE 9-bus | Algorithm 2 | 55 | 70 | 1 | 18 | 34 | 75 | 40 |
| | Brute-force search | 1 434 | 834 | 1 | 18 | 154 | 835 | 235 |
| IEEE 14-bus | Algorithm 2 | 1 | 35 | 39 | 2 | 42 | 71 | 107 |
| | Brute-force search | 1 | 35 | 36 | 2 | 73 | 629 | 7 108 |

| Test systems | Bus number | 8 | 9 | 10 | 11 | 12 | 13 |
|---|---|---|---|---|---|---|---|
| IEEE 9-bus | Algorithm 2 | 25 | – | – | – | – | – |
| | Brute-force search | 25 | – | – | – | – | – |
| IEEE 14-bus | Algorithm 2 | 72 | 109 | 123 | 141 | 135 | 183 |
| | Brute-force search | 630 | 7 140 | 6 235 | 6 236 | 6 237 | 7 109 |

subsets obtained from two algorithms to protect every single-state variable from being targeted by adversaries. The whole system protection subsets shown in Table 11.1 are the unions of the protection subsets for protecting single variable.

From both plots in Figure 11.6, it can be seen that in nearly all cases, the proposed algorithm can find a protection subset having the same size as that found by brute-force method. The size differences are small when the two methods find subsets with a different number of elements, i.e., the difference are 1 for IEEE 9-bus system and the maximum difference is 2 for IEEE 14-bus system. In IEEE 14-bus system, only once the difference is 2, which is quite small compared with the total number of 33 measurements in IEEE 14-bus. Therefore, the proposed algorithm is an applicable replacement of brute-force method to find small protection subset.

Table 11.2 compares the complexities of the two algorithms in terms of the number of feasibility testing times when generating small protection subsets. The results correspond with the simulation shown in Figure 11.6 in which measurement protection subsets are searched for protecting every single-state variable. It is obvious that when the size of protection subset exceeds 3, the difference of the two methods becomes significant. This difference is more significant when the size of protection subset is bigger as the brute-force search needs to exhaust all subset combinations with smaller sizes. The numbers for the proposed algorithm are quite small, indicating that this fast method is feasible to find measurement protection subset to defend against stealth attacks in a large power system, in which the brute-force method is infeasible.

It is also clear that in a larger power system the difference is much larger for two algorithms to find a subset with the same size as that in a smaller system. The testing times of the proposed algorithm increase only slightly when the size of protection subset and the system scale grow, which is also described by (11.30). In a real power system, while brute-force method is infeasible because of the combinatorial complexity, the proposed method instead is fast and practical to search a quasi-optimal measurement protection subset to defend against the stealth attacks satisfying (11.9).

### 11.3.3 Robust detection

Traditional residue testing based false data detection methods cannot provide protection of state estimation from carefully designed stealth attacks. Therefore, new detection methods need to be designed to detect random errors as well as stealth attacks. It is shown that a series of measurement data exhibit low rank and sparse structure, which can be employed in anomaly detection method [28]. In practice, measurements tend to be contaminated with noise. Additionally, it also may happen that some of the measurements are lost due to the measurement device failures or disrupted communication links. This section deals with these situations.

Considering a time interval $T$, the power system obtains a series of measurements $[\mathbf{z_a}_1, \mathbf{z_a}_2, \ldots, \mathbf{z_a}_T]$ at the time instants $t_1, t_2, \ldots, t_T$. These measurements can form a matrix $\mathbf{Z_a} \in \mathbb{R}^{M\times T}$, which can be decomposed as follows:

$$\mathbf{Z_a} = \mathbf{Z} + \mathbf{A} + \mathbf{E} \tag{11.31}$$

where $\mathbf{Z} \in \mathbb{R}^{M\times T}$ is the block of true measurements with each column $\mathbf{z}_i$ representing true measurements at time $t_i$, $\mathbf{A} \in \mathbb{R}^{M\times T}$ denotes the attack matrix formed by all instant sparse attacks and $\mathbf{E}$ represents the noise.

It is known that fast system dynamics are usually well damped in the power system. This implies that the system states would change gradually in a small period $T$, making the measurements highly correlated and thus the matrix $\mathbf{Z}$ typically low rank. Additionally, malicious injection data matrix $\mathbf{A}$ tends to be sparse since the attack $\mathbf{a}$ on measurement at each time instant is sparse. This is due to the fact that some of the measurements may be protected and also because attackers would launch attacks with least effort. Given corrupted measurement matrix $\mathbf{Z_a}$, it is possible to recover low-rank matrix $\mathbf{Z}$ and sparse attack matrix $\mathbf{A}$ by performing low rank and sparse decomposition, which is well discussed in the robust principle component analysis (PCA) problem [29], which solves:

$$\min \|\mathbf{Z}\|_* + \lambda\|\mathbf{A}\|_1 \quad \text{s.t. } \|\mathbf{Z_a} - \mathbf{Z} - \mathbf{A}\|_F \le \delta \tag{11.32}$$

where $\|\cdot\|_*$ denotes the nuclear norm. The nuclear norm of a matrix $\mathbf{A} \in \mathbb{R}^{m\times n}$ is defined as $\|\mathbf{A}\|_* = trace(\sqrt{\mathbf{A}^*\mathbf{A}}) = \sum_i^{\min\{m,n\}} \sigma_i$, where $\mathbf{A}^*$ denotes the conjugate transpose of matrix $\mathbf{A}$ and $\sigma_i$ is the singular value. $\|\cdot\|_F$ denotes the Frobenius norm. The Frobenius norm of a matrix $\mathbf{A} \in \mathbb{R}^{m\times n}$ is defined as $\|\mathbf{A}\|_F = \sqrt{\sum_i^m \sum_i^n |a_{ij}|^2} = \sqrt{trace(\mathbf{A}^*\mathbf{A})} = \sqrt{\sum_i^{\min\{m,n\}} \sigma_i^2}$, where $\mathbf{A}^*$ denotes the conjugate transpose of matrix $\mathbf{A}$ and $\sigma_i$ is the singular value. $\delta$ represents a small positive noise bound and $\lambda$ is the regulation parameter. This problem is also addressed in compressive sensing and matrix completion [30] literature. Thus, true measurements $\mathbf{Z}$ can be recovered and the sparse perturbations including malicious attacks and other false data can also be identified.

Unlike coordinated malicious attacks, the missing data in the measurements can result in residue changes in (11.6). These incomplete measurement data, as well as the measurements with errors, would be identified as bad data by traditional BDD algorithms. The proposed algorithm can not only detect the missing and inaccurate measurement data, but also detect the carefully designed stealth attacks, which is undetectable to traditional methods. More importantly, the proposed algorithm can recover the true measurements from the incomplete measurements.

In order to address the problem that only noise-contaminated partial measurements are collected, the PCA problem in (11.32) can be extended to the following form with element-wise error constraints:

$$\min \|\mathbf{Z}\|_* + \lambda\|\mathbf{A}\|_1 \quad \text{s.t.} \ |\mathscr{P}_\Omega(\mathbf{Z_a}) - \mathscr{P}_\Omega(\mathbf{Z}+\mathbf{A})| \preceq \varepsilon \tag{11.33}$$

where $\preceq$ represents element-wise inequality and $\varepsilon$ is the matrix of entry-wise error bounds. $\mathscr{P}_\Omega(\cdot)$ denotes a projection operation, in which all elements outside the set $\Omega$ are forced to 0:

$$[\mathscr{P}_\Omega(\mathbf{A})]_{ij} = \begin{cases} a_{ij} & ij \in \Omega \\ 0 & ij \notin \Omega \end{cases} \tag{11.34}$$

It is demonstrated in Reference 31 that the above problem in (11.33) is equivalent to the following problem:

$$\min \|\mathbf{Z}\|_* + \lambda\|\mathscr{T}(\mathbf{A}, \tilde{\varepsilon})\|_1 \quad \text{s.t.} \ \mathbf{Z_a} = \mathbf{Z} + \mathbf{A} \tag{11.35}$$

where $\tilde{\varepsilon}$ has the same value as $\varepsilon$ in the projection set $\Omega$ and infinite outside set $\Omega$, and the soft thresholding operation $\mathscr{T}_\varepsilon(a_{ij})$ is defined as

$$\mathscr{T}(a_{ij}, \varepsilon) = \text{sign}(a_{ij}) \cdot \max\{|a_{ij}| - \varepsilon, 0\} \tag{11.36}$$

A variant of the augmented Lagrangian method (ALM), which is also known as the alternating direction method of multipliers (ADMM) algorithm [32], is used to solve the problem defined by (11.35). The Lagrangian corresponding to this problem is

$$\mathscr{L}(\mathbf{Z}, \mathbf{A}, \mathbf{Y}, \mu) = \|\mathbf{Z}\|_* + \lambda\|\mathscr{T}(\mathbf{A}, \tilde{\varepsilon})\|_1 + \langle \mathbf{Y}, \mathbf{H} \rangle + \frac{\mu}{2}\|\mathbf{H}\|_2^2 \tag{11.37}$$

where $\langle\cdot\rangle$ denotes the Frobenius product, $\mathbf{H} = \mathbf{Z_a} - \mathbf{Z} - \mathbf{A}$, and $\mu > 0$. The regulation parameter $\lambda$ can be set to $\sqrt{M/|\Omega|}$. We further define the singular-value thresholding operation as follows:

$$\mathscr{D}(\mathbf{X}, \tau) = \mathbf{U}\mathscr{T}(\Sigma, \tau)\mathbf{V}^T \tag{11.38}$$

where $\tau$ is the threshold and $\mathbf{X} = \mathbf{U}\Sigma\mathbf{V}^T$. It is notable that ADMM updates $\mathbf{Z}, \mathbf{A}, \mathbf{Y}$ separately only once in each iteration thus it is efficient. The convergence of the whole algorithm is analyzed in Reference 32, which states that the condition for convergence requires $\sum_1^\infty \mu_k^{-1} = +\infty$ where $\mu_k$ denotes the value of $\mu$ in the $k$th iteration. The whole process of solving (11.35) is shown in Algorithm 4.

It is notable that when incomplete measurements are collected, Algorithm 4 will take the missing data to be anomalies and it can also recover the low-rank true measurement matrix and sparse anomaly matrix. However, the recovery accuracy would be impacted as the sparsity is changed. The recovered sparse attack matrix can ignore those injected data outside the observation set. Thus, it is more difficult to identify all malicious attacks with partial observations.

**Algorithm 4** RPCA with entry wise constraints

Input: $\mathbf{Z_a^p} = \mathscr{P}_\Omega(\mathbf{Z_a}) \in \mathbb{R}^{\mathbf{M}\times\mathbf{T}}$, $\tilde{\varepsilon} \in \mathbb{R}^{M\times T}$, $\lambda$.
Initialize $\mathbf{Z} = \mathbf{0}$, $\mathbf{A} = \mathbf{0}$, $\mathbf{Y} = \mathbf{0}$, $\mu > 0$, $\rho > 1$, $k = 0$.
**while** not converged
1. Update the value of low-rank matrix $\mathbf{Z_{k+1}}$: $\mathbf{Z_{k+1}} = \mathscr{D}\left(\mathbf{Z_a^p} - \mathbf{A}_k + \frac{\mathbf{Y}_k}{\mu_k}, \mu_\mathbf{k}^{-1}\right)$.
2. Compute the value of sparse matrix $\mathbf{A_{k+1}}$ by minimizing:
$F(\mathbf{A}) = \frac{\lambda}{\mu}\|\mathscr{T}(\mathbf{A}, \tilde{\varepsilon})\|_1 - \mathrm{tr}\left(\frac{\mathbf{Y}_k}{\mu_k}(\mathbf{A} - (\mathbf{Z_a^p} - \mathbf{Z}_k))\right) + \frac{1}{2}\|\mathbf{A} - (\mathbf{Z_a^p} - \mathbf{Z}_k)\|_\mathbf{F}$.
3. Update the Lagrange multiplier $\mathbf{Y}$: $\mathbf{Y_{k+1}} = \mathbf{Y_k} + \mu_\mathbf{k}(\mathbf{Z_a^p} - \mathbf{Z_{k+1}} - \mathbf{A_{k+1}})$.
4. Update $\mu_{k+1} = \rho \cdot \mu_k$.
5. Update $k = k + 1$.
**end while**
**Return** $\mathbf{Z}$, $\mathbf{A} = \mathscr{T}(\mathbf{A}, \tilde{\varepsilon})$.
Output: $\mathbf{Z}$, $\mathbf{A}$.

### *11.3.4 Numerical results*

The performance of the detection algorithm is tested on IEEE 14-bus system and IEEE 57-bus system. The malicious attack vectors are constructed using the proposed Algorithm 1. In order to obtain sparsity in the rows of the attack block matrix **A**, different column vectors in the null-space in Algorithm 1 are utilized. The SR of the attack vectors is chosen as 15%. In Figure 11.3(b), it is shown that when SR = 0.15 traditional residual testing based algorithms will not be able to detect those attacks. Thus, in the simulation, the algorithm is not compared with traditional methods. Additionally, recently proposed algorithms such as Reference 27 do not deal with the problem of partial observations. These algorithms do not address the problem of error-contaminated measurements as well. The detection method discussed in this chapter addresses both problems. Most importantly, it can not only detect anomalies but also recover the true measurements from partially contaminated observations.

We use the false alarm rate (FAR) which is the probability of positive alarms when there are no attacks. The noise performance of the algorithm compared to robust PCA with Frobenius constraints in (11.32) has been extensively studied in Reference 31. In this chapter, we focus on identifying anomalies in different scenarios when undetectable attacks are injected in power systems.

Figure 11.7 shows the error tolerance performance in the IEEE 14-bus system. It is shown that when FAR exceeds 10%, the algorithm can identify attacks with high probabilities which are approaching 100%. This probability is still quite high in the presence of high noise (around 95% when SNR = 5 dB). When FAR decreases, the system will absorb more noise and detection probability decreases. It can be seen that there is still a high chance of detecting anomalies, more than 90%, when FAR decreases to an extremely low level (2.5%) under SNR = 10 dB. Therefore, carefully

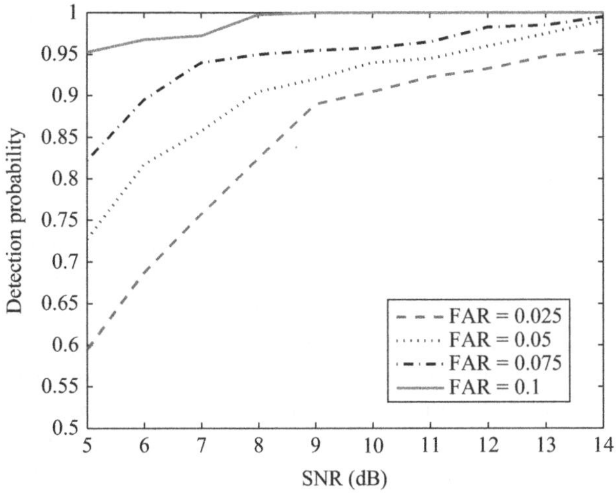


*Figure 11.7 Sparsity of* **a** *under different attack conditions*

*Table 11.3 Detection probability and measurements deviation with partial observations*

| Observations (%) | Detection probability (%) | Variable deviation (%) |
|---|---|---|
| 100 | 100 | 0.53 |
| 95 | 92 | 2.06 |
| 90 | 56.3 | 3.7 |
| 85 | 44.7 | 5.3 |

designed stealth attacks cannot bypass detection in state estimator when utilizing the proposed sparse attack detection algorithm.

In the case where partial measurements are collected, the missing data are regarded as sparse anomalies having the values equal to the negative real measurement data in Algorithm 4. Additionally, non-zero entries in sparse matrix **A** can only be confirmed as attacks when they are located in the observation set. This makes identifications of attacks more difficult. Algorithm 4 can circumvent this problem since it also recovers the block of true measurements. We evaluate the attack detection probabilities as well as the deviation rate of the recovered measurement variables, which is defined as $\|\mathbf{z} - \mathbf{z}'\|_2/\|\mathbf{z}\|_2$, where $\mathbf{z}'$ denotes the recovered measurements. Table 11.3 shows the results when incomplete measurements are collected based on the IEEE 57-bus system. The FAR equals 0.05 and SNR is set to 8 dB in the simulations. It can be seen that the attack detection probability declines greatly with increasing missing observations. However, the recovered measurement variables experience only small deviations. Therefore, the proposed algorithm can successfully verify the true measurements while detecting carefully designed malicious stealth attacks, even in the situation that only partial measurements are observed.

## 11.4 Conclusions

In this chapter, we looked into the problem of malicious FDIAs in power grid state estimation. We proposed stealth attack construction strategies for different scenarios and also introduced the countermeasures. The results demonstrate that the proposed random attack construction algorithm can generate extremely sparse attack vectors with high probabilities with consideration of the noise in measurements. Traditional successful attacks tend to compromise a number of measurements, which exceeds a certain value. The proposed algorithm can construct undetectable attacks which only compromise a much smaller numbers of measurements than this known value. The targeted attack construction method is evaluated considering different percentages of state variables are targeted and different number of measurements are protected. The results show that attack vectors in this scenario cannot be extremely sparse, unless only an extremely small number of state variables are targeted. It also demonstrates that targeted stealth attack vectors do not exist when a number of measurements are protected from being modified.

An efficient protection scheme is proposed in this chapter to find an effective measurement protection subset to defend from the stealth attacks. The simulation results have demonstrated that the proposed algorithm can find protection subsets with the same size as that from brute-force method in nearly all cases. More importantly, the algorithm is quick, and thus feasible in practice when the power system is large. Additionally, a detection algorithm is introduced to detect the stealth attacks as well as other false data. This algorithm considers the case in which only partial measurements are collected in the presence of noise. The performance is demonstrated via the simulation results based on IEEE test power systems.

## Bibliography

[1] W. Wang and Z. Lu, "Cyber security in the smart grid: Survey and challenges," *Computer Networks*, vol. 57, no. 5, 7, pp. 1344–1371, Apr. 2013.

[2] Y. Liu, P. Ning, and M. K. Reiter, "False data injection attacks against state estimation in electric power grids," in *Proceedings of the ACM Conference on Computer and Communication Security*, Chicago, IL, USA, pp. 21–32, 2009.

[3] M. Ozay, I. Esnaola, F. T. Y. Vural, S. R. Kulkarni, and H. V. Poor, "Sparse attack construction and state estimation in the smart grid: Centralized and distributed models," *Selected Areas in Communications, IEEE Journal on*, vol. 31, no. 7, pp. 1306–1318, Jul. 2013.

[4] "Power systems test case archive," University of Washington, Seattle [Online]. Available from http://www.ee.washington.edu/research/pstca/. July 2016.

[5] R. D. Zimmerman, C. E. Murillo-Sanchez, and R. J. Thomas, "Matpower: Steady-state operations, planning and analysis tools for power systems research and education," *IEEE Transactions on Power Systems*, vol. 26, no. 1, pp. 12–19, Feb. 2011.

[6] A. Abur and A. G. Exposito, *Power System State Estimation Theory and Implementation*, CRC Press, Boca Raton, FL, 24 Mar. 2004.

[7] M. A. Rahman and H. Mohsenian-Rad, "False data injection attacks with incomplete information against smart power grids," in *Global Communications Conference (GLOBECOM), 2012 IEEE*, Anaheim, California, USA, pp. 3153–3158, 3–7 Dec. 2012.

[8] A. Monticelli, "Electric power system state estimation," *Proceedings of the IEEE*, vol. 88, no. 2, pp. 262–282, Feb. 2000.

[9] E. Handschin, F. C. Schweppe, J. Kohlas, A. Fiechter, "Bad data analysis for power system state estimation," *Power Apparatus and Systems, IEEE Transactions on*, vol. 94, no. 2, pp. 329–337, Mar. 1975.

[10] F. C. Schweppe, J. Wildes, and D. P. Rom, "Power system static state estimation – parts I, II, III," *IEEE Transactions on Power Apparatus and Systems*, vol. PAS-89, pp. 120–135, 1970.

[11] T. T. Kim and H. V. Poor, "Strategic protection against data injection attacks on power grids," *Smart Grid, IEEE Transactions on*, vol. 2, no. 2, pp. 326–333, Jun. 2011.

[12] H. Sandberg, A. Teixeira, and K. H. Johansson, "On security indices for state estimators in power networks," in *First Workshop on Secure Control Systems, CPSWEEK 2010*, Stockholm, Sweden, 2010.

[13] O. Kosut, L. Jia, R. J. Thomas, and L. Tong, "Malicious data attacks on the smart grid," *Smart Grid, IEEE Transactions on*, vol. 2, no. 4, pp. 645–658, Dec. 2011.

[14] M. G. Adamiak, A. P. Apostolov, M. M. Begovic, *et al.*, "Wide area protection – Technology and infrastructures," *Power Delivery, IEEE Transactions on*, vol. 21, no. 2, pp. 601–609, Apr. 2006.

[15] E. J. Candès and M. B. Wakin, "An introduction to compressive sampling," *Signal Processing Magazine, IEEE*, vol. 25, no. 2, pp. 21–30, Mar. 2008.

[16] M. Grant and S. Boyd. "CVX: MATLAB software for disciplined convex programming" [Online]. Available from http://cvxr.com/cvx, July 2016.

[17] X. Bei, Y. J. Yoon, and A. Abur, "Optimal placement and utilization of phasor measurements for state estimation," *PSERC Publication*, pp. 5–20, 2005.

[18] J. Chen and A. Abur, "Placement of PMUs to enable bad data detection in state estimation," *Power Systems, IEEE Transactions on*, vol. 21, no. 4, pp. 1608–1615, 2006.

[19] A. Giani, E. Bitar, M. Garcia, M. McQueen, P. Khargonekar, and K. Poolla, "Smart grid data integrity attacks," *Smart Grid, IEEE Transactions on*, vol. 4, no. 3, pp. 1244–1253, 2013.

[20] A. Phadke, J. Thorp, R. Nuqui, and M. Zhou, "Recent developments in state estimation with phasor measurements," in *Power Systems Conference*

*and Exposition, 2009. PSCE'09. IEEE/PES.* IEEE, Piscataway, NJ, pp. 1–7, 2009.

[21] R. B. Bobba, K. M. Rogers, Q. Wang, H. Khurana, K. Nahrstedt, and T. J. Overbye, "Detecting false data injection attacks on dc state estimation," in *Preprints of the First Workshop on Secure Control Systems, CPSWEEK*, Stockholm, Sweden, vol. 2010, 2010.

[22] G. Dán and H. Sandberg, "Stealth attacks and protection schemes for state estimators in power systems," in *Smart Grid Communications (SmartGridComm), 2010 First IEEE International Conference on*. IEEE, Piscataway, NJ, 2010, pp. 214–219.

[23] G. Granelli and M. Montagna, "Identification of interacting bad data in the framework of the weighted least square method," *Electric Power Systems Research*, vol. 78, no. 5, pp. 806–814, 2008.

[24] S. Cui, Z. Han, S. Kar, T. T. Kim, H. V. Poor, and A. Tajer, "Coordinated data injection attack and detection in the smart grid: A detailed look at enriching detection solutions," *Signal Processing Magazine, IEEE*, vol. 29, no. 5, pp. 106–115, 2012.

[25] Y. Huang, H. Li, K. A. Campbell, and Z. Han, "Defending false data injection attack on smart grid network using adaptive CUSUM test," in *Information Sciences and Systems (CISS), 2011 45th Annual Conference on*. IEEE, Piscataway, NJ, pp. 1–6, 2011.

[26] Y. Huang, M. Esmalifalak, H. Nguyen, *et al.*, "Bad data injection in smart grid: Attack and defense mechanisms," *Communications Magazine, IEEE*, vol. 51, no. 1, pp. 27–33, 2013.

[27] M. Esmalifalak, N. T. Nguyen, R. Zheng, and Z. Han, "Detecting stealthy false data injection using machine learning in smart grid," in *Global Communications Conference (GLOBECOM), 2013 IEEE*. IEEE, Piscataway, NJ, 2013, pp. 808–813.

[28] L. Liu, M. Esmalifalak, and Z. Han, "Detection of false data injection in power grid exploiting low rank and sparsity," in *Communications (ICC), 2013 IEEE International Conference on*. IEEE, Piscataway, NJ, 2013, pp. 4461–4465.

[29] E. J. Candès, X. Li, Y. Ma, and J. Wright, "Robust principal component analysis?" *Journal of ACM*, vol. 58, no. 3, pp. 11:1–11:37, Jun. 2011.

[30] E. Candès and B. Recht, "Exact matrix completion via convex optimization," *Communications of the ACM*, vol. 55, no. 6, pp. 111–119, Jun. 2012.

[31] R. Paffenroth, P. du Toit, R. Nong, L. Scharf, A. P. Jayasumana, and V. Bandara, "Space-time signal processing for distributed pattern detection in sensor networks," *Selected Topics in Signal Processing, IEEE Journal of*, vol. 7, no. 1, pp. 38–49, Feb. 2013.

[32] S. Boyd, N. Parikh, E. Chu, B. Peleato, and J. Eckstein, "Distributed optimization and statistical learning via the alternating direction method of multipliers," *Foundations of Trends in Machine Learning*, vol. 3, no. 1, pp. 1–122, Jan. 2011.

*Chapter 12*

# Overview of research in the ADVANTAGE project

*Marko Angjelichinoski[1], Mirsad Cosovic[2], Charalampos Kalalas[3], Ruben Lliuyacc[4], Mehdi Zeinali[5], Jesus Alonso-Zarate[3], Juan Manuel Mauricio[4], Petar Popovski[1], Cedomir Stefanovic[1], John S. Thompson[5] and Dejan Vukobratovic[6]*

## 12.1 Introduction

The European Marie Curie Project ADVANTAGE (Advanced Communications and Information processing in smart grid systems) was launched in 2014 [1]. It represents a major inter-disciplinary research project into the topic of Smart Grid technology. A key aspect of the project is to bring together and train 13 early stage researchers from the traditionally separate fields of power systems and communications engineering research. This chapter describes some of the initial research results that have arisen from the project and to highlight some of the key advances and developments that are being studied in the project.

The major research focus of the ADVANTAGE project is on advancing technologies for the smart grid, providing architectural solutions and developing innovative information and communications technology (ICT) solutions to support its operation. More specifically, the researchers within ADVANTAGE are organised to make research advances in four main thematic areas:

- **Smart homes:** This work will develop novel ICT solutions and applications focussing on home energy management solutions for smart grid household consumers. Novel approaches to integrate communications technology and new ways to model the energy consumption of houses are being considered.

[1]Department 8 – Department of Electronic Systems, Aalborg University, Aalborg, Denmark
[2]Schneider Electric Dms Ns, Novi Sad, Serbia
[3]Centre Tecnològic de Telecomunicacions de Catalunya (CTTC), Parc Mediterrani de la Tecnologia, Castelldefels, Spain
[4]University of Seville, Electrical Engineering, Seville, Spain
[5]Institute for Digital Communications, School of Engineering, University of Edinburgh, Edinburgh, UK
[6]Faculty of Technical Sciences, University of Novi Sad, Novi Sad, Serbia

- **Neighbourhood/industrial area networks:** This research will provide efficient neighbourhood area ICT solutions for advanced smart grid data exchange, gathering and processing. It will investigate a variety of communications technologies for this purpose, including both wireless and wireline solutions.
- **Microgrids:** The focus here is on developing ICT solutions for challenging problem of communications, processing and distributed control of microgrids. These are small to medium power networks which include both power generation and energy consumers and which may be either connected or switched off (islanded) from the main power grid.
- **Intelligent distribution networks:** The researchers will study advanced and intelligent large-scale distribution networks. A major focus is on supporting efficient power distribution and developing new approaches to efficient balance supply and demand such as load clustering, demand side management and distributed microgrid control.

In this chapter, we will now provide initial examples of research carried out in the ADVANTAGE project. Five different topics will be addressed, covering both communications and power engineering topics. Section 12.2 will review device-to-device (D2D) communications, which is a promising emerging technology for communication between different entities in the smart grid. In Section 12.3, we will discuss PowerTalk, an innovative communications protocol for microgrid systems which uses changes in electrical power properties, such as voltage and frequency, to communicate information about the state of the microgrid network. Section 12.4 will study communication of electrical information within the smart grid, focussing particularly on data from smart meters and how it may be compressed efficiently to reduce the data load on communications networks. Section 12.5 discusses the wide area power grid and novel techniques that can be used to monitor grid behaviour and stability using ideas from communications systems. Finally, control methods for distributed generators operating within the smart grid are discussed and evaluated in Section 12.6. Section 12.7 presents brief conclusions to this chapter.

## 12.2 Cellular-enabled D2D communication for smart grid neighbourhood area networks

This section provides a comprehensive discussion on the applicability of D2D communication in cellular networks as a key enabling technology for the fundamental operations of smart grid neighbourhood area networks (NANs). Smart grid NANs constitute the communication infrastructure of the electric power distribution systems and involve communication between diverse electric devices, e.g., intelligent electronic devices (IEDs) and metering data aggregation points, which are deployed in large and potentially complex geographical areas [2].

A variety of communication technologies have been proposed for NAN services to realise the full potential of smart grid [3]. While the use of wired communication solutions can be economically and/or physically prohibitive in NAN applications, e.g., advanced metering infrastructure (AMI) and distribution automation (DA), current wireless technologies, e.g., cellular long-term evolution (LTE), need to be enhanced with a wider range of use case characteristics and adapted to a much more complex range of access requirements in order to efficiently support distribution grid services [4]. LTE network-assisted D2D communication constitutes a promising technology to satisfy the stringent requirements of advanced smart grid NAN operations and overcome the shortcomings of the current communication capabilities of power systems. Direct connectivity and information exchange among devices in the smart grid, with network control provided by a cellular base station, offers key benefits in the performance, from both communications and power perspective.

### *12.2.1 Limitations of LTE technology*

Compared to other wireless communication technologies, e.g., IEEE 802.15.4, IEEE 802.11 and IEEE 802.16, LTE networks offer extended coverage and enhanced security mechanisms for reliable smart grid data delivery in licensed spectrum bands [5,6]. LTE provides improved performance (low delay, high data rates, and interference management) for the realisation of demanding DA operations, such as fault detection and outage management, and massive metering data transfer in AMI systems. Utilising the existing cellular communication infrastructure, the operational cost for distribution system operators (DSOs) is kept at minimum, allowing cellular networks to play an important role in the day-to-day business of energy utilities [7]. However, conventional LTE networks were originally designed for human-type communication (HTC); therefore, important technical challenges and design requirements arise while integrating traffic related to smart grid functionalities in shared LTE networks. In particular:

- Smart grid data traffic characteristics are fundamentally different than those generated by human mobile services originally thought for LTE, e.g., voice, video and web browsing. While metering and monitoring data follow a periodic traffic pattern, control and protection messages are event-driven with sporadic nature [8]. Smart grid data packets are relatively short compared to HTC and metering traffic is uplink-based, in contrast to HTC applications which are mainly downlink-dominant.
- The envisioned number of connected devices in the smart grid, bringing together IEDs and metering devices, is order of magnitudes higher than the number of simultaneous HTC users in traditional LTE networks. Such a high density of devices introduces significant challenges for the accommodation of the generated traffic volume and the avoidance of network congestion due to massive access attempts [9].
- DA operations, e.g., protection-related message exchange and inter-substation communication, are often associated with strict end-to-end latencies (in the order of few milliseconds) and high reliability figures for packet delivery, which current LTE standard configurations cannot guarantee [10,11].
- In shared LTE networks, smart grid traffic needs to coexist with regular HTC without jeopardising the quality-of-service (QoS) levels. This coexistence calls for a differentiated radio resource management strategy to efficiently handle the resource allocation to the competing links. The design of traffic prioritisation mechanisms requires a sophisticated understanding of the interaction between the smart grid and traditional communication networks while the performance of conventional cellular users should not be adversely affected [12].

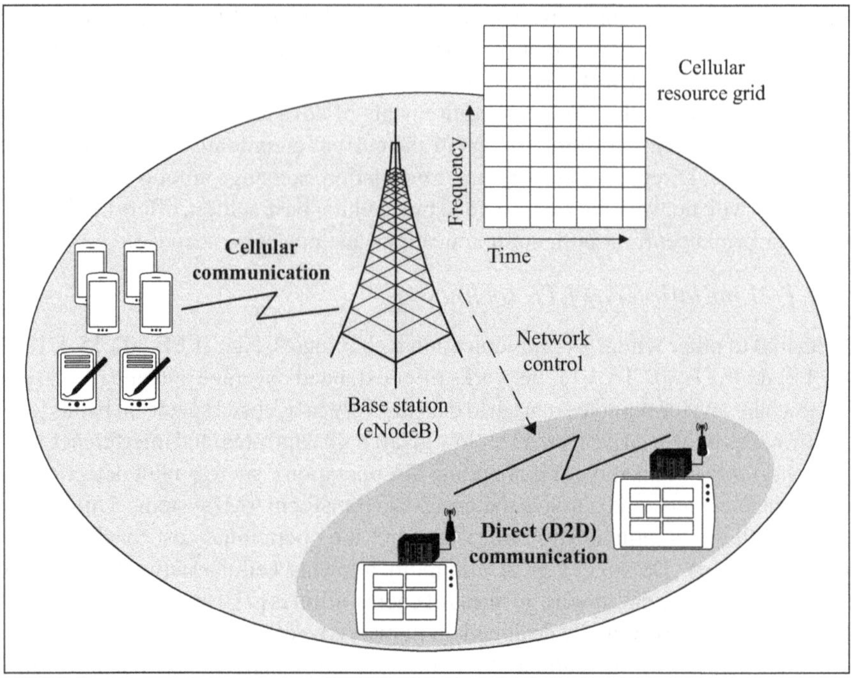


*Figure 12.1 LTE network-assisted D2D communication scheme. D2D refers to a radio technology that allows smart grid devices, such as smart meters, circuit breakers and, in general, IEDs, to autonomously communicate with each other in a direct manner, bypassing the transmission to a central base station. Network control and resource management among D2D links and cellular users remain under the responsibility of the base station*

### *12.2.2 A promising approach: LTE-D2D communication*

LTE network-assisted D2D communication constitutes a recently introduced radio technology that enhances standardised cellular networks to overcome some of LTE limitations when supporting smart grid NAN services [13]. As illustrated in Figure 12.1, D2D communication in the context of cellular networks refers to the direct communication between entities in a centralised cellular network without the need to route data transmissions through a base station. In this scheme, smart grid devices can exchange information utilising licensed cellular resources over a direct link, allowing for a decentralised and fully automated power system operation.

The key benefits of LTE-D2D communication as the enabling technology for smart grid NAN operations can be highlighted as follows:

- Latency and reliability improvements. By using D2D communications, time-critical DA functionalities related with the protection and control of the power

distribution grid can be performed without the additional delay imposed by the core cellular network. In addition to the immediate end-to-end delay gain achieved by shifting from a two-hop communication model (via the base station) to a single-hop direct communication, a combination of both D2D and infrastructure-based communications can lead to an increased reliability by means of multi-path diversity.
- Radio resource management. The localised nature of the D2D transmissions allows for the reuse of cellular radio resources while maintaining acceptable interference levels outside a certain spatially limited area around each transmitting node in the system. Thus, resource sharing with conventional cellular users becomes more efficient [14].
- Network offload. By offloading traffic onto direct D2D links, base stations and other LTE network components are relieved of the extensive infrastructure network load, e.g., large volume of metering data in AMI systems.
- Energy efficiency. The shorter communication path among devices compared to the distance between a device and its serving base station improves energy efficiency and increases device lifetime [15].

Besides the benefits in communication, LTE-D2D technology brings fundamental advancements to the current ageing distribution grid resulting in a dramatic improvement of the overall power system operation. In particular, LTE-D2D networks allow for a decentralised structure of the power grid with an automated system management that efficiently coordinates the diversified functions, e.g., microgrid distributed management across-the-network components [16,17]. By exchanging information in their local networks, control and real-time status monitoring of all IEDs in the distribution level can be possible, leading to fast detection of faults and reduced system response times, thus, full support of mission-critical functionalities, e.g., substation automation [18]. In addition, bidirectional communication among smart meters and utility centres allows consumers to become aware of the timing and quantity of their personal electricity usage and timely respond to the electricity pricing information delivery [19].

### *12.2.3 State of the art – open challenges*

Before LTE-D2D networks successfully support smart grid NAN operations in the distribution grid, there are many research challenges that need to be resolved. A topic with growing research interest nowadays refers to the efficient radio resource management in LTE-D2D networks that accommodate both cellular HTC and D2D communication among smart grid devices. Based on spectrum utilisation, D2D can be generally classified into two categories: in-band and out-of-band. In-band refers to D2D utilising the same spectrum (uplink or downlink resources) used for cellular communications, while out-of-band refers to D2D utilising bands other than cellular band (e.g., 2.4 GHz ISM band). The integration of LTE with short-range technologies has been also proposed as a means for achieving enhanced reliability and availability. Figure 12.2 illustrates a taxonomy of approaches for resolving radio

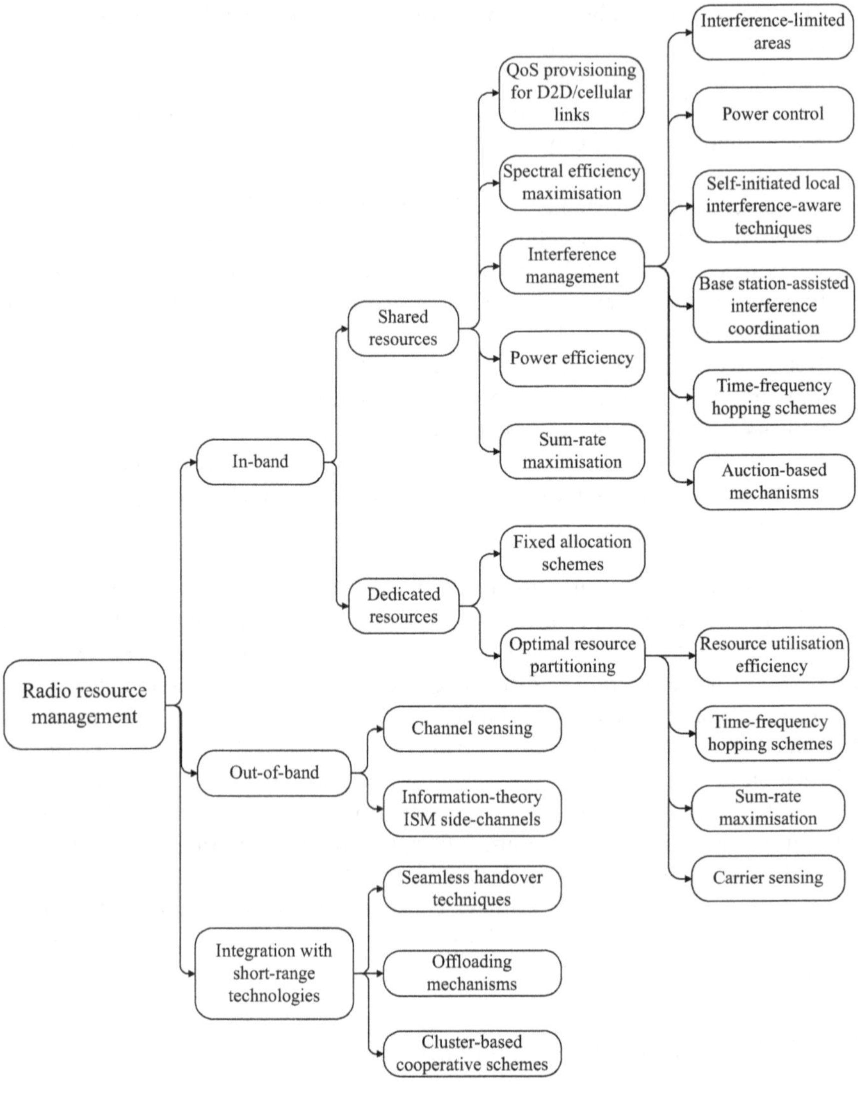


*Figure 12.2 Diagram of approaches for resolving radio resource management in D2D-enabled cellular networks*

resource management in D2D-enabled cellular networks. The scope and limitations of the existing works in the literature are thoroughly compared in Table 12.1.

In in-band LTE-D2D networks, cellular links can either share resources with D2D links (shared resources) or utilise mutually orthogonal parts of the cellular spectrum (dedicated resources). In a network with shared resources, the main challenge refers to the efficient management of intra-cell interference experienced by both cellular and

*Table 12.1 Comparison of proposals for resolving radio resource management in D2D-enabled cellular networks*

| Proposal | | | | Analytical tool | | | | | | | Metrics | | | | | | | | | | Validation | | Resources | |
|---|---|---|---|---|---|---|---|---|---|---|---|---|---|---|---|---|---|---|---|---|---|---|---|---|
| Type | Sub-type | Approach | Reference | Optimisation theory | Game theory | Stochastic geometry | Graph theory | Probability theory | Geometry | Fuzzy logic | Throughput | Coverage probability | Fairness | Transmit power | Interference | Latency | System efficiency | Energy efficiency | Spectrum efficiency | Handover metrics | Simulation | Theory | Uplink | Downlink |
| In-band | Shared resources | QoS provisioning for D2D/cellular links | [45] | ✓ | | | | | | | ✓ | | | | | | | | | | ✓ | | ✓ | ✓ |
| | | | [42] | ✓ | | | | | | | ✓ | | | | | | | | | | ✓ | | ✓ | ✓ |
| | | | [43] | ✓ | | | | | | | ✓ | | | | | | | | | | ✓ | | | ✓ |
| | | | [44] | ✓ | | | | ✓ | | | ✓ | ✓ | | | | | | | | | ✓ | | ✓ | |
| | | Interference-limited areas | [27] | | | | | ✓ | ✓ | | ✓ | | ✓ | | | | | | | | ✓ | | ✓ | |
| | | | [28] | | | | | | ✓ | | ✓ | | | | ✓ | | | | | | ✓ | | | ✓ |
| | | Power control | [20] | | | | | ✓ | ✓ | | ✓ | | | | ✓ | ✓ | | | | | ✓ | ✓ | ✓ | |
| | | | [21] | – | | | | | | | ✓ | | | | ✓ | | | | | | ✓ | | ✓ | |
| | | | [23] | – | | | | | | | ✓ | ✓ | | | | | | | | | ✓ | | ✓ | ✓ |
| | | | [24] | ✓ | | ✓ | | ✓ | | | ✓ | ✓ | | | | | | | | | ✓ | ✓ | ✓ | |
| | | | [22] | – | | | | | | | ✓ | | | | | | | | | | ✓ | | ✓ | ✓ |
| | | | [25] | ✓ | | | ✓ | | | | ✓ | | | ✓ | | | | | | | ✓ | | Not specified | |
| | | | [26] | ✓ | ✓ | | | | | | ✓ | | ✓ | ✓ | | | | | | | ✓ | | ✓ | |
| | | Self-initiated local interference-aware techniques | [29] | – | | | | | | | ✓ | ✓ | | | | | | | | | ✓ | | ✓ | ✓ |
| | | | [30] | ✓ | | | | | | | ✓ | | | | | | | | | | ✓ | | ✓ | ✓ |
| | | | [31] | ✓ | | | ✓ | | | | ✓ | | | | | | | | | | ✓ | | | ✓ |
| | | | [32] | – | | | | | | | ✓ | | | | | | | | | | ✓ | | ✓ | |
| | | Base station-assisted interference coordination | [37] | – | | | | | | | ✓ | | | | | | | | | | ✓ | | ✓ | ✓ |
| | | | [35] | | | | ✓ | | | | ✓ | | | | | | | | | | ✓ | | Not specified | |
| | | | [36] | | | | ✓ | | | | ✓ | | | | ✓ | | | | | | ✓ | | ✓ | |
| | | | [34] | ✓ | ✓ | | | | | | ✓ | | | | | | | | | | ✓ | | ✓ | |
| | | | [33] | ✓ | ✓ | | | ✓ | | | ✓ | | | | ✓ | | | | | | ✓ | ✓ | ✓ | |
| | | Time-frequency hopping schemes | [41] | ✓ | | ✓ | | ✓ | | | ✓ | ✓ | | | | | | | | | ✓ | ✓ | | ✓ |
| | | | [40] | | | ✓ | | ✓ | | | ✓ | ✓ | | ✓ | | | | | ✓ | | ✓ | ✓ | ✓ | |
| | | Auction-based mechanisms | [38] | | ✓ | | | | | | ✓ | | | | | | ✓ | | | | ✓ | | | ✓ |
| | | | [39] | | ✓ | | | | | | ✓ | | ✓ | | | | ✓ | | | | ✓ | | | ✓ |
| | | Power efficiency | [46] | ✓ | | | | | | | | | | ✓ | | | | | | | ✓ | | | ✓ |
| | | | [47] | ✓ | | | | | | | | | ✓ | ✓ | | | | ✓ | ✓ | | ✓ | | ✓ | |
| | | Sum-rate maximisation | [66] | ✓ | | | ✓ | | | | ✓ | | ✓ | | | | | | | | ✓ | | Not specified | |
| | | Spectral efficiency maximisation | [48] | | | ✓ | | ✓ | | | ✓ | ✓ | | | | | | | ✓ | | ✓ | ✓ | ✓ | |
| | Dedicated resources | Fixed allocation schemes | [67] | | | | | ✓ | | | | | | | | ✓ | | | | | ✓ | ✓ | ✓ | |
| | | | [37] | – | | | | | | | ✓ | | | | | | | | | | ✓ | | ✓ | ✓ |
| | | | [50] | | | | | ✓ | | | | | | | | ✓ | | ✓ | ✓ | | ✓ | ✓ | ✓ | |
| | | | [51] | | | | | ✓ | | | | | | | | ✓ | | | | | ✓ | | ✓ | |
| | | | [49] | | | | ✓ | | | | ✓ | ✓ | | | | | | | | | ✓ | | ✓ | |
| | | Carrier sensing | [54] | ✓ | | ✓ | | ✓ | | | ✓ | ✓ | | | | | | | ✓ | | ✓ | | ✓ | |
| | | Resource utilisation efficiency | [68] | ✓ | | | | ✓ | | | | | | | | | | | ✓ | | ✓ | | ✓ | |
| | | Sum-rate maximisation | [40] | | | ✓ | | ✓ | | | ✓ | ✓ | | ✓ | | | | | ✓ | | ✓ | ✓ | ✓ | |
| | | | [53] | | | ✓ | | ✓ | | | ✓ | ✓ | ✓ | | | | | | ✓ | | ✓ | ✓ | ✓ | |
| | | Time-frequency hopping schemes | [41] | ✓ | | ✓ | | ✓ | | | ✓ | ✓ | | | | | | | | | ✓ | ✓ | | ✓ |
| | | | [52] | ✓ | | ✓ | | ✓ | | | ✓ | ✓ | | | | | | | | | ✓ | ✓ | | ✓ |
| Out-of-band | | Channel sensing | [55] | | | | | ✓ | | | ✓ | | | | | | | | ✓ | | ✓ | | Not specified | |
| | | Information-theory ISM side-channels | [56] | – | | | | | | | ✓ | | | | ✓ | | | | | | ✓ | | ✓ | ✓ |
| Integration with short-range technologies | | Seamless handover techniques | [58] | – | | | | | | | ✓ | | | | | | | | | ✓ | ✓ | | ✓ | |
| | | | [59] | – | | | | | | | | | | | | | | | | ✓ | ✓ | | Not specified | |
| | | | [61] | – | | | | | | | ✓ | | | | | | | | | ✓ | ✓ | | | ✓ |
| | | | [62] | – | | | | | | | ✓ | ✓ | | | | | | ✓ | | ✓ | ✓ | | ✓ | ✓ |
| | | | [60] | | | | | | | ✓ | | ✓ | | | | | | | | ✓ | ✓ | | Not specified | |
| | | Cluster-based cooperative schemes | [63] | – | | | | | | | | | | | | | | ✓ | | | ✓ | | | ✓ |
| | | Offloading mechanisms | [64] | – | | | | | | | ✓ | | | | | | | ✓ | | | ✓ | | ✓ | ✓ |
| | | | [65] | – | | | | | | | ✓ | | | | | ✓ | | | ✓ | | ✓ | | | ✓ |

D2D links. Various resource allocation strategies have been proposed in the literature to deal with interference mitigation. Power control schemes, where D2D pairs dynamically modify their transmission power levels to minimise the interference effects to cellular users, have been studied in References 20–26. Other works focus on the definition of interference-limited areas for resource sharing, where D2D and cellular links cannot exploit the same frequencies [27,28]. Local interference-aware mechanisms, driven by the D2D terminals to maximise their performance while guaranteeing the quality of cellular links, have been proposed in References 29–31. In an effort to avoid the acquisition of channel state information for every potential link and thus reduce the computational complexity of the formulated optimisation problems, the authors in Reference 32 propose a location-based resource sharing scheme that instead relies on network planning.

Interference coordination mechanisms with base station assistance have been also considered in the literature [33–37]. In particular, the works in References 33,34 employ game-theoretic techniques for resource allocation with fictional pricing mechanisms optimised by the base stations and transmitted to the D2D pairs which in turn compete to maximise their individual utility functions. Similar modelling frameworks that rely on auction mechanisms are proposed in References 38,39. Leveraging stochastic geometry tools, the authors in References 40,41 present analytical frameworks for the analysis and design of D2D spectrum sharing and make use of time-frequency hopping schemes for interference management and transmit power control. Resource allocation schemes with advanced optimisation techniques for QoS provisioning of cellular and D2D links have been studied in References 42–45 and the objective lies in the maximisation of the total system throughput. Heuristic power and resource allocation schemes aiming at an enhanced power efficiency with QoS guarantees for cellular and D2D links are proposed in References 46,47, whereas the maximisation of spectral efficiency for D2D links is the objective in Reference 48.

In a network with orthogonal spectrum allocation for D2D communication, part of the available cellular resources is subtracted from the general pool and is exclusively used by D2D links, instead of allocating the entire resource grid to both cellular and D2D links. In this case, interference among cellular and D2D links is not the primal concern and the main objective resides in the efficient and fair partition of the cellular resources to achieve increased spectrum efficiency, while satisfying the QoS requirements and traffic demands of the competing entities. The option for dedicated resources for D2D communication has been introduced in Reference 37.

Resource partitioning can be either fixed or dynamically determined. A static allocation scheme based on graph theory is presented in Reference 49, to avoid the interference caused among D2D pairs in a network with orthogonal resources. In Reference 50, a contention-based LTE mechanism is proposed where dedicated resources are utilised for direct data transmission, avoiding the signalling overhead required for network access. A similar method for improving access latency is presented in Reference 51, where the authors propose detailed LTE physical layer enhancements to address the occurring collisions in overload conditions. In References 41,52, the authors investigate the optimal resource partitions between D2D and cellular networks and apply time-frequency hopping schemes to achieve interference randomisation. Efficient spectrum sharing strategies that allow a relatively fair and interference-aware partition of cellular resources between cellular users and devices have been also proposed in References 40,53. In Reference 54, the authors present a distributed mechanism for spectrum allocation using a carrier-sensing threshold for self-interference control among D2D communication pairs.

In out-of-band LTE-D2D networks, D2D links utilise unlicensed spectrum in an effort to eliminate interference between D2D and cellular radio connections. The use of other frequency bands, non-overlapping with the cellular spectrum, introduces complexity in coordinating the communication over the two different bands. Out-of-band D2D communication may also suffer from the uncontrolled nature of unlicensed spectrum. Despite its potential gains, out-of-band D2D approaches have received less attention in the literature to date. In Reference 55, the use of unlicensed ISM band for communication among D2D pairs is proposed. D2D pairs form resource-contention groups depending on their QoS/bandwidth requirements and a group-wise channel sensing technique is then applied. The use of cellular-controlled ISM bands to mitigate intra-cell interference management and increase the achieved cellular network capacity has been also studied in Reference 56.

Another hybrid approach encountered in the literature, refers to the integration of LTE with short-range radio technologies, e.g., WiFi or ZigBee, forming the so-called capillary networks [57]. In this heterogeneous network deployment, reliability and availability could be improved by exploiting the transmission diversity with simultaneous radios used for the same purpose. Various seamless handover techniques for optimal network selection among the available ones have been proposed in the literature, mainly aiming at low handover delays [58–60], QoS preservation [60,61] and energy efficiency [62]. The authors in Reference 63 present a reliable multicast scheme with cooperative retransmissions in LTE-WiFi networks for reducing both the traffic load and the energy consumption of devices. Offloading mechanisms in LTE-WiFi networks are proposed in References 64,65, along with a performance evaluation of energy consumption and resource utilisation efficiency, respectively.

### *12.2.4 Conclusions and outlook*

The transformation of the existing power distribution grid into a smart grid requires an efficient and reliable communication technology that would support advanced power grid functionalities. LTE-D2D communication constitutes a promising approach to enhance current cellular technology and ensure the performance requirements for reliable connectivity in the next-generation power grid. However, many challenges remain open and research efforts are required to facilitate the use of LTE-D2D as an enabler for the true smart grid of the future, solving open challenges and uncovering opportunities for still unforeseen automated functionalities in the distribution energy networks.

## 12.3 Power talk in DC MicroGrids: merging primary control with communication

Power talk is a novel method for communication among voltage source converters (VSCs) in DC MicroGrids (MGs), achieved through variations of the supplied power incurred by modulation of the parameters of the primary control. This chapter discusses general aspects of power talk and identifies its main challenges from communication perspective, via a simple framework of one-way communication where one VSC unit sends control message to another VSC unit. The important concepts of signalling space (where the input power-talk symbol constellations are designed) and detection space (where demodulation of the symbols is performed) are defined and illustrated. To conclude, the performance of the power-talk detector for representative symbol-constellations in the signalling space of arbitrary order is evaluated in terms of symbol-error probability.

### *12.3.1 Why power talk?*

MGs are clusters of distributed energy resources (DERs) and loads connected to common buses through *power electronic converters* (see Figure 12.3) which span relatively small geographical areas and are able to operate connected to the main grid or autonomously, in islanded, self-sustaining mode [69]. To support, optimise and enhance MG control, information exchange between units in the MG system is required, especially on secondary and tertiary control levels [70,71] that drive value-added control applications, such as voltage restoration, unit commitments (UCs) and optimal dispatch [71,72]. Recent approaches for MG control avoid

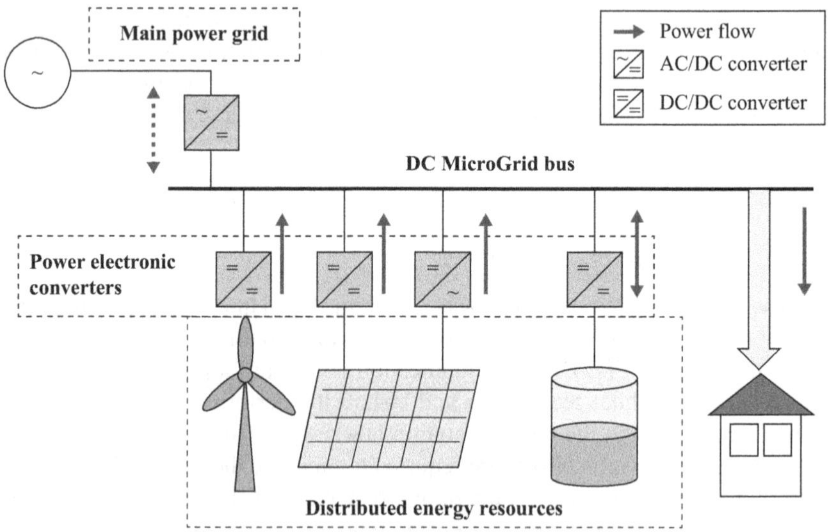


*Figure 12.3 Low-voltage DC MicroGrid*

reliance on external communication networks [73–76] due to the availability and reliability requirements, and advocate the use of the capabilities of the power electronic interfaces, foremost the capability of digital signal processing, as the potential communication enabler.

Power talk is a novel communication technique designed specifically for control applications in DC MGs [77,78] that fits in the framework of autonomous, self-sustaining MG framework. The key idea is to modulate information in the parameters of the *primary control loops*, implemented in the power electronic converters and regulating the voltage of the MG buses. This incurs bus voltage deviations that can be detected by other MG units, resulting in information transfer; the effective physical medium upon which the power talk communication channel is established is the level of the voltage bus, jointly maintained by all units in the system through primary control. The achievable rates of the power talk are limited by the response times of primary control loops, which are at the order of tens of milliseconds. Nevertheless, such rates may prove sufficient for the communication requirement of the MG control applications. On the other hand, power talk is, in essence, as reliable as the MG power-control system itself, since all units connected to the same bus measure the bus voltage.

## *12.3.2 Embedding information in primary control loops*

To illustrate the main idea of power talk, consider the steady-state model of standalone single-bus DC MG system with $K$ DER units, shown in Figure 12.4, where all units and loads are connected to the same bus and observe the same bus voltage [79]. The equivalent load is denoted with $r_{eq} = r_{eq}(r, P_{CP})$ and includes resistive $r$ and non-linear components, such as constant power loads $P_{CP}$; it represents the variable power demand that changes randomly through time. The common bus voltage is denoted with $v^*$ and is jointly regulated by all converters in the system that are operated as voltage sources, i.e., VSCs using *droop control* technique [70,71,79]:

$$v^* = v_k - r_{d,k} i_k, \tag{12.1}$$

where $i_k$ is the output current of the converter, while $v_k$ and $r_{d,k}$ are the droop parameters, *reference voltage* and *virtual resistance*, respectively. The droop parameters are

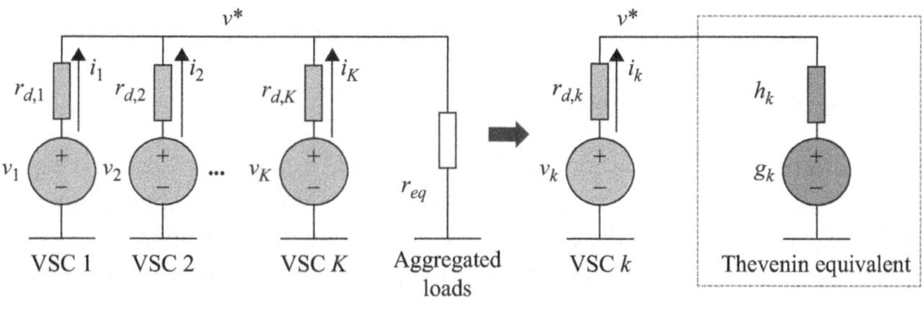


*Figure 12.4 Steady-state model of single-bus droop-controlled DC MicroGrid*

controllable; their values are set according to the current power demand such that the proportional power sharing is enabled. Specifically, the bus voltage level in steady state is regulated to the following value using the droop control:

$$v^* = \frac{\sum_{k=1}^{K} \frac{v_k}{r_{d,k}} + \sqrt{\left(\sum_{k=1}^{K} \frac{v_k}{r_{d,k}}\right)^2 - 4P_{CP}\left(\frac{1}{r} + \sum_{k=1}^{K} \frac{1}{r_{d,k}}\right)}}{2\left(\frac{1}{r} + \sum_{k=1}^{K} \frac{1}{r_{d,k}}\right)}. \tag{12.2}$$

The key idea of power talk is to modulate information into $v_k$ and $r_{d,k}$, thus incurring subtle deviations of the bus voltage (12.2), which can be detected by the units in the system. The main impairments of the power talk communication channel are the sporadic, random variations of the loads in the system, causing unpredictable shifts of the voltage that can lead to erroneous reception. Additionally, by inspecting (12.2), two important issues arise. First, the bus voltage as output of the power talk communication channel is non-linear function of the droop control parameters as the inputs. This non-linear feature implies that existing transmission and receptions techniques, primarily developed for linear channels cannot be directly applied to power talk. Second, solving (12.2) necessitates detailed knowledge of the physical configuration of the underlying system, such as the parameters of other non-transmitting units, the values of the loads and impedances of transmission lines. Such knowledge is *not* available on primary control level, since primary control operates only on the locally available voltage and current measurements, making the design of a receiver based on (12.2) complicated. The above discussion suggests use of system-configuration agnostic approach to design power-talk solutions.

### *12.3.3 One-way power talk communication*

In the rest of the chapter, we focus on simple DC MG with two VSC units, denoted by VSC *a* and VSC *b*, that supply a common load $r_{eq}$. We develop modulation and demodulation techniques for one-way power talk communication, in which VSC *a* is the transmitter and VSC *b* the receiver. We assume that the time axis is slotted and the transmitter and the receiver are slot synchronised. The duration of each slot is denoted with $T_s$.

#### 12.3.3.1 Model

In nominal mode, the droop control parameters of the units are denoted with

$$\mathbf{x}_a^{\mathrm{n}} = (v_a^{\mathrm{n}}, r_{d,a}^{\mathrm{n}}), \quad \mathbf{x}_b^{\mathrm{n}} = (v_b^{\mathrm{n}}, r_{d,b}^{\mathrm{n}}), \tag{12.3}$$

and the nominal supplied powers with $P_a^{\mathrm{n}}$ and $P_b^{\mathrm{n}}$. VSC *b*, acting as receiver, keeps the parameters $v_b$ and $r_{d,b}$ fixed to the nominal values, while VSC *a* modulates information into the DC voltage level of the common bus by varying the controllable steady-state parameters $v_a$ and $r_{d,a}$ in each slot. Thus, the input in the communication channel is

$$\mathbf{x}_a = (v_a, r_{d,a}). \tag{12.4}$$

Assume modulation of arbitrary order $Q$. Then, there are $Q$ different input symbols $\mathbf{x}_a^q$, $q = 0, \ldots, Q-1$; every input symbol represents a combination of $\log_2 Q$ transmitted bits, denoted with $\mathbf{b}_a^q \in \{0,1\}^{\log_2 Q}$. Assume that the Hamming weights of the transmitted bit combinations $\mathbf{b}_a^q$ satisfy

$$w_H(\mathbf{b}_a^0) < \cdots < w_H(\mathbf{b}_a^{Q-1}), \tag{12.5}$$

and assume that VSC $a$ chooses $\mathbf{x}_a^q$, $q = 0, \ldots, Q-1$, to satisfy the following:

$$P_a(\mathbf{x}_k = \mathbf{x}_a^0) < \cdots < P_a(\mathbf{x}_a^{Q-1}), \tag{12.6}$$

where $P_a(\mathbf{x}_a^q)$ is the supplied, i.e., output power by VSC $a$ that corresponds to input $\mathbf{x}_a^q$. The output bus voltage observed by VSC $b$ can be written as follows:

$$y_b \equiv v^* = v^* = \frac{\frac{v_a}{r_{d,a}} + \frac{v_b}{r_{d,b}} + \sqrt{\left(\frac{v_a}{r_{d,a}} + \frac{v_b}{r_{d,b}}\right)^2 - 4P_{CP}\left(\frac{1}{r} + \frac{1}{r_{d,a}} + \frac{1}{r_{d,b}}\right)}}{2\left(\frac{1}{r} + \frac{1}{r_{d,a}} + \frac{1}{r_{d,b}}\right)} + z_b, \tag{12.7}$$

where $r$ and $P_{CP}$ are the resistive and constant power components of the equivalent load, and $z_b \sim \mathcal{N}(0, \sigma_b^2)$ is the measurement noise. The receiver samples the bus voltage with frequency $f_0$ and obtains $N$ samples $y_b[n]$, $n = 1, \ldots, N$, per slot. Finally, the equivalent load $r_{eq} = r_{eq}(r, P_{CP})$ is modelled as a random variable:

$$R_{eq} \sim p_{R_{eq}}(r_{eq}), \tag{12.8}$$

that changes slowly with respect to the signalling rate $1/T_s$.

#### 12.3.3.2 Receiver: detection space

As noted, solving (12.7) by VSC $b$ in order to obtain input of VSC $a$, besides being difficult to perform analytically due to non-linearity, also necessitates knowledge of the current configuration of the system, including the values of the loads, which is not available on primary control level. To address this issue, we introduce a detector based on *Thevenin equivalent* (see Figure 12.5) through which VSC $b$ sees the rest of the system. The parameters of the equivalent representation are denoted with $h_b$ (the equivalent system resistance) and $g_b$ (the equivalent voltage source) and they are

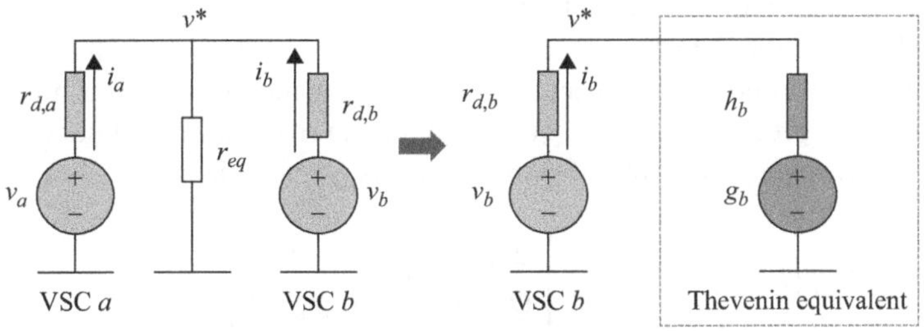


*Figure 12.5 Steady-state model of single-bus DC MG with two droop-controlled VSC units*

functions of $r_{eq}$, the droop parameters of the transmitter VSC $a$, as well as the rest of the system. Then, the bus voltage level as output of the communication channel and observed by VSC $b$ (12.7) can be written as follows:

$$v^* = \frac{v_b r_{d,b}^{-1} + g_b h_b^{-1}}{r_{d,b}^{-1} + h_b^{-1}}. \tag{12.9}$$

When VSC $a$ transmits, the parameters $h_b$ and $g_b$ change. The physical properties of the system are such that, if VSC $a$ supplies *more* power than nominally, the Thevenin equivalent as seen from VSC $b$ also supplies more power and the bus voltage *increases*. The increase in the bus voltage results in *decrease* of the output current $i_b$, see (12.1), and correspondingly, in *decrease* in the output power $P_b$. Conversely, if VSC $a$ supplies less power, the output power $P_b$ increases. Thus, VSC $b$ can determine the bit combination sent by VSC $a$ by observing its local output power $P_b = v^* i_n$ as illustrated in Figure 12.6.

The set of all possible outputs $(v^*, i_b)$ observed (measured) by VSC $b$ in power talk mode is referred as *detection space*. The points in the detection space of the receiver VSC $b$ lie on the line determined by the local droop law $v^* = v_b - r_{d,b} i_b$. When VSC $a$ transmits binary inputs, i.e., $Q = 2$, the two possible combinations $(v^*, i_b)$ that VSC $b$ can observe are located at the intersections of the local droop law (12.1) and the bus voltage law determined by the Thevenin equivalent (12.9); see Figure 12.6(a). The same holds for higher-order modulations; see Figure 12.6(b) for $Q = 4$, as well as for the arbitrary number of units communicating in all-to-all setup. The bit-to-output power mappings in the detection space of the receiver can be learned in a training phase, where the transmitters insert all possible input bit combinations in a predetermined manner, allowing the receiver to observe and learn them. This is equivalent to channel estimation phase in standard communication system; in power talk the channel estimation phase amounts to determining the Thevenin equivalent of the rest of the system as seen by the receiver. We note that the parameters of the Thevenin equivalent can also vary due to load variations. In such case, the combinations $(v^*, i_b)$ in the detection space slide on the droop law line; see Figure 12.6(c). Assuming that the loads vary infrequently on in time compared to the signalling rate of power talk, then the values of the parameters remain fixed for multiple time slots and can be efficiently (re-)estimated using training sequences. However, a change of the load might necessitate initiation of the training phase.

Finally, once the detection space is constructed, and the set of all possible output combinations known, the receiver can employ standard detection mechanisms for detection in presence of noise. When employing maximum likelihood detection (MLD), the average probability of symbol error is given with

$$P_\varepsilon = \mathbb{E}_{R_{eq}} \{P(\varepsilon | r_{eq})\} = \mathbb{E}_{R_{eq}} \left\{ \frac{2}{Q} \sum_{i=2}^{Q} \mathcal{Q}\left( \frac{v_{i-1}^* - v_i^*}{\sigma_b \sqrt{2}} \right) \right\}, \tag{12.10}$$

where $v_q^*$ is the bus voltage value corresponding to a specific input bit combination $\mathbf{b}_a^q$ when $\mathbf{x}_a^q$, $q = 0, \ldots, Q-1$ is inserted by the transmitter.

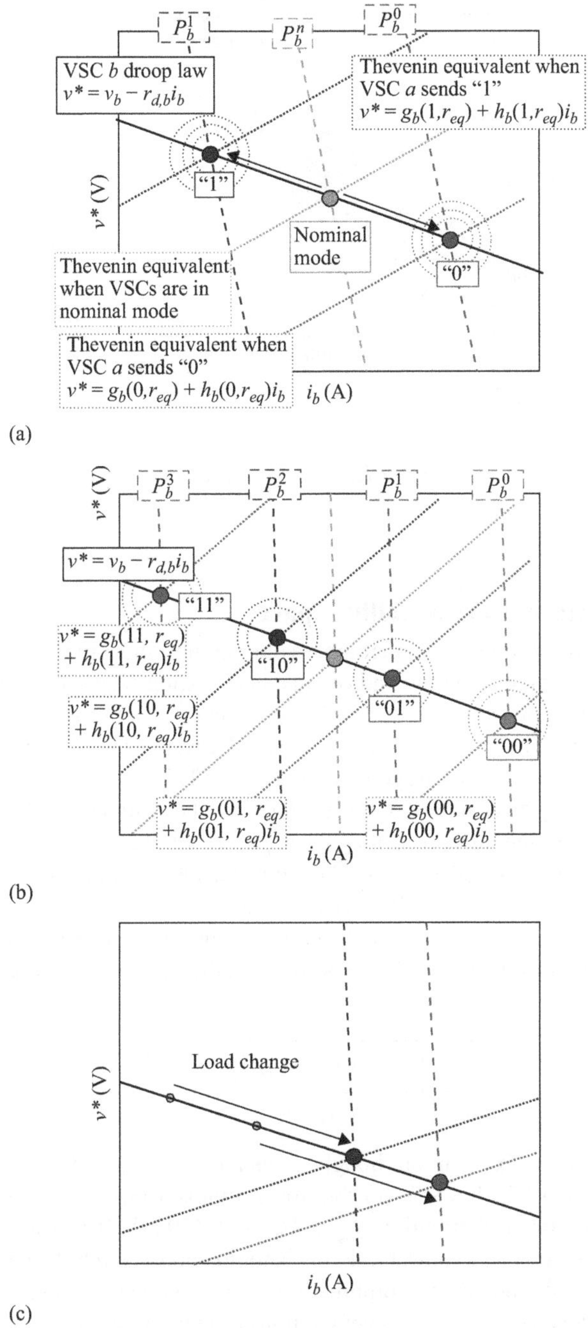


*Figure 12.6 Detection space. (a) $Q = 2$, (b) $Q = 4$ and (c) $Q = 2$, load change*

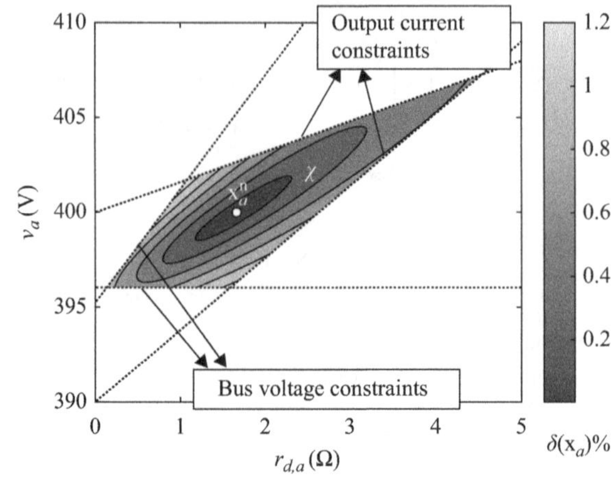


*Figure 12.7 Signalling space*

#### 12.3.3.3 Transmitter: signalling space

The MG as a power supply system imposes operational constraints, such as maximum allowable bus voltage deviation and/or current deviation. The set of operational constraints $\mathcal{C}$ defines the *signalling space* $\mathcal{X}$, as the set of all input symbols $\mathbf{x}_a$ that do not violate constraints in $\mathcal{C}$. The signalling space for the system in Figure 12.4 and $K = 2$ is illustrated in Figure 12.7 for the constraint set $\mathcal{C} = (390\,\text{V} \le v^* \le V_{max} = 400\text{ V}, 0 \le i_a \le 6\text{ A}, 0 \le i_b \le 4\text{ A})$ and $r_{eq} \in [50\ \Omega, 250\ \Omega]$. The boundaries of $\mathcal{X}$ represent the output voltage $v^*$ and current $i_a$ at $r_{eq} = R_{eq,min} = 50\ \Omega$ and $r_{eq} = R_{eq,max} = 250\ \Omega$, and they are easily obtained using the Thevenin equivalent for maximum and minimum load $h_a(R_{eq,min})$, $h_a(R_{eq,max})$, $g_a(R_{eq,min})$ and $g_a(R_{eq,max})$.

Each symbol $\mathbf{x}_a \in \mathcal{X}$ results in different output power supplied to the load $P(\mathbf{x}_a)$ for each $r_{eq}$. We introduce the relative power deviation in respect to the nominal mode of operation:

$$\delta(\mathbf{x}_a) = \frac{\sqrt{\mathbb{E}_{R_{eq}}\left\{[P(\mathbf{x}_a, r_{eq}) - P(\mathbf{x}_a^{\mathrm{n}}, r_{eq})]^2\right\}}}{\mathbb{E}_{R_{eq}}\left\{P(\mathbf{x}_a^{\mathrm{n}}, r_{eq})\right\}}, \tag{12.11}$$

averaged over $R_{eq}$. $\delta(\mathbf{x}_a)$ can be thought of as a cost assigned to each input symbol $\mathbf{x}_a \in \mathcal{X}$. Figure 12.7 shows $\delta$ (in %) for symbols from $\mathcal{X}$, assuming the non-informative, uniform distribution $R_{eq} \sim \mathcal{U}[50\ \Omega, 250\ \Omega]$. If one aims to minimise the power deviation due to power talk, one should choose symbols in the close vicinity of the nominal mode droop combination $\mathbf{x}_a^{\mathrm{n}}$, i.e., with small $\delta(\mathbf{x}_a)$. In this respect, we introduce a constraint on the average power deviation per input symbol $\mathbf{x}_a$ as an additional criterion for designing input constellations:

$$\delta(\mathbf{x}_a) \le \gamma. \tag{12.12}$$

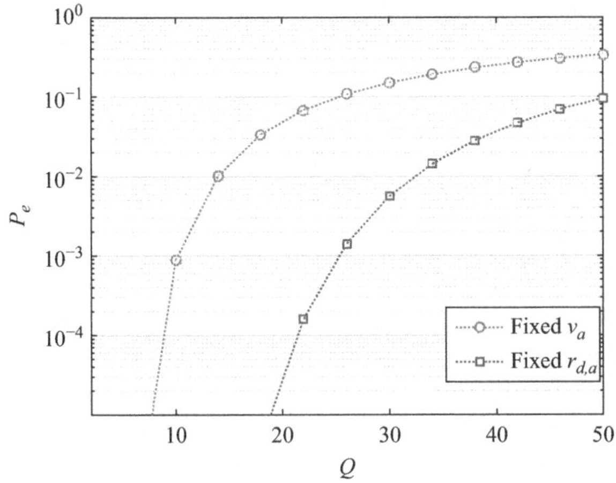


*Figure 12.8 The error probability versus the order of the modulation ($\sigma = 0.01$ V)*

#### 12.3.3.4 Performance: probability of symbol error

We evaluate the performance of the above power talk modulation and demodulation techniques using the system shown in Figure 12.5. The signalling space is shown in Figure 12.7. We fix the average power deviation constraint at $\gamma = 0.004$, and investigate the increase in the error probability with the order of the constellation. Two symbol constellations in the signalling space are investigated: (i) fixed $v_a$ constellation, where the reference voltage is fixed and the data is modulated in the virtual resistances and (ii) fixed $r_{d,a}$ constellation, where the virtual resistance is fixed and the data is modulated in the reference voltages. The results are shown in Figure 12.8. Evidently, in this example the fixed $r_{d,a}$ constellation performs better in terms of average error probability than the fixed $v_a$.

### *12.3.4 Conclusions and outlook*

For more deeper development and discussion of power talk and its applications in general DC MGs with multiple control units, we refer the interested reader to the recent works [77,78,80]. Particularly, in Reference 77, we develop rudimentary power talk communication protocol leveraging on the role of the droop combination in nominal mode playing the role of a pilot and exploiting communication techniques for simple binary asymmetric and erasure channels. In Reference 78, we provide deeper analysis on the design of symbol constellation in the signalling space, evaluate their performance in terms of error probability and show that in certain channel states (i.e., certain values of the load) significant performance gain can be achieved by modulating information in the two droop parameters. Finally, in Reference 80, we extend the power talk concept to a general single-bus DC MG with multiple VSC units that communicate in all-to-all scenario, develop noise-robust receiver based on detection spaces, introduce strategies to cope with sporadic load changes based on training sequences and design communication protocols based on Time Division

Multiple Access and Full Duplex strategies. Part of the ongoing work includes theoretical characterisation of the power talk channel and defining specific control applications using power talk as communication enabler.

## 12.4 Compression techniques for smart meter data

### *12.4.1 Introduction*

Smart meters are electronic devices which are coordinated through communication networks and are used to measure energy consumption in homes, offices and factories. The meters can provide energy consumption data using bidirectional communication between consumers and utility companies. Effective use of the meters can inform consumers to achieve energy saving along with reducing meter reading and billing costs. Smart meters can also improve the stability of electricity production through reducing the need for excess energy production through real-time coordination of supply and demand. The market for smart meters in Europe is predicted to achieve significant growth due to the various national- and European-level initiatives underway; it is expected that smart meters will be operating in about 80% of the continent by 2020 [81].

There are three methods suggested for smart meter communication in the existing literature. The first method is to use the existing cellular networks for smart meter data communication by adding a wireless module to smart meters, e.g., using Global System for Mobile Communications (GSM) or third generation (3G) or LTE technology. This will increase the hardware complexity and the communication cost. A second method is applying wireless sensor network technology which build low-frequency communication link trough mesh network with collecting data in one gateway. A third distinct approach is to communicate the data directly over power lines to local substations. No matter the technology used, as the smart meters are deployed, the data volume will increase dramatically [82].

However, reducing the network load of smart meter data will decrease the communication costs. Due to the very large data volumes expected for smart meters in future, data compression is a necessary technique to be applied. Data compression is thus needed to maintain the data fidelity while reducing the data volume for smart metering. Although powerful compression algorithms exist for personal computers, they require more memory and computational load than is typically available in smart meter devices. The most important reasons for compression are as follows:

1. **Reducing data volume:** This is required for billing purposes, as very small amounts of data are produced by individual users but for smart grid networks, the data volume increases dramatically. For example, one study of Holland predicted that the total data per year produced by the grid could be up to several hundred terabytes of raw data; see Figure 2 of Reference 83. With the application of compression algorithms, it can be reduced by up to 70%–90% compared to the original data volume which can be stored more easily in the data centre in addition to lower traffic volumes for the communication system.

2. **Communication bandwidth efficiency:** Smart meters will be connected to the communication systems through a low-bandwidth communication links. By using compression techniques, we can reduce the required data rates for communication and we can increase the reliability and number of connected smart meters without significant packet losses.
3. **Energy efficiency:** By using compression methods to reduce the volume of data transmitted by smart meters data can improve the energy efficiency of communication, especially when considering the large density of meters that is expected in future.

### *12.4.2 Basic concepts of data compression*

The basic principles of data compression are to reducing the size of data packets for transmission by encoding data more efficiently. In both wired and wireless communications equipment, modems make use of compression schemes to improve communications throughput and efficiency. In order to do this, most compression schemes take advantage of the fact that most data packets contain at least some repetition [84]. Two important compression concepts here are lossy and lossless compression:

- **Lossy compression:** In this type of compression, loss of some data could be acceptable and it does not make a big problem in content of information required. A good example of this is in videoconferencing applications where users can accept a certain proportion of packet losses in order to maintain an overall real-time communication.
- **Lossless compression:** With lossless compression, data is compressed without any loss of data fidelity. It is required to recover all of the compressed data without any loss of information. For example, to compress important financial data files, it is required to apply lossless compression.

Smart meter data types should be compressed using *lossless compression*. This means the compressed file will be restored to its original values with no loss of information during the decompression process. Lossless compression algorithms use data analysis techniques to reduce the amount of repetition of characters within in a packet. Some of the methods used may include deleting blank characters, replacing a sequence of repeating characters by one character or substituting recurring characters with shorter bit sequences. These methods are used to reduce the data size for storing and transmitting information [85]. Compression evaluation measures the reduction in data size. The most useful metric for measuring compression in terms of performance is the compression ratio:

$$\text{Compression ratio} = \frac{\text{Compressed data size} - \text{Uncompressed data size}}{\text{Compressed data size}}. \tag{12.13}$$

For example, if compression reduces the smart meter data volume from 2 to 1 kbyte, then the compression ratio is 50%. In this work, we present lossless compression algorithms that are suitable to be implemented in smart meter hardware without considering the specific features of the collected data. Because of that they are compatible with a wide range of smart meter data formats. We assess these algorithms using two different sets of real smart meter data and compare the compression results.

Another important factor that is always considered when performing a compression operation is complexity of compression algorithm. Different compression algorithms can achieve widely different results even for the same data sets. In this section, two compression methods are studied and evaluated in terms of both compression and processing time performance.

1. **Lempel–Ziv–Welch (LZW):** If we look at almost any data file and examine consecutive characters, we would often see that the same characters are repeated. The LZW method is one algorithm that takes advantage of this repetition. The original version of the method was proposed by Lempel and Ziv in 1978, with enhancements by Welch in 1984, leading to the algorithm's name. The idea is to (1) begin with an initial compression model, (2) read in data character by character and (3) then update the model and encode the data on the fly [86–88]. The LZW method is a compression algorithm based around a dictionary, which stores information about typical patterns observed in the data. Therefore, to encode a character sequence, only a single character, corresponding to that substring's index in the dictionary, needs to be written to the output file. It generally performs best on files with repeated patterns, such as text files.
2. **Adaptive Huffman (AH):** Huffman coding is a well-known technique for lossless compression but it suffers from the fact that the algorithm which recovers the data needs to have some knowledge of the probabilities of occurrence the characters in the compressed files. This adds to the bits needed to encode the file and if this information is unavailable, compressing the file requires two stage – the first is to find the likelihood of each character in order to construct the Huffman encoding tree and a second stage to actually compress the data. Faller and Gallagher, and later Knuth and Vitter, developed a method to perform the Huffman compression in a single-stage process [89–91]. It follows four simple steps: (1) start with a flat code tree; (2) encode some symbols from the source and at the same time count how often each symbol appears; (3) update the Huffman tree with these probability values and (4) repeat the process from step 2 until all data are encoded.

### *12.4.3 Smart meter data and communication scenario*

In order to simulate the idea of smart meter data compression, we analysed the different data sets. In this work, we have used the low-frequency Massachusetts Institute of Technology (MIT) Reference Energy Disaggregation Data Set (REDD) [92]. The MIT data consists of a set of 116 load profiles, where each profile contains average power readings of one individual circuit from one of six different houses. The data is sampled at intervals of 1 s with a precision of 0.01 W. Also in our simulations, we worked with a communication scenario assumptions which compressed smart meter

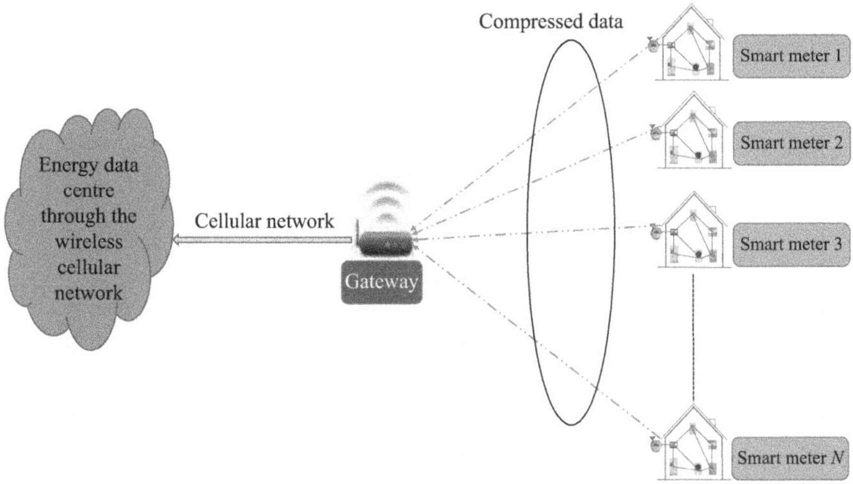


*Figure 12.9 Smart meter data compression from each consumer to the data centre*

data are sent to the gateway and then sent to the Energy Data Centre (EDC) through wireless cellular networks. Each smart meter transmits measured data in different intervals of 15 min, 30 min, 1 h or 2 h. This scenario is shown in Figure 12.9.

Simulation results for the above-mentioned communication scenario are shown for the LZW algorithm in Figure 12.10 and for the AH method in Figure 12.11. Each figure shows four plots – scatter plots of compression rates for individual packets of different sizes (top left), average compression rates (top right), average compression processing times (bottom left) and decompression times (bottom right).

We have worked on LZW and AH compression algorithms for smart meter data and have evaluated their compression gains and processing time. Our study shows that smart meter data is well compressed by compression algorithms, since the data mainly comprises meter readings, where a limited number of symbols occur frequently in data string and the overall data has low entropy. The effects of compression method performance have been investigated by the variety for different reading types and data set sizes. Figures 12.10 and 12.11 depict the distribution of compression rates. The algorithms achieve average compression rates of 74%–88% but exhibit large variations in the case of the LZW algorithm. If extremely low execution time is mandatory, the AH algorithm is the best choice, still achieving approximately 74% compression rate. When considering data packet size it can be seen easily that the LZW method has a very significant performance advantage in compression of larger data packets because of its dictionary base specification that could encode all input patterns. Another important factor for evaluation of compression algorithm is required hardware to implement on the smart meter, gateway and EDC and this investigation is currently ongoing. The trade-off between compression rate, processing time and hardware requirement can lead us to the best selection of compression algorithm for each part of our communication scenario.

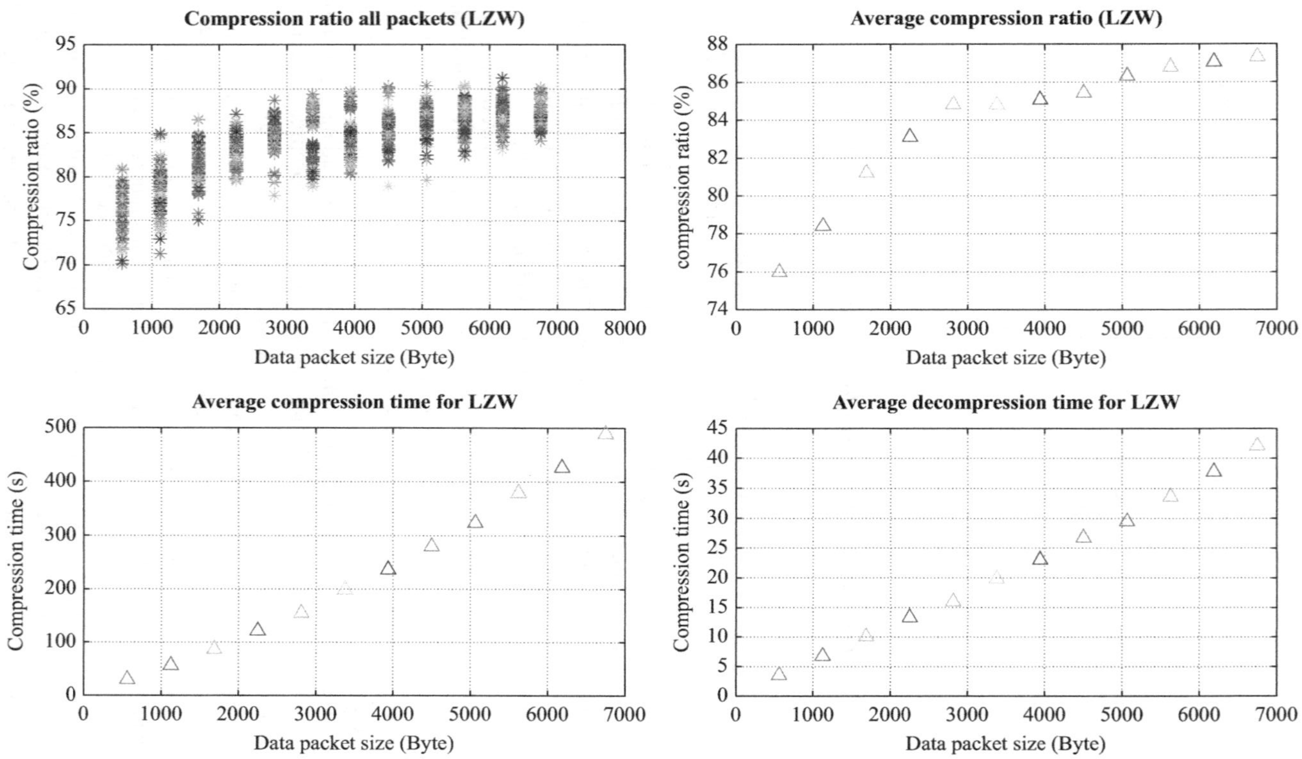


*Figure 12.10 Compression and processing time results for the LZW compression algorithm on the REDD*

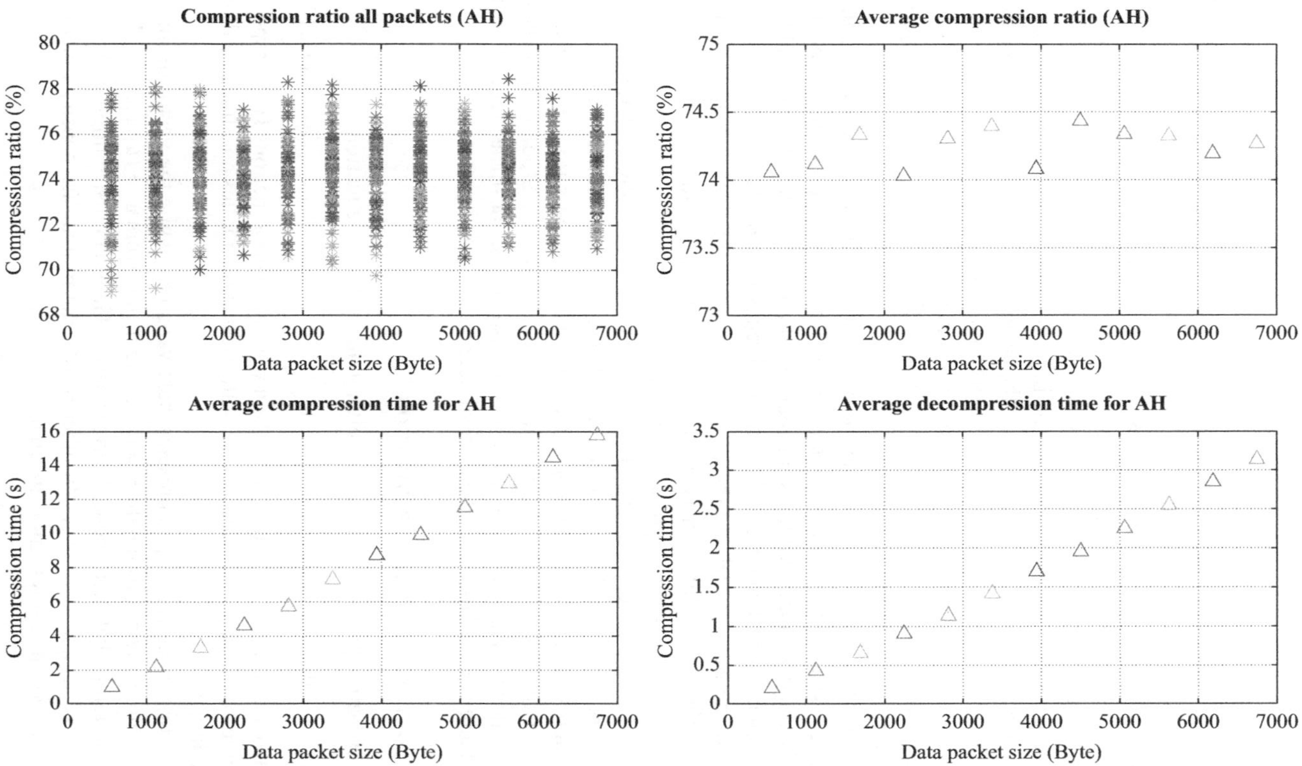


*Figure 12.11 Compression and processing time results for the AH compression algorithm on the REDD*

## 12.5 State estimation in electric power distribution system with belief propagation algorithm

### 12.5.1 Introduction

The performance of energy management systems (EMSs) critically depends on the quality and timeliness of measurements collected across the power system. The measurement devices measure their values with errors incurred by noise, while the communication links from measurement devices may be prone to transmission errors or may even be disconnected, leaving the system operator without the knowledge of smaller or larger part of the system [93]. The imperfections in performing and communicating measurements will greatly affect the accuracy of the state estimation: the one of the most important functions of the real-time EMS. In addition to the state estimation, the topology processing, observability analysis and bad data analysis are the key inputs for other control and management functions, such as security analysis, optimal power flow or economic control.

The state estimation uses real-time measurements to estimate the state of the electric power system. It may be considered as a filter for measurement errors and for level of redundant meter readings. In addition, it allows the determination of the power flows in parts of the network that are not directly metered [94].

Given the network model and the gathered measurements from the system, the main task of the state estimation is to provide an optimal estimate of the system variables. In the state estimation problem, the state variables are the voltage magnitude and voltage phase at buses, the magnitude of turns ratio and angle of turns ratio in transformers. On the other hand, the set of measurements consists of voltage magnitude and voltage angle at buses, active and reactive power flow in branches, active and reactive power injection into buses and current magnitude in branches. Besides real-time measurements, state estimation also use pseudo-measurements and virtual measurements of mentioned electrical quantities [94,95].

### 12.5.2 Conventional state estimation

The state estimation problem reduces to solving the system of equations:

$$\mathbf{z} = \mathbf{f}(\mathbf{x}) + \mathbf{u}. \tag{12.14}$$

Linear state estimation or DC state estimation implies linear functions $\mathbf{f}(\mathbf{x})$, respectively, the non-linear state estimation or AC state estimation implies non-linear and/or linear functions $\mathbf{f}(\mathbf{x})$.

The vector $\mathbf{x} = (x_1, \dots, x_n)$ represents state variables, while the vector of independent measurements is given with $\mathbf{z} = (z_1, \dots, z_n)$ and errors $\mathbf{u} = (u_1, \dots, u_n)$. The usual assumption is that measurement errors have a Gaussian distribution with zero mean $u_k \sim \mathcal{N}(0|\Sigma_k)$, $k = 1, \dots, n$. In other words, the probability density function associated with $k$th measurement is given as:

$$\mathcal{N}(z_k|\mathbf{x}, \Sigma_k) = \frac{1}{\sqrt{2\pi\,\Sigma_k}} \exp\left\{\frac{[z_k - f_k(\mathbf{x})]^2}{2\Sigma_k}\right\}, \quad k = 1, \dots, n, \tag{12.15}$$

where $z_k$ is the value of measurement with variance $\Sigma_k$ and the function $f_k(\mathbf{x})$ connects the vector of state variables with the value of $k$th measurement.

With the above probabilistic measurement error model, the problem of the state estimation reduces to the maximisation of the likelihood function:

$$\max_{\mathbf{x}} \mathcal{L}(\mathbf{z}|\mathbf{x}) = \prod_{k=1}^{n} \mathcal{N}(z_k|\mathbf{x}, \Sigma_k). \tag{12.16}$$

Finding the maximum likelihood solution is equivalent to solving the weighted least-square problem:

$$\min_{\mathbf{x}} [\mathbf{z} - \mathbf{f}(\mathbf{x})]^{\mathrm{T}}\mathbf{W}[\mathbf{z} - \mathbf{f}(\mathbf{x})]. \tag{12.17}$$

The matrix $\mathbf{W}$ is a diagonal positive definite matrix whose elements are the measurement weights, equal to the inverses of measurement variances. The solution of the linear weighted least-square problem corresponding to DC state estimation can be obtained as follows:

$$\hat{\mathbf{x}} = (\mathbf{H}^{\mathrm{T}}\mathbf{W}\mathbf{H})^{-1}\mathbf{H}^{\mathrm{T}}\mathbf{W}\mathbf{z}, \tag{12.18}$$

where the matrix $\mathbf{H}$ defines the linear functions $\mathbf{f}(\mathbf{x})$.

Further, the non-linear weighted least-square problem corresponding to the AC state estimation can be solved using Gauss–Newton method. The non-linear weighted least-square problems are usually non-convex and may have several local minima. Therefore, the solution of the non-linear weighted least-square problem strongly depends of a priori knowledge of an initial point [96]. The Gauss–Newton iterative scheme is given as follows:

$$\begin{aligned} &\mathbf{x}^{\nu+1} = \mathbf{x}^{\nu} + \Delta\mathbf{x}^{\nu}, \\ &\mathbf{H}(\mathbf{x}^{\nu})^{\mathrm{T}}\mathbf{W}\mathbf{H}(\mathbf{x}^{\nu})\Delta\mathbf{x}^{\nu} = \mathbf{H}(\mathbf{x}^{\nu})^{\mathrm{T}}\mathbf{W}\mathbf{r}(\mathbf{x}^{\nu}), \end{aligned} \tag{12.19}$$

where $\mathbf{r}(\mathbf{x}) = \mathbf{z} - \mathbf{f}(\mathbf{x})$ is the residual of measurements and $\mathbf{H}(\mathbf{x})$ is the Jacobian matrix.

### *12.5.3 Belief propagation algorithm in electric power distribution system*

The distribution grid of the electric power system has a high ratio between resistance and reactance. Therefore, decoupling between the active power and voltage phase, and reactive power and voltage magnitude is not allowed. Further, the quality of the state estimator depends the selection of initial states of state variables, while in distribution network, these initial states might be difficult to obtain. In the conventional state estimation, the gain matrix $\mathbf{G} = \mathbf{H}^{\mathrm{T}}\mathbf{W}\mathbf{H}$ can become almost singular (ill-conditioned) causing numerical instability of the algorithm. These are just some of the problems that may arise in conventional state estimation applied to distribution network.

A probabilistic model based on probabilistic graphical modelling, in particular, the factor graphs, possess a potential to bypass many problems of conventional state estimation. The algorithm for exact inference on probabilistic graphical models without loops is known as belief propagation algorithm [97–99]. Using the belief propagation algorithm, it is possible to efficiently calculate marginal distributions or mode of the joint distribution of the system of random variables which is decomposable, i.e., can be factored, into a product of local functions. Belief propagation algorithm can be also applied to graphical models with loops (loopy belief propagation), although in that case, the solution is not guaranteed to converge to correct marginals/modes of the joint distribution. For the purposes of solving inference problems, it is often convenient to convert both directed and undirected graphs into a different representation called a factor graph.

The input to the belief propagation algorithm is a factorised joint distribution. In the context of state estimation, that is likelihood function which is formed as the product of individual probability density function of each measurement [100]:

$$\mathcal{L}(\mathbf{z}|\mathbf{x}) = \prod_{k=1}^{n} \mathcal{N}(z_k|\mathbf{x}, \Sigma_k). \tag{12.20}$$

This equation can be visualised as a factor graph. The factor graph contains variable nodes for each variable and function nodes for each local factorisation term in the joint probability distribution:

$$\mathcal{L}(\mathbf{z}|\mathbf{x}) = \prod_{k=1}^{n} f_k(z_k, \mathbf{x}). \tag{12.21}$$

The sum–product belief propagation algorithm is applied to obtain exact marginals for each variable; the reader can find step-by-step sum–product algorithm on factor graphs in References 99,101,102.

In order to demonstrate the concept, one small radial feeder with metering devices is shown in Figure 12.12. The bus and branch data are given in Tables 12.2 and 12.3, representing the set of measurements generated using power-flow analysis, additionally corrupted by Gaussian white noise as described in Table 12.4.

The radial system according to available measurements can be easily translated into the factor graph, Figure 12.13. The factor node $f_1(x_1)$ represents the slack bus, each other factor node represents voltage magnitude and voltage phase at bus and branch active power measurements.

An initial model involves a discretisation of continuous state variables and computing the marginal distribution over the discrete states according to the values of

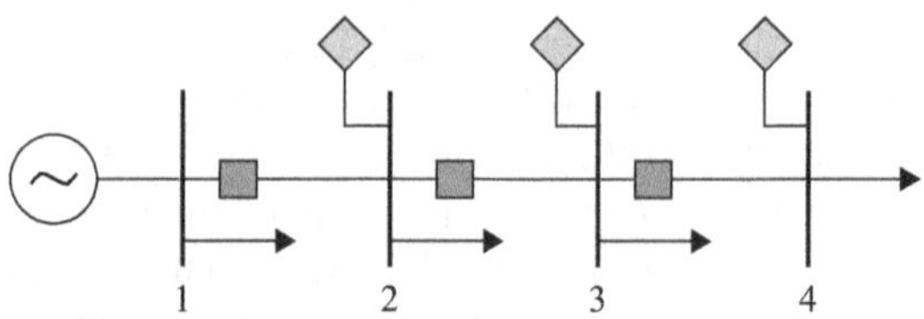


*Figure 12.12 Radial feeder*

*Table 12.2 The bus data*

| Bus no. $i$ | Bus type | Bus voltage | | Load | |
|---|---|---|---|---|---|
| | | $V_i$ (pu) | $\theta_i$ (rad) | $P_{li}$ (pu) | $Q_{li}$ (pu) |
| 1 | Slack | 1.00 | 0 | 0.15 | 0.20 |
| 2 | Demand | 1.00 | 0 | 0.67 | 1.10 |
| 3 | Demand | 1.00 | 0 | 10.45 | 7.50 |
| 4 | Demand | 1.00 | 0 | 5.30 | 3.41 |

*Table 12.3 The branch data*

| From bus $i$ | To bus $j$ | Resistance $r_{ij}$ (pu) | Reactance $x_{ij}$ (pu) | Charging susceptance $y_i, y_j$ (pu) |
|---|---|---|---|---|
| 1 | 2 | 0.0016981 | 0.0052976 | 0 |
| 2 | 3 | 0.0003016 | 0.0060302 | 0 |
| 3 | 4 | 0.0015998 | 0.0003016 | 0 |

*Table 12.4 The measurement data*

| Type | Unit | Measurement | Standard deviation |
|---|---|---|---|
| $P_{12}$ | pu | 18.0769 | 0.3000 |
| $P_{23}$ | pu | 16.6015 | 0.3000 |
| $P_{34}$ | pu | 4.7338 | 0.3000 |
| $V_2$ | pu | 0.8075 | 0.2000 |
| $V_3$ | pu | 0.8265 | 0.2000 |
| $V_4$ | pu | 0.5977 | 0.2000 |
| $\theta_2$ | rad | −0.0751 | 0.0035 |
| $\theta_3$ | rad | −0.2139 | 0.0035 |
| $\theta_4$ | rad | −0.2063 | 0.0035 |

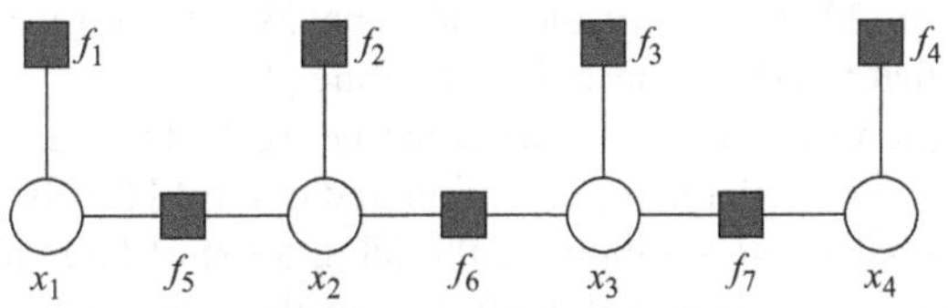


*Figure 12.13 Factor graph*

*Table 12.5 The simulation results*

| Bus no. $i$ | Power flow | | State estimation | | Belief propagation | |
|---|---|---|---|---|---|---|
| | $V_i$ (pu) | $\theta_i$ (∘) | $V_i$ (pu) | $\theta_i$ (∘) | $V_i$ (pu) | $\theta_i$ (∘) |
| 1 | 1.0000 | 0.0000 | 1.0192 | 0.0000 | 1.0000 | 0.0000 |
| 2 | 0.8682 | −4.0757 | 0.9116 | −4.2980 | 0.8800 | −4.3100 |
| 3 | 0.7677 | −12.0399 | 0.7253 | −12.2563 | 0.8100 | −12.1600 |
| 4 | 0.7551 | −11.6588 | 0.7134 | −11.8204 | 0.8000 | −11.9200 |

measurements. This approach results in a non-iterative AC estimator that avoids computational complexity of matrix operations (such as forming the Jacobian matrix and factorisation); however, the quality of estimator strongly depends of discretisation, what is a further research problem.

The results from AC power flow will be used to compare two different approaches, conventional state estimation and belief propagation estimation.

From the results presented in Table 12.5, the belief propagation estimator is obviously a good competitor to conventional estimator. It is very important to emphasise that the results are obtained under a discretisation scheme which is not optimised, still the algorithm is able to find accurate solution. The promising results we obtained across several distribution network topologies motivate us to pursue further investigation of applications of belief propagation algorithm in both discretised and continuous model of state variables in state estimation.

## 12.6 Research and design of novel control algorithms needed for the effective integration of distributed generators

### *12.6.1 Overview*

Variable energy resource (VER) generation such as wind and solar are widely spread in many power systems all over the world. The level of VER penetration is so significant for many countries that it can be accounted over 25% of the installed capacity in some cases [103]. Moreover, their penetration is expected to be still increasing by 3.2%/year on average due to technological advances and favourable policy incentives [104]. Lot of VER penetration are located into distribution networks as distributed generation (DG). In 2012, 39% of the total capacity addition was invested in DG installations, and that trend is expected to continue [105].

In the integration of DGs into distribution networks there are many challenges from the point of view of planning, operation and control [106,107]. The microgrid concept aims to simplify the solution for the integration of large penetration of DG by decentralising the operation and control from the main grid to the microgrids. A microgrid can be defined as a cluster of DGs, storage systems and loads (lately loads tend to be controllable). The operation of a microgrid depends not only on the type of technologies of DGs, storage systems and loads, but also on the condition of the microgrid whether it is connected or not to the host grid.

The microgrid can be operated in two modes: grid connected and stand alone. So far, when a microgrid is disconnected from the host grid (due to faults, maintenances, etc.), all DGs within the microgrid are disconnected due to the inexistence of a proper frequency control that can handle generation and consumption balance during autonomous operation. Thus, renewable generation is wasted. During grid- connected mode, it is desired to control the active and reactive power flow between the microgrid and the host grid in order to improve the energy management in the whole system. However, the transition between the two modes must be handled properly in both directions: connection or disconnection from the host grid. Therefore, implementing control strategies for microgrids will support the integration of DG into the grid, exploit the production of renewable energy and make more efficient use of electricity.

This project seeks to solve problems related to the control and power management for DGs within a microgrid in both operational modes by proposing a novel control scheme for frequency/voltage regulation. In order to achieve so a hierarchical control structure is adopted as a platform for solving those problems. Then, the next subsection presents an overview of different levels of the control hierarchy, and finally, some conclusions are drawn in the last subsection.

## *12.6.2 Hierarchical control of a microgrid*

The control structure adopted in this work is based on hierarchical approach which consists of four levels, namely zero-level, primary, secondary and tertiary [108–112]. Primary control is responsible to stabilise frequency and voltage and control the distribution of active and reactive power among DGs. Secondary control deals with the restoration of frequency and voltage to their nominal values. At last, tertiary control ensures optimal operation in both operating modes.

### 12.6.2.1 Zero-level control

DGs are usually connected to the grid thought power electronic interfaces that include DC/AC converters. These converters can be operated in two modes: current source inverter (CSI) or voltage source inverter (VSI).

In CSI mode, the inverter supplies a given active and reactive power reference points. Active power set point is subject to power availability from the primary resource (wind, the Sun, etc.), while reactive power is predefined either locally or through a central control [113]. In VSI mode, the inverter seeks to control predefined frequency and voltage, which is the reason it is usually connected to a storage systems. When a VSI is connected to the host grid, where frequency and voltage are fixed, VSI can supply desired active and reactive power by adjusting both grid angular frequency and voltage magnitude references through active and reactive gains [114]. Nevertheless, it is usual to have a converter in CSI mode when it is in grid-connected mode, while VSI is more needed in stand-alone mode to keep the frequency and voltage in a microgrid [110].

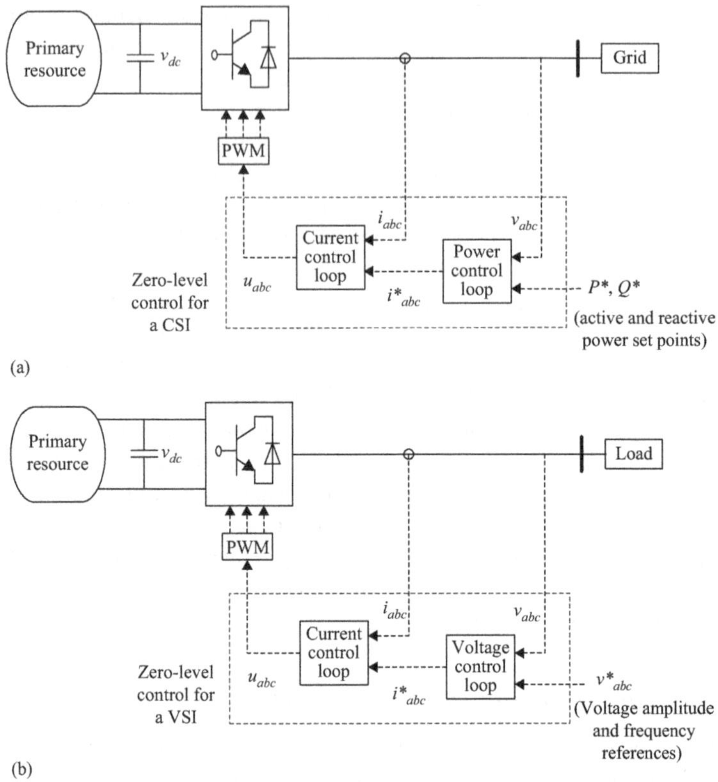


*Figure 12.14 Basic control structure of power converters. (a) CSI and (b) VSI*

Both modes of operation are based on inner control loops, also referred to as zero-level control. Figure 12.14 shows the control structure of a converter for (a) CSI and (b) VSI, when they are operating in grid-connected mode and stand-alone mode, respectively.

#### 12.6.2.2 Primary control

In order to avoid circulating current when more than one VSI are operating in parallel two strategies have been adopted [115]:

*A. Active load sharing*

This method was first thought for paralleling Uninterruptible Power Supply (UPS) units that operate close to each other [13]. The strategy consists of making the output current of each converter to be equal or at least proportional to its nominal power rating. The current reference point is determined through different approaches such as centralised [116,117], master–slave [114] and average load sharing [118,119]. In a microgrid, where its elements might be separated by several kilometres, using this method might require large communication infrastructure through all the grids which

*Table 12.6 Typical line impedances values, following [113]*

| Voltage level | $R$ (Ω/km) | $X$ (Ω/km) | $R/X$ (pu) |
|---|---|---|---|
| High voltage | 0.06 | 0.191 | 0.31 |
| Medium voltage | 0.161 | 0.190 | 0.85 |
| Low voltage | 0.642 | 0.083 | 7.7 |

make this application unfeasible for some cases. To overcome this problem, the droop characteristic method was proposed.

*B. Droop characteristic*

If the droop control method is based on local measurements, then it does not require communication infrastructure between DGs. The idea of this method is to control the power distribution among different DGs within a microgrid by emulating the droop control characteristic of a synchronous generator. The grid impedance influences on the droop control method by its inductive and resistive components. High-voltage networks are inductive, medium voltage is a mix between inductive and resistive, while low voltage are mainly resistive, as Table 12.6 shows.

First, if the resistive component is neglected and only inductive is taken into account, frequency decreases with the increase of output active power, and voltage amplitude decreases with the increase of the reactive output power. This principle can be integrated in a VSI by using the following mathematical formulation [111]:

$$\begin{aligned} f &= f^* - k_P(P - P^*), \\ E &= E^* - k_Q(Q - Q^*), \end{aligned} \tag{12.22}$$

where $f$ and $E$ are the frequency and voltage amplitude of the VSI's output, $f^*$ and $E^*$ are their references, $P$ and $Q$ are active and reactive power delivered, $P^*$ and $Q^*$ are their references, and $k_P$ and $k_Q$ are the droop coefficients which are based on the converter power rating and maximum voltage and frequency deviation. Nevertheless, these coefficients can also be optimally designed by using some heuristic techniques such as particle swarm optimisation or genetic algorithms [112].

$$\begin{aligned} k_P &= \Delta f / P_{max}, \\ k_Q &= \Delta V / 2Q_{max}. \end{aligned} \tag{12.23}$$

On the contrary, if only resistive component is taken into account, the frequency decreases with the increase of the output reactive power, and the voltage amplitude decreases with the increase of the active output power. This can be expressed as follows:

$$\begin{aligned} E &= E^* - k_P(P - P^*), \\ f &= f^* + k_Q(Q - Q^*). \end{aligned} \tag{12.24}$$

In the general case, that is taken into account both resistive and inductive components, the droop control can be modified by using park transformation [111], then it can be written as follows:

$$\begin{aligned} f &= f^* - k_P X/Z(P - P^*) + k_Q R/Z(Q - Q^*), \\ E &= E^* - k_P R/Z(P - P^*) - k_Q X/Z(Q - Q^*), \end{aligned} \tag{12.25}$$

where $R$ and $X$ are the resistive and reactive component of the Thevenin equivalent impedance from the VSI's eyes.

As it has seen above, conventional $f$–$P$ and $v$–$Q$ droop control strategies cannot be applied to medium-low voltage level, where resistive component cannot be neglected. Potential solutions have been discussed in the literature to overcome this issue. The virtual power method consists of an orthogonal linear transformation matrix which transfer active and reactive power ($P$,$Q$) to a new reference frame ($P'$,$Q'$) where resistive and inductive components are decoupled [120,121]. A similar method was presented using frequency and voltage frame transformation [122,123]. Overviews of some of the aforementioned methods are presented in Reference 108,112. However, an outstanding solution is the introduction of the called "virtual impedance" within the primary control [124,125]. The virtual impedance's function is to regulate power sharing among different VSI by modifying their voltage references as the next equation shows [111]:

$$v^*_{new,ref} = v^*_{ref} - Z_v i_{grid}. \tag{12.26}$$

It is important to point out that the value of $Z_v$ should be larger than the actual line impedance so that its effect ensures inductive behaviour within the inner control loops [113]. The implementation of virtual impedance within the primary control for a VSI is shown in Figure 12.15.

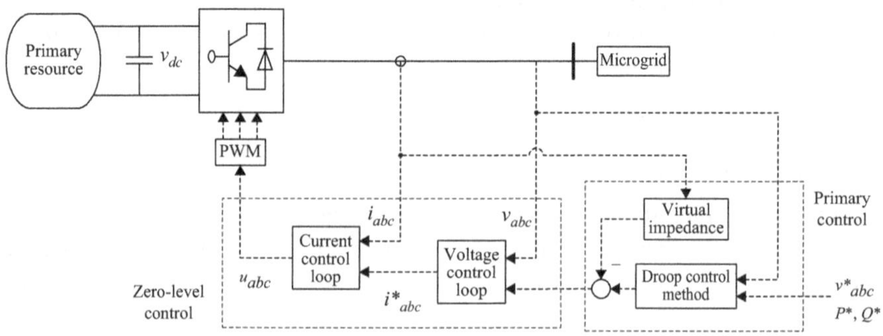


*Figure 12.15 Block diagram of the virtual output impedance loop within the primary control in the VSI*

#### 12.6.2.3 Secondary control

The secondary control is a centralised control related to each microgrid. This control level is expected to restore frequency and voltage to their nominal values subsequently primary control was happening, thus, the dynamic response of secondary control is slower than primary. The structure for secondary control has been proposed in Reference 111. Figure 12.16 shows the block diagram for a microgrid composed by one VSI and multiples CSI, all they connected though the microgrid network (in the stand-alone operation of a microgrid), there must be at least one VSI so that frequency can be set [113].

It is seen in Figure 12.16 that frequency and voltage output of VSI unit are compared with the corresponding references values, $w^*$ and $E^*$, respectively (in grid-connected mode these references are set by the main grid). $G_w$ and $G_e$ represent Proportional, Integral (PI) controllers which will process their input signals as follows [111]:

$$\Delta w = k_{P_w}(w^* - w) + k_{Iw} \int (w^* - w)dt + \Delta w_s, \tag{12.27}$$

$$\Delta E = k_{PE}(E^* - E) + k_{IE} \int (E^* - E)dt. \tag{12.28}$$

The coefficients $k_{P_w}$, $k_{Iw}$, $k_{PE}$ and $k_{IE}$ are the controller's parameters and, $\Delta w_s$ is a synchronisation term that is zero during stand-alone operation, while during

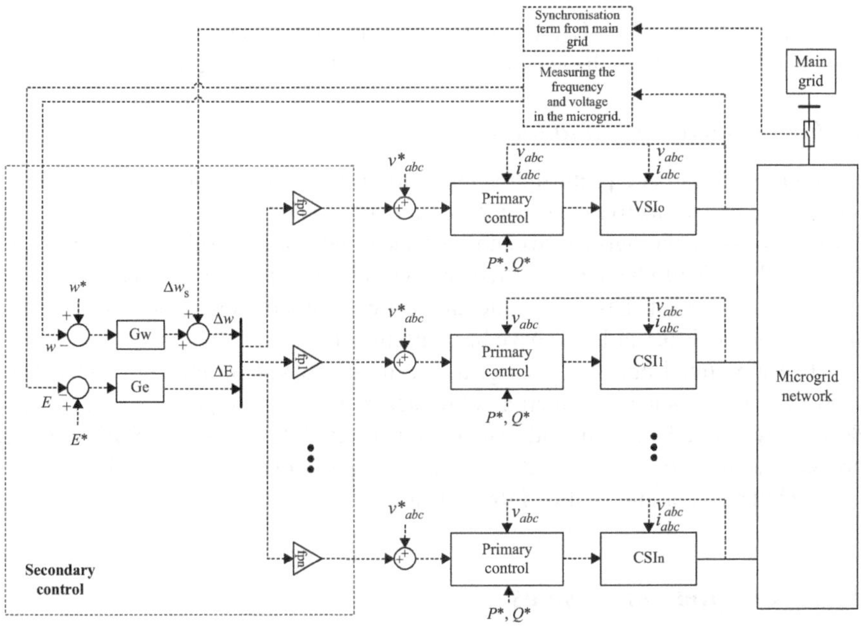


*Figure 12.16 Block diagram of the hierarchical control of primary and secondary controls of an AC MG*

synchronisation to the main grid it provides an additional input to the secondary control in order to facilitate the synchronisation process. The introduction of participation factors $(pf_0, pf_1, \ldots, pf_n)$ will allow desired distribution of power among the DGs.

#### 12.6.2.4 Tertiary control

Tertiary control is responsible for the economical and reliable operation of the microgrid, it is also referred as EMS, which consists of finding the optimal generation schedule for available DG units for both operating modes.

In grid-connected mode the concern lay on maximising revenues by finding optimal set points of DGs according to the requirement of the main grid, bids and market prices [24]. Normally, wind and solar generation power within the microgrid are totally used, whereas mismatches within the microgrid are supplied by the main grid. Some works have proposed UC for a microgrid operating in stand-alone mode, where stochastic techniques were used to deal with intermittent nature from non-dispatchable generators [126–128].

As seen in Figure 12.17, when the microgrid is in grid-connected mode, the power is controlled by adjusting the frequency and voltage amplitude, inputs of the secondary control, through PI controller based on the following expressions:

$$w^* = k_{PP}(P_{ref} - P) + k_{IP} \int (P_{ref} - P)dt, \tag{12.29}$$

$$E^* = k_{PQ}(Q_{ref} - Q) + k_{IQ} \int (Q_{ref} - Q)dt. \tag{12.30}$$

The coefficients $k_{Pw}$, $k_{Iw}$, $k_{PE}$ and $k_{IE}$ are the controller's parameters. Whereas, in stand-alone operation the generation dispatch is the result of the UC process.

### *12.6.3 Conclusions and outlook*

Microgrids appear as a promising solution towards the integration of DG based on VERs. A reliable/effective control and power management are required in both operational modes of a microgrid: grid connected and stand alone; as well as in the transition between them. This chapter presented an overview of different aspects related to the control of microgrids based on a hierarchical control structure, where the role and function of the different levels have been discussed.

It was pointed out that microgrids are mostly based on medium/low voltage networks, where generation inertia is low, active and reactive power control are not decoupled, and unbalanced loads cannot be neglected. In that sense, further studies are required to be done taking into account these particularities that involve not only control but also communication infrastructure.

## 12.7 Chapter conclusions

In this chapter, we have discussed several important research topics in relation to the smart grid that have arisen from the European ADVANTAGE project. The chapter

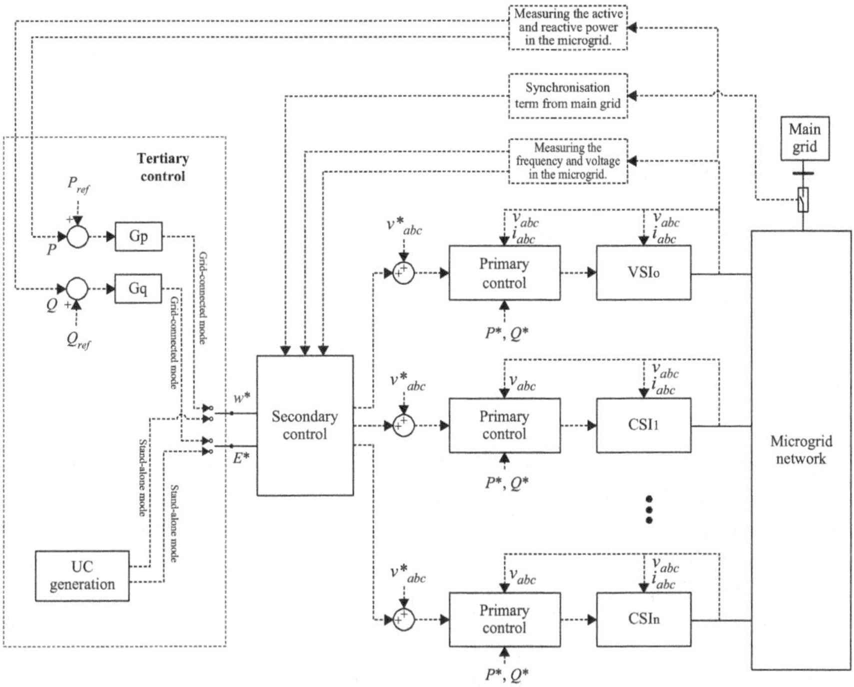


*Figure 12.17 The overview for the hierarchical control schemes for a microgrid in the two operation modes*

began by discussing D2D communications technologies, which offer a simple and energy efficient approach to forming neighbourhood communication networks to exchange smart grid data. A number of research challenges have also been identified on radio spectrum resource management for these networks. The chapter then moved on to discuss Power Talk, which is a novel communications approach exploiting the properties of the mains voltage, particularly for communication between different entities in local power grid networks and small scale microgrids. The performance of a simple two-node PowerTalk system was analysed considering both voltage and resistance changes as a means to communicate information. The research outlook and open problems for this promising communications technology have also been explored. The next topic to be discussed concerns data compression techniques to reduce the volume of smart grid data arising from smart meters. Two different compression algorithms have been tested using real data sets and their performance compared. The AH approach is simpler to implement, but better performance is achieved using the LZW method for larger data packet sizes at the cost of extra complexity. The fourth subject concerns monitoring of wide area power networks, where the electrical properties of different parts of the power network are estimated. A novel approach using the concept of belief propagation is evaluated for this purpose and is shown to achieve promising results. Finally, control methods for ensuring the stability of power grids are considered. Different levels of control are described with particular application to microgrids. It is hoped that these topics will give the reader a clear impression of research progress to date within the ADVANTAGE project.

## Acknowledgements

This project has received funding from the European Union's Seventh Framework Programme for research, technological development and demonstration under grant agreement no. 607774.

## Bibliography

[1] Project ADVANTAGE Website. Available from http://www.fp7-advantage.eu, 2016. Accessed on 18-July-2016.

[2] V. Gungor, D. Sahin, T. Kocak, *et al.*, "Smart grid technologies: Communication technologies and standards," *Industrial Informatics, IEEE Transactions on*, vol. 7, pp. 529–539, Nov. 2011.

[3] W. Meng, R. Ma, and H.-H. Chen, "Smart grid neighbourhood area networks: A survey," *Network, IEEE*, vol. 28, pp. 24–32, Jan. 2014.

[4] Q.-D. Ho, Y. Gao, and T. Le-Ngoc, "Challenges and research opportunities in wireless communication networks for smart grid," *Wireless Communications, IEEE*, vol. 20, pp. 89–95, Jun. 2013.

[5] S. Sesia, I. Toufik, and M. Baker, *LTE: The UMTS Long Term Evolution from Theory to Practice*. John Wiley & Sons Ltd., New York, NY, 2009.

[6] P. Cheng, L. Wang, B. Zhen, and S. Wang, "Feasibility study of applying LTE to smart grid," in *Smart Grid Modelling and Simulation (SGMS), 2011 IEEE First International Workshop on*, Brussels, Belgium, pp. 108–113, Oct. 2011.

[7] Ericsson White Paper, "LTE for utilities, supporting smart grids," Sep. 2013.

[8] P. Kansal and A. Bose, "Bandwidth and latency requirements for smart transmission grid applications," *Smart Grid, IEEE Transactions on*, vol. 3, pp. 1344–1352, Sep. 2012.

[9] A. Laya, L. Alonso, and J. Alonso-Zarate, "Is the random access channel of LTE and LTE-A suitable for M2M communications? A survey of alternatives," *Communications Surveys Tutorials, IEEE*, vol. 16, pp. 4–16, First 2014.

[10] IEC, "IEC 61850-90-1: Communication networks and systems for power utility automation – Part 90-1: Use of IEC 61850 for the communication between substations," 2010.

[11] V. Gungor, D. Sahin, T. Kocak, *et al.*, "A survey on smart grid potential applications and communication requirements," *Industrial Informatics, IEEE Transactions on*, vol. 9, pp. 28–42, Feb. 2013.

[12] C. Kalalas, "Enabling LTE for control system applications in a smart grid context," KTH Royal Institute of Technology, Stockholm, Sweden, Aug. 2014.

[13] 3GPP TS 22.368 v12.1.0, "Service requirements for machine-type communications (MTC) stage 1," Release 12, Dec. 2012.

[14] P. Phunchongharn, E. Hossain, and D. Kim, "Resource allocation for device-to-device communications underlying LTE-advanced networks," *Wireless Communications, IEEE*, vol. 20, pp. 91–100, Aug. 2013.

[15] G. Fodor, E. Dahlman, G. Mildh, *et al.*, "Design aspects of network assisted device-to-device communications," *Communications Magazine, IEEE*, vol. 50, pp. 170–177, Mar. 2012.

[16] A. Timbus, M. Larsson, and C. Yuen, "Active management of distributed energy resources using standardized communications and modern information technologies," *Industrial Electronics, IEEE Transactions on*, vol. 56, pp. 4029–4037, Oct. 2009.

[17] Q. Yang, J. Barria, and T. Green, "Communication infrastructures for distributed control of power distribution networks," *Industrial Informatics, IEEE Transactions on*, vol. 7, pp. 316–327, May 2011.

[18] M. Eriksson, M. Armendariz, O. Vasilenko, A. Saleem, and L. Nordstrom, "Multiagent-based distribution automation solution for self-healing grids," *Industrial Electronics, IEEE Transactions on*, vol. 62, pp. 2620–2628, Apr. 2015.

[19] N. Cheng, N. Lu, N. Zhang, T. Yang, X. Shen, and J. Mark, "Vehicle-assisted device-to-device data delivery for smart grid," in *IEEE Transactions on Vehicular Technology*, vol. 65, no. 4, pp. 2325–2340, April 2016.

[20] B. Kaufman and B. Aazhang, "Cellular networks with an overlaid device to device network," in *Signals, Systems and Computers, 2008 42nd Asilomar Conference on*, Pacific Grove, USA, pp. 1537–1541, Oct. 2008.

[21] T. Peng, Q. Lu, H. Wang, S. Xu, and W. Wang, "Interference avoidance mechanisms in the hybrid cellular and device-to-device systems," in *Personal, Indoor and Mobile Radio Communications, 2009 IEEE 20th International Symposium on*, Tokyo, Japan, pp. 617–621, Sep. 2009.

[22] K. Vanganuru, S. Ferrante, and G. Sternberg, "System capacity and coverage of a cellular network with D2D mobile relays," in *Military Communications Conference, 2012 – MILCOM 2012*, Orlando, USA, pp. 1–6, Oct. 2012.

[23] P. Janis, C. Yu, K. Doppler, *et al.*, "Device-to-device communication underlaying cellular communications systems," *International Journal of Communications, Network and System Sciences*, vol. 2, no. 3, pp. 169–178, 2009.

[24] N. Lee, X. Lin, J. Andrews, and R. Heath, "Power control for D2D underlaid cellular networks: Modelling, algorithms, and analysis," *Selected Areas in Communications, IEEE Journal on*, vol. 33, pp. 1–13, Jan. 2015.

[25] J. Han, Q. Cui, C. Yang, M. Valkama, and X. Tao, "Optimized power allocation and spectrum sharing in device to device underlaying cellular systems," in *Wireless Communications and Networking Conference (WCNC), 2014 IEEE*, Istanbul, Turkey, pp. 1332–1337, Apr. 2014.

[26] F. Wang, L. Song, Z. Han, Q. Zhao, and X. Wang, "Joint scheduling and resource allocation for device-to-device underlay communication," in

*Wireless Communications and Networking Conference (WCNC), 2013 IEEE*, Shanghai, China, pp. 134–139, Apr. 2013.

[27] H. Min, J. Lee, S. Park, and D. Hong, "Capacity enhancement using an interference limited area for device-to-device uplink underlaying cellular networks," *Wireless Communications, IEEE Transactions on*, vol. 10, pp. 3995–4000, Dec. 2011.

[28] X. Chen, L. Chen, M. Zeng, X. Zhang, and D. Yang, "Downlink resource allocation for device-to-device communication underlaying cellular networks," in *Personal Indoor and Mobile Radio Communications (PIMRC), 2012 IEEE 23rd International Symposium on*, Sydney, Australia, pp. 232–237, Sep. 2012.

[29] P. Janis, V. Koivunen, C. Ribeiro, J. Korhonen, K. Doppler, and K. Hugl, "Interference-aware resource allocation for device-to-device radio underlaying cellular networks," in *Vehicular Technology Conference, 2009. VTC Spring 2009. IEEE 69th*, Barcelona, Spain, pp. 1–5, Apr. 2009.

[30] M. Zulhasnine, C. Huang, and A. Srinivasan, "Efficient resource allocation for device-to-device communication underlaying LTE network," in *Wireless and Mobile Computing, Networking and Communications (WiMob), 2010 IEEE 6th International Conference on*, Niagara Falls, Canada, pp. 368–375, Oct. 2010.

[31] R. Zhang, X. Cheng, L. Yang, and B. Jiao, "Interference-aware graph based resource sharing for device-to-device communications underlaying cellular networks," in *Wireless Communications and Networking Conference (WCNC), 2013 IEEE*, Shanghai, China, pp. 140–145, Apr. 2013.

[32] M. Botsov, M. Klugel, W. Kellerer, and P. Fertl, "Location dependent resource allocation for mobile device-to-device communications," in *Wireless Communications and Networking Conference (WCNC), 2014 IEEE*, Istanbul, Turkey, pp. 1679–1684, Apr. 2014.

[33] Q. Ye, M. Al-Shalash, C. Caramanis, and J. Andrews, "Distributed resource allocation in device-to-device enhanced cellular networks," *Communications, IEEE Transactions on*, vol. 63, pp. 441–454, Feb. 2015.

[34] C. Xia, S. Xu, and K. S. Kwak, "Resource allocation for device-to-device communication in LTE-A network: A Stackelberg game approach," in *Vehicular Technology Conference (VTC Fall), 2014 IEEE 80th*, Vancouver, Canada, pp. 1–5, Sep. 2014.

[35] K. Zheng, F. Hu, W. Wang, W. Xiang, and M. Dohler, "Radio resource allocation in LTE-advanced cellular networks with M2M communications," *Communications Magazine, IEEE*, vol. 50, pp. 184–192, Jul. 2012.

[36] B. Guo, S. Sun, and Q. Gao, "Graph-based resource allocation for D2D communications underlying cellular networks in multiuser scenario," *International Journal of Antennas and Propagation*, vol. 2014, Article ID 783631, pp1-6, Aug. 2014.

[37] K. Doppler, M. Rinne, C. Wijting, C. Ribeiro, and K. Hugl, "Device-to-device communication as an underlay to LTE-advanced networks," *Communications Magazine, IEEE*, vol. 47, pp. 42–49, Dec. 2009.

[38] C. Xu, L. Song, Z. Han, D. Li, and B. Jiao, "Resource allocation using a reverse iterative combinatorial auction for device-to-device underlay cellular networks," in *Global Communications Conference (GLOBECOM), 2012 IEEE*, Anaheim, USA, pp. 4542–4547, Dec. 2012.

[39] C. Xu, L. Song, Z. Han, Q. Zhao, X. Wang, and B. Jiao, "Interference-aware resource allocation for device-to-device communications as an underlay using sequential second price auction," in *Communications (ICC), 2012 IEEE International Conference on*, Ottawa, Canada, pp. 445–449, Jun. 2012.

[40] X. Lin, J. Andrews, and A. Ghosh, "Spectrum sharing for device-to-device communication in cellular networks," *Wireless Communications, IEEE Transactions on*, vol. 13, pp. 6727–6740, Dec. 2014.

[41] Q. Ye, M. Al-Shalash, C. Caramanis, and J. Andrews, "Resource optimization in device-to-device cellular systems using time-frequency hopping," *Wireless Communications, IEEE Transactions on*, vol. 13, pp. 5467–5480, Oct. 2014.

[42] M.-H. Han, B.-G. Kim, and J.-W. Lee, "Subchannel and transmission mode scheduling for D2D communication in OFDMA networks," in *Vehicular Technology Conference (VTC Fall), 2012 IEEE*, Quebec City, Canada, pp. 1–5, Sep. 2012.

[43] L. Su, Y. Ji, P. Wang, and F. Liu, "Resource allocation using particle swarm optimization for D2D communication underlay of cellular networks," in *Wireless Communications and Networking Conference (WCNC), 2013 IEEE*, Shanghai, China, pp. 129–133, Apr. 2013.

[44] D. Feng, L. Lu, Y. Yuan-Wu, G. Li, G. Feng, and S. Li, "Optimal resource allocation for device-to-device communications in fading channels," in *Global Communications Conference (GLOBECOM), 2013 IEEE*, Atlanta, USA, pp. 3673–3678, Dec. 2013.

[45] L. B. Le, "Fair resource allocation for device-to-device communications in wireless cellular networks," in *Global Communications Conference (GLOBECOM), 2012 IEEE*, Anaheim, USA, pp. 5451–5456, Dec. 2012.

[46] X. Xiao, X. Tao, and J. Lu, "A QoS-aware power optimization scheme in OFDMA systems with integrated device-to-device (D2D) communications," in *Vehicular Technology Conference (VTC Fall), 2011 IEEE*, San Francisco, USA, pp. 1–5, Sep. 2011.

[47] M. Belleschi, G. Fodor, and A. Abrardo, "Performance analysis of a distributed resource allocation scheme for D2D communications," in *GLOBECOM Workshops (GC Wkshps), 2011 IEEE*, Houston, USA, pp. 358–362, Dec. 2011.

[48] Z. Chen and M. Kountouris, "Distributed SIR-aware opportunistic access control for D2D underlaid cellular networks," in *Global Communications Conference (GLOBECOM), 2014 IEEE*, Austin, USA, pp. 1540–1545, Dec. 2014.

[49] C. Lee, S.-M. Oh, and A.-S. Park, "Interference avoidance resource allocation for D2D communication based on graph-coloring," in *Information and Communication Technology Convergence (ICTC), 2014 International Conference on*, Busan, South Korea, pp. 895–896, Oct. 2014.

[50] S. Andreev, A. Larmo, M. Gerasimenko, *et al.*, "Efficient small data access for machine-type communications in LTE," in *Communications (ICC), 2013 IEEE International Conference on*, Budapest, Hungary, pp. 3569–3574, Jun. 2013.

[51] K. Zhou, N. Nikaein, R. Knopp, and C. Bonnet, "Contention based access for machine-type communications over LTE," in *Vehicular Technology Conference (VTC Spring), 2012 IEEE 75th*, Yokohama, Japan, pp. 1–5, May 2012.

[52] Q. Ye, M. Al-Shalash, C. Caramanis, and J. Andrews, "A tractable model for optimizing device-to-device communications in downlink cellular networks," in *Communications (ICC), 2014 IEEE International Conference on*, Sydney, Australia, pp. 2039–2044, Jun. 2014.

[53] X. Lin and J. Andrews, "Optimal spectrum partition and mode selection in device-to-device overlaid cellular networks," in *Global Communications Conference (GLOBECOM), 2013 IEEE*, Atlanta, USA, pp. 1837–1842, Dec. 2013.

[54] B. Cho, K. Koufos, and R. Jantti, "Spectrum allocation and mode selection for overlay D2D using carrier sensing threshold," in *Cognitive Radio Oriented Wireless Networks and Communications (CROWNCOM), 2014 9th International Conference on*, Oulu, Finland, pp. 26–31, Jun. 2014.

[55] B. Zhou, S. Ma, J. Xu, and Z. Li, "Group-wise channel sensing and resource pre-allocation for LTE D2D on ISM band," in *Wireless Communications and Networking Conference (WCNC), 2013 IEEE*, Shanghai, China, pp. 118–122, Apr. 2013.

[56] J. Bai, C. Liu, and A. Sabharwal, "Increasing cellular capacity using ISM band side-channels: A first study," in *Proceedings of the Fourth Workshop on All Things Cellular: Operations, Applications and Challenges*, AllThingsCellular'14 (New York, NY), pp. 9–14, ACM, 2014.

[57] J. Sachs, N. Beijar, P. Elmdahl, J. Melen, F. Militano, and P. Salmela, "Capillary networks – a smart way to get things connected," *Ericsson Review*, September 9, 2014.

[58] T.-S. Kim, R. Oh, S.-J. Lee, S.-H. Yoon, C.-H. Cho, and S.-W. Ryu, "Vertical handover between LTE and wireless LAN systems based on common resource management (CRRM) and generic link layer (GLL)," in *Proceedings of the Second International Conference on Interaction Sciences: Information Technology, Culture and Human*, ICIS'09 (New York, NY), pp. 1160–1166, ACM, 2009.

[59] L. Nithyanandan and I. Parthiban, "Seamless vertical handoff in heterogeneous networks using IMS technology," in *Communications and Signal Processing (ICCSP), 2012 International Conference on*, Chennai, India, pp. 32–35, Apr. 2012.

[60] A. Miyim, M. Ismail, R. Nordin, and G. Mahardhika, "Generic vertical handover prediction algorithm for 4G wireless networks," in *Space Science and Communication (IconSpace), 2013 IEEE International Conference on*, Melaka, Malaysia, pp. 307–312, Jul. 2013.

[61] S. J. Bae, M. Y. Chung, and J. So, "Handover triggering mechanism based on IEEE 802.21 in heterogeneous networks with LTE and WLAN," in *Information Networking (ICOIN), 2011 International Conference on*, Kuala Lumpur, Malaysia, pp. 399–403, Jan. 2011.

[62] J. M. Rodríguez Castillo, "Energy-efficient vertical handovers," KTH Royal Institute of Technology, Stockholm, Sweden, 2013.

[63] A. Laya, L. Alonso, P. Chatzimisios, and J. Alonso-Zarate, "Reliable machine-to-machine multicast services with multi-radio cooperative retransmissions," *Mobile Networks and Applications*, vol. 20, No. 6, pp. 734–744, Dec. 2015.

[64] W. Yoon and B. Jang, "Enhanced non-seamless offload for LTE and WLAN networks," *Communications Letters, IEEE*, vol. 17, pp. 1960–1963, Oct. 2013.

[65] F. Rebecchi, M. Dias de Amorim, and V. Conan, "Flooding data in a cell: Is cellular multicast better than device-to-device communications?" in *Proceedings of the Ninth ACM MobiCom Workshop on Challenged Networks*, CHANTS'14 (New York, NY), pp. 19–24, ACM, 2014.

[66] Y. Li, D. Jin, F. Gao, and L. Zeng, "Joint optimization for resource allocation and mode selection in device-to-device communication underlaying cellular networks," in *Communications (ICC), 2014 IEEE International Conference on*, Sydney, Australia, pp. 2245–2250, Jun. 2014.

[67] A. Gotsis, A. S. Lioumpas, and A. Alexiou, "Analytical modelling and performance evaluation of realistic time-controlled M2M scheduling over LTE cellular networks," *Transactions on Emerging Telecommunications Technologies*, vol. 24, no. 4, pp. 378–388, 2013.

[68] B. Zhou, H. Hu, S.-Q. Huang, and H.-H. Chen, "Intracluster device-to-device relay algorithm with optimal resource utilization," *Vehicular Technology, IEEE Transactions on*, vol. 62, pp. 2315–2326, Jun. 2013.

[69] R. Lasseter, "Microgrids," in *Power Engineering Society Winter Meeting, 2002. IEEE*, New York, USA, vol. 1, pp. 305–308, 2002.

[70] J. Guerrero, J. Vasquez, J. Matas, L. de Vicua, and M. Castilla, "Hierarchical control of droop-controlled AC and DC MicroGrids: A general approach toward standardization," *Industrial Electronics, IEEE Transactions on*, vol. 58, pp. 158–172, Jan 2011.

[71] J. Guerrero, M. Chandorkar, T. Lee, and P. Loh, "Advanced control architectures for intelligent microgrids – Part I: Decentralized and hierarchical control," *Industrial Electronics, IEEE Transactions on*, vol. 60, pp. 1254–1262, Apr. 2013.

[72] H. Liang, B. J. Choi, A. Abdrabou, W. Zhuang, and X. Shen, "Decentralized economic dispatch in microgrids via heterogeneous wireless networks," *Selected Areas in Communications, IEEE Journal on*, vol. 30, pp. 1061–1074, Jul. 2012.

[73] J. Schonberger, R. Duke, and S. Round, "DC-bus signalling: A distributed control strategy for a hybrid renewable nanogrid," *Industrial Electronics, IEEE Transactions on*, vol. 53, pp. 1453–1460, Oct. 2006.

[74] D. Chen, L. Xu, and L. Yao, "DC voltage variation based autonomous control of DC MicroGrids," *Power Delivery, IEEE Transactions on*, vol. 28, pp. 637–648, Apr. 2013.

[75] T. Vandoorn, B. Renders, L. Degroote, B. Meersman, and L. Vandevelde, "Active load control in islanded microgrids based on the grid voltage," *Smart Grid, IEEE Transactions on*, vol. 2, pp. 139–151, Mar. 2011.

[76] K. Sun, L. Zhang, Y. Xing, and J. Guerrero, "A distributed control strategy based on dc bus signaling for modular photovoltaic generation systems with battery energy storage," *Power Electronics, IEEE Transactions on*, vol. 26, pp. 3032–3045, Oct. 2011.

[77] M. Angjelichinoski, C. Stefanovic, P. Popovski, H. Liu, P. Loh, and F. Blaabjerg, "Power talk: How to modulate data over a dc micro grid bus using power electronics," *2015 IEEE Global Communications Conference (GLOBECOM)*, San Diego (USA), 2015, pp. 1–7.

[78] M. Angjelichinoski, C. Stefanovic, P. Popovski, and F. Blaabjerg, "Power talk in dc micro grids: Constellation design and error probability performance," *2015 IEEE International Conference on Smart Grid Communications (SmartGridComm)*, Miami (USA), 2015, pp. 689–694.

[79] T. Dragicevic, J. Guerrero, J. Vasquez, and D. Skrlec, "Supervisory control of an adaptive-droop regulated DC MicroGrid with battery management capability," *Power Electronics, IEEE Transactions on*, vol. 29, pp. 695–706, Feb. 2014.

[80] M. Angjelichinoski, C. Stefanovic, P. Popovski, H. Liu, P. Loh, and F. Blaabjerg, "Multiuser communication through power talk in DC Micro-Grids," in *IEEE Journal on Selected Areas in Communications*, vol. 34, no. 7, pp. 2006–2021, July 2016.

[81] Frost & Sullivan, "Smart meter market: Frost & Sullivan forecasts 109% growth in the UK," 2011. Available from http://tinyurl.com/3u9hhzr. Accessed on 18-July-2016.

[82] A. Unterweger and D. Engel, "Resumable load data compression in smart grids," *IEEE Transactions on Smart Grid*, vol. 6, pp. 919–929, Mar. 2015.

[83] M. Aiello and G. Pagani, "The smart grid's data generating potentials," in *Computer Science and Information Systems (FedCSIS), 2014 Federated Conference on*, Warsaw, Poland, pp. 9–16, Sep. 2014.

[84] T. Sheldon, "Web article describing data compression," 2001. Available from http://tinyurl.com/ze5fmwr. Accessed on 18-July-2016.

[85] M. Ringwelski, C. Renner, A. Reinhardt, A. Weigel, and V. Turau, "The hitchhiker's guide to choosing the compression algorithm for your smart meter data," in *IEEE International Energy Conference and Exhibition (ENERGYCON)*, Florence, Italy, pp. 935–940, Sep. 2012.

[86] S. Bhat, "Web article on LZW data compression," 2008. Available from http://tinyurl.com/gujevl2. Accessed on 18-July-2016.

[87] J. Ziv and A. Lempel, "A universal algorithm for sequential data compression," *Information Theory, IEEE Transactions on*, vol. 23, pp. 337–343, May 1977.

[88] T. Welch, "A technique for high-performance data compression," *Computer*, vol. 17, pp. 8–19, Jun. 1984.

[89] J. Low, "Web article on adaptive Huffman coding," 2008. Available from http://tinyurl.com/gr7g37p. Accessed on 18-July-2016.

[90] D. Huffman, "A method for the construction of minimum-redundancy codes," *Proceedings of the IRE*, vol. 40, pp. 1098–1101, Sep. 1952.

[91] J. S. Vitter, "Design and analysis of dynamic Huffman codes," *Journal of the Association for Computing Machinery*, vol. 34, pp. 825–845, Oct. 1987.

[92] J. Z. Kolter and M. J. Johnson, "REDD: A public data set for energy disaggregation research," in *In Proceedings of the SustKDD Workshop on Data Mining Applications in Sustainability*, San Diego, USA, pp. 1–6, 2011.

[93] A. Wood and B. Wollenberg, *Power Generation, Operation, and Control.* A Wiley-Interscience Publication, Wiley, New York, NY, 1996.

[94] A. Monticelli, *State Estimation in Electric Power Systems: A Generalized Approach*. Kluwer International Series in Engineering and Computer Science, Springer US, New York, NY, 1999.

[95] A. Abur and A. Expósito, *Power System State Estimation: Theory and Implementation*. Power Engineering (Willis), Taylor & Francis, Oxfordshire, 2004.

[96] P. C. Hansen, V. Pereyra, and G. Scherer, *Least Squares Data Fitting with Applications*. Johns Hopkins University Press, Baltimore, MD, 2013.

[97] Judea Pearl "Probabilistic Reasoning in Intelligent Systems: Networks of Plausible Inference (2nd ed.)," San Francisco (USA) Morgan Kaufmann. ISBN 1-55860-479-0, 1988.

[98] S. L. Lauritzen and D. J. Spiegelhalter, "Local Computations with Probabilities on Graphical Structures and Their Application to Expert Systems," *Journal of the Royal Statistical Society. Series B (Methodological)*, Vol. 50, No. 2, 1988, pp. 157–224, Blackwell Publishing for the Royal Statistical Society. [online] Available from: http://www.jstor.org/stable/2345762

[99] C. M. Bishop, *Pattern Recognition and Machine Learning*. Springer, Berlin, 2006.

[100] Y. Weng, R. Negi, and M. Ilic, "Graphical model for state estimation in electric power systems," in *Smart Grid Communications (SmartGridComm), 2013 IEEE International Conference on*, Vancouver, Canada, pp. 103–108, Oct. 2013.

[101] D. Barber, *Bayesian Reasoning and Machine Learning*. Cambridge University Press, New York, NY, 2012.

[102] F. Kschischang, B. Frey, and H.-A. Loeliger, "Factor graphs and the sum-product algorithm," *Information Theory, IEEE Transactions on*, vol. 47, pp. 498–519, Feb. 2001.

[103] "Global wind energy report," Technical report, Global Wind Energy Council, 2013. Available from http://www.gwec.net. Accessed on 18-July-2016.

[104] "Annual energy outlook," Technical report, Energy Information Administrational, 2014.

[105] B. Owens, "The rise of distributed power," Technical report, General Electric Company, 2014.

[106] E. J. Coster, J. M. Myrzik, B. Kruimer, and W. L. Kling, "Integration issues of distributed generation in distribution grids," *Proceedings of the IEEE*, vol. 99, no. 1, pp. 28–39, 2011.

[107] N. Roy and H. Pota, "Current status and issues of concern for the integration of distributed generation into electricity networks," in *IEEE Systems Journal*, vol. 9, no. 3, pp. 933–944, Sept. 2015.

[108] A. Bidram and A. Davoudi, "Hierarchical structure of microgrids control system," *Smart Grid, IEEE Transactions on*, vol. 3, no. 4, pp. 1963–1976, 2012.

[109] D. E. Olivares, A. Mehrizi-Sani, A. H. Etemadi, *et al.*, "Trends in microgrid control" in *IEEE Transactions on Smart Grid*, vol. 5, no. 4, pp. 1905–1919, July 2014.

[110] J. M. Guerrero, M. Chandorkar, T.-L. Lee, and P. C. Loh, "Advanced control architectures for intelligent microgrids, part I: Decentralized and hierarchical control," *IEEE Transactions on Industrial Electronics*, vol. 60, no. 4, pp. 1254–1262, 2013.

[111] J. M. Guerrero, J. C. Vasquez, J. Matas, L. G. De Vicuña, and M. Castilla, "Hierarchical control of droop-controlled AC and DC MicroGrids: A general approach toward standardization," *Industrial Electronics, IEEE Transactions on*, vol. 58, no. 1, pp. 158–172, 2011.

[112] N. Yang, D. Paire, F. Gao, and A. Miraoui, "Power management strategies for microgrid: A short review," in *Industry Applications Society Annual Meeting, 2013 IEEE*, pp. 1–9, IEEE, Piscataway, NJ, 2013.

[113] J. Rocabert, A. Luna, F. Blaabjerg, and P. Rodriguez, "Control of power converters in ac microgrids," *Power Electronics, IEEE Transactions on*, vol. 27, no. 11, pp. 4734–4749, 2012.

[114] J. Peas Lopes, C. Moreira, and A. Madureira, "Defining control strategies for microgrids islanded operation," *Power Systems, IEEE Transactions on*, vol. 21, no. 2, pp. 916–924, 2006.

[115] J. M. Guerrero, L. Hang, and J. Uceda, "Control of distributed uninterruptible power supply systems," *Industrial Electronics, IEEE Transactions on*, vol. 55, no. 8, pp. 2845–2859, 2008.

[116] N. Pogaku, M. Prodanovic, and T. C. Green, "Modelling, analysis and testing of autonomous operation of an inverter-based microgrid," *Power Electronics, IEEE Transactions on*, vol. 22, no. 2, pp. 613–625, 2007.

[117] T. Iwade, S. Komiyama, Y. Tanimura, M. Yamanaka, M. Sakane, and K. Hirachi, "A novel small-scale ups using a parallel redundant operation system," in *Telecommunications Energy Conference, 2003. INTELEC'03. The 25th International*, pp. 480–484, IEEE, Piscataway, NJ, 2003.

[118] C.-L. Chen, Y. Wang, J.-S. Lai, Y.-S. Lee, and D. Martin, "Design of parallel inverters for smooth mode transfer microgrid applications," *Power Electronics, IEEE Transactions on*, vol. 25, no. 1, pp. 6–15, 2010.

[119] M. Prodanovic and T. C. Green, "High-quality power generation through distributed control of a power park microgrid," *Industrial Electronics, IEEE Transactions on*, vol. 53, no. 5, pp. 1471–1482, 2006.

[120] Y. Li and Y. W. Li, "Power management of inverter interfaced autonomous microgrid based on virtual frequency-voltage frame," *Smart Grid, IEEE Transactions on*, vol. 2, no. 1, pp. 30–40, 2011.

[121] K. De Brabandere, B. Bolsens, J. Van den Keybus, A. Woyte, J. Driesen, and R. Belmans, "A voltage and frequency droop control method for parallel inverters," *Power Electronics, IEEE Transactions on*, vol. 22, no. 4, pp. 1107–1115, 2007.

[122] C. Lee, C. Chu, P. Cheng, *et al.*, "A new droop control method for the autonomous operation of distributed energy resource interface converters," *Power Electronics, IEEE Transactions on*, vol. 28, no. 4, pp. 1980–1993, 2013.

[123] Y. Li and Y. W. Li, "Virtual frequency-voltage frame control of inverter based low voltage microgrid," in *Electrical Power and Energy Conference (EPEC), 2009 IEEE*, pp. 1–6, IEEE, Piscataway, NJ, 2009.

[124] J. M. Guerrero, J. Matas, L. G. de Vicuña, M. Castilla, and J. Miret, "Decentralized control for parallel operation of distributed generation inverters using resistive output impedance," *Industrial Electronics, IEEE Transactions on*, vol. 54, no. 2, pp. 994–1004, 2007.

[125] J. He, Y. W. Li, J. M. Guerrero, F. Blaabjerg, and J. C. Vasquez, "An islanding microgrid power sharing approach using enhanced virtual impedance control scheme," *IEEE Transactions on Power Electronics*, vol. 28, no. 11, pp. 5272–5282, 2013.

[126] D. E. Olivares, C. A. Cañizares, and M. Kazerani, "A centralized energy management system for isolated microgrids," in *IEEE Transactions on Smart Grid*, vol. 5, no. 4, pp. 1864–1875, July 2014.

[127] A. H. Hajimiragha, M. Dadash Zadeh, and S. Moazeni, "Microgrids frequency control considerations within the framework of the optimal generation scheduling problem," in *IEEE Transactions on Smart Grid*, vol. 6, no. 2, pp. 534–547, March 2015.

[128] B. Zhao, Y. Shi, X. Dong, W. Luan, and J. Bornemann, "Short-term operation scheduling in renewable-powered microgrids: A duality-based approach," *Sustainable Energy, IEEE Transactions on*, vol. 5, no. 1, pp. 209–217, 2014.

## Chapter 13

# Big data analysis of power grid from random matrix theory

*Robert C. Qiu[1,2], Xing He[1], Lei Chu[1] and Qian Ai[1]*

Data will become a strategic resource and even prime driving force for future grid. Essentially, rather than massive data themselves, we are much more interested in the potential contained in the data. In other words, how to mine the value from the 4Vs data (data with features of volume, variety, velocity, and veracity) within tolerant resources (time, hardware, human, etc.) is the key challenge.

This chapter studies the methodology of applying big data analytics to power grids. First, the definition of big data and random matrix theories (RMTs), as well as related system mapping framework and data processing methods are introduced as foundations. Especially, some mathematical contents, such as random matrix models (RMMs), probability in high dimension, and linear eigenvalue statistics (LES), are discussed in detail. Then, a series of functions related to situation awareness (SA) of power grids, including early event detection (EED), fault diagnosis and location, correlation analysis, high-dimensional indicator system and its visualization (i.e., auxiliary 3D power-map), are developed as concrete applications.

In this way, a typical data-driven methodology, mainly based on RMT, is proposed to cognize power grids. Three main procedures are essential: (1) big data model—to build the RMMs with raw data; (2) big data analysis—to conduct high-dimensional analyses to construct the indicator system via statistical transformations; and (3) engineering interpretation—to visualize and interpret the statistical results to human beings. This methodology is a more precise and natural way to gain insight into the large-scale interconnected systems. Furthermore, the indicator system will build a new epistemology to reveal the physical systems; it will open a new ear for the SA.

[1]Department of Electrical Engineering, Research Center for Big Data Engineering Technology, State Energy Smart Grid Research and Development Center, Shanghai Jiaotong University, Shanghai 200240, China

[2]Department of Electrical and Computer Engineering, Tennessee Technological University, Cookeville, TN 38505, USA

## 13.1 Background for conduct SA in power grid with big data analytics

This section gives the basic descriptions about the smart grids, the big data analytics, and the SA, as well as their relationship.

### *13.1.1 Smart grid—an essential big data system with 4Vs data*

Data become a strategic resource for smart grids; as shown in Figure 13.1, data are readily accessible caused by developments of various technologies and devices [67]. Hence, data with features of volume, velocity, variety, and veracity (i.e., 4Vs data) [31], as well as the curse of dimensionality [45], are inevitably generated and daily aggregated in power systems. A smart grid, especially the one large in scale, is a complex big data system essentially [23,71]. For such a system, it is a big challenge to mine the value from the 4Vs data within tolerable resources.

Particularly, the "4Vs" are elaborated as follows:

- *Volume.* For a system with 1,000 phasor measurement units (PMUs), we obtain up to nearly 402 gigabytes of PMU data per day; for a conventional Wide Area Measurement System (WAMS) system with 1,000 sampled points, the volume of annual data is Petabytes (PB) level [54].
- *Velocity.* Massive data must be processed within a fraction of second to sever online decision-makings.
- *Variety.* The data are often derived from diverse departments and in various formats. Besides, in the view of data management, the requirements for the access frequency and the processing speeds, as well as the expected performances are always distinct.
- *Veracity.* Inevitably, for a massive data system, there exist bad data (e.g., the incomplete, the inaccurate, the unsynchronized, and the unavailable data). Challenges in classical data processing (often via a low-dimensional statistical method), such

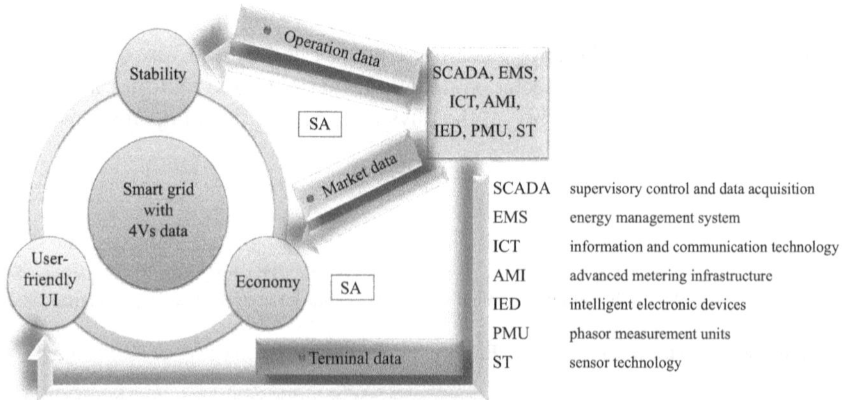


*Figure 13.1 Smart grid with 4Vs data and its SA*

as error accumulations and spurious correlations, will almost lead to bad results; however, for system operations, the final results should be highly reliable.

### *13.1.2 Smart grid and its stability, control, and SA*

The complexity of power systems is increasing, owing to following phenomena: (1) the evolution of the system network (particularly the expansion in size), (2) the penetration of renewable/distributed resources and controllable electronic components, and (3) the revolution of the operation mechanism. Meanwhile, the financial, the environmental, and the regulatory constrains are pushing the system towards its stability limits. The notions of *power system stability* and *power system control* are closely related; greater reliance is, therefore, being placed on the control strategy to enhance system security, to facilitate economic design, and to provide greater flexibility of system operation [38].

The SA is a key ingredient for the emergency control strategy; it detects and identifies anomaly patterns by continuously monitoring and processing 4Vs data [47] (Figure 13.1). In other words, SA aims to tell *signals* from *noises*—we treat trivial or even tolerable sample errors and irregular fluctuations of distributed generators and small loads as noises; whereas system faults, network reconfigurations, and dramatic loads/generators changes (often unintended) as signals.

### *13.1.3 Approach to SA—big data analytics and unsupervised learning mechanism*

The integration of *big data analytics* and *unsupervised learning mechanism* is an effective approach to SA [26]. *For the former*, big data analytics is a scientific trend of complex data processing [46,56]. It is a data-driven tool and aims to work out the statistical characteristics (especially correlations) indicated by statistical parameters (e.g., eigenvalues of matrix, LES) [54]. That means, it conducts high-dimensional data processing, rather than builds and analyzes physical models, to help understand and gain insight into the systems [52]. Big data analytics has already been successfully applied in numerous phenomena, such as quantum systems [8], financial systems [39], biological systems [29], and wireless communication networks [40,52,69]; we believe that it will also have a wide applied scope in power systems [32,36,54]. *For the latter*, the supervised learning methods are prevailing in data processing. The key parts are the inferred functions and empirical models; these functions/models, produced via an artificial training procedure (often in low dimensions), lead to a determinate parameter as the indicator [44]. However, for a complex system, it is hard to find a convincing training way to ensure the validity of this deterministic indicator; besides, it is impossible to create an ergodic event identification framework which is robust enough to manage all the scenarios. In contrast, the proposed unsupervised method, in a mathematical manner, utilizes the raw data (often derived from diverse departments of power system) in the form of RMMs to seek the statistical solution. The proposed approach is more suitable to SA of smart grids; this topic will be elaborated in Section 13.7.

### 13.1.4 RMM and probability in high dimension

RMM is selected as the most fundamental tool for mapping the power grids. RMM is a typical high-dimensional probabilistic tool which always takes the form of non-asymptotic probabilistic inequalities. It means that we are not concerned with limit theorems, but rather with explicit estimates that are either dimension free, or that capture precisely the dependence of the problem on the relevant dimensional parameters [59].

The explicit nature of non-asymptotic estimates makes RMM suit the grid very well. For a particular large-scale interconnected grid, there may exist several heterologous parameters of interest. In asymptotic results, one must take all these parameters to the limit in a fixed relation to one another (often based on assumptions and simplifications). However, the non-asymptotic viewpoint, deriving from the raw data directly, allows us to express the interrelation between these parameters in a much more flexible and precise way (even if the ultimate result of interest is asymptotic in nature).

## 13.2 Three general principles related to big data analytics

The nomenclature is given in Table 13.1.

All our high-dimensional theories are organized around three fundamental principles: (1) Concentration, (2) Suprema, and (3) Universality. Rather than corresponds to one particular theorem or admits a precise mathematical description, each principle

*Table 13.1 Some frequently used notations in the theories*

| Notations | Meanings |
|---|---|
| $\mathbf{X}, \mathbf{x}, x, x_{i,j}$ | a matrix, a vector, a single value, an entry of a matrix |
| $\hat{\mathbf{X}}, \hat{\mathbf{x}}, \hat{x}$ | hat: raw data |
| $\tilde{\mathbf{X}}, \tilde{\mathbf{x}}, \tilde{x}, \tilde{\mathbf{Z}}$ | tilde: intermediate variables, formed by normalization |
| $N, T, c$ | the numbers of rows and columns; $c=N/T$ |
| $\mathbb{C}^{N\times T}$ | $N\times T$-dimensional complex space |
| $\mathbf{X}_u$ | the singular value equivalent of $\tilde{\mathbf{X}}$ |
| $\mathbf{M}$ | another form of covariance matrix of $\mathbf{X}$: $\mathbf{M}=c\mathbf{S}$ |
| $\mathbf{Z}$,$L$ | $L$-independent matrices product: $\mathbf{Z}=\prod_{i=1}^{L}\mathbf{X}_{u,i}$ |
| $\lambda_{\mathbf{S}}, \lambda_{\tilde{\mathbf{Z}}}, \lambda_{\mathbf{M}}$ | the eigenvalue of matrix $\mathbf{S}$, $\tilde{\mathbf{Z}}$, $\mathbf{M}$ |
| $\lambda_{\mathbf{S},i}$ | the $i$th eigenvalue of matrix $\mathbf{S}$ |
| $r$ | the circle radius on the complex plane of eigenvalues |
| $\tau$ | linear eigenvalue statistics |
| $\tau_{\mathrm{MSR}}$ | mean value of radius for all eigenvalues of $\tilde{\mathbf{Z}}$: $\mu(\mathbf{r}_{\lambda_{\tilde{\mathbf{Z}}}})$ |
| $\varphi, \widehat{\varphi}$ | the test function and its Fourier transformation |
| $X$ | random variable |
| $\mathbb{E}(X)$, $\mathbb{D}(X)$ | expectation, variance for $X$ |
| $\mu(\mathbf{x})$, $\sigma^2(\mathbf{x})$ | mean, variance for $\mathbf{x}$ |
| $X^\circ$ | $X-\mathbb{E}(X)$ |
| $\kappa_i$ | $i$th cumulant of a random variable $X$ |
| $[\zeta(\theta)]_{\theta=\theta_2}^{\theta=\theta_1}$ | $\zeta(\theta_1)-\zeta(\theta_2)$ |

encompasses a family of conceptually related results that appear in different guises in different settings [59].

### *13.2.1 Concentration*

If $X_1, \ldots, X_n$ are independent identically distributed (i.i.d.) random variables, then

$$\frac{1}{n}\sum_{k=1}^{n} X_k - \mathbb{E}\left[\frac{1}{n}\sum_{k=1}^{n} X_k\right] \to 0 \quad \text{as} \quad n \to \infty \tag{13.1}$$

by the law of Large Numbers. In other words, if we define the function $f(x_1, \ldots, x_n) = \frac{1}{n}\sum_{k=1}^{n} x_k$, then for large $n$ the random variable $f(X_1, \ldots, X_n)$ is close to its expectation (i.e., its fluctuations are quite small). This principle can be stated informally as follows.

**Theorem 13.2.1.** *If $X_1, \ldots, X_n$ are independent (or weakly dependent) random variables, then the random variable $f(X_1, \ldots, X_n)$ is "close" to its expectation $\mathbb{E}[f(X_1, \ldots, X_n)]$ provided that the function $f(x_1, \ldots, x_n)$ is not too "sensitive" to any of the coordinates $x_i$.*

### *13.2.2 Suprema*

The suprema principle is concerned with the value of $\mathbb{E}[f(X_1, \ldots, X_n)]$ itself. It can be stated informally as follows.

**Theorem 13.2.2.** *If the random process $\{X_t\}_{t\in\mathbf{T}}$ is "sufficiently continuous", then the magnitude of the supremum $\sup_{t\in\mathbf{T}} X_t$ is controlled by the "complexity" of the index set* **T**.

### *13.2.3 Universality*

By the central limit theorem (CLT), we have a precise description of the distribution of the fluctuations, as

$$\frac{1}{\sqrt{n}}\sum_{k=1}^{n} \{X_k - \mathbb{E}X_k\} \approx \frac{1}{\sqrt{n}}\sum_{k=1}^{n} \{G_k - \mathbb{E}G_k\} \tag{13.2}$$

where $G_k$ are independent Gaussian random variables with the same mean and variance of $X_k$ (here $\approx$ denotes closeness of the distribution).

This principle can be stated informally as follows.

**Theorem 13.2.3.** *If $X_1, \ldots, X_n$ are independent (or weakly dependent) random variables, then the expectation $\mathbb{E}[f(X_1, \ldots, X_n)]$ is "insensitive" to the distribution of $X_1, \ldots, X_n$ when the function $f$ is "sufficiently smooth".*

Akin to CLTs, universality [52] refers to the phenomenon that the asymptotic distributions of various covariance matrices (such of eigenvalues and eigenvectors) are identical to those of Gaussian covariance matrices. These results let us calculate the exact asymptotic distributions of various test statistics without restrictive distributional assumptions of matrix entries.

The data of real systems can be viewed as a spatial and temporal sampling of the random graph. Randomness is introduced by the uncertainty of spatial locations and the system uncertainty. Under real-life applications, we cannot expect the matrix entries follow i.i.d. distribution. Numerous studies based on both simulations [23] and experiments, however, demonstrate that the Ring law and M-P law are universally followed. In such cases, universality properties provide a crucial tool to reduce the proofs of general results to those in a tractable special case—the i.i.d. case.

## 13.3 Fundamentals of random matrices

### 13.3.1 Types of matrices

In this section, we provide a brief review of the mathematical results that are corresponding to the analysis of RMM arising in power grid. We are inspired by problems of engineering interest. We start by providing the definitions for three classes of random matrices mostly related to the power grid: Gaussian, Wigner, and Wishart. These matrices help us build the RMM for the high-dimensional system.

**Definition 13.3.1.** *Let* $\mathbf{X}$ *be standard Gaussian* $M \times N$ *matrix. It has independent identical distribution* (i.i.d.) *zero mean Gaussian entries* $X_{ij}$ *with identical variance* $\sigma^2 = \frac{1}{M}$. *The probability density function (p.d.f.) of* $\mathbf{X}$ *is*

$$(\pi\sigma^2)^{-MN} \exp\left(-\frac{\mathrm{Tr}\left(\mathbf{X}\mathbf{X}^{\mathrm{H}}\right)}{\sigma^2}\right) \tag{13.3}$$

*where* $\mathrm{Tr}(\cdot)$ *is the trace operator.*

Gaussian matrices play an important role in RMT and attract enormous attention. Inspired by practical problems in power grid, we look into other two special Gaussian matrices.

**Definition 13.3.2.** *Let* $\mathbf{W}$ *be* $N \times N$ *Wigner matrix or so-called Gaussian unitary ensemble* GUE, *and* $\mathbf{W} = \{w_{ij}\}_{1 \le i,j \le N}$. $\mathbf{W}$ *satisfies:*

1. *The entries of* $\mathbf{W}$ *are* i.i.d *Gaussian variables.*
2. *For* $1 \le i \le j \le N$, $\mathrm{Re}(W_{ij})$ *and* $\mathrm{Re}(W_{ij})$ *are* i.i.d. *with distribution* $N(0, \frac{1}{2}\sigma^2)$.
3. *For any* $i,j$ *in* $\{1, 2, \ldots, N\}$, $W_{ij} = \bar{W}_{ji}$.
4. *The diagonal entries of* $\mathbf{W}$ *are real random variable with distribution* $N(0, \sigma^2)$.

For convenience, we denote GUE as $\mathbf{W} = \frac{1}{2}(\mathbf{X} + \mathbf{X}^H)$. Thus, the joint p.d.f. of ordered eigenvalues of GUE $(\lambda_1 \ge \lambda_2 \ge \cdots \ge \lambda_N)$ is [55,63]

$$2^{-N/2}\pi^{-N^2/2} \exp\left[-\frac{\mathrm{Tr}\mathbf{W}^2}{2}\right] \tag{13.4}$$

**Definition 13.3.3.** *Let* $\{X_{ij}\}_{1\le i\le M, 1\le j\le N}$ *be* i.i.d. *Gaussian random variables with* $\mathbb{E}(X_{ij}) = 0$ *and* $\mathbb{E}X_{ij}^2 = \frac{1}{2}(1+\delta_{ij})$. *The so-called Wishart matrix or Laguerre unitary ensemble* LUE *can be expressed as* $\mathbf{W} = \frac{1}{N}\mathbf{X}\mathbf{X}^{\mathrm{H}}$. *The* p.d.f. *of* $\mathbf{W}$ *for* $N \ge M$ *is [53–55]*

$$\frac{\pi^{-M(M-1)/2}}{\det \sum \prod_{i=1}^{M}(N-i)!}\exp\left[-\mathrm{Tr}\{\mathbf{W}\}\right]\det \mathbf{W}^{N-M} \tag{13.5}$$

We also conclude some non-asymptotic prosperities of LUE matrices.

**Lemma 13.3.4.** ([55,60]). *For a* $M \times N$ LUE *matrix* $\mathbf{W}$ *with* $N > M$,

$$\begin{aligned}
&\mathbb{E}[\mathrm{Tr}\{\mathbf{W}\}] = MN\\
&\mathbb{E}[\mathrm{Tr}\{\mathbf{W}^2\}] = MN(M+N)\\
&\mathbb{E}[\mathrm{Tr}^2\{\mathbf{W}\}] = MN(MN+1)\\
&\mathbb{E}[\mathrm{Tr}\{\mathbf{W}^{-1}\}] = \tfrac{M}{N-M}\\
&\mathbb{E}[\mathrm{Tr}\{\mathbf{W}^{-2}\}] = \tfrac{MN}{(N-M)^3-(N-M)}\\
&\mathbb{E}[\mathrm{Tr}^2\{\mathbf{W}^{-1}\}] = \tfrac{M}{N-M}\left(\tfrac{N}{(N-M)^2-1}+\tfrac{M-1}{N-M+1}\right)\\
&\mathbb{E}[\det\{\mathbf{W}^k\}] = \prod_{j=0}^{m-1}\tfrac{\Gamma(N-j+k)}{\Gamma(N-j)}
\end{aligned}$$

*where* $\Gamma(\cdot)$ *is Gamma function*

$$\Gamma(x) = \int_0^{\infty} t^{x-1}e^{-t}dt \tag{13.6}$$

## *13.3.2 Central limiting theorem*

Theorems for Wigner's semicircle law and Marchenko–Pastur law can be viewed as random matrix analogues of the law of Large Numbers from classical probability theory. Thus, a CLT for or fluctuations of LES is a natural second step to study the eigenvalue distribution of the random matrices. Here, we only give the result for a sample covariance matrix.

For each $N \ge 1$, let $\mathbf{A} = \frac{1}{N}\mathbf{X}^H\mathbf{X}$ be a real sample covariance matrix of size $N$, where $\mathbf{X} = \{X_{ij}\}_{1\leqslant i,j\leqslant N}$, and $\{X_{ij} : 1 \leqslant i,j \leqslant N\}$ is a collection of real independent random variables with zero mean and unit variance (i.e., $\mu = 0, \delta^2 = 1$). The eigenvalues are ordered as $\lambda_1(\mathbf{A}), \lambda_2(\mathbf{A}), \ldots, \lambda_N(\mathbf{A})$. The test function $f$ from the space $\mathcal{H}_s$ has the norm

$$\|f\|_s^2 = \int (1+2|\omega|)^{2s}|F(\omega)|^2 d\omega$$

for some $s > 3/2$, where $F(\omega)$ is the Fourier transform defined by

$$F(\omega) = \frac{1}{\sqrt{2\pi}}\int e^{j\omega t}f(t)dt \tag{13.7}$$

We note that if $f$ is a real-valued function with $f \in \mathcal{H}_s$, the both $f$ and its derivative $f'$ are continuous and bounded almost everywhere [30]. It means $f$ is Lipschitz.

Suppose that $\mathbb{E}[X_{ij}^4] = m_4$ for all $1 \le i,j \le N$ $(N \ge 1)$. We assume there exists $\varepsilon > 0$ such that

$$\sup_{N \geqslant 1} \sup_{1 \leqslant i,j \leqslant N} \mathbb{E}|X_{ij}|^{4+\varepsilon} < \infty$$

Let $f$ be a real-valued function with $\|f\|_s < \infty$ for some $s > 3/2$. Then

$$\sum_{i=1}^{N} f(\lambda_i(\mathbf{A})) - \mathbb{E}\sum_{i=1}^{N} f(\lambda_i(\mathbf{A})) \to \mathcal{N}(0, v^2[f])$$

in distribution as $N \to \infty$, where the variance $v^2[f]$ is a function defined as

$$v^2[f] = \frac{1}{2\pi^2}\int_0^4\int_0^4 \left(\frac{f(x)-f(y)}{x-y}\right)^2 \frac{(4-(x-2)(y-2))}{\sqrt{4-(x-2)^2}\sqrt{4-(y-2)^2}}dxdy + \frac{m_4-3}{4\pi^2}\left(\int_0^4 \frac{x-2}{\sqrt{4-(x-2)^2}}dx\right)^2 \tag{13.8}$$

In addition, we are interested in the performance of this algorithm in different scales of matrix sizes $N$. The variance of the linear eigenvalue statistics does not grow to infinity in the limit $N \to \infty$ for sufficiently smooth test functions. These points to very effective cancellations between different terms of sum and a rigidity property [13] for the distribution of the eigenvalues.

### 13.3.3 Limit results of GUE and LUE

Here, we review some useful asymptotic results of GUE and LUE including the limiting spectral distribution, transforms, and characteristic function.

For GUE and LUE, remarkable results exist that describe the empirical spectral distribution (ESD). We are interested in the limiting behavior of marginal eigenvalue probability density $\rho_{N\to\infty}(x)$, in terms of the true eigenvalue probability density $\rho(x)$. For $N \times N$ GUE matrix and $M \times N$ LUE matrix, let $M, N \to \infty$ and $\beta = \frac{M}{N}$, the relationship between $\rho_{N\to\infty}(x)$ and $\rho(x)$ can be described as

$$\rho(x) := \lim_{N\to\infty} \rho_N(x) = \begin{cases} \frac{1}{2\pi}\sqrt{4-x^2}, x \in [-2,2], & \text{GUE} \\ \frac{1}{2\pi\beta x}\sqrt{(x-a)(b-x)}, x \in [-a,b], & \text{LUE} \end{cases} \tag{13.9}$$

where $a = (1-\sqrt{\beta})^2, b = (1+\sqrt{\beta})^2$. The first form in (13.9) is referred to as Wigner semicircle law, and the second as Marchenko–Pastur law [50,54,63].

Another important ESD of Gaussian random matrices is Ring law [58]: consider a $L$-independent matrices product $\mathbf{Z} = \prod_{i=1}^{L} \mathbf{X}_{u,i}$, where $\mathbf{X}_u \in \mathbb{C}^{N\times N}$ is the singular value equivalent [33] of $\tilde{\mathbf{X}}$ (see (13.12)); $\tilde{\mathbf{X}}$ is obtained directly from raw data $\hat{\mathbf{X}} \in \mathbb{C}^{N\times T}$

(see (13.11)). Furthermore, the matrices product $\mathbf{Z}$ is converted into $\tilde{\mathbf{Z}}$ (see (13.13)). Thus, the ESD of $\tilde{\mathbf{Z}}$ converges almost surely to the same limit given by

$$\rho_{ring}(\lambda) = \begin{cases} \frac{1}{\pi c L}|\lambda|^{(2/L-2)}, & (1-c)^{L/2} \leqslant |\lambda| \leqslant 1 \\ 0, & \text{otherwise} \end{cases} \tag{13.10}$$

as $N, T \to \infty$ with the ratio $c = N/T \in (0, 1]$.

$$\tilde{\mathbf{x}}_i = \frac{\sigma(\tilde{\mathbf{x}}_i)}{\sigma(\hat{\mathbf{x}}_i)}(\hat{\mathbf{x}}_i - \mu(\hat{\mathbf{x}}_i)) + \mu(\tilde{\mathbf{x}}_i), \quad 1 \leqslant i \leqslant N \tag{13.11}$$

where $\hat{\mathbf{x}}_i = (\hat{x}_{i,1}, \hat{x}_{i,2}, \cdots, \hat{x}_{i,T})$ and $\mu(\tilde{\mathbf{x}}_i) = 0$, $\sigma^2(\tilde{\mathbf{x}}_i) = 1$.

$$\mathbf{X}_u = \sqrt{\tilde{\mathbf{X}}\tilde{\mathbf{X}}^H}\mathbf{U} \tag{13.12}$$

where $\mathbf{U} \in \mathbb{C}^{N\times N}$ is a Haar unitary matrix; $\mathbf{X}_u \mathbf{X}_u^H \equiv \tilde{\mathbf{X}}\tilde{\mathbf{X}}^H$.

$$\tilde{\mathbf{z}}_i = \mathbf{z}_i / (\sqrt{N}\sigma(\mathbf{z}_i)), \quad 1 \leqslant i \leqslant N \tag{13.13}$$

where $\mathbf{z}_i = (z_{i1}, z_{i2}, \cdots, z_{iN})$, $\mathbf{Z} = \prod_{i=1}^{L} \mathbf{X}_{u,i}$.

It is often the case that we can obtain the limiting spectrum from the transforms of its distribution. In this section, we will review the useful transforms including Stieltjes transform, R transform, and S transform.

**Definition 13.3.5.** *Let* $\mathbf{X}$ *be a random matrix with distribution* $F(\cdot)$. *Its Stieltjes transform is defined as*

$$G(z) = \left\{\frac{1}{N}\mathrm{Tr}\left[(z\mathbf{I} - \mathbf{X})^{-1}\right]\right\} = \int_{\mathbb{R}} \frac{1}{x - z} dF(x), \quad z \in \mathbb{C} \setminus \mathbb{R} \tag{13.14}$$

*where* $\rho(x)$ *is the empirical spectral density of* $\mathbf{X}$ *and* $\mathbf{I}$ *represents the identity matrix of dimension* $N$.

For GUE and LUE matrices, the corresponding Stieltjes transforms are shown in the following examples.

**Example 13.3.6.** *Let* $\mathbf{X}$ *be a* GUE *matrix and its limit spectral density is defined in the first form in (13.9). The related Stieltjes transform is*

$$G(z) = \frac{1}{2\pi}\int_{-2}^{2} \frac{\sqrt{4-x^2}}{x-z} dx = \frac{1}{2}\left[z - \sqrt{z^2 - 4}\right] \tag{13.15}$$

**Example 13.3.7.** *Let* $\mathbf{X}$ *be a* LUE *matrix and its limit spectral density is defined in the second form in (13.9). The related Stieltjes transform is*

$$G(z) = \frac{1}{2\pi}\int_{a}^{b} \frac{\sqrt{(x-a)(b-x)}}{\beta x(x-z)} dx = \frac{\beta + z - 1 - \sqrt{z^2 - 2(\beta+1)z + (\beta-1)^2}}{2\beta z} \tag{13.16}$$

An important application of Stieltjes transform is to calculate the eigenvalue distribution of $\mathbf{X}$.

**Theorem 13.3.8.** *Let* $\mathbf{X}$ *be an* $N \times N$ *random Hermitian matrix and its Stieltjes transform is* $G(z)$*, the corresponding eigenvalue density* $\rho(x)$ *is expressed as*

$$\rho(x) = -\frac{1}{\pi} \lim_{\mathrm{Im} z \to 0} \mathrm{Im}\left\{G(z)\right\} \tag{13.17}$$

Another two handy transforms which we elaborate in the following are R transform and S transform. The key point of these two transforms is that R/S transform enables the characterization of the limiting spectrum of a sum/product of random matrices from their own limiting spectra.

**Definition 13.3.9.** *Let* $G(z)$ *denote the Stieltjes transform of* $\mathbf{X}$*, the R transform is defined as [61]*

$$R(G(z)) = z - G^{-1}(z) \tag{13.18}$$

**Example 13.3.10.** *R transform of the semicircle law and Marchenko–Pastur law as defined in* (13.9) *are*

$$R(z) = \begin{cases} z, & \text{GUE} \\ \frac{1}{1-\beta z}, & \text{LUE} \end{cases} \tag{13.19}$$

Two important prosperities of R transform are shown in the following.

**Theorem 13.3.11.** *Additivity law: let* $R_{\mathbf{A}}(z)$*,* $R_{\mathbf{B}}(z)$*, and* $R_{\mathbf{A}+\mathbf{B}}(z)$ *be the R transforms of matrices* $\mathbf{A}$*,* $\mathbf{B}$*, and* $\mathbf{A}+\mathbf{B}$*, respectively. We have*

$$R_{\mathbf{A}+\mathbf{B}}(z) = R_{\mathbf{A}}(z) + R_{\mathbf{A}}(z) \tag{13.20}$$

**Theorem 13.3.12.** *For any* $\alpha > 0$,

$$R_{\alpha\mathbf{X}}(z) = \alpha R_{\mathbf{X}}(\alpha z) \tag{13.21}$$

Additivity law 13.3.11 can be easily understood in terms of Feynman diagrams [54,61]. Theorems 13.3.11 and 13.3.12 enable us to conduct linear calculation for the asymptotic spectrum of random matrices. Another important transform of engineering sense in RMT is the S transform. S transform is related to the R transform which is defined in the following.

**Definition 13.3.13.** *[61]*

$$S(z) = \frac{1}{R(zS(z))} \tag{13.22}$$

**Example 13.3.14.** *S transform of Marchenko–Pastur law as defined in* (13.9) *is*

$$S(z) = \frac{1}{1 + \beta x} \tag{13.23}$$

**Theorem 13.3.15.** *Multiplication law: let $S_{\mathbf{A}}(z)$, $S_{\mathbf{B}}(z)$, and $S_{\mathbf{AB}}(z)$ be the S transforms of matrices* **A**, **B**, *and* **AB**, *respectively. We have*

$$S_{\mathbf{AB}}(z) = R_{\mathbf{A}}(z)S_{\mathbf{B}}(z) \tag{13.24}$$

For multiplication law, we refer readers to Reference 9 for technical details.

Moreover, for the above two ensemble matrices, let $\rho(x)$ be the density of the population spectral distribution of a random matrix and let $f(t)$ be its characteristic function. Haagerup and Thorbjørnsen [21] give some analytical proofs of a number of classical results on the asymptotic behavior of GUE and LUE matrices including the characteristic functions. The characteristic functions of GUE and LUE matrices are shown in the following. For $N \times N$ GUE matrices, we have

$$f_{\mathrm{GUE}}(t) = \exp\left\{-t^2/2N\right\} \Phi\left(1 - N, 2, t^2/N\right) \tag{13.25}$$

where $\Phi\left(\alpha, \beta, s\right)$ is the confluent hypergeometric function defined as

$$\Phi\left(\alpha, \beta, s\right) = 1 + \frac{\alpha s}{\beta 1!} + \frac{\alpha\left(\alpha + 1\right) s^2}{\beta\left(\beta + 1\right) 2!} + \frac{\alpha\left(\alpha + 1\right)\left(\alpha + 2\right) s^3}{\beta\left(\beta + 1\right)\left(\beta + 2\right) 3!} + \cdots \tag{13.26}$$

Also shown in Reference 21 that $\Phi(\alpha, \beta, s)$ satisfies

$$s\frac{d^2\Phi}{ds^2} - \left(\beta - s\right)\frac{d\Phi}{ds} - \alpha\Phi = 0 \tag{13.27}$$

Based on (13.27), we can derive a differential equation for $\rho(x)$.

**Theorem 13.3.16.** *Let $\rho(x)$ be the expected spectral distribution of an $N \times N$ GUE random matrix, and it satisfies*

$$\left(4 - x^2\right)\rho'(x) + x\rho(x) + \frac{1}{N^2}\rho'''(x) = 0 \tag{13.28}$$

For LUE matrices defined in (13.5), Haagerup and Thorbjørnsen [21] prove

$$f_{\mathrm{LUE}}(t)' = i\left(1 - \frac{it}{M}\right)^{-(M+N)} F\left(1 - M, 1 - N, 2; -t^2/m^2\right) \tag{13.29}$$

where the hypergeometric function $F\left(\alpha, \beta, \gamma, s\right)$ is defined as

$$\begin{aligned} F\left(\alpha, \beta, \gamma, s\right) &= 1 + \frac{\alpha\beta s}{\gamma 1!} + \frac{\alpha\left(\alpha + 1\right)\beta\left(\beta + 1\right) s^2}{\gamma\left(\gamma + 1\right) 2!} \\ &\quad + \frac{\alpha\left(\alpha + 1\right)\left(\alpha + 2\right)\beta\left(\beta + 1\right)\left(\beta + 2\right) s^3}{\gamma\left(\gamma + 1\right)\left(\gamma + 2\right) 3!} + \cdots \end{aligned} \tag{13.30}$$

Let $\kappa(s) = F(\alpha, \beta, \gamma, s)$, $F(\alpha, \beta, \gamma, s)$ satisfies

$$s\left(1 - s\right)\frac{d^2\kappa}{ds^2} + \left[\gamma - \left(\alpha + \beta + 1\right) s\right]\frac{d\kappa}{ds} - \alpha\beta\kappa = 0 \tag{13.31}$$

Similarly, let $g(x) := x\rho_{\mathrm{LUE}}(x), \varphi(t) = -i(f_{\mathrm{LUE}}(t))', \eta = \frac{N}{M}$, we can derive the differential function of $g(x)$ from (13.31) as follows.

**Theorem 13.3.17.** *The function $g(x)$ satisfies*

$$(x-a)(b-x)g'(x) + \left(x - \frac{a+b}{2}\right)g(x) + \frac{xg''(x) + x^2 g''(x)}{M^2} = 0 \qquad (13.32)$$

*where* $a = (1-\sqrt{\eta})^2, b = (1+\sqrt{\eta})^2$.

We refer interested readers to Reference 21 for details. Characteristic function is useful for many applications in RMT. In this chapter, we use it to study the asymptotic properties of GUE and LUE in Section 13.3.5.

So far, we provide some basic results of the random matrices which help us model the data in power grid. In addition, we are concerned that how small the dimension of data is that the RMM is still valid for data analysis. Inspired by this, we will further study some asymptotic results.

### *13.3.4 Asymptotic expansion for the Stieltjes transform of GUE*

Since the landmark work [60] by Voiculescu, the asymptotics for sets of massive, independent GUE random matrices has become an important analytic tool in the RMT. Here, we study the asymptotic expansion for the Stieltjes transform of GUE random matrix.

We focus on the asymptotic expansion of the mean $\mathbb{E}\left(\frac{1}{N}\mathrm{Tr}\left(f(\mathbf{X})\right)\right)$ and covariance $\mathrm{cov}\left\{\frac{1}{N}\mathrm{Tr}\left(f(\mathbf{X})\right), \frac{1}{N}\mathrm{Tr}\left(g(\mathbf{X})\right)\right\}$ of the Stieltjes Transform of GUE, where $f(x) = \frac{1}{\lambda - x}$ and $g(x) = \frac{1}{\mu - x}$, $\{\lambda, \mu\} \subseteq \mathbb{C}\backslash\mathbb{R}$ are compactly supported $C^\infty$ functions.

Let $f : \mathbb{R} \to \mathbb{C}$ be a $\mathbb{C}^\infty$ function that all derivations $f^{(k)}, k \in \mathbb{N}_0$ are bounded on $\mathbb{R}$. Ercolani and McLaughlin established asymptotic expansions of the expectation value $\mathbb{E}\left(\frac{1}{N}\mathrm{Tr}\left(f(\mathbf{X})\right)\right)$ for GUE random matrix $\mathbf{X}$ [12]. Their proof is based on Riemann–Hilbert techniques and rather involved. Alternatively, Haagerup and Thorbjørnsen prove this asymptotic expansion by an analytical approach [22]. Specifically,

$$\mathbb{E}\left\{\frac{1}{N}\mathrm{Tr}\left(f(\mathbf{X})\right)\right\} = \frac{1}{2\pi}\int_{-2}^{2} f(x)\sqrt{4-x^2}dx + \sum_{j=1}^{k}\frac{\alpha_j(f)}{N^{2j}} + O(N^{-2k-2}) \qquad (13.33)$$

where $k \in \mathbb{N}^+$, and $\mathrm{Tr}(\cdot)$ is the trace operator.

In the GUE case, let $f(x) = \frac{1}{z-x}$, based on the differential function as introduced in (13.29). Haagerup and Thorbjørnsen shown in Reference 22 that the one-dimensional Stieltjes transform can be expressed in the following explicit form:

$$G(z) = \eta_0(z) + \frac{\eta_1(z)}{N^2} + \cdots + \frac{\eta_k(z)}{N^{2k}} + O(N^{-2k-2}), \quad z \in \mathbb{C}\setminus\mathbb{R}$$

where

$$\eta_0(z) = \frac{z}{2} - \frac{(z^2-4)^{\frac{1}{2}}}{2} \quad \text{and} \quad \eta_j(z) = \sum_{l=2j}^{3j-1} C_{j,l}(z^2-4)^{-l-\frac{1}{2}} \qquad (13.34)$$

and with $j \geq 1$, $2j + 2 \leq l \leq 3j + 2$, $C_{j,l}$ can be obtained from a recursion formula:

$$C_{j+1,l} = \frac{(2r-3)(2r-1)}{r+1}((r-1)C_{j+1,l-2} + (4r-10)C_{j,l-3}) \tag{13.35}$$

Furthermore, we establish the asymptotic expansion for two-dimensional Stieltjes transform in the form $\mathrm{cov}\left\{\frac{1}{N}\mathrm{Tr}\left[f(\mathbf{X})\right], \frac{1}{N}\mathrm{Tr}\left[g(\mathbf{X})\right]\right\}$. Based on the inspiring formula in Reference 41 which is given by

$$\mathrm{cov}\left\{\frac{1}{N}\mathrm{Tr}\left[f(\mathbf{X})\right], \frac{1}{N}\mathrm{Tr}\left[g(\mathbf{X})\right]\right\} = \int_{\mathbb{R}^2}\left(\frac{f(x)-f(y)}{x-y}\right)\left(\frac{g(x)-g(y)}{x-y}\right)\varphi(x,y)\,dxdy$$

where the kernel function $\varphi(x,y)$ is defined as

$$\varphi(x,y) = \frac{N}{4}\left[\Lambda_N\left(\sqrt{\frac{N}{2}}x\right)\Lambda_{N-1}\left(\sqrt{\frac{N}{2}}y\right) - \Lambda_{N-1}\left(\sqrt{\frac{N}{2}}x\right)\Lambda_N\left(\sqrt{\frac{N}{2}}y\right)\right]^2 \tag{13.36}$$

with $\Lambda_N(x)$ the $N$th Hermite function. As shown in Reference 22, the kernel function satisfies

$$\varphi(x,y) = \frac{1}{4}\left[\tilde{\rho}(x)\tilde{\rho}(y) - 4\rho'(x)\rho'(y) - \frac{1}{N^2}\rho''(x)\rho''(y)\right] \tag{13.37}$$

where $\tilde{\rho}(x) = \rho(x) - x\rho'(x)$.

For GUE matrices, let $f(x) = \frac{1}{\lambda - x}$ and $g(x) = \frac{1}{\mu - x}$, based on (13.36) and (13.37), the asymptotic expansion of covariance of the Stieltjes transform is

$$\mathrm{cov}\left\{\mathrm{Tr}\left[\frac{1}{N}(\lambda\mathbf{I}-\mathbf{X})^{-1}, \frac{1}{N}(\mu\mathbf{I}-\mathbf{X})^{-1}\right]\right\} = \frac{1}{2(\lambda-\mu)^2}\sum_{j=0}^{k}\frac{\Gamma_j(\lambda,\mu)}{N^{2j}} + O(N^{-2k-2}) \tag{13.38}$$

where the leading term $\frac{\Gamma_0(\lambda,\mu)}{2(\lambda-\mu)^2}$ is identified as

$$\int_{R^2}\left(\frac{(\lambda-x)^{-1}-(\lambda-y)^{-1}}{x-y}\right)\left(\frac{(\mu-x)^{-1}-(\mu-y)^{-1}}{x-y}\right)\varphi(x,y)\,dxdy \tag{13.39}$$

with the limiting kernel function

$$\varphi(x,y) = \frac{1}{4\pi^2}\frac{4-xy}{\sqrt{4-x^2}\sqrt{4-y^2}}, \quad |x| \leq 2, |y| \leq 2 \tag{13.40}$$

In the special case, for any $\lambda = \mu$, $k \in \mathbb{N}$, we have that

$$G(\lambda,\lambda) = \frac{1}{4}\left[\Upsilon_0(\lambda) + \frac{\Upsilon_1(\lambda)}{N^2} + \cdots + \frac{\Upsilon_k(\lambda)}{N^{2k}} + O(N^{-2k-2})\right]$$

where

$$\Upsilon_0(\lambda) = (\lambda^2-4)\eta_0''(\lambda)$$

and for $j = 1, 2, \ldots, k$

$$\Upsilon_j(\lambda) = (\lambda^2 - 4)\sum_{i=0}^{j-1} \eta_i''(\lambda)\eta_{j-i}''(\lambda) - \sum_{i=0}^{j-1} \eta_i'''(\lambda)\eta_{j-1-i}'''(\lambda) \tag{13.41}$$

### 13.3.5 The rate of convergence for spectra of GUE and LUE

In this section, we investigate the spectral asymptotics for GUE and LUE matrices. We are motivated by the practical problems introduced in References 10, 54. Let $F(x)$ be the ESD function of GUE or LUE matrices and $G(x)$ be the distribution function of the limit law (semicircle law for GUE matrices and Marchenko–Pastur law for LUE matrices). Here, we study the convergence rate of expected empirical distribution function $\mathbb{E}F(x)$ to $G(x)$. Specially, the bound $\Delta$

$$\Delta = |\mathbb{E}F(x) - G(x)| \tag{13.42}$$

is mainly concerned in the following.

The rate of convergence for the expected spectral distribution of GUE matrices has attracted numerous attention due to its increasingly appreciated importance in applied mathematics and statistical physics. Wigner initially looked into the convergence of the spectral distribution of GUE matrices [42]. Bai [2,5] conjectured that the optimal bound for $\Delta$ in GUE case should be of order $n^{-1}$. Bai *et al.* in Reference 3 proved that $\Delta = O(N^{-1/3})$. Götze and Tikhomirov in Reference 17 improve the result in Reference 3 and proved that $\Delta = O(N^{-1/2})$. Bai *et al.* in Reference 4 also shown that $\Delta = O(N^{-1/2})$ on the condition that the eighth moment of $\mathbf{X}$ satisfies $\sup E|X_{ij}|^8 < \infty$. Girko in Reference 14 stated as well that $\Delta = O(N^{-1/2})$ assuming uniform bounded fourth moment of $\mathbf{X}$. Recently, Götze and Tikhomirov prove an optimal bound as follows:

**Theorem 13.3.18.** *There exists a positive constant C such that, for any $N \geq 1$,*

$$\Delta \leq CN^{-1} \tag{13.43}$$

Similarly, the convergence of the density (denoted by $g(x)$) of standard semicircle law to the expected spectral density $p(x)$ is also provided in the following theorem.

**Theorem 13.3.19.** *There exists a positive constant $\varepsilon$ and C such that, for any $x \in [-2 + N^{-\frac{1}{3}}\varepsilon, 2 - N^{-\frac{1}{3}}\varepsilon]$,*

$$|p(x) - g(x)| \leq \frac{C}{N(4 - x^2)} \tag{13.44}$$

Here, we present a simulation result of Theorem 13.3.19 shown in Figure 13.2. It is noted that the experimental data is generated by Matpower toolbox according to IEEE 118 bus standard. The normal case and abnormal case mean that the power system is under normal condition and abnormal condition, respectively. The results in Figure 13.2 demonstrate the correctness of the method that we model the data in power system as GUE matrices. Besides, Theorems 13.3.18 and 13.3.19 can be used to do anomaly detection in power system.

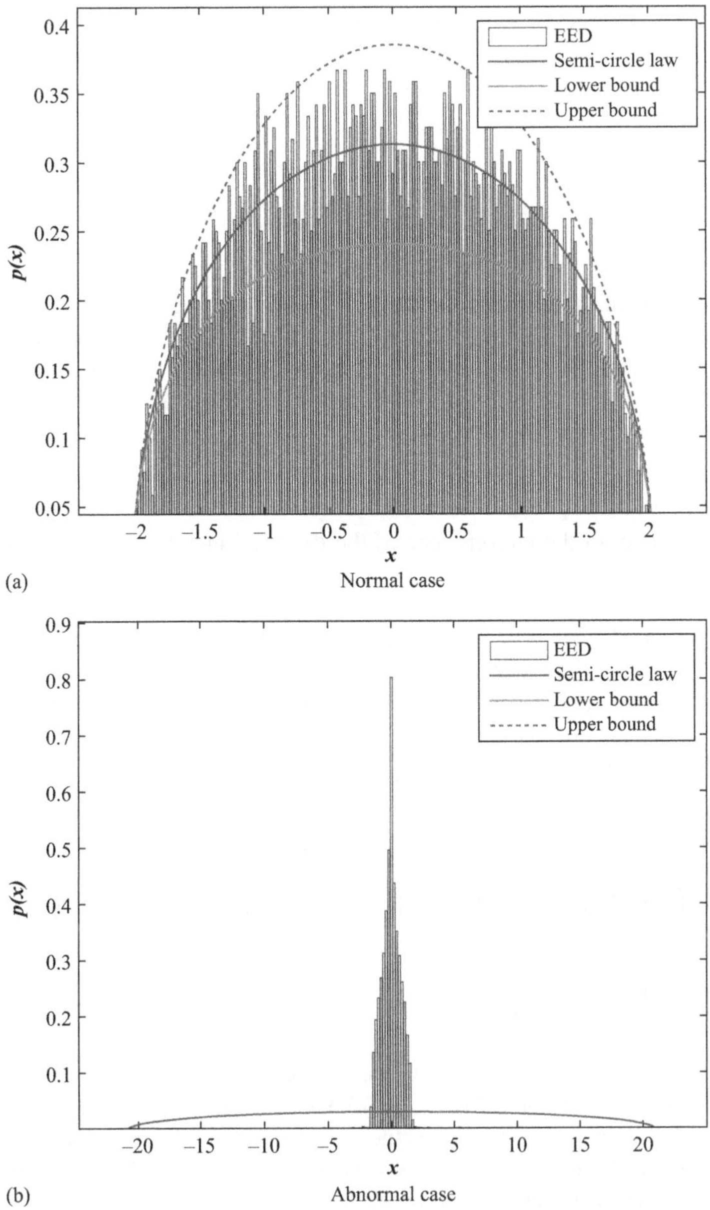


*Figure 13.2 Modeling the data in power system as GUE matrices. (a) Normal case and (b) abnormal case*

For LUE matrix $\mathbf{W}$ with spectral distribution function $F(x)$, let $\beta = \frac{N}{M}$ as $N, M \to \infty$, it is well known that $\mathbb{E}F(x)$ convergences to the Marchenko–Pastur law $H(x)$ with density $h(x)$

$$h(x) = \frac{1}{2\pi\beta x}\sqrt{(x-a)(b-x)} \tag{13.45}$$

where $a = (1-\sqrt{\beta})^2$, $b = (1+\sqrt{\beta})^2$ (see 13.9). The bound

$$\Delta = |\mathbb{E}F(x) - H(x)| \tag{13.46}$$

for the convergence rate is shown in the following theorems.

**Theorem 13.3.20.** *For $\beta = \frac{N}{M}$, there exists some positive constant $\beta_1$ and $\beta_2$ such that $0 < \beta_1 \leq \beta \leq \beta_2 < 1$, for all $N \geq 1$. Then there exists a positive constant $C$ depending on $\beta_1$ and $\beta_2$ and for any $N \geq 1$*

$$\Delta \leq CN^{-1} \tag{13.47}$$

With respect to the convergence of the density function, the main result is as follows:

**Theorem 13.3.21.** *For $\beta = \frac{N}{M}$, there exists some positive constant $\beta_1$ and $\beta_2$ such that $0 < \beta_1 \leq \beta \leq \beta_2 < 1$, for all $N \geq 1$. Then there exists a positive constant $C$ and $\varepsilon$ depending on $\beta_1$ and $\beta_2$ and for any $N \geq 1$ and $x \in [a + N^{-\frac{2}{3}}\varepsilon, b - N^{-\frac{2}{3}}\varepsilon]$*

$$|p(x) - h(x)| \leq \frac{C}{N(x-a)(b-x)} \tag{13.48}$$

For the special case $M = N$, we have the following result:

**Theorem 13.3.22.** *There exists a positive constant $C$ such that, for any $N \geq 1$ and any $x \in [N^{-2}\varepsilon, 4 - N^{-2}\varepsilon]$*

$$|p(x) - h(x)| \leq \frac{C}{Nx(4-x)} \tag{13.49}$$

We refer the interested readers to Reference 18 for technical details. Here, we present a simulation result of Theorem 13.3.21 shown in Figure 13.3. The results in Figure 13.3 demonstrate the correctness of the method that we model the data in power system as LUE matrices. Besides, Theorems 13.3.20 and 13.3.22 can be also used to do anomaly detection in power system.

## 13.4 From power grid to RMM

According to the big data definition, power system is a typical big data system. For such a system, we assume $t$ times observation for $n$-dimensional vectors $\hat{\mathbf{x}}_1, \hat{\mathbf{x}}_2, \ldots, \hat{\mathbf{x}}_t$ $\left(\hat{\mathbf{x}}_j \in \mathbb{C}^{n\times 1}, j = 1, 2, \ldots, t\right)$; hence, a data source, denoted as $\mathbf{\Omega}$ (in size of $n \times t$), is obtained. $\mathbf{\Omega}$ is in a high-dimensional space but not an infinite one

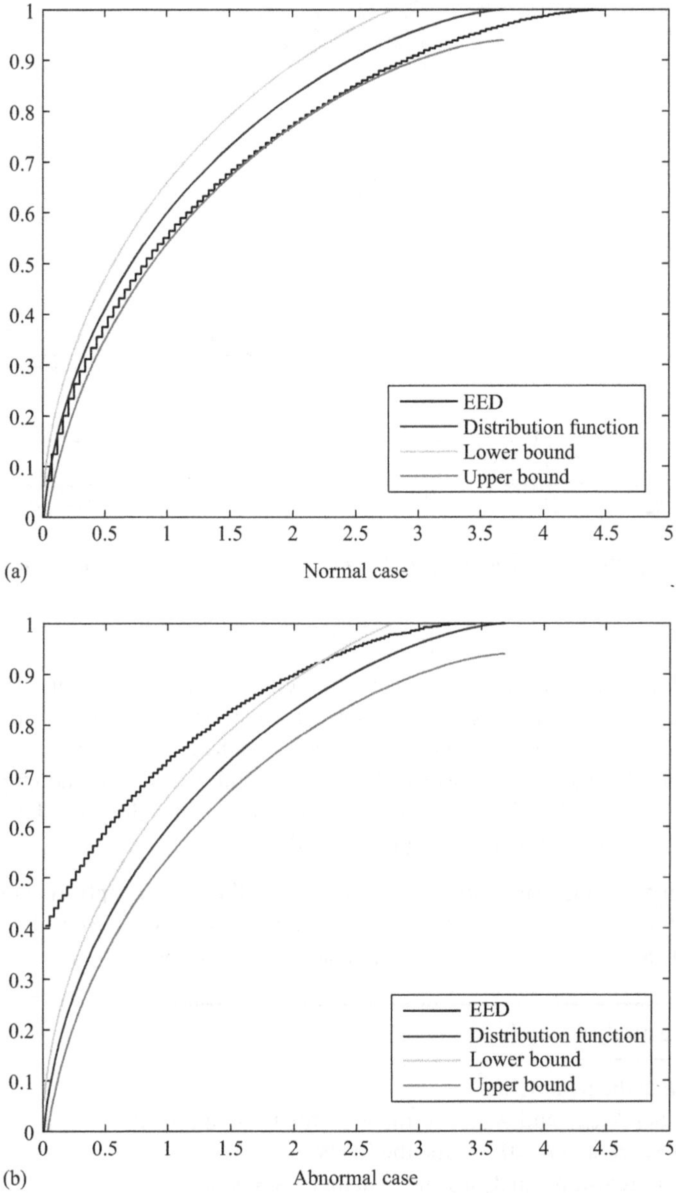


*Figure 13.3 Modeling the data in power system as LUE matrices. (a) Normal case and (b) abnormal case*

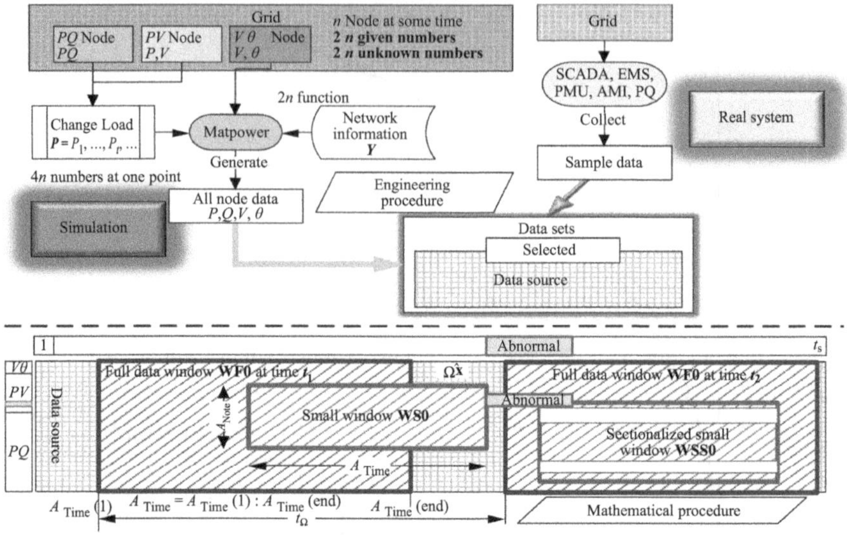


*Figure 13.4 The designed big data architecture for smart grids*

(or more explicitly, we are interested in the practical regime in which $n = 100$–$10{,}000$, and $t$ is sufficient large); this disables most classical tools. In the other side, arbitrary data are able to be selected—both *in temporal dimensions* ($T$ from $t$) and *in spatial dimensions* ($N$ from $n$)—to form an $\hat{\mathbf{X}} \in \mathbb{C}^{N\times T}$ naturally, $\hat{\mathbf{X}}$ is a random matrix due to the presence of ubiquitous noises. Figure 13.4 elaborates the procedures.

Here, we would like to highlight two modes:

(1) Real-time analysis mode: under this mode, the moving split-window (MSW) sets its right edge with the current time (i.e., $\mathbf{A}_{\text{Time}}(\text{end}) = t_0$); the MSW will slide as the time goes by. The work steps are as follows:

---

**Real-time analysis mode**

---

(1) Initialize the parameters
  (1a) Set $\mathbf{A}_{\text{Time0}}$ and $\mathbf{A}_{\text{Node0}}$ to focus on the first data window
  (1b) Set $t_\Omega$ and $k = 0$ to slide the MSW
(2) Conduct real-time analyses and visualize the results
  (2a) $\mathbf{A}_{\text{Time}} = \mathbf{A}_{\text{Time0}} + k$, and form $\hat{\mathbf{X}}$ with $\mathbf{A}_{\text{Time}}$
  (2b) Conduct high-dimensional analysis
  (2c) Visualize the statistical indicators
(3) Judge as time goes by:
  (3a) $k < t_\Omega \Rightarrow k{+}{+}$; back to *step (2)*
  (3b) $k \geqslant t_\Omega \Rightarrow$ END

---

(2) Blocking analysis mode: this mode deal with $\mathbf{A}_{\text{Node}}$ but not $\mathbf{A}_{\text{Time}}$. It can focus on a smaller window consisting of data only in designated dimensions, even those disjunct ones; it is a natural way to decouple the interconnected system.

The both modes give the $\hat{\mathbf{X}}$ as the row data; furthermore, we convert $\hat{\mathbf{X}}$ into a normalized matrix $\tilde{\mathbf{X}}$ row-by-row. Thus, for a system with any designated spatial areas and temporal period, the RMM is built for further analyses.

## 13.5 LES and related research

### *13.5.1 Definition of LES*

The LES of a matrix as defined via the continuous test function $\varphi : \mathbb{R} \to \mathbb{C}$

$$\mathcal{N}_N[\varphi] = \sum_{i=1}^{N} \varphi(\lambda_i) \tag{13.50}$$

### *13.5.2 Law of Large Numbers*

The law of Large Numbers is the first step in studies of eigenvalue distributions for a certain random matrix ensemble. The result, for the Wigner ensemble, obtained initially in Reference 64, was improved in Reference 51, where the Stieltjes transformation was introduced and the famous semicircle law was shown under the minimal conditions on the distribution of **W** (the Lindeberg-type conditions) [57]. The law of Large Numbers tells us that $N^{-1}\mathcal{N}_N[\varphi]$ converges in probability to the limit

$$\lim_{N\to\infty} \frac{1}{N}\mathcal{N}_N[\varphi] = \int \varphi(\lambda)\rho(\lambda)\,d\lambda \tag{13.51}$$

where $\rho(\lambda)$ is the p.d.f. of the eigenvalues.

### *13.5.3 CLTs of LES*

CLTs, as the natural second step, aim to study the LES fluctuations. Lots of papers devote to proofs of CLTs for different random matrix ensembles (see References 1,6,20,35,41,49,57). CLTs for LESs with polynomial test functions of some generalizations for the Wigner and covariances matrices were proved in Reference 1 via moment methods. In contrast, CLTs for LES with real analytic test functions of the Wigner and covariances matrices were established in Reference 6 under additional assumptions that

$$\left| \begin{array}{ll} \mathbb{E}(W_{i,i}{}^2)=2, \mathbb{E}(W_{i,j}{}^4)=3\mathbb{E}^2(W_{i,j}{}^2)=3 & \text{Wigner} \\ \mathbb{E}(X_{i,j}{}^4)=3\mathbb{E}^2(X_{i,j}{}^2) & \text{Covariance} \end{array} \right. \tag{13.52}$$

In the recent paper [41], CLTs for LESs of the Wigner and covariances matrices were proved under assumptions that $\mathbb{E}(W_{i,i}{}^2)=2$, the third and the forth moments of all entries are the same, but $\mathbb{E}({W_{i,j}}^4)$ is not necessary 3. Moreover, the test functions are not supposed to be real analytic. It was assumed that the Fourier transformation $\widehat{\varphi}$ satisfies the inequality

$$\int (1+|k|^5)|\widehat{\varphi}(k)|\,dk < \infty \tag{13.53}$$

which means that $\varphi$ has more than five bounded derivatives.

### 13.5.4 CLT for covariance matrices

In this section, we will study the CLT for LUE as defined in Section 13.3.1. The CLT for **M** is given as follows [57]:

**Theorem 13.5.1** [57]. *Consider a rectangular $N \times T$ non-Hermitian random matrix* $\mathbf{X}$, *with entries* $X_{i,j}$; $\mathbf{M}$ *is the covariance matrix. Let the real-valued test function $\varphi$ satisfies condition $\|\varphi\|_{3/2+\varepsilon} < \infty$ $(\varepsilon > 0)$. Then $\mathcal{N}_N{}^{\circ}[\varphi]$, in the limit $N, T \to \infty, c = N/T \leq 1$, converges in the distribution to the Gaussian random variable with zero mean and the variance:*

$$V_{SC}[\varphi] = \frac{2}{c\pi^2} \iint_{-\frac{\pi}{2}<\theta_1,\theta_2<\frac{\pi}{2}} \psi^2(\theta_1,\theta_2)(1-\sin\theta_1 \sin\theta_2)d\theta_1 d\theta_2 + \frac{\kappa_4}{\pi^2}\left(\int_{-\frac{\pi}{2}}^{\frac{\pi}{2}} \varphi\left(\zeta\left(\theta\right)\right) \sin\theta d\theta\right)^2 \tag{13.54}$$

*where* $\psi(\theta_1,\theta_2) = \frac{[\varphi(\zeta(\theta))]_{\theta=\theta_2}^{\theta=\theta_1}}{[\zeta(\theta)]_{\theta=\theta_2}^{\theta=\theta_1}}$, *and* $\zeta(\theta) = 1 + 1/c + 2/\sqrt{c}\sin\theta$; $\kappa_4 = \mathbb{E}(X^4) - 3$ *is the fourth cumulant of entries of* $\mathbf{X}$.

For Gauss variable $X$, $\mathbb{E}(X)=0$, $\mathbb{E}(X^2)=1$, and $\mathbb{E}(X^4)=3$. A typical scenario is assumed: $N=118$ and $T=240$, thus $c = N/T = 0.4917$.

### 13.5.5 LES for Ring law

MSR is a special LES[2]; it is defined as follows:

$$\tau_{\text{MSR}} = \sum_{i=1}^{N} \frac{1}{N} |\lambda_{\widetilde{\mathbf{Z}},i}| \tag{13.55}$$

where $\lambda_{\widetilde{\mathbf{Z}},i}$ $(i=1,2,\ldots,N)$ are the eigenvalues of $\widetilde{\mathbf{Z}}$, and $|\lambda_{\widetilde{\mathbf{Z}},i}|$ is the radius of $\lambda_{\widetilde{\mathbf{Z}},i}$ on the complex plane.

[2]Since $\lambda_{\widetilde{\mathbf{Z}},i}$ are highly correlated random variables (each one is a complicated function of the raw random matrices $\hat{\mathbf{X}}$), $\tau_{\text{MSR}}$ is a random variable.

According to (13.51), the theoretical expectation of $r$ when $N \to \infty$ ($\mathbb{E}(\tau_{\text{MSR}})$), are calculated as follows:

$$\mathbb{E}(\tau_{\text{MSR}}) = \iint_{\text{Area}} \mathbb{P}(r) \times r \cdot r\, dr\, d\theta = \int_0^{2\pi} \int_{\sqrt{1-c}}^{1} \frac{1}{c\pi} r \cdot r\, dr\, d\theta = 0.8645 \quad (13.56)$$

where $\mathbb{P}(r)$ is given in (13.10), and $c = 0.4917$ for this scenario.

### 13.5.6 LES for covariance matrices

1. Chebyshev polynomials $\varphi(\lambda) = T_2 = 2x^2 - 1$

$$\tau_{T_2} = \sum_{i=1}^{N} (2\lambda_i{}^2 - 1) \quad (13.57)$$

For the scenario, according to (13.51), we get $\mathbb{E}(\tau_{T_2})$:

$$\mathbb{E}(\tau_{T_2}) = N \int \varphi(\lambda) \rho_{mp2}(\lambda)\, d\lambda = 6600 \quad (13.58)$$

and according to (13.54), we get $\mathbb{D}(\tau_{T_2})$:

$$\mathbb{D}(\tau_{T_2}) = 1080 \quad (13.59)$$

Similarly, for a certain RMM $\mathbf{X}$, we can also design other test functions $\varphi(\lambda)$ to obtain diverse LESs $\tau$ , as well as their theoretical values. Here, we list some classical test functions:

2. Chebyshev polynomials: $T_3 = 4x^3 - 3x$
3. Chebyshev polynomials: $T_4 = 8x^4 - 8x^2 + 1$
4. Determinant: $\text{DET} = \ln(x)$
5. Likelihood-ratio function: $\text{LRF} = x - \ln(x) - 1$

The test function is similar to a filter somehow. With the LESs, the high-dimensional indicator system is built; it gives us a multiple view angle to learn from the system via the medium RMM $\mathbf{X}$. About the indicator system, we will make a further discussion in Section 13.7.

## 13.6 Data preprocessing—data fusion

*How to choose data* and *how to deal with heterogeneous data* are two key questions for data preprocessing. This section we only take account of the second one. Data fusion is related to the augmentation, the blocking, the sum, and the product of matrices. Comparing to the LES designs, which aim to define the LES $\tau$ via the test functions $\varphi(\lambda)$ for a determinate $\mathbf{X}$, data fusion manages to handle multiple data sources (i.e., $\mathbf{X}_1, \mathbf{X}_2, \ldots$), even with distinct features (e.g., in different size).

### 13.6.1 Augmented matrix method for power systems

Diverse **causing factors** affect the **system status** differently. Assuming that there are $N$ status variables and $M$ factors ones; they are all measurable. In a fixed study period $t_i$ ($i = 1, 2, \ldots, T$), sampling data of status variables consist of a matrix $\mathbf{B} \in \mathbb{C}^{N \times T}$ (i.e., status matrix), and the factor ones consist of $\mathbf{c}_j \in \mathbb{C}^{1 \times T}$ ($j = 1, 2, \ldots, M$) (i.e., factor vector).

As we all know, two matrices with the same length can be put together and an augmented matrix is formed; in such a way, we are able to obtain a new matrix $\mathbf{A}$ using the status matrix $\mathbf{B}$ and factor matrix $\mathbf{c}_j$.

In order to balance the proportion (enlarge the statistic correlation), we form a factor matrix for each factor vector. First, for the factor $\mathbf{c}_j$, we duplicate it for $K$ times[3] to construct a matrix $\mathbf{D}_j$, formulated by

$$\mathbf{D}_j = \left[\mathbf{c}_j{}^T \ \mathbf{c}_j{}^T \ \cdots \ \mathbf{c}_j{}^T\right]^T \in \mathbb{C}^{K \times T} \tag{13.60}$$

Then, we introduce white noise into $\mathbf{D}_j$ to release the inner correlations. Thus, the factor matrix is formulated by

$$\mathbf{C}_j = \mathbf{D}_j + m_{e,j}\mathbf{R} \quad (j = 1, 2, \ldots, m) \tag{13.61}$$

where $m_{e,j}$ is related to the signal-to-noise ratio (SNR), and the entries $R_{i,j}$ of the matrix $\mathbf{R}$ is Gaussian random variable.

The SNR of the factor matrix $\mathbf{C}_j$ is defined via the trace function $\mathrm{Tr}(\cdot)$

$$\rho_j = \frac{\mathrm{Tr}\left(\mathbf{D}_j\mathbf{D}_j^H\right)}{\mathrm{Tr}\left(\mathbf{R}\mathbf{R}^H\right) \times m_{e,j}^2} \quad (j = 1, 2, \ldots, m) \tag{13.62}$$

Thus, in parallel, we can construct the augmented matrix for each factor $\mathbf{c}_j$, formulated by

$$\mathbf{A}_j = \begin{bmatrix} \mathbf{B} \\ \mathbf{C}_j \end{bmatrix} \quad (j = 1, 2, \ldots, m) \tag{13.63}$$

The relationships between causing factors and effective status can be revealed by the new RMM $\mathbf{A}_j$.

The operating status of power systems can be estimated by various kinds of status variables, such as frequencies, voltages, currents, and power flows. For system analysis, we often choose bus voltage magnitudes for the following reasons: (1) the voltage magnitude is a fundamental and directly measurable parameter which is independent from the topology; (2) it is easily measurable for every bus; that means the measurement has considerable redundancy and accuracy.

The status is mainly affected by following factors:

(a) Electrical factors: include nodal loads, nodal distributed generations, etc.
(b) Climatic factors: include temperature, wind speed, light intensity, etc.
(c) Economic factors: include electricity price, gross domestic product, etc.

[3] $K$ is appropriated to $0.4 \times N$.

The augmented model is compatible with different units and magnitudes for each variable (in the form of **A**'s row), owing to the normalization during the data processing. For data source with different sampling frequency, some simple mathematical methods, e.g., interpolation, are able to apply as a solution.

### *13.6.2 Another kind of data fusion*

Data fusion is very common and meaningful in engineering. The theories about data fusion are deep and novel: Götze *et al.* and Kösters, and Tikhomirov, in References 15, 16,37, have already studied the performance of the matrices in the form of:

$$\mathbf{F}_n^{(1)} + \cdots + \mathbf{F}_n^{(m)} \tag{13.64}$$

where $\mathbf{F}_n^{(i)} = \mathbf{X}_n^{(i,1)} \cdots \mathbf{X}_n^{(i,k)}$, and $\mathbf{X}_n^{(0,0)}, \ldots, \mathbf{X}_n^{(i,j)}, \ldots, \mathbf{X}_n^{(m,k)}$ are independent $n \times n$ matrices with independent entries.

We will not go so far and just give a typical example: Zhang and Qiu [70], using the data from a 70 nodes network testbed, validated the data fusion in the field of signal detection for Massive MIMO systems.

## 13.7 A new methodology and epistemology for power systems

RMM, which is built based on RMT, is a new methodology to map the systems. This processing procedure for RMM, driven by data, leads to the LESs. LESs are high-dimensional indicators, which provide a novel high-dimensional perspective for the systems. Or even we can go further—these new indicators are able to result in a new epistemology which is, in our opinion, much more suitable to smart grids.

### *13.7.1 The evolution of power systems and group-work mode*

Some brief but novel introductions about the development of power grids, including new situations and challenges, and the managing modes, are given as the related background for applying big data into smart grids. Generally, the power grid's evolutions are summarized as three generations—G1, G2, and G3 [71]. Their own network structures are depicted in Figure 13.5 [43]. Meanwhile, their data flows and energy flows (Figure 13.6), as well as corresponding work modes, are quite different [24]. In the following discussions, we will come to a conclusion that the group-work mode is the precondition for data-driven analysis, and also is the trend for smart grids.

Small-scale isolated grids were developed from around 1900 to 1950, featured by small-scale isolated grids. For G1, components interchange energy, as well as data, in a balance state within the isolated grid. These components are fully controlled by decentralized control system and operating under individual-work mode. It means that each apparatus collects designated data, and makes corresponding decisions only with its own application, shown in the above part of Figure 13.7. The individual-work mode works with an easy logic and little information communication. Whereas, it means few advanced functions and inefficient utilization of resources. It is only suitable for a small isolated place, e.g., an island or a mothership.

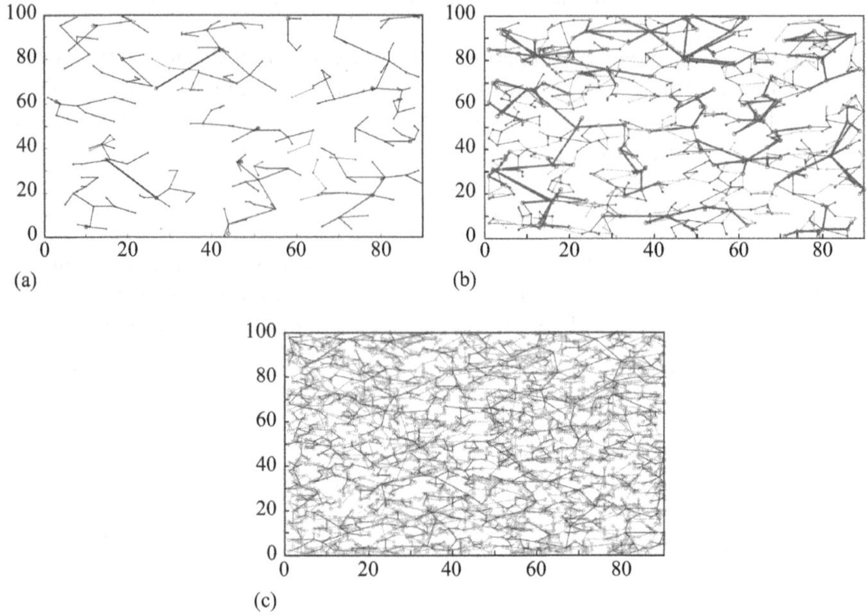


*Figure 13.5* *Simulation of network structures. The above twos are G1—small-scale isolated grids, and G2—large-scale interconnected power grids; and the below one is G3—smart grids, which have complex network structures without clear-cut partitioning. Reproduced from Reference 43*

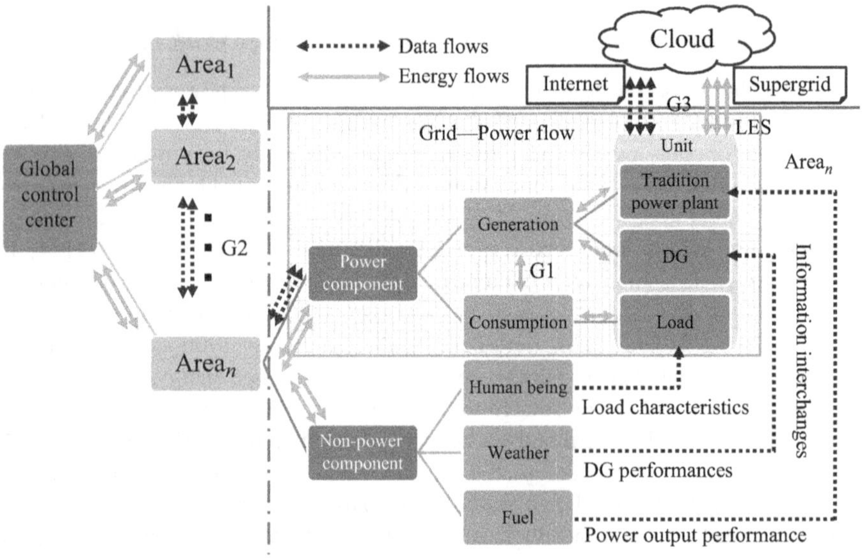


*Figure 13.6* *Data flows and energy flows for three generations of power systems. The single lines, double lines, and triple lines indicate the flows of G1, G2, and G3, respectively. Reproduced from Reference 23*

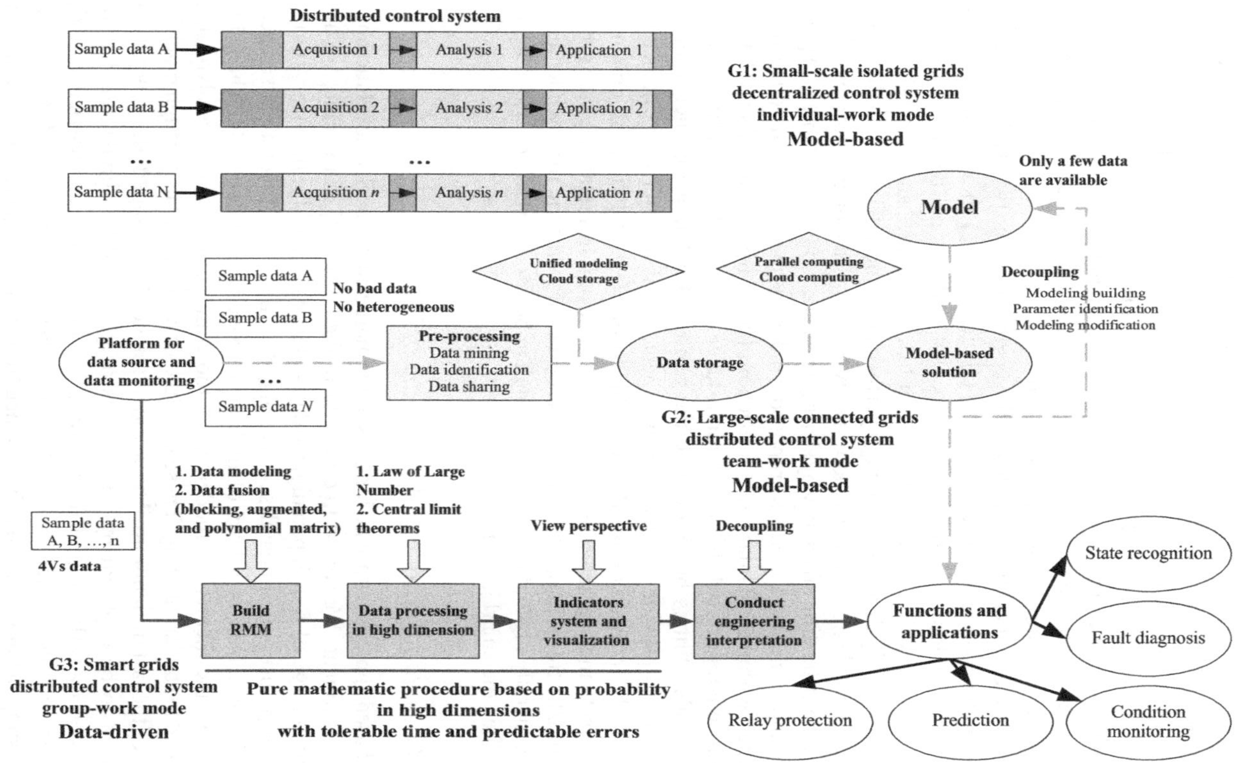


*Figure 13.7 Data utilization methodology for SA of power systems. The above, middle, and below parts indicate the data processing procedures and the work modes for G1, G2, and G3, respectively. Reproduced from Reference 23*

Large-scale interconnected grids were developed from about 1960 to 2000, featured by zone-dividing large-scale interconnected grids. For G2, utilities interchange energy and data within the adjacent ones. The components are dispatched by control center and operating under team-work mode. The regional team leaders, like local dispatching centers, substations, or microgrid control centers, aggregate their own team members (i.e., components in the region) into a standard black box model. These standard models will be further aggregated by the global control center for the control or prediction purposes. The above two aggregations are achieved by four steps, which are data monitoring, data preprocessing, data storage, and data processing. The above work mode is in a hierarchical model-based way.

The development of G3 was launched at the beginning of the twenty-first century and expected to be completed around 2050 [71]. Figure 13.5(c) shows that the clear-cut partitioning is no longer suitable for G3, as well as the model-based work mode which is based on the hierarchical structure. For G3, the abilities and demands of the individual units are highlighted: these high performance and self-control individuals result in much more flexible flows for both energy exchange and data communication; they improve the utilization efficiency by sharing resources among the whole grid [28]. Accordingly, the *group-work mode* is proposed. Under this mode, the individuals play a dominant part within the fundamental restrictions [24]. Virtual power plants (VPPs) [34] and multi-microgrids (MMGs) [25], for instance, are typically G3 utilities. These group-work mode utilities break through the regional limitation for energy flows and data flows and provide a relaxed environment. It has a better performance by benefiting both the individuals and the grids: the former (i.e., individuals), driven by their own interests and characteristics, are able to create or join a relatively free group to benefit mutually from sharing the superior resources of their own; meanwhile, these utilities are generally big and controllable enough to be good customers or managers to the latter (i.e., the smart grids).

### *13.7.2 The methodology of SA for smart grids*

The model-based approaches are no longer suit to smart grids; to seek a novel methodology of SA is urgent. We would like to refer to the book [27] as the clue; we are now entering the age of fourth paradigm—data-intensive scientific discovery. Originally, there was just experimental science, and then there was theoretical science, with Kepler's laws, Newton's laws of motion, Maxwell's equations, etc. Then, for many problems, the theoretical models grew too complicated to solve analytically, and people had to start simulating. These simulations have carried us through much of the last half of the last millennium. The world of science has changed; the new model is for the data to be captured by instruments or generated by simulations before being processed by software and for the resulting information or knowledge [19]. The techniques and technologies for such data-intensive science are so different that it is worth distinguishing data-intensive science from computational science as a new, fourth paradigm for scientific exploration [7].

The second and third paradigms are typically model based—they use the equations, expressions, formulas, or the aggregated simulations to describe the operation regulations and interaction mechanisms; in this way, it gets insight into the objects and even the world. These descriptions are always in low dimensions, i.e., they are fully depended on a few model parameters and chosen sampling data. The massive data of the systems and classical data processing methods (e.g., principal component analysis (PCA)) merely serve the models' determinant factors via training or iteration; in these paradigms, the massive data, in essence, play as a server but not a king. Accordingly, back to the power system, we summarize the classical decision-making approaches (always model based) and design the data-driven ones (Figure 13.7).

The classical approaches in power grids are always based on models. For instance, (1) when the initial value of PQ nodes, PV nodes, $V\theta$ nodes, and the network admittance matrix $\mathbf{Y}$ are given, the grid's static state is estimated via the power flow equations; (2) ideally, wind power is proportional to the cube of its speed; and (3) the Lyapunov exponent which estimates the transient stability is decided by the parameters of the grid network and the generating units. They all have one thing in common: the value is fully dependent upon only a few parameters, owing to the low-dimensional model (e.g., $y = ax^2 + bx + c$ is a three-dimensional model—the relationship between $x$ and $y$ depends on $a$, $b$, and $c$). In other words, model-based work mode is not able to turn massive data into driven force; we can hardly conduct SA more precisely as the data growth. Even worse, the more data mean the more bad ones; if we take the bad ones into the fixed models, we will almost surely obtain bad results.

The prevailing data-processing methods, essentially, are serving under model-based mode, e.g., (1) the procedure of PCA mainly aims to form a proper medium (i.e., principal component) for taking into the model and (2) most massive data processing algorithms just aim to modify the model and its parameters by the supervised training with a large number of samples. For concrete discussion, we take the PCA for an example. PCA is a prevailing data-driven method [65,66]; the steps are listed as follows:

---

**Steps of PCA**

---

(1) Select data $\mathbf{Y}=[\mathbf{y}_1,\mathbf{y}_2,\cdots,\mathbf{y}_N]$, $\mathbf{y}_i=[y_{1,i},y_{2,i},\ldots,y_{n,i}]^T$.
(2) $\mathbf{C}_\mathbf{Y}=\mathbf{Y}^H\mathbf{Y}$, calculate $\lambda(\mathbf{C}_\mathbf{Y})$.
(3) Rearrange and select the top $m$ eigenvalues: $\lambda_1\rightarrow\mathbf{p}_1,\ldots,\lambda_m\rightarrow\mathbf{p}_m$.
(4) Form $m$-dimensional principal component subspace $\mathbb{L}(\mathbf{p}_1,\mathbf{p}_2,\cdots,\mathbf{p}_m)$ and project the original $N$ variables onto it: select $m' \leqslant m$ vector-based variables as the pilot PMUs from $N$ PMUs to form the linear basis matrix $\mathbf{Y}_B=[\mathbf{y}_{b1},\mathbf{y}_{b2},\cdots,\mathbf{y}_{bm'}]$. The selected $m'$ variables should be as orthogonal to each other as possible, which means $\min(\cos\theta)=(\mathbf{y}_{bi}\cdot\mathbf{y}_{bj})/(|\mathbf{y}_{bi}||\mathbf{y}_{bi}|)\quad (i,j=1,2,\ldots,m';\ i\neq j)$.
(5) Represent non-pilot PMUs $\mathbf{y}_{ci}\quad (i=1,2,\ldots,N-m')$ for training: let $\mathbf{v}_{ci}=[v_{1,ci},v_{2,ci},\ldots,v_{m',ci}]^T$ be the vector of regression coefficients:

$$\mathbf{y}_{ci}=\langle(\underbrace{\mathbf{y}_{b1},\mathbf{y}_{b2},\cdots,\mathbf{y}_{bm'}}_{\text{Basis}})\cdot(v_{1,ci},v_{2,ci},\ldots,v_{m',ci})\rangle=\mathbf{Y}_B\mathbf{v}_{ci}$$

$$\Rightarrow \mathbf{v}_{ci}=(\mathbf{Y}_B^H\mathbf{Y}_B)^{-1}\mathbf{Y}_B^H\mathbf{y}_{ci};$$

(6) Train at $t_s = s-1$ :

$$\mathbf{v}_{ci}{}^{(s-1)} \Leftarrow f(\mathbf{Y}_B{}^{(s-1)}, \mathbf{y}_{ci}{}^{(s-1)});$$

(7) Judge at $t = s$ :

$$\|\mathbf{y}_{ci}{}^{s}, \mathbf{Y}_B^s \mathbf{v}_{ci}{}^{(s-1)}\|.$$

---

*Steps (6) and (7)* are the executive parts based on data processing procedure—*steps (1)–(5)*. The key steps are *step (4)* and *step (5)*; they constitute the training procedure. By choosing $m'$ PMUs (as pilot PMUs) from total $N$ PMUs, the procedure tags the system in a reduced subspace ($N \to m'$); in this way, the function $\mathbf{v}_{ci}$ is inferred.

Furthermore, we take Ring Law Analysis as an example. The Ring Law Analysis conducts EED with following steps:

---

**Steps of Ring Law Analysis**

(1) Select arbitrary raw data (or all available data) as the data source $\mathbf{\Omega}$.
(2) Forming RMM $\hat{\mathbf{X}}$ at a certain time $t_i$.
(3) Obtain $\tilde{\mathbf{Z}}$ by variable transformations ($\hat{\mathbf{X}} \to \tilde{\mathbf{X}} \to \mathbf{X}_u \to \mathbf{Z} \to \tilde{\mathbf{Z}}$).
(4) Calculate eigenvalues $\lambda_{\tilde{\mathbf{Z}}}$ and plot the Ring on the complex plane.
(5) Conduct high-dimensional analysis.
  (4a) Observe the experimental ring and compare it with the reference one.
  (4b) Calculate $\tau_{\text{MSR}}$ as the experimental value.
  (4c) Compare $\tau_{\text{MSR}}$ with the theoretical value $N\mathbb{E}(\tau_{\text{MSR}})$.
(6) Repeat (2)–(5) at the next time point ($t_i = t_i + 1$).
(7) Visualize $\tau_{\text{MSR}}$ on the time series.
(8) Make engineering explanations.

---

*Steps (2)–(7)* conduct high-dimensional analysis only using raw data, and then visualize the indicator $\tau$; they are unsupervised statistical proceedings without assumptions and simplifications. In *step (2)*, for different purposes, arbitrary raw data, even ones from distributed nodes or intermittent time periods are able to be focused on to form the RMM $\hat{\mathbf{X}}$. It is also an online data-driven method requiring no knowledge of the physical models/topologies. In addition, the size of $\hat{\mathbf{X}}$ is controllable during *step (2)*; it relieves the curse of dimensionality in some ways.

### *13.7.3 Novel indicator system and its advantages*

In Section 13.5, we have studied *how to design various LESs and calculate their values*. These indicators are derived from a common RMM $\mathbf{X}$ and generated via distinct statistical procedures; they are in high dimensions and without introducing system errors (assumption and simplification). In other words, they learn the real systems from multiple perspectives objectively; they consist of a indicator system

*Table 13.2 Indicator system for power system*

| Statistical indicator | Classical indicator |
|---|---|
| Supported by novel guidelines | Supported by tradition laws |
| Unclear defined engineering concept | Clear defined one |
| Sensors and network of high quality are required | Sensors and network are okay enough |
| Non-asymptotic probabilistic value | Asymptotic accurate one |
| Obtained by data-driven methodology | Obtained by model-based one (often in low dimensions) |
| The value relies on massive (all) data | The value relies on a few data (less than model's dimensions) |
| Robust against bad data and insensitive to individual data | Sensitive to sample selection |
| Pure statistical procedure without system errors | System errors are inevitable |
| Random errors can be estimated by the model size $(N, T)$ | The errors depend on the model building procedure |
| Compatible with diverse data | Only for the assigned ones (rely on the model) in a fix/inflexible form |
| Ready transformation in statistical space (preprocessing) | |
| Naturally decoupling the interconnected system | Decoupling based on assumptions and simplifications |

which may lead to a new epistemology. The novel statistical indicator is shown in Table 13.2.

As in Table 13.2, the statistical indicator system provides a much better way than the classical one to utilize the massive data for a certain system. The former to the latter, in some sense, is just like the quantum physics to the classical physics. Instead of an accuracy measurement (for the mass, position, velocity, etc.), the latter only gives a statistical law to describe the system, owing to the complex of the operation rules and interaction mechanism for the quantum world. And the quantum physics is one of the greatest achievements in the history of physics (even the most greatest one ever); it still influences and puzzles us deeply nowadays [11].

So, in conclusion, RMT is a data-driven methodology and based on which we can turn the real power grid into RMM. With pure mathematical procedure, a statistical indicator system is formed as a new epistemology for the grid. This data-driven solution has following advantages:

(1) For a smart grid, it is hard to build an acceptable model (enough veracity, tolerable time, hardware resources, etc.).
(2) The online sampling data is redundant and reliable than the rating one (e.g., conductor resistance).

(3) RMM is a statistical model but not a physical one; it requires no knowledge of the grid topologies and operation mechanism.
(4) Some RMM operations (matrix augmentation, matrix blocking, etc.) perfectly match traditional challenges (the heterogeneous data combination, interconnected systems decoupling, etc.). Moreover, these operations do not bring in system errors.
(5) Some parallel calculation (e.g., status matrix augmented with multiple factor matrix simultaneously) together with matrix blocking will shorten the processing times greatly and releases the dimensionality curse somehow.
(6) This mode is mainly conducted in high dimensions; it is robust against some problems (such as error accumulations, spurious correlations, and even bad data in core area) which are hard to be solved in low dimensions.
(7) This model gives a statistical description for correlation analysis. Especially for a complex system, the statistical description performs better than the causal one. Take pairs of sunlight and system for an example, it is hard to make a considerate correlation analysis via the causal models (light intensity ⇒ solar cells ⇒ power system is a typical logic, so is light intensity ⇒ outdoor temperature ⇒ air-conditioner ⇒ power system). In contrast, only using the sampling data, we can easily obtain the correlation coefficient.

## 13.8 Case studies

In this section, we use both simulated data and real ones to validate the proposed approach. For the simulated case, we adopt the standard IEEE 118-bus system (Figure 13.8). Detailed information about the simulation is referred to the *case118.m* in *Matpower package* and *Matpower 4.1 User's Manual* [72]. There are generally three scenarios for the system's input: (1) only white noises (i.e., $\mathbf{R}$), e.g., small random fluctuations of loads and Gaussian sample errors; (2) signals to be detected and noises (i.e., $\mathbf{R}+\mathbf{S}$), it means that there are also sudden changes, or even serious faults; (3) signals already known, signals to be detected, and noise (i.e., $\mathbf{R}+\mathbf{A}+\mathbf{S}$). For the real case, we use a PMU database of some power grid. Generally, the event detect may be modeled as binary hypothesis testing: normal hypothesis $\mathcal{H}_0$ (no signal $\mathbf{S}$ present) and abnormal hypothesis $\mathcal{H}_1$ (signal $\mathbf{S}$ present).

### *13.8.1 Case 1: anomaly detection and statistical indicators designing using simulated 118-bus system*

According to *From Physical System to Random Matrix* in Section 13.3, the data source $\Omega_{\mathbf{V}}=\hat{v}_{i,j}$ ($n=118$, $t=1{,}500$) is obtained to map the simulated system.

For details about this work, see our previous work [23]. Figure 13.9 shows the results.

At sampling time $t_s=600$ s, the RMM $\hat{\mathbf{X}}$ includes a time period $\mathbf{A}_{\mathrm{Time}}=361:600$ s; the noises play a dominant part during the period. Figure 13.9(a) shows that the distribution of $\lambda_{\tilde{\mathbf{Z}}}$ is more closely to the reference ring (when $L=1$); and Figure 13.9(c) shows that the distribution of $\lambda_{\mathbf{M}}$ (in bars) meets the M- P distribution (in solid line)

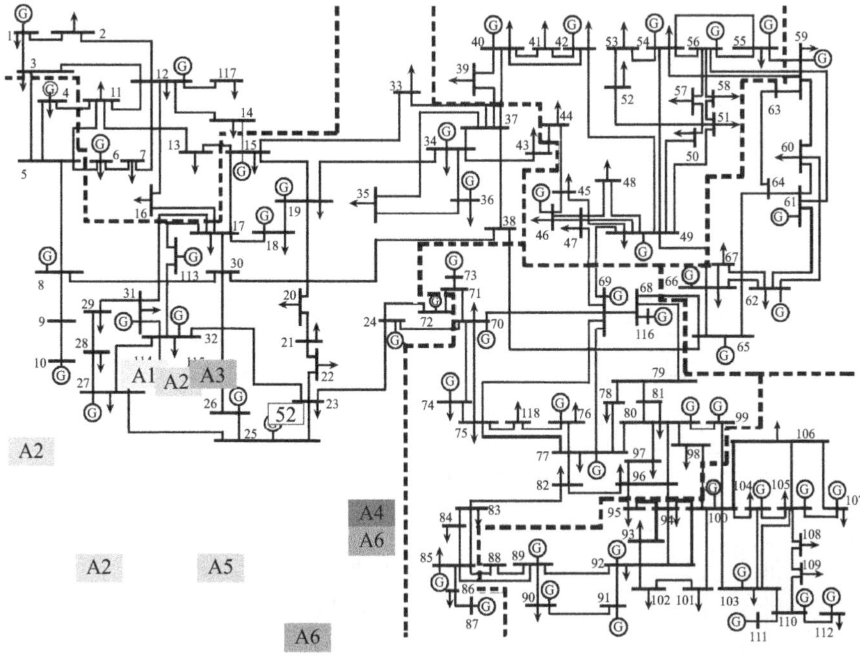


*Figure 13.8 Partitioning network for the IEEE 118-bus system. There are six partitions, i.e., A1, A2, A3, A4, A5, and A6*

quite well. Whereas, at sampling time $t_s = 601$ s, Figure 13.9(b) shows that the eigenvalue points collapse to the center point of the circle, which means that the correlations of the data have been enhanced somehow, and Figure 13.9(d) also shows the deviation between the experimental distribution and the theoretical one.

Furthermore, we can plot $\tau_{\text{MSR}}$-$t$ curve (Figure 13.9(e)). It is observed that $\tau_{\text{MSR}}$ starts to decrease $(0.8665, 0.6308, \ldots, 0.4927)$ at $t = 600$ s, just when the event (*sudden change of* $P_{Bus\text{-}52}$) occurs as the signal. The influence lasts for full-time length ($T = 240$ s) and the decreasing lasts for half (120 s); thus, a "U"-shaped curve is observed. In this way, we conduct anomaly detection; the time for the beginning point of "U" ($t = 600$ s for this case) is right the anomaly start time.

Designing LESs is a major target for the indicator system; here, we study diverse LESs with distinct test functions (see Table 13.3). Keeping $T = 240$ s, we can divide the temporal space into five subspaces (stages) according to the status of $P_{\text{Bus-52}}$ as follows:

| | **Time areas (time length)** | | | | **Description** |
|---|---|---|---|---|---|
| **S1** | 241 s | – | 600 s | (360 s) | Fluctuations around 0 M |
| **S2** | 601 s | – | 840 s | (240 s) | A step signal |
| **S3** | 841 s | – | 1,200 s | (360 s) | Fluctuations around 300 M |
| **S4** | 1,201 s | – | 1,306 s | (106 s) | A ramp signal |
| **S5** | 1,307 s | – | 1,500 s | (194 s) | Static voltage collapse |

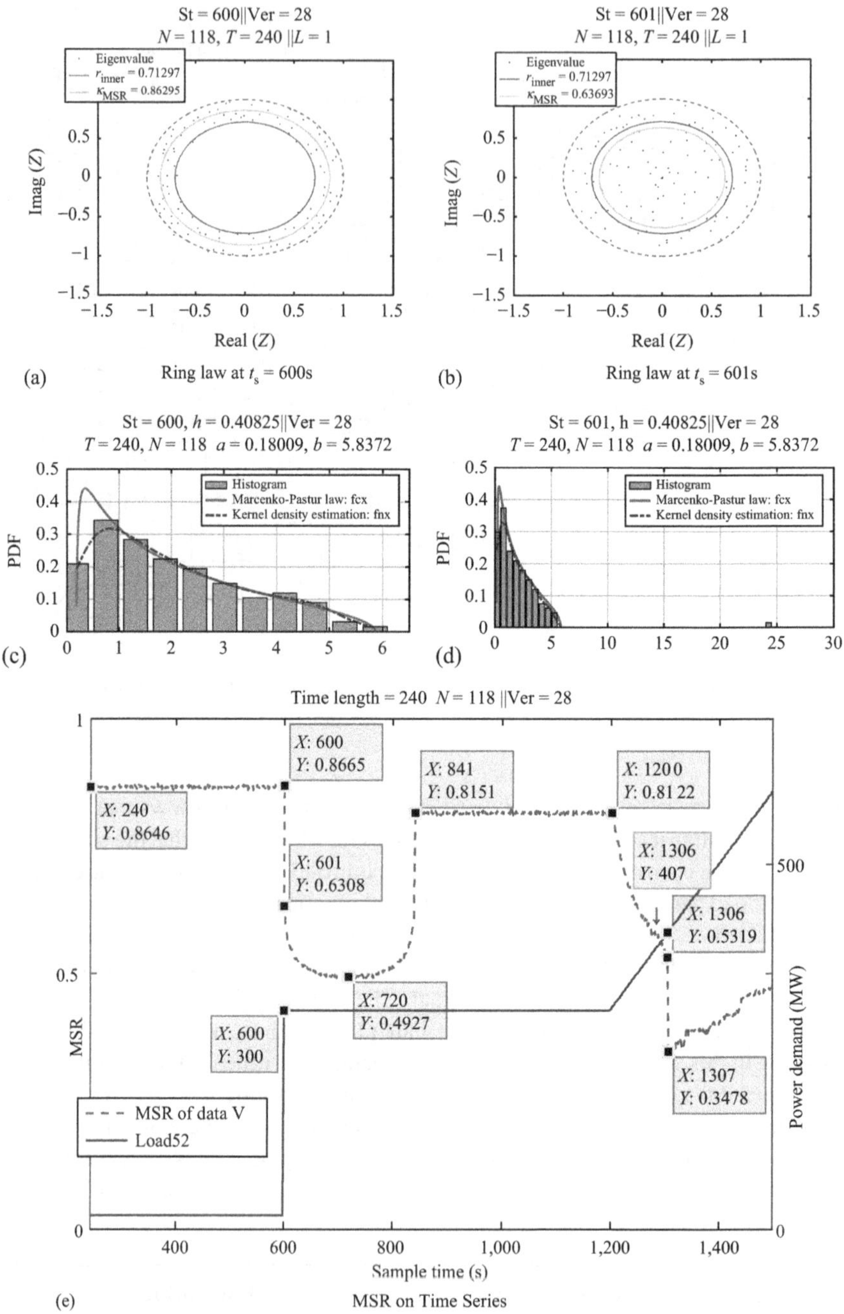


*Figure 13.9 Ring law, M-P law, and MSR for simulated data* **V**. *(a) Ring law at $t_s = 600$ s, (b) Ring law at $t_s = 601$ s, (c) M-P law at $t_s = 600$ s, (d) M-P law at $t_s = 601$ s, and (e) MSR on time series. Reproduced from Reference 26*

*Table 13.3 LESs and their values. Reproduced from Reference 26 with permission of ©IEEE*

| | MSR | $T_2$ | $T_3$ | $T_4$ | DET | LRF |
|---|---|---|---|---|---|---|
| **E0**: Theoretical value | | | | | | |
| $\mathbb{E}(\tau)$ | 0.8645 | 1.34E3 | 1.01E4 | 8.35E4 | 48.3 | 73.68 |
| $\mathbb{D}(\tau)$ | – | 6.65E2 | 9.35E4 | 1.30E7 | 1.32 | 1.42 |
| $c_v$ | – | 0.0193 | 0.0304 | 0.0432 | 0.0238 | 0.0162 |
| **S1**: Only small fluctuations around 0 MW | | | | | | |
| $\mu(\tau)$ | 0.8648 | 1.33E3 | 9.93E3 | 8.19E4 | 73.68 | 73.3 |
| $\sigma^2(\tau)$ | 0.0080 | 6.53E1 | 2.20E4 | 4.67E6 | 0.406 | 0.322 |
| **S2**: A step signal ($P_{\text{Bus-52}}$: 0 MW → 300 MW) is included | | | | | | |
| $\mu(\tau)$ | 0.5149 | 1.29E4 | 1.92E6 | 3.04E8 | −174 | 295 |
| $\sigma^2(\tau)$ | 0.0788 | 3.30E6 | 1.51E11 | 5.66E15 | 890 | 893 |
| **S3**: Only small fluctuations around 300 MW | | | | | | |
| $\mu(\tau)$ | 0.8141 | 1.54E3 | 1.73E4 | 2.89E5 | 27.9 | 93.1 |
| $\sigma^2(\tau)$ | 0.0250 | 1.81E2 | 3.30E5 | 4.20E8 | 0.507 | 0.419 |
| **S4**: A ramp signal as the system incoming | | | | | | |
| $\mu(\tau)$ | 0.6448 | 6.43E3 | 6.40E5 | 7.54E7 | −61.4 | 182 |
| $\sigma^2(\tau)$ | 0.0571 | 7.20E6 | 1.68E11 | 3.29E15 | 1.74E3 | 1.73E3 |
| **S5**: Static voltage collapse for the system | | | | | | |
| $\mu(\tau)$ | 0.4136 | 7.49E3 | 6.48E5 | 7.25E7 | −598 | 719 |
| $\sigma^2(\tau)$ | 0.1076 | 7.47E6 | 2.99E11 | 1.02E15 | 4.16E4 | 4.16E4 |

*$c_v = \sqrt{\mathbb{D}(\tau)}/\mathbb{E}(\tau)$ is the coefficient of variation.

The above results in Table 13.3 validate that it is feasible to analyze the power grid with these statistical indicators. The upper layer should be ontology and we do not go that far; here, we just make some engineering statements/analyses:

1. Independent of the system models/topologies, we can design the LES $\tau$. For a RMM $\mathbf{X}$ with determinate size $N \times T$, if the test function $\varphi(\lambda)$ is given, the LES $\tau$ is obtained; some related theoretical values (the expectation $\mathbb{E}(\tau)$, the variance $\mathbb{D}(\tau)$, and the coefficient of variation $c_v$) are able to be calculated as well.
2. Among $\varphi(\lambda)$ given above, LRF performs best in the view of $c_v$ (low $c_v$ means high precision and repeatability of the assay [48]; here, we regard 12% as the upper bound).

*Table 13.4 Events assume for Case 2*

| | | | | |
|---|---|---|---|---|
| Event 1 | $t$/s | 1:350 | 351:650 | 651:1,000 |
| | $P_{bus\_117}$/MW | 20 | 120 | 20 |
| Event 2 | $t$/s | 1:300 | 301:700 | 701:1,000 |
| | $P_{bus\_59}$/MW | 113 | $113 + (t - 300)$ | 213 |
| Event 3 | $t$/s | 1:350 | 351:650 | 651:1,000 |
| | $P_{bus\_117}$/MW | 20 | 120 | 20 |
| | $t$/s | 1:300 | 301:700 | 701:1,000 |
| | $P_{bus\_59}$/MW | 113 | $113 + (t - 300)$ | 213 |

3. We can make a summary about the performance of the experimental values (mean $\mu(\tau)$ and variance $\sigma^2(\tau)$) comparing to the theoretical ones (expectation $\mathbb{E}(\tau)$ and variance $\mathbb{D}(\tau)$):
   (a) During **S1**, $\mu(\tau)$ is close to $\mathbb{E}(\tau)$; $\sigma^2(\tau)$ is much less than $\mathbb{D}(\tau)$.
   (b) During **S3**, $\mu(\tau)$ has a little bias (i.e., $|\mu(\tau) - \mathbb{D}(\tau)|$), but more than **S1**; $\sigma^2(\tau)$ is acceptable ($\sqrt{\sigma^2(\tau)}/\mu(\tau) < 12\%$).
   (c) For **S2**, **S4**, and **S5**, $\mu(\tau)$ has much more bias; $\sigma^2(\tau)$ is always too big to be accepted.
   (d) Variance $\sigma^2(\tau)$ is much more sensitive than mean $\mu(\tau)$.

   Then, we can conjecture that:

   (a) The more stable the system is, the more effective the theoretical values become ($\mu(\tau)$ is close to $\mathbb{E}(\tau)$; $\sigma^2(\tau)$ is less than $\mathbb{D}(\tau)$).
   (b) Different test functions $\varphi(\lambda)$ have distinct characteristics and functions. In this sense, we can balance the reliability and sensitivity for the anomaly detection in a special system.
   (c) In addition, a test function is akin to a filter in some sense; it has the potential to trace a specific anomaly (conduct EED under input model **R** + **A** + **S**).
   (d) For a special purpose, e.g., the lowest $c_v$ or the lowest bias, there should exist an optimal combination of the Chebyshev Polynomials as the test function.

### *13.8.2 Case 2: correlation analysis for single factor using simulated 118-bus system*

We assumed the load of each bus is the factor, and a change of the factor is designed as the signal. Similar to the first case, we arrange three distinct events in Table 13.4.

Let $n = 118$, $m = 118$, $t = 1{,}000$, $N = 118$, $T = 240$; the results are shown in Figure 13.10: Events 1 and 2 validate the effectiveness of the correlation analysis method when signals of single factor are added into the system; and Event 3 shows the situation of a scene of a mixed signals.

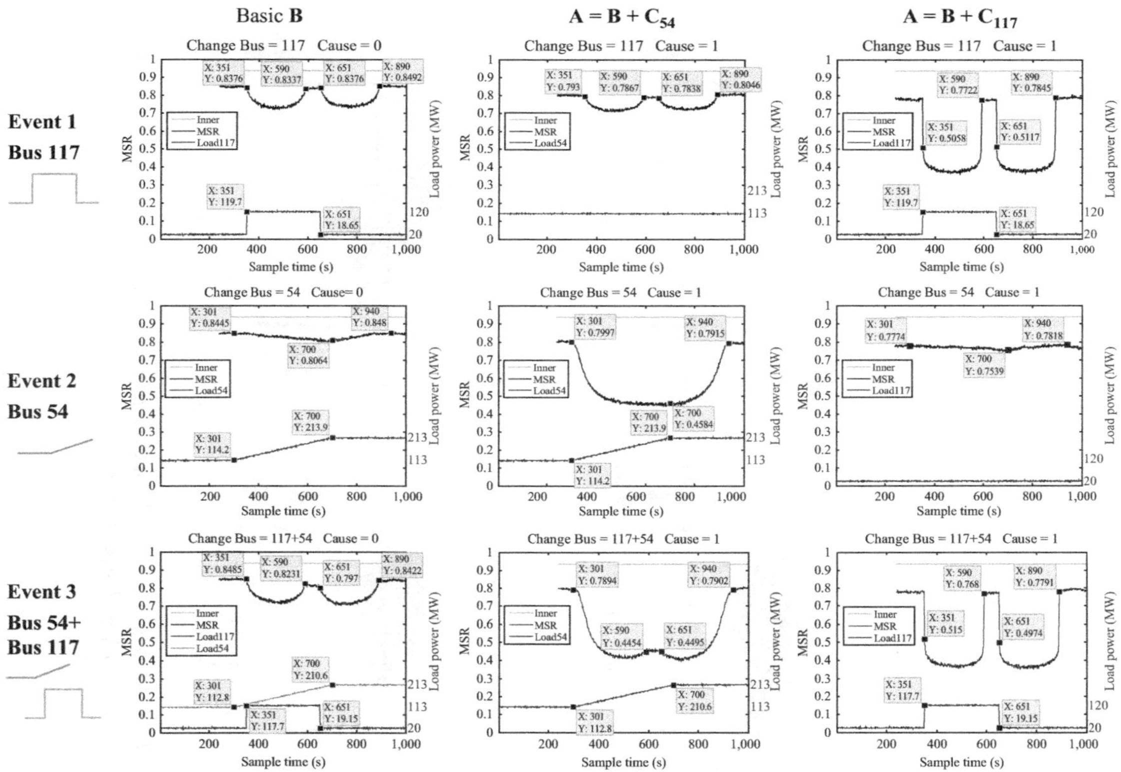


*Figure 13.10 Augmented matrix method for correlation analyses. Reproduced from Reference 68*

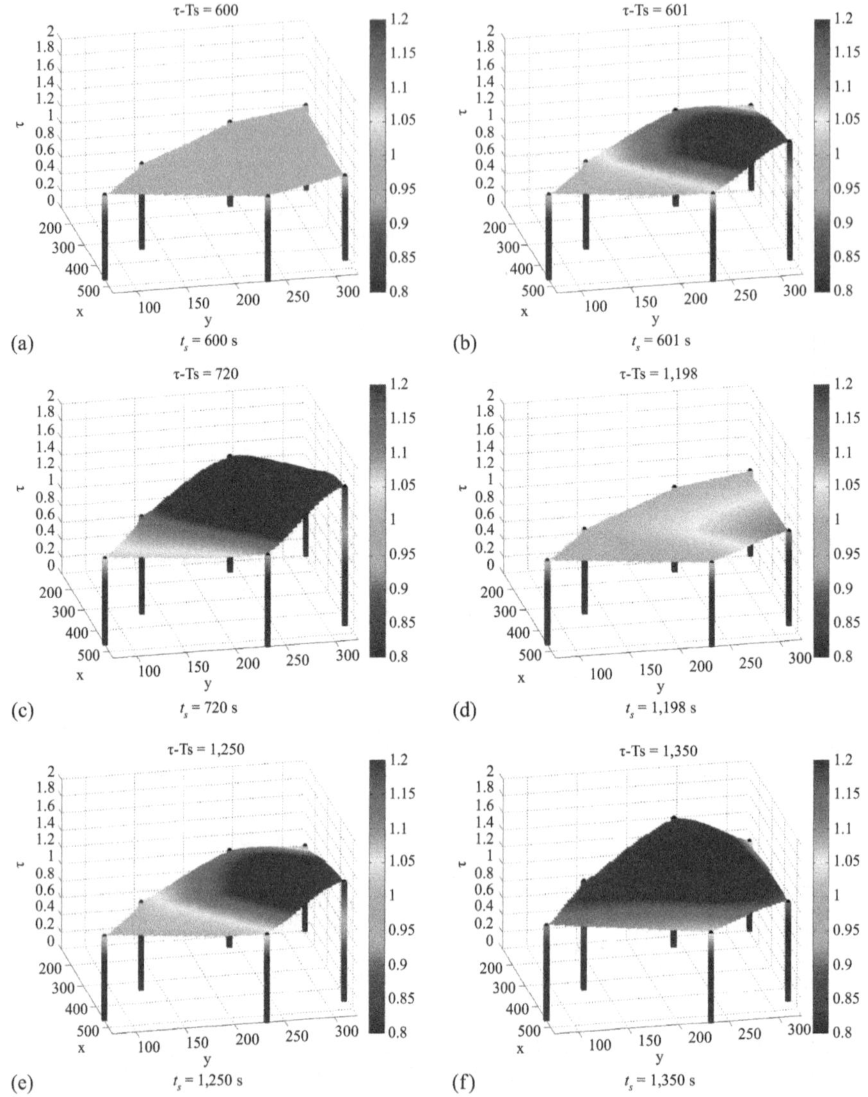


*Figure 13.11 Visualization of the high-dimensional indictor $\eta$ with full data sets. Reproduced from Reference 26*

### *13.8.3 Case 3: advantages of LES and visualization using 3D power-map*

LES $\tau$, as a statistical indicator, is sensitive and robust against bad data; the advantages will be validated via the visualization of $\tau$ using 3D power-map. We denote

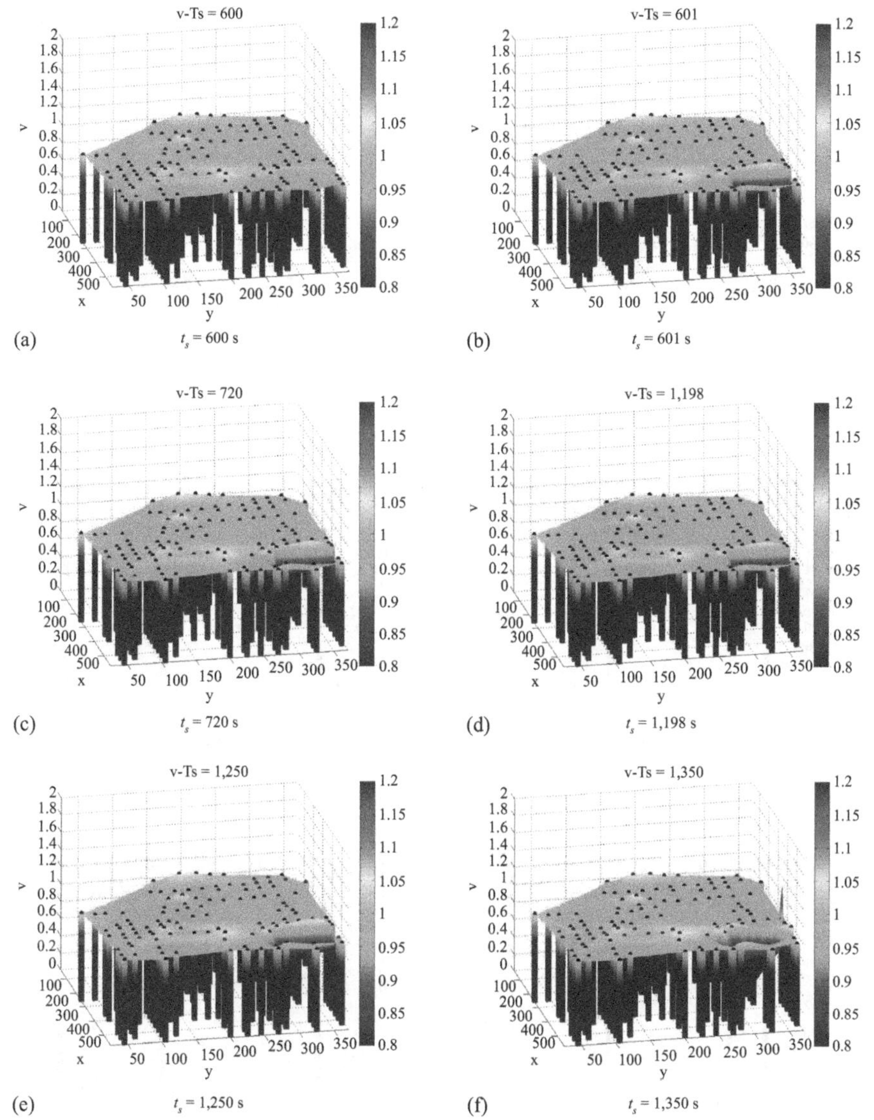


*Figure 13.12 Visualization of the voltage* **V** *with full data sets. Reproduced from Reference 26*

$\eta = \tau_{\text{LRF}}/\mathbb{E}(\tau_{\text{LRF}})$ as the high-dimensional indicator. For each time point, with an interpolation method [62], a 3D map is able to be plotted. Figures 13.11 and 13.13 depict some key frames of the 3D power-map animation with $\eta$, whereas Figures 13.12 and 13.14 with the raw data **V**.

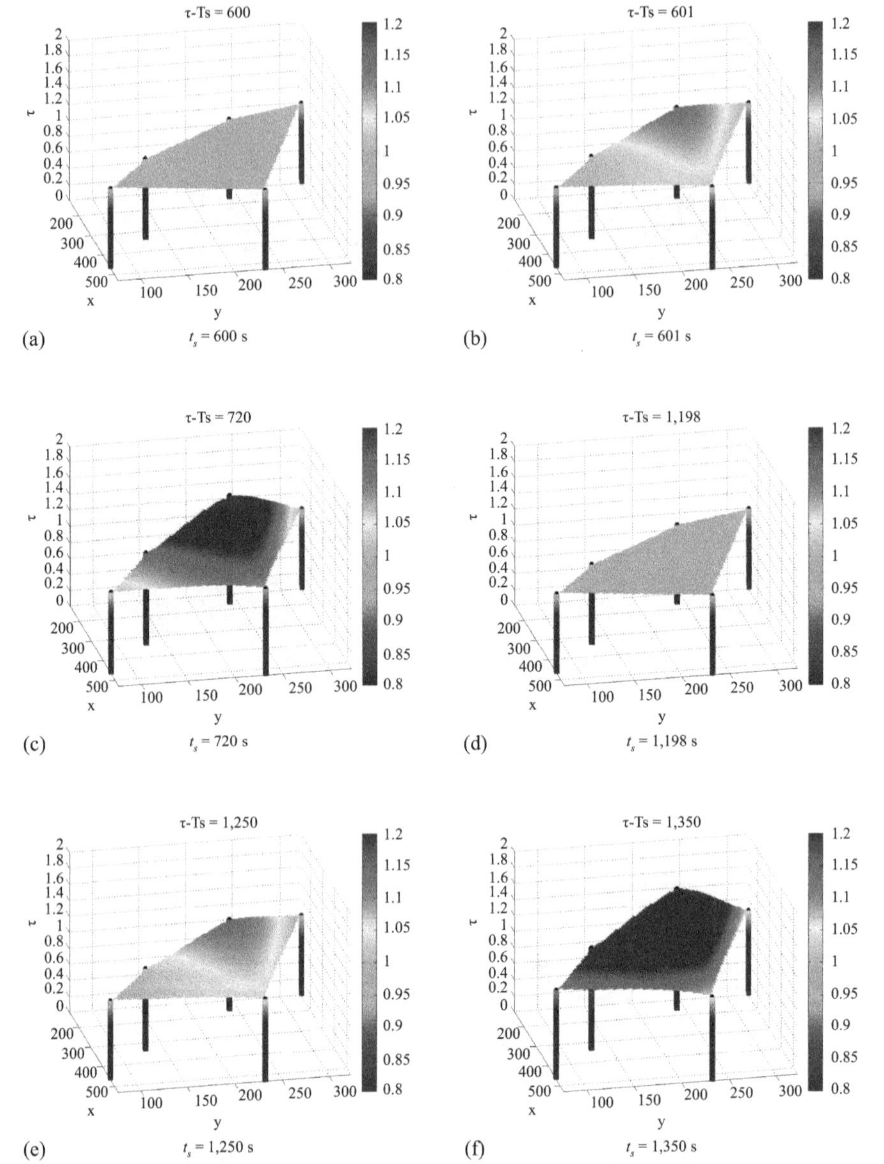


*Figure 13.13 Visualization of $\eta$ without data set of A3. Reproduced from Reference 26*

(a) For Figure 13.11, at time $t=601$ s, $\eta$ of the area around A3 changes relative dramatically; this trend last for the next $T-2=238$ sampling points (i.e., $t_s=602:839$ s) in the animation. Therefore, we conjecture that some event occurs in A3; even we can go further that the event is influential to A1, A2, A4, and A5,

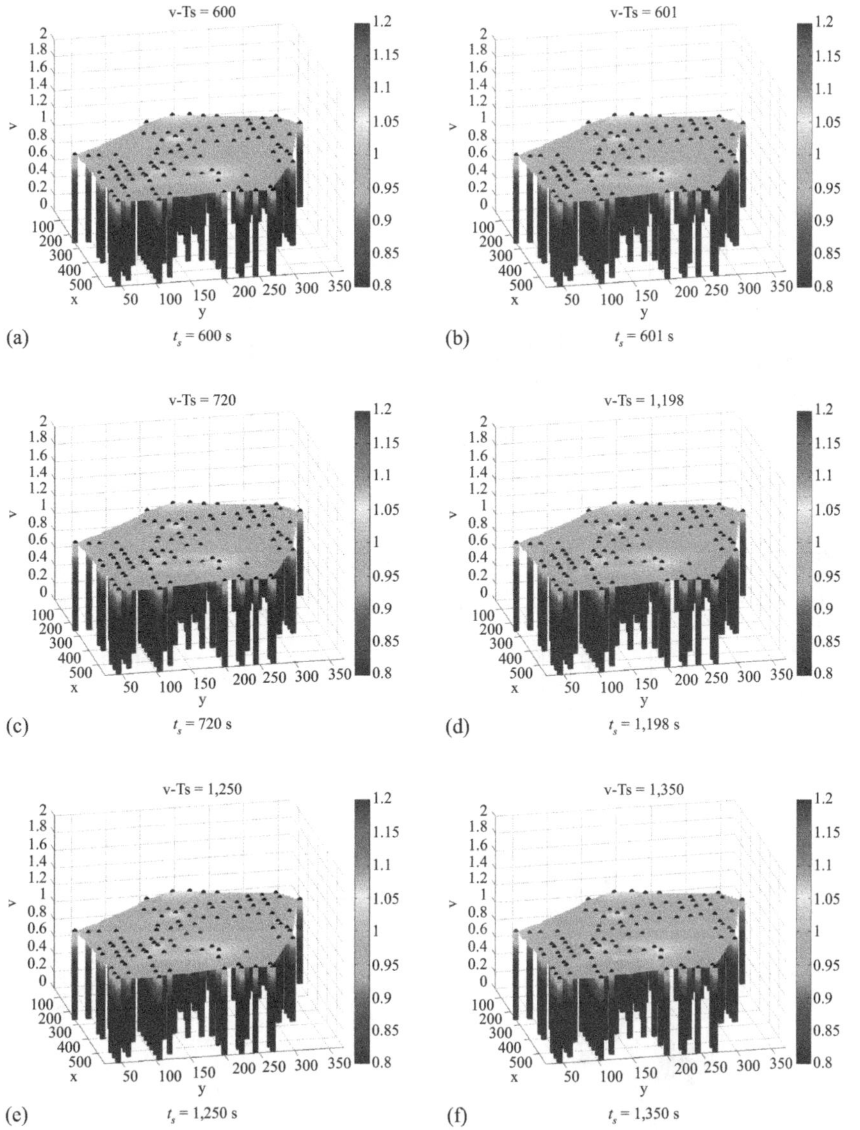


*Figure 13.14 Visualization of* **V** *without data set of A3. Reproduced from Reference 26*

and has little impact on A6. These conjectures, in a reasoning way, coincide with the common sense that there is a sudden change in A3 at $t = 601$ s.

(b) For Figure 13.11, with sustainable growth of power demand at some bus ($P_{\text{Bus-52}}$ for this case), the whole system becomes more and more vulnerable. The

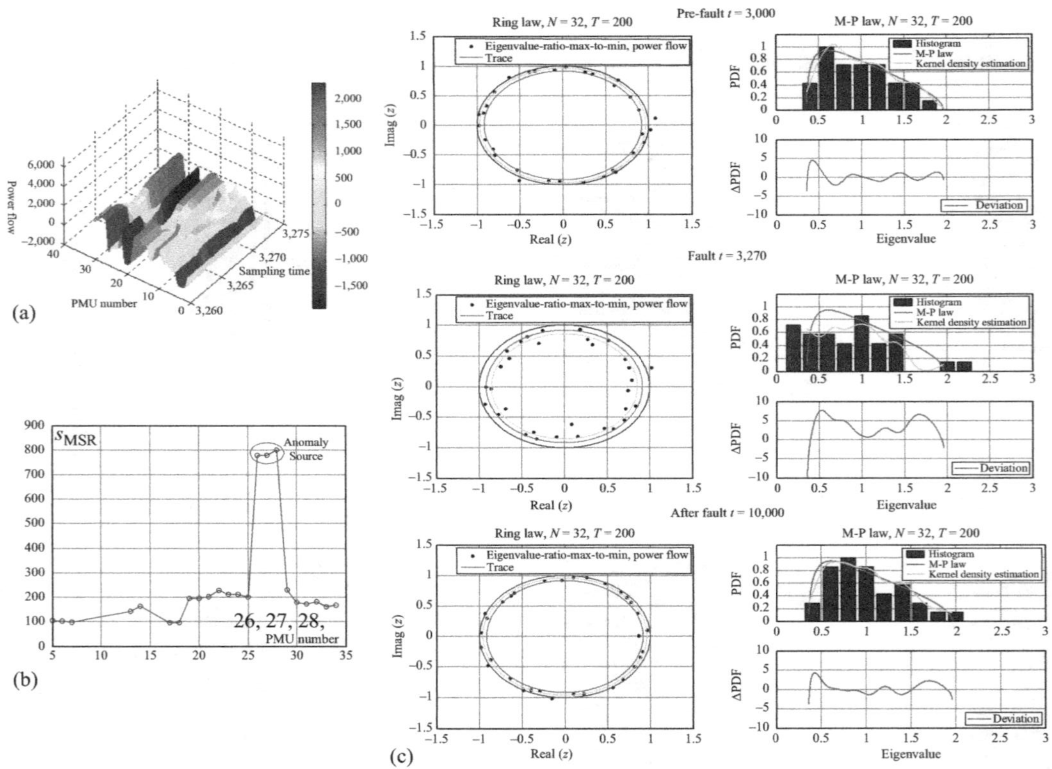


*Figure 13.15 Situation awareness using real data. (a) PMU data, (b) correlation analysis, and (c) anomaly detection*

vulnerability can be estimated, before the system has a breakdown due to voltage collapse, via the visualization of $\eta$.

(c) Moreover, if the most related data (data of A3 for this case) are lost somehow, hardly any information can be gotten by $\mathbf{V}$ in Figure 13.14, whereas the proper judgments can still be achieved by $\eta$ in Figure 13.13.

In general, the combination of high-dimensional indicators (e.g., $\eta$) and 3D power-map is really a novel and feasible approach to conduct SA.

### *13.8.4 Case 4: SA using real data*

With the PMU data shown in Figure 13.15(a) (32 PMUs, 14,200 Sampling Times), we validate the anomaly detection and correlation analysis in Figure 13.15(c) and 13.15(b)—The fault is occurring at $t=3{,}270$ and is very related to the PMU 26, 27, and 28.

## Bibliography


[1] Greg W Anderson and Ofer Zeitouni. "A CLT for a band matrix model". *Probability Theory and Related Fields*, 134(2):283–338, 2006.

[2] ZD Bai. "Convergence rate of expected spectral distributions of large random matrices. Part I. Wigner matrices". *Annals of Probability*, 21(2): 625–648, 1993.

[3] ZD Bai, Baiqi Miao, and Jhishen Tsay. "A note on the convergence rate of the spectral distributions of large random matrices". *Statistics & Probability Letters*, 34(1):95–101, 1997.

[4] ZD Bai, Baiqi Miao, and Jian-Feng Yao. "Convergence rates of spectral distributions of large sample covariance matrices". *SIAM Journal on Matrix Analysis and Applications*, 25(1):105–127, 2003.

[5] Zhidong D Bai. "Convergence rate of expected spectral distributions of large random matrices. Part II. Sample covariance matrices". *Annals of Probability*, 21(2): 649–672, 1993.

[6] Zhidong D Bai and Jack W Silverstein. "CLT for linear spectral statistics of large-dimensional sample covariance matrices". *Annals of Probability*, 32(1):553, 2004.

[7] Gordon Bell, Tony Hey, and Alex Szalay. "Beyond the data deluge". *Science*, 323(5919):1297–1298, 2009.

[8] Tómas A Brody, J Flores, J Bruce French, PA Mello, A Pandey, and Samuel SM Wong. "Random-matrix physics: spectrum and strength fluctuations". *Reviews of Modern Physics*, 53(3):385–480, 1981.

[9] Z Burda, RA Janik, and MA Nowak. "Multiplication law and S transform for non-Hermitian random matrices". *Physical Review E*, 84(6):061125, 2011.

[10] Yingshuang Cao, Long Cai, C Qiu, *et al.* "A random matrix theoretical approach to early event detection using experimental data". *arXiv preprint arXiv:1503.08445*, 2015.

[11] Tianyuan Capo. *History of Quantum Physics*. Liaoning Education Press, Shenyang, 2008.

[12] Nicholas M Ercolani and KDT-R McLaughlin. "Asymptotics of the partition function for random matrices via Riemann–Hilbert techniques and applications to graphical enumeration". *International Mathematics Research Notices*, 2003(14):755–820, 2003.

[13] László Erdős, Horng-Tzer Yau, and Jun Yin. "Rigidity of eigenvalues of generalized Wigner matrices". *Advances in Mathematics*, 229(3):1435–1515, 2012.

[14] VL Girko. "Extended proof of the statement: Convergence rate of the expected spectral functions of symmetric random matrices $\xi$ n is equal to o (n1/2) and the method of critical steepest descent". *Random Operators and Stochastic Equations*, 10(3):253–300, 2002.

[15] F Götze, H Kösters, and A Tikhomirov. "Asymptotic spectra of matrix-valued functions of independent random matrices and free probability[J]". *Random Matrices: Theory and Applications*, 2015, 4(02): 1550005.

[16] F Götze, A Naumov, and A Tikhomirov. "Distribution of linear statistics of singular values of the product of random matrices". *ArXiv e-prints*, Dec. 2014.

[17] F Götze and A Tikhomirov. "Rate of convergence to the semi-circular law". *Probability Theory and Related Fields*, 127(2):228–276, 2003.

[18] Friedrich Götze and Alexander Tikhomirov. "The rate of convergence for spectra of GUE and LUE matrix ensembles". *Open Mathematics*, 3(4):666–704, 2005.

[19] Jim Gray. *Jim Gray on eScience: A Transformed Scientific Method[J]*. The Fourth Paradigm: Data-Intensive Scientific Discovery, 2009, 1.

[20] Alice Guionnet. "Large deviations upper bounds and central limit theorems for non-commutative functionals of Gaussian large random matrices". *Annales de l'IHP Probabilités et Statistiques*, 38:341–384, 2002.

[21] Uffe Haagerup and Steen Thorbjørnsen. "Random matrices with complex Gaussian entries". *Expositiones Mathematicae*, 21(4):293–337, 2003.

[22] Uffe Haagerup and Steen Thorbjørnsen. "Asymptotic expansions for the Gaussian unitary ensemble". *Infinite Dimensional Analysis, Quantum Probability and Related Topics*, 15(01):1250003, 2012.

[23] Xing He, Qian Ai, Robert C Qiu, Wentao Huang, Longjian Piao, and Haichun Liu. "A big data architecture design for smart grids based on random matrix theory[J]". 2015. DOI = "10.1109/TSG.2015.2445828".

[24] Xing He, Qian Ai, Zhiwen Yu, Yiting Xu, and Jian Zhang. "Power system evolution and aggregation theory under the view of power ecosystem". *Power System Protection and Control*, 42(22):100–107, Nov. 2014.

[25] Xing He, Qian Ai, Peng Yuan, and Xiaohong Wang. "The research on coordinated operation and cluster management for multi-microgrids". In *International Conference on Sustainable Power Generation and Supply (SUPERGEN 2012)*, pages 1–3, Hangzhou, Sep. 2012. London: IET.

[26] Xing He, Robert C Qiu, Qian Ai, Yinshuang Cao, Jie Gu, and Zhijian Jin. "A random matrix theoretical approach to early event detection in smart grid[J]". *arXiv preprint arXiv:1502.00060*, 2015.

[27] Anthony JG Hey, Stewart Tansley, Kristin Michele Tolle, *et al. The Fourth Paradigm: Data-Intensive Scientific Discovery*, vol. 1. Microsoft Research, Redmond, WA, 2009.

[28] Zhang Hong, Dongmei Zhao, Chenghong Gu, Furong Li, and Bo Wang. "Economic optimization of smart distribution networks considering real-time pricing". *Journal of Modern Power Systems and Clean Energy*, 2(4):350–356, 2014.

[29] Doug Howe, Maria Costanzo, Petra Fey, *et al.* "Big data: The future of biocuration". *Nature*, 455(7209):47–50, 2008.

[30] John K Hunter and Bruno Nachtergaele. *Applied Analysis*. World Scientific, Singapore, 2001.

[31] Barwick H. The "four Vs" of Big Data. Implementing Information Infrastructure Symposium[J]. 2012-10-02. http://www.computerworld.com.au/article/396198/iiis_four_vs_big_data, 2012.

[32] IS Group. "Managing big data for smart grids and smart meters." IBM Corporation, whitepaper (May 2012) (2012). https://www.dataclaritycorp.com/search-resources/item/managing-big-data-for-smart-grids-and-smart-meters.html.

[33] Jesper R Ipsen and Mario Kieburg. "Weak commutation relations and eigenvalue statistics for products of rectangular random matrices". *Physical Review E*, 89(3), 2014. Art. ID 032106.

[34] Yang Ji. "Multi-agent system based control of virtual power plant and its application in smart grid". Master's thesis, School of Electronic Information and Electrical Engineering, Shanghai Jiaotong University, Shanghai, 2011.

[35] K Johansson. "On fluctuations of eigenvalues of random Hermitian matrices". *Duke Mathematical Journal*, 91(1):151–204, 1998.

[36] Mladen Kezunovic, Le Xie, and Santiago Grijalva. "The role of big data in improving power system operation and protection". In *Bulk Power System Dynamics and Control-IX Optimization, Security and Control of the Emerging Power Grid (IREP), 2013 IREP Symposium*, pages 1–9, Rethymno, Aug. 2013. Piscataway, NJ: IEEE.

[37] H. Kösters and A. Tikhomirov. "Limiting spectral distributions of sums of products of non-Hermitian random matrices". *ArXiv e-prints*, Jun. 2015.

[38] Prabha Kundur, Neal J Balu, and Mark G Lauby. *Power System Stability and Control*, volume 7. McGraw-Hill, New York, NY, 1994.

[39] Laurent Laloux, Pierre Cizeau, Marc Potters, and Jean-Philippe Bouchaud. "Random matrix theory and financial correlations". *International Journal of Theoretical and Applied Finance*, 3(3):391–397, Jul. 2000.

[40] Xia Li, Feng Lin, and Robert C Qiu. "Modeling massive amount of experimental data with large random matrices in a real-time UWB-MIMO system". *ArXiv e-prints*, Apr. 2014.

[41] A Lytova and L Pastur. "Central limit theorem for linear eigenvalue statistics of random matrices with independent entries". *Annals of Probability*, 37(5):1778–1840, 2009.

[42] Madan Lal Mehta. *Random Matrices*, volume 142. Academic Press, San Diego, CA, 2004.

[43] Shengwei Mei, Yuan Gong, and Feng Liu. "The evolution model of three generation power systems and characteristic analysis". *Proceedings of the CSEE*, 34(7):1003–1012, Mar. 2014.

[44] Mehryar Mohri, Afshin Rostamizadeh, and Ameet Talwalkar. *Foundations of Machine Learning*. MIT Press, Cambridge, MA, 2012.

[45] LS Moulin, AP Alves da Silva, MA El-Sharkawi, Robert J Marks. "Support vector machines for transient stability analysis of large-scale power systems". *IEEE Transactions on Power Systems*, 19(2):818–825, 2004.

[46] Lynch C. "Big data: How do your data grow?[J]". Nature, 2008, 455(7209): 28–29.

[47] Daniel B Neill and Gregory F Cooper. "A multivariate Bayesian scan statistic for early event detection and characterization". *Machine Learning*, 79(3): 261–282, 2010.

[48] Forrest W Nutter Jr and Patricia M Schultz. "Improving the accuracy and precision of disease assessments: Selection of methods and use of computer-aided training programs". *Canadian Journal of Plant Pathology*, 17(2): 174–184, 1995.

[49] Guangming Pan, QiMan Shao, and Wang Zhou. "Universality of sample covariance matrices: CLT of the smoothed empirical spectral distribution". *ArXiv e-prints*, Nov. 2011.

[50] L Pastur and V Vasilchuk. "On the law of addition of random matrices". *Communications in Mathematical Physics*, 214(2):249–286, 2000.

[51] Leonid A Pastur. "On the spectrum of random matrices". *Theoretical and Mathematical Physics*, 10(1):67–74, 1972.

[52] Robert Qiu and Michael Wicks. *Cognitive Networked Sensing and Big Data*. Springer, Berlin, 2013.

[53] Robert Qiu and Michael Wicks. *Cognitive Networked Sensing and Big Data*. Springer, Berlin, 2014.

[54] Robert C Qiu and Paul Antonik. *Smart Grid and Big Data: Theory and Practice*. John Wiley & Sons, New York, NY, 2015.

[55] Robert Caiming Qiu, Zhen Hu, Husheng Li, and Michael C Wicks. *Cognitive Radio Communication and Networking: Principles and Practice*. John Wiley & Sons, New York, NY, 2012.

[56] Jonathan T O and Gerald A M. "Special online collection: dealing with data[J]". Science, 2011, 331(6018): 639–806.

[57] Mariya Shcherbina. "Central limit theorem for linear eigenvalue statistics of the Wigner and sample covariance random matrices". *ArXiv e-prints*, Jan. 2011.

[58] Terence Tao, Van Vu, Manjunath Krishnapur, *et al.* "Random matrices: Universality of ESDs and the circular law". *Annals of Probability*, 38(5):2023–2065, 2010.

[59] van Handel R. "Probability in high dimension[R]". PRINCETON UNIV NJ, 2014.

[60] Dan Voiculescu. "Limit laws for random matrices and free products". *Inventiones Mathematicae*, 104(1):201–220, 1991.

[61] Dan V Voiculescu, Ken J Dykema, and Alexandru Nica. *Free Random Variables*. Number 1. American Mathematical Society, Providence, RI, 1992.

[62] James D Weber and Thomas J Overbye. "Voltage contours for power system visualization". *IEEE Transactions on Power Systems*, 15(1):404–409, 2000.

[63] Rudolf Wegmann. "The asymptotic eigenvalue-distribution for a certain class of random matrices". *Journal of Mathematical Analysis and Applications*, 56(1):113–132, 1976.

[64] Eugene P Wigner. "On the distribution of the roots of certain symmetric matrices". *Annals of Mathematics*, 67(2):325–327, Mar. 1958.

[65] L Xie, Y Chen, and PR Kumar. "Dimensionality reduction of synchrophasor data for early event detection: Linearized analysis". *IEEE Transactions on Power Systems*, 29(6):2784–2794, 2014.

[66] Le Xie, Yang Chen, and Huaiwei Liao. "Distributed online monitoring of quasi-static voltage collapse in multi-area power systems". *IEEE Transactions on Power Systems*, 27(4):2271–2279, 2012.

[67] Wilsun Xu and Jing Yong. "Power disturbance data analytics: New application of power quality monitoring data". *Proceedings of the CSEE*, 33(19):93–101, Jul. 2013.

[68] Xinyi Xu, Xing He, Qian Ai, and Caiming Qiu. "A correlation analysis method for power systems based on random matrix theory[J]". 2015. DOI: 10.1109/TSG.2015.2508506.

[69] C Zhang and RC Qiu. "Data modeling with large random matrices in a cognitive radio network testbed: Initial experimental demonstrations with 70 nodes". *ArXiv e-prints*, Apr. 2014.

[70] Changchun Zhang and Robert Qiu. 'Massive MMO as a big data system: Random matrix models and testbed". *Access, IEEE*, 3(4):837–851, 2015. Included in Special Section in IEEE Access: Big Data for Green Communications and Computing.

[71] Xiaoxin Zhou, Shuyong Chen, and Zongxiang Lu. "Review and prospect for power system development and related technologies: A concept of three-generation power systems". *Proceedings of the CSEE*, 33(22):1–11, Aug. 2013.

[72] Zimmerman R D and Murillo-Sanchez C E. "MATPOWER 4.1 A MATLAB Power System Simulation Package-User's Manual Ithaca, NY: Cornell Univ[J]". 2011.

*Chapter 14*

# A model-driven evaluation of demand response communication protocols for smart grid

*Emad Ebeid[1], Sergi Rotger-Griful[1], Søren Aagaard Mikkelsen[1] and Rune Hylsberg Jacobsen[1]*

## 14.1 Introduction

The current electrical grid faces some emerging challenges. These challenges cover a continuous increase of electricity demand, stopping the increase of greenhouse emissions from electricity production, and maintaining a reliable supply energy [1]. The smart grid is the future electrical grid that addresses these issues. A higher penetration of green energy sources is expected in the envisioned smart grid, especially in the end points of the grid. This will require an Information and Communication Technology (ICT) platform to handle Distributed Energy Resources (DER). The benefits of the smart grid include postponement of grid investments, an increase of efficiency, and higher reliability of the overall system [2].

There are several challenges that need to be solved for the smart grid to be a success. One of them is grid balancing. Balancing the grid is a continuous process, which purpose is to equate the energy production with the consumption of electricity in the grid. With society gradually moving toward the integration of renewable energy sources for lowering the greenhouse emissions, generation of energy becomes more fluctuating. With an energy generation side that is less controllable, there is a need for solutions that manages the consumption side. However, since the consumption side consists of customers of the grid, it demands for willingness to adapt to more optimal grid conditions.

Demand response can support the stability of the grid by incorporating the customer side. It targets reducing the peak of electricity consumption, since this can relieve stress in the grid and postpone grid investments (usually the capacity of the electrical grid is based on peak consumption). Recent advancements in demand response technology have focused on the standardization of control protocols for

[1]Department of Engineering, Aarhus University, Aarhus, Denmark

home appliances. These protocols provide a flexible electricity demand for offering intelligent automation services in the smart grid [3].

The Association of Home Appliance Manufacturers (AHAM) did an exhaustive survey of smart grid communications protocols that could enable large deployment of smart home appliances [4]. Among all reviewed protocols, two communication protocols received the top score: OpenADR and Smart Energy Profile 2.0 (SEP2). OpenADR was created by the OpenADR Alliance based on the OASIS Energy Operation Standard [5]. In contrast, the ZigBee Alliance and HomePlug Powerline Alliance have published the SEP2 Application Protocol Standard [6] which has recently been adopted by the IEEE 2030 standard [7]. Both protocols strive to standardize and simplify demand response.

This chapter presents: (I) a formal way to describe, model, and synthesize a demand response communication protocol combined with a user scenario and the demand response strategy, (II) a unique method for modeling, simulation, and evaluation of demand response protocols, (III) a structured method for fine-tuning protocol parameters, and (IV) a way to evaluate the performance of a demand response communication protocol.

This chapter extends the work performed in Reference 8 which proposes a design methodology to model, simulate, and evaluate demand response communication protocols together with demand response strategies for smart grids applications. The previous work is extended with a more detailed background description, enhancement of Unified Modeling Language (UML) models to be synthesized, and an additional case study.

The methodology design flow is shown in Figure 14.1, where alphabetic labels indicate different stages. The first stage of the methodology begins with the natural

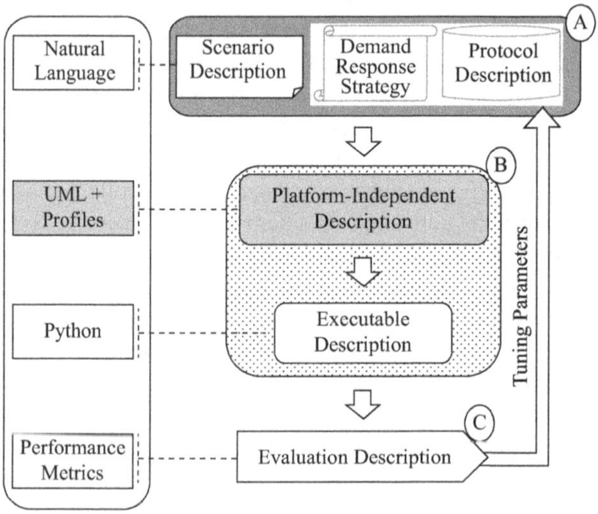


*Figure 14.1 The model-driven evaluation methodology for demand response communication protocols [8]*

description of a user scenario (e.g., user_1 turns ON the lights), the chosen demand response strategy, and a protocol under test (label A). These descriptions are then formalized, modeled, and simulated to validate the overall system functionality (label B). From the simulation results, the demand response protocol is then evaluated based on the defined performance metrics. In order to improve the protocol performance, the evaluation description results (label C) can be used to adjust the protocol's tuning parameters. This work attempts to provide a systematic way of studying the interplay between demand response strategies and the actual protocol implementations. System developers and power grid engineers can benefit from using the methodology by shortening the design and deployment time of demand response programs. Furthermore, demand response strategies can be validated before launched and the results can be applied to ensure reliability and stability of the power grid. It also allows for a heuristic way to couple end user behavior with the system performance.

## 14.2 State of the art

This section gives an overview about the related literature work on modeling, simulation, techniques, methods, and protocols of the smart grid.

Simulators represent an essential tool for gaining knowledge about possible outcomes when constructing new communication systems, power systems, and control strategies in the smart grid. Furthermore, it provides a useful tool during deployment in order to verify results when tuning parameters within these domains. Generally, there exist two approaches for a unified simulation of a complex system. A bottom-up approach, where either a new or existing simulation tools are adapted to a given domain by domain experts. A top-down approach where a simulation is based on a modeling language that can be synthesized to executable software.

The Mosaik framework [9] performs a large-scale analysis by using existing simulators to simulate a given smart grid scenario. This allows for a unified evaluation of grid topologies, control strategies, and scenarios by taking all domains into account. Originally, the co-simulation framework did not take the communication infrastructure into account, however recently a system architecture for integrating OMNeT++ with the Mosaik framework has been proposed [10]. The implementation details and evaluation are unknown. Furthermore, the approach of adding new simulators will increase the development complexity, since it would require the software developer to be familiar with all simulators. Thus, a unified modeling approach seems more viable.

The use of model-driven development methods for simulating smart grid scenarios is currently limited. Andren *et al.* [11] propose a semantic-driven design method using IEC standards, such as Common Information Model (CIM), IEC 61850, and IEC 61499. These cover the domains of control, communication, power grid, and application. Similarly, Nieße *et al.* [12] propose a methodology that approach the smart grid with a system engineering view. It focuses on closing the gap between research and industry by changing the development methodology. For both proposals, the exact outcome of changing practice is unknown and they do not specifically address demand response protocols.

Abstraction and platform-independent models become a solution in many fields to cope with design complexity. UML is a high-level modeling language broadly applied for modeling and specification of software systems [13]. Different studies have proven that UML is also applicable for hardware/software/network co-modeling and design [14]. However, there are several studies that make use of UML and derive network simulation models from its description. As an example, de Miguel *et al.* [15] introduce UML extensions for the representation of temporal requirements and resource usage for real-time systems. They propose tools to generate executable models from UML for the OPNET simulator.[2] In the same context, Hennig *et al.* [16] describe a UML-based simulation framework for early performance assessment of Software/Hardware (SW/HW) systems described as UML deployment and sequence diagrams. Their proposed simulator is based on the discrete event simulation package OMNet++.[3] Therefore, validation by simulation is considered one of suitable solutions to verify high-level models. As a result, several simulators have been combined to perform co-simulation for smart grid applications [17]. Nevertheless, UML profiles such as the Modeling and Analysis of Real-Time and Embedded Systems (MARTE) [18] are broadly used by system-level designer to enhance the UML models by embedded devices semantics.

For demand response to be a successful product there has been an important work on standardization communication protocols. OpenADR 2.0 is an application layer protocol created ease communication between service providers, grid operators, and energy management systems in consumer premises on demand response actions like load shifting. SEP2 is also an application layer protocol with a broader scope than OpenADR 2.0 but also suitable for demand response applications [6]. Both protocols have received the highest scores by the AHAM.[4] The specifications of both protocols contain different UML diagrams to describe the functionality of the protocols. For a comparison at specification level between these two protocols the reader is addressed to Reference 19. It is important to highlight that the methodology presented in this chapter is protocol agnostic.

Demand response protocols define the communications mechanism to exchange necessary information to deploy diverse control strategies. Possible control strategies may target actuators in residential appliances while others may target energy management systems that can control a set of loads. Common strategies are based on establishing simple operation rules (e.g., run appliance if power lower than threshold) [20]. Other strategies are based on multi-agent based systems to support on the decision-making [21] while others use models of the systems under control predict the best control action [2]. See Reference 2 for an overview of different strategies to regulate energy usage in a demand response scenario.

This chapter bases on a model-driven approach using UML diagrams for smart grid research. The chapter provides a methodology to evaluate demand response

[2]http://www.opnet.com
[3]http://www.omnetpp.org
[4]http://www.aham.org/

protocols in a holistic way that takes into account protocol specifics in combination with a chosen demand response strategy.

## 14.3 Background

This section introduces a reference architecture and gives a description about demand response programs, demand response protocols, modeling languages, and tools that will be used in this chapter.

### *14.3.1 Demand response reference architecture*

A reference architecture is needed to better understand the different actors that interact in a demand response scenario. Part of the standardization efforts toward the smart grid is related to define such architecture. In 2012, the Smart Grid Coordination Group presented a reference architecture for smart grids [22]. This architecture is very extensive and overly complicated for the purpose of this chapter. The proposed demand response architecture in this chapter is inspired from Reference 4 and is displayed in Figure 14.2.

In the upper level of Figure 14.2, there are three actors: *Market*, *Energy Service Provider*, and *Utility Operation*. In the *Market*, the different *Energy Service Providers* and *Utility Operators* can trade electricity and services (e.g., demand response).

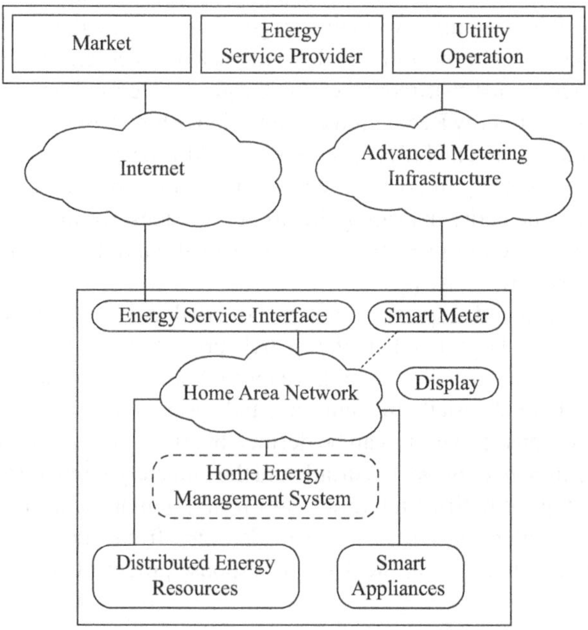


*Figure 14.2 Reference architecture of demand response systems*

An example of a *Market* could be Nord Pool Spot,[5] an example of an *Energy Service Provider* could be a balance responsible party or a demand response providers, and an example of *Utility Operation* could be a Distribution System Operator (DSO) or Transmission System Operator (TSO). These three top-level actors communicate with the different consumer households through the *Internet* or an *Advanced Metering Infrastructure*. The *Advanced Metering Infrastructure* is used to communicate with the smart meter in the consumer household (e.g., power line communication) while the *Internet* is used to communicate with the *Energy Service Interface* within the consumer household (e.g., demand response messages). In the consumer household, all the actors can communicate via the *Home Area Network*. The *Home Energy Management System* can control the local *Distributed Energy Resources* and *Smart Appliances* to respond to petition from *Energy Service Providers*. Alternatively, the *Energy Service Provider* can directly control *Distributed Energy Resources* and *Smart Appliances*. Furthermore, information can be shown to the user by means of the *Display*. This reference architecture is used throughout the chapter.

### *14.3.2 Demand response programs*

In the United States, customers can typically provide demand response to the regional Independent System Operator (ISO) or Regional Transmission Operator (RTO) by subscribing to different programs through the wholesale electricity market. Examples of ISOs that offer demand response programs are: New York Independent System Operator (NYISO), Pacific Gas and Electric (PG&E), San Diego Gas and Electric (SDG&E), and Southern California Edison (SCE) [23]. Demand response programs can usually be classified into two main categories: price-based programs and event-based programs [24]. In the former, the end user of electricity is provided with a varying electricity price that motivates the usage of electricity in periods with low prices. The Federal Energy Regulation Commission (FERC) uses the term time-based for this program type and includes programs like real-time pricing, critical peak price, and time-of-use rate [25]. In event-base programs, known by the FERC as incentive-based programs, the entity managing the program request for direct power changes. In this category there are programs like direct load control, demand bidding, and emergency demand response [25].

In Europe, demand response is not as extended as in the United States and is being targeted by a large number of research projects [26]. Compared to United States, European grids usually do not have an ISO or a RTO but a TSO. The TSO is responsible for both electricity transmission and system operation.

Demand response programs can be defined by: the notification time, the curtailment time, the amount of power to curtail, and the monetary remuneration [24]. In the United States, typical notification times vary from 30 minutes prior event start until 3.5 hours. The length of the curtailment can also vary, from 1 hour intervention up to 14 hours. Furthermore, most of the demand response programs are called during day

[5] http://www.nordpoolspot.com

peaks 14:00–18:00 [23]. There are other characteristics that define demand response programs. For example, the NYISO offers two types of demand response programs. The first program of the participation is voluntary and the client is paid after provision. The second program of the participation is mandatory but the client is paid upon enrollment [23].

By participating in a demand response program, the consumer needs to change their normal electricity usage. In some circumstances, this change may be compensated by the monetary remuneration provided by participating in the program but in others it may not. It is therefore crucial to consider different types of programs depending on the different characteristics of the consumers. This is one of the reasons why demand response providers usually target industrial consumers rather than residential consumer: they can provide more demand response but they are also more flexible in their electricity usage.

### *14.3.3 Demand response protocols*

#### 14.3.3.1 Smart Energy Profile 2.0

Smart Energy Profile 2.0 (SEP2)[6] [7] is an open standard released under IEEE (IEEE 2030.5-2013) and is initially the result of a joint collaboration between the ZigBee Alliance and the HomePlug® Alliance. Future work has continued within the IEEE 2030.5 technical group and multiple partners have supported the work, including National Institute of Standards and Technology (NIST) and Internet Engineering Task Force (IETF). Thus, at NIST the SEP2 is identified as a standard to address common smart grid use cases [27].

SEP2 is an evolution of the ZigBee Smart Energy 1.x standard design to be interoperable with Internet-based technologies that supports the Internet Protocol (IP) as seen in Figure 14.3. It follows the Representational State Transfer (REST) architectural style and is an HTTP-based application protocol. Clients use HTTP methods to connect with servers that host resources. The definition of the role of being a client or server relies upon whether the device initiates the connection or hosts a resource, respectively. Furthermore, it is independent of the protocol of the physical and medium layer in the Open Systems Interconnection (OSI) model, but requires that they are compatible with the Internet protocol stack. This makes it possible for the SEP2 to support a range of wireless and wired mediums, e.g., 802.15.4, 802.11, and Ethernet. However, to minimize the storage and memory requirements for resource-constrained devices to support the full IP stack, the SEP2 is intended to run on top of ZigBee IP [28]. This layer is compliant with a regular IP stack. SEP2 aims specifically for the communication between smart grid applications end-to-end and conforms to the CIM. CIM is a data model that addresses different parts of the electric system and is typical used in enterprises that work with electricity. The structure of the payloads are specified with the XML Schema Definition (XSD) and can be encoded with the Efficient XML Interchange (EXI) format. Moreover, since SEP2 is based on IP,

[6]The ZigBee Smart Energy Profile 2.0 is now known as IEEE 2030.5™-2013.

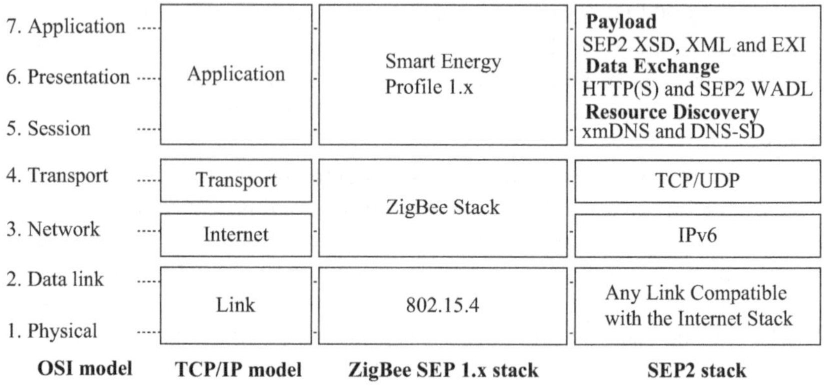


*Figure 14.3 Comparison of protocol stacks. While the ZigBee SEP 1.x stack is targeting 802.15.4 links with a customized network and transport layer, the SEP2 stack is similar to the TCP/IP model and uses IETF compatible protocols*

it supports multiple backhaul networks that different smart grid use cases demand for, e.g., smart metering with an Advanced Metering Infrastructure (AMI), customer premises with Energy Service Interface (ESI) to connect customer energy resources to external systems using the existing Internet connection, or in-home monitoring typically with battery-driven metering devices.

The SEP2 protocol is primarily targeting customer premises and its purpose is to inform and request action in order to assist the electric grid. Since the SEP2 protocol is client initiated, the concept is based on in-home devices that control their energy consumption, and request permission before they start. It does that either directly by carrying on and off requests from the in-home devices or indirectly by carrying electricity tariffs advertisements for end devices to display for the customer.

To ease the use, the SEP2 has an automatic service discovery and configuration. It uses the extended multicast Domain Name System (xmDNS) and DNS-Service Discovery (DNS-SD) protocol to discover SEP2 compliant devices on the local network [29]. This enables devices to automatically request for resources from other devices on the Local Area Network (LAN) in order to obtain additional information. This could be a heat pump that requests for a price tariff in the LAN. The smart meter might have this information and response with its location of the price tariff. Then the heat pump will store this location and retrieve the information when necessary.

To ensure secure communication between devices, it transmits HTTP requests over the Transport Layer Security (TLS) protocol; also known as HTTPS. HTTPS provides confidentiality, authentication, and integrity of the information transmitted and facilitates end-to-end security. The confidentiality property ensures that the information is kept secret and integrity ensures tamper-resistant transactions. The authentication protects against Man-in-the-Middle (MITM) attacks, but requires a certificate management system for authenticating trustworthy devices. The SEP2 therefore relies on a Public Key Infrastructure (PKI), where SEP2 device manufacturers issues RFC 5290 compliant PKI certificates to the devices at manufacturing time [30]. These certificates are intended to be persistent during the entire life cycle of the device. While compliant SEP2 devices must support processing of manufacturer PKI certificates, support for additional certificates is optional.

Using the REST ontology, the SEP2 standard classifies things into function sets and resources. A function set is a logical grouping of resources that implement related features. The SEP2 standard specifies resources and function sets within the area of **Support Resources**, **Common Resources**, and **Smart Energy Resources**. Depending on a given scenario, function sets from different areas can be applied.

In the process of evaluating SEP2 as a demand response protocol, the Demand Response and Load Control (DRLC) function set in the **Smart Energy Resources** is used. It targets client devices that support load control, where server devices are envisioned to be proxies for upstream demand response systems. For instance, a client device could be a heat pump, while a server device could be a Home Energy Management System (HEMS) receiving a demand response program. The server exposes load control events through resources called End Device Controls (EDC). EDC instances have attributes for devices types that allow devices to respond if they are within this device category. Furthermore, it includes features as randomization of duration and start time to mitigate the rebound effect.

#### 14.3.3.2 OpenADR

OpenADR is an application layer protocol used to securely exchange demand response messages and tariffs information between demand response service providers, customers, and customers' automation system equipment. OpenADR can be used for specific demand response task like power reduction, electricity shifting, or load shedding.

The development of OpenADR started in 2002 in Lawrence Berkeley National Laboratory [19]. The research and development of the standard proceed until its commercialization and release of OpenADR 1.0 in 2008. In 2010, an alliance between different industries and research institutions was formed to deepen the development of this protocol. OpenADR 2.0 is the latest version of the protocol and was released in August 2013 [5]. More recently, in November 2015 OpenADR and Universal Smart Energy Framework (USEF) have signed a strategic memorandum of understanding seeking a high cooperation in interconnecting and optimizing smart energy systems. This memorandum can be a key point for enabling a larger deployment and success of OpenADR.

OpenADR presents a client–server architecture with two types of nodes: Virtual Top Node (VTN) and Virtual End Node (VEN). The communication always happens between this two different types of nodes, i.e., VTN–VTN communication and VEN–VEN is not possible. However, one same entity can have both a VTN and VEN. Typically, a demand response event would be announced by a VTN to several VENs, which would be responsible for providing the load shed or load shift.

OpenADR has two different operation modes: PUSH mode and PULL mode. In the PUSH mode, the VTN initiates all communications and sends them to the VEN. In the PULL model, the VEN periodically pulls events from the VTN.

OpenADR provides four main types of services: parties registration, event communication, event opting, and data reporting. Before a VTN and VEN can collaborate in a demand response event, it is necessary that these entities identify each other. The *EiRegister Party* service is responsible for this task. The core functionality of OpenADR is provided through the *EiEvent* service. This service is responsible for communicating information about specific demand response events. Some event messages do not require reply from the VEN while others do. When event reply is required, the opting service is used (*EiOpt*). A VEN uses this service to show its availability and unavailability in specific time windows to participate in demand response events. The last service is the data reporting where a VTN can ask for historical data to a VEN or a VEN can set periodic data reporting to a VTN.

To transport OpenADR messages, this protocol offers two types of configurations: Hypertext Transfer Protocol (HTTP) over TLS or Extensible Messaging and Presence Protocol (XMPP). As data format Extensible Markup Language (XML) is used.

#### 14.3.3.3 Other protocols

There are other communication protocols that have more general-purpose application but are also suitable for demand response. Some of these protocols are listed in the exhaustive analysis that AHAM conducted to identify suitable protocols to be used by residential customers and their appliances in a Smart Grid scenario [4].

BACnet is a ASHRAE, ANSI, and ISO standard communication protocol specially designed to interact with building automation equipment and control systems [31]. The latest version of BACnet includes a couple of objects that can be used in a demand response scenario: *load control object* and *accumulator object* [32]. The former details controllability requirements (e.g., shed duration, shed level, and start time) while the latter provides monitoring of electricity usage.

OPC-Unified Architecture (OPC-UA) is a machine-to-machine communication protocol developed by OPC Foundation. OPC-UA present a client–server architecture and present an abstract specification divided in 13 different parts to enable high interoperability [33]. OPC-UA can be used in a demand response application domain as in Reference 34, where the authors propose a web service for demand response.

### *14.3.4 Modeling languages and tools*

This section presents the modeling languages and tools used in this chapter.

#### 14.3.4.1 UML

Unified Modeling Language (UML) [13] is a standardized general-purpose modeling language with a widespread application in the field of object-oriented software engineering. It became the de facto standard for modeling software intensive systems. UML includes a set of graphic notation techniques to create visual models of object-oriented SW-intensive systems. UML 2.x has 14 types of diagrams divided into two categories: *structural diagrams*, which emphasize the components that must be present in the system being modeled (e.g., Class, Deployment, and Profile diagrams), and *behavioral diagrams*, which emphasize what must happen in the system being modeled (e.g., State machine, Activity, and Sequence diagrams). Figure 14.4 shows an overview of these diagrams. A particularity of UML is the possibility to extend its semantics by a generic extension mechanism named profile definition. Profiles are defined using stereotypes, tag definitions, and constraints that are applied to specific model elements. Profiles create a lightweight semantic extension of standard UML to cover various domains.

In this work, Profiles, Class, and Object structure diagrams have been used to depict the static part of smart grid application, and Activity and Sequence diagrams have been used to depicts the functionalities of such applications (see the double-boxed shapes in Figure 14.4).

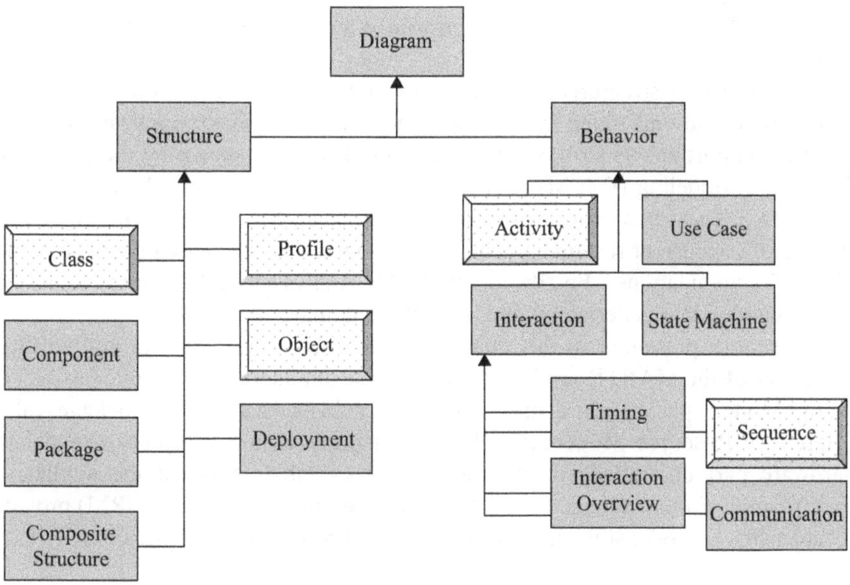


*Figure 14.4 UML2.x diagrams [13], the double-boxed shapes show the used diagrams in this work*

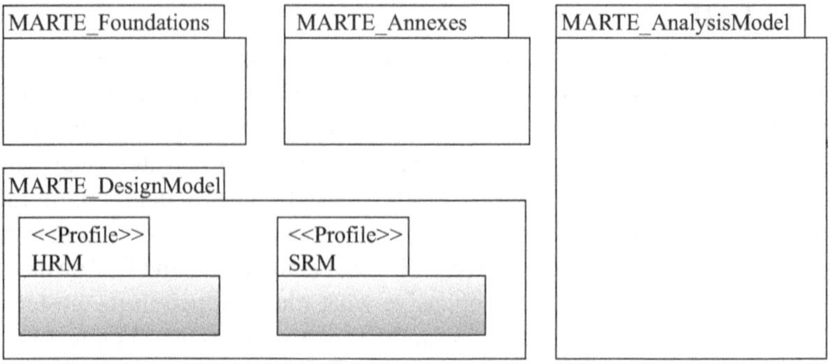


*Figure 14.5 Architecture of the MARTE profile [18]*

#### 14.3.4.2 MARTE profile

Modeling and Analysis of Real-Time and Embedded systems (MARTE) [18] is a UML profile standardized by Object Management Group (OMG).[7] It provides support for specification, design, and verification/validation stages in real-time embedded system development. It is designed to allow an easy specification of real-time and embedded systems. It provides some sub-profiles, like Non-Functional-Properties (NFPs) that allow to describe the "fitness" of the system behavior (e.g., performance, memory usage, and power consumption). Comparing with other OMG standard profiles, MARTE is the most appropriate profile to model the ICT aspects of smart grid system.

The profile is structured around two main concerns, one to model the features of real-time and embedded systems and the other to annotate application models so as to support analysis of system properties. These are shown by the Real-Time Embedded Modeling (RTEM) package named "MARTE_DesignModel" shown in Figure 14.5. These two major parts share common concerns with describing time and the use of concurrent resources, which are contained in the shared package named "MARTE_Foundations." Finally, the "MARTE_AnalysisModel" features are broken into a foundational generic part and an analysis domains part [18].

The work presented in this chapter benefits from the hardware and software semantics of the MARTE profile to model home appliances. It exploits "MARTE_DesignModel" package and uses two of its sub packages. One package called "Hardware Resource Modeling" (HRM) that provides mechanisms to model the hardware part of embedded systems, that is essential to fulfill the application specification. Other package called "Software Resource Modeling" (SRM) provides mechanisms to model software real-time and embedded (RTE) applications.

[7] http://www.omg.org

#### 14.3.4.3 Papyrus/Acceleo

Papyrus[8] is a graphical editing tool for UML2 as defined by OMG. It offers a very advanced support of UML profiles that enable users to define editors for domain-specific modeling based on the UML2 standard. In this work, Papyrus will be used to formalize the natural language (see Figure 14.1 label B box, "Platform-Independent Description") into UML models. Acceleo[9] is a pragmatic implementation of the OMG Meta-Object Facility (MOF) model to text language standard. It combines tooling, simple syntax, and efficient code generation. In this work, Acceleo will be used to generate executable Python code from UML models (see Figure 14.1 label B box, "Executable Description").

#### 14.3.4.4 Python/SimPy

Python is chosen as a programming language for the model evaluation through simulation. Python is a high-level programming language specially suited for fast prototyping with a large set of libraries that eases usage in different applications.

The Python library SimPy is used to perform the simulations. A SimPy simulation is composed by *Processes*, which are active components (e.g., washing machine), *Environments*, where *Processes* live, and *Events*, which are interaction between *Processes* and the *Environment*.

### *14.3.5 Evaluation metrics*

In a demand response provisioning scenario, there are many metrics that can be considered. Some relevant metrics are described in this section but the reader is referred to References 35,36 for more details on key performance indicators to evaluate demand response.

The Peak-to-Average Ratio (PAR) is a metric that is defined as the coefficient of the maximum power divided by the average power over a time window [35]. In a peak clipping scenario, it is expected that the PAR would be significantly reduced. Another typical index is the Root Mean Square (RMS) of the difference between the reference demand curve and the real demand curve. For a successful load following case, it is desired that this RMS is kept relatively low. Thanos *et al.* also propose to look at the variance of the power [35]. In this case, a lower variance implies a more even distributed of the electricity load.

In control engineering, there are a set of performance metrics that can be used for demand response assessments [37]. Most of these metrics focus on certain operations on the error signal, being the difference between the reference load and the real load. The Integral of the Absolute Error (IAE) provides an index on how well a system is following a certain reference by calculating the integral of the absolute value of the error over a time window. Alternatively, the Integral of the Error (IE) is also often used.

[8]http://www.eclipse.org/papyrus
[9]http://www.eclipse.org/acceleo

In a case where the load is always below or above the reference, the IE value will be the same as the IAE value.

## 14.4 The methodology

The viability of a demand response protocol in a real-world setup depends on multiple factors each having an impact on the performance of the protocol. For instance, a given user behavior impacts on the grid control. Likewise, a change of demand response strategy will influence on how the user is requested to change behavior. The presented approach allows modeling the user behavior while also taking into account a demand response scheduling algorithm.

In order to assess if a given demand response protocol is suitable for a smart grid environment, evaluation parameters associated with the demand response protocol and the resulting consumption pattern of all households are considered. The evaluation parameters for the protocol can include transmission overhead and time responsiveness between the device client and the demand response server. These metrics allow demand response protocol developers to benchmark the protocols against timing requirements, but also for making comparisons between them. Furthermore, the specification of demand response protocols often gives the possibility of adjusting parameters of the protocol. Tuning these parameters may have an impact on the protocol evaluation, but also on how successful the demand response strategy is in shifting the electricity consumption. Figure 14.6 details each part of the methodology and is labeled (A, B, and C) referring to Sections 14.4.1–14.4.3.

### *14.4.1 Describing household scenarios, demand response strategy, and protocol*

The description of household scenarios accounts for the majority of the dynamic behavior in the system. Their compositions are essential for the evaluation but also the alignment with the real world. Descriptions should be written in a natural language such that a non-technical person can understand and create these. The household scenarios should only include information relevant for the scope of the simulation. However, the order of execution is obligatory for the formal transformation. To gain better insights into the temporal progress, details about the execution time can be included.

Protocols are generally implemented based on their specifications, i.e., a document that contains communication requirements and protocol parameters which can be adjusted for conforming the needs of the domain. Together with a schema (or a data format hereof), these form the building blocks for the communication. These specifications can occasionally be found modeled in UML (e.g., in SEP2).

Demand response is a temporary adjustment on an electricity consumption to provide flexibility to the power grid. A goal of a demand response strategy is to reduce or shift load. The former implies a reduction on electricity usage while the

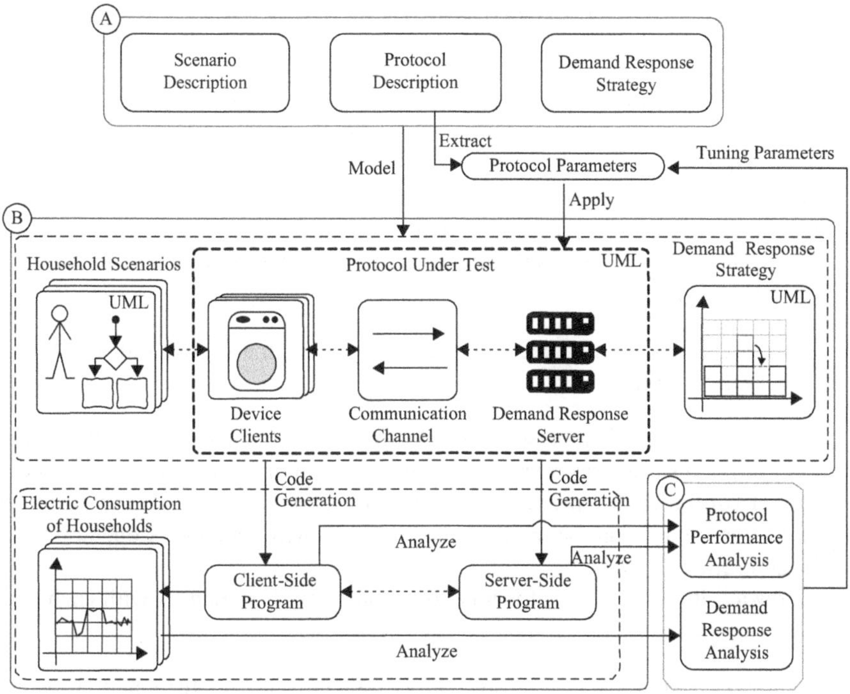


*Figure 14.6 Detailed overview of the methodology for evaluating and tuning demand response protocols*

latter entails changing the electricity consumption to a more suitable time period. The creation of a demand response strategy is typically to secure the delivery of electricity and provide a better integration of renewable energy sources. These are first formulated as mathematical expressions that later on can be modeled.

## 14.4.2 Platform-independent and executable descriptions

The core of the methodology is the UML with profiles (Figure 14.1) that provides a standard and interoperable representations of the scenario, demand response strategy, and protocol descriptions. Therefore, a combination of UML structural and behavioral diagrams has been used to model these descriptions.

*UML class diagrams* capture the structure of the whole system. *Sequence diagrams* depict the interactions between the consumer and the appliances and describe the demand response protocols (e.g., SEP2). *Activity diagrams* capture the behavioral aspect of smart grid components (e.g., electric vehicle). A novel Smart Grid profile has been developed as an extension to MARTE profile [18] to overcome the missing smart grid semantics in MARTE profile. Thus, the profile enables system simulation by code generation from UML models.

The extended MARTE stereotypes are as follows:

- **HWSensor**: it represents a device that measures a physical quantity and converts it into a signal which can be read by an observer or by an instrument.
- **HWActuator**: it is a device that transforms an input signal (mainly an electrical signal) into motion.
- **HwResource**: it denotes any hardware entity that provides one or many services and may require some services from other resources.
- **SwResource**: it models software structural entities provided to the user by execution supports.

New stereotypes have been created to model smart grid semantics. The stereotypes are as follows:

- **ElectricityMeter**: it represents an electronic device that periodically records consumption of electric energy and is able to communicate with other devices.
- **SmartAppliance**: it represents an appliance that has an embedded smart meters and actuators for automatizing its operations. Smart appliance's operations can be controlled.
- **Gateway**: it is a device that is able to connect smart appliances with each other and with the Internet.

Figure 14.7 shows the developed UML profile diagram for smart grids that extends MARTE profile with smart grid semantics. Figure 14.8 shows an example of a home automation application with the smart grid profile annotations.

Model validation by simulation is a complementary approach to validate high-level models. One approach is the Models-to-Text transformation which aims to synthesize such high-level models by generating textual artifacts from them. Acceleo, Eclipse plug-in tool, can be used by developing a generator to transform UML models to code (see Section 14.3.4.3).

### *14.4.3 Evaluating demand response strategy and protocol*

The evaluation process of a demand response strategy relies on a set of evaluation metrics. These metrics are different depending on the perspective chosen. On the consumer side, important metrics to consider can be indoor comfort, economic savings, and inconveniences caused (e.g., waiting time for appliances to run). On the grid side, relevant metrics include among others power peak reduction, time responsiveness, and economic benefits [35].

Time responsiveness indicates the response time of a certain electrical load to be controlled. In some circumstances, it is required to have fast time responsiveness (e.g., for frequency regulation) while in others the speed of response is not as critical (e.g., day-ahead demand response).

Demand response is a change on the electricity usage. This change usually leads to shift the electricity usage over time. The amount of electricity shifted is an important metric for an energy service provider because the monetary remuneration for the

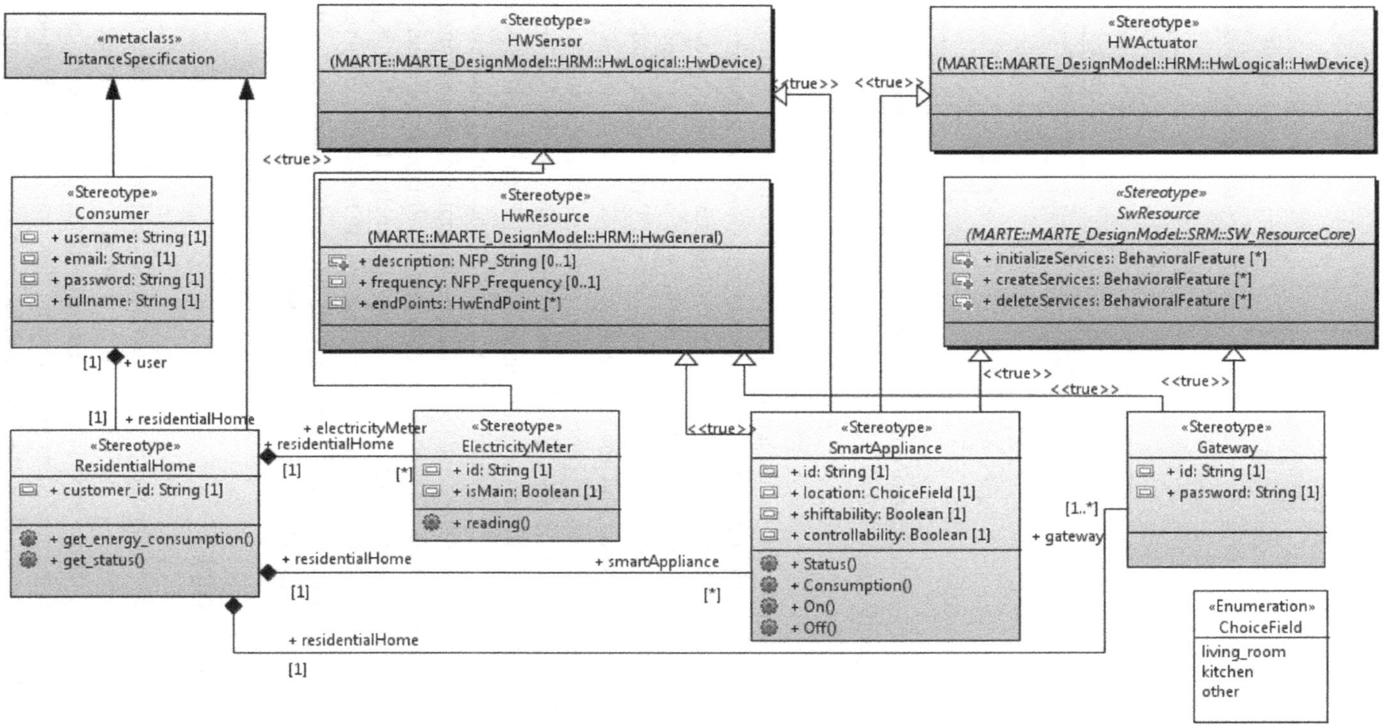


Figure 14.7 UML profile diagram for smart grid as an extension to MARTE profile

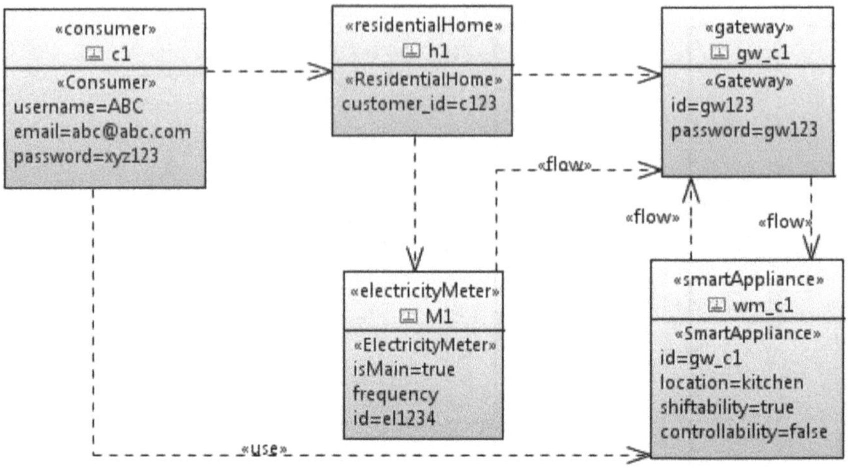


*Figure 14.8 UML object diagram of a home automation application with smart grid profile annotations*

consumer may depend on it. Predicting a demand response baselines is a challenging task currently attracting the attention of many researchers. Common techniques for baseline establishment use historical data analysis to find days with similar conditions and use them as baseline [38]. Important metrics for the grid operator are those that reflect the power peak. Examples of these indexes are the PAR and the RMS of difference between actual power and power reference.

In a client–server architecture, communication protocols can be evaluated by counting the number of packets being exchanged, by measuring the latency on the communication, and by studying the communication overhead. For a demand response scenario, it would also be relevant to consider the metrics like the waiting time between the client request to run an appliance and the actual time the appliance start running. It is important to consider such metric because the consumers' willingness to participate in a demand response even may depend on that. Another metric to analyze is the presence of rebound peaks due to massive reconnection of loads after a demand response event [39]. Some communication protocols have some parameters within their specification to avoid such problems (e.g., like SEP2).

## 14.5 Proof of concept

This section introduces the mapping rules to generate executable Python code from UML diagrams. First, a scenario, a demand response strategy, and a protocol description will be captured by UML models and then mapped to Python source code. In this work, Python has been used as an executable description of the high-level models. Tables 14.1 and 14.2 show the correspondences between UML diagrams, XML Schema description of such diagrams, and the corresponding Python syntax.

*Table 14.1 Mapping between UML structure diagrams and Python static elements*

| UML entity | XML description | Python source code |
|---|---|---|
| Class:<br>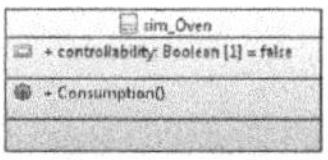 | `<packagedElement`<br>`↪ xmi:type="uml:Class"`<br>`↪ name="sim_Oven">`<br>`<ownedAttribute`<br>`↪ xmi:type="uml:Property"`<br>`↪ name="controllability">`<br>`<type xmi:type=`<br>`↪ "uml:PrimitiveType"/>`<br>`<defaultValue value=false/>`<br>`</ownedAttribute>`<br>`<ownedOperation xmi:type=`<br>`↪ "uml:Operation"`<br>`↪ name="Consumption"/>`<br>`</packagedElement>` | `class sim_Oven():`<br>`def __init__(self,`<br>`↪ control=False):`<br>`self.controllability`<br>`↪ = control`<br>`def Consumption(self,`<br>`↪ sec):`<br>`pass` |
| Object:<br>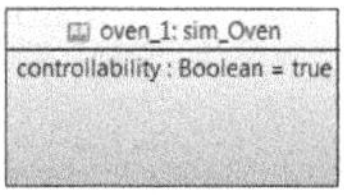 | `<packagedElement xmi:type=`<br>`↪ "uml:InstanceSpecification"`<br>`↪ name="oven_1">`<br>`<slot xmi:type="uml:Slot"`<br>`↪ definingFeature=`<br>`↪ "controllability">`<br>`<value xmi:type=`<br>`↪ "uml:LiteralBoolean"`<br>`↪ value="true"/>`<br>`</slot>`<br>`</packagedElement>` | `oven_1 = sim_Oven(`<br>`↪ controllability=True)` |

*Table 14.2 Mapping between UML behavioral diagrams and Python elements*

| UML entity | XML description | Python source code |
|---|---|---|
| Sequence:<br>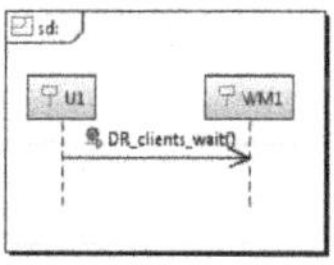 | `<fragment xmi:type="uml:Message`<br>`OccurrenceSpecification"`<br>`↪ name="MessageSend"`<br>`↪ message="DR_clients_wait()"/>` | `# Function call`<br>`DR_clients_wait()` |
| Activity:<br>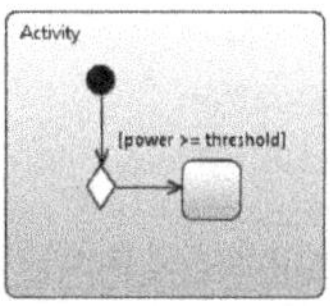 | `<guard`<br>`↪ xmi:type="uml:Expression"`<br>`↪ name="g1">`<br>`<expr`<br>`↪ xmi:type="uml:LiteralString"`<br>`↪ value="power>=threshold"/>`<br>`</guard>` | `# Conditional`<br>`if power >=`<br>`↪ threshold:`<br>`pass` |

The structural diagrams (i.e., the class and object diagrams) are directly mapped into Python's class syntax. The behavioral diagrams (i.e., sequence and activity diagrams) are first formalized and then synthesized into executable code. One way to formalize sequence diagrams is to use finite state machines that can be converted into

*IF-THEN-ELSE* statements. More details about the high-level synthesis step can be found in Reference 40.

## 14.6 Experimental results

The methodology presented here is demonstrated in a couple of case studies. The first case consists of a scenario with an individual household and a control strategy to provide demand response from certain loads from the household. In the second case, a load aggregation scenario is considered where demand response is provided by coordinating the operation of several appliances in different households. In both cases, the scenario and control strategy are described and modeled, and SEP2 is selected as a protocol to be analyzed. The obtained models are used to manually generate executable Python code and simulations are subsequently done. The simulation results are then analyzed using performance metrics to tune different parameters of the SEP2 protocol.

### *14.6.1 Case 1: individual household*

#### 14.6.1.1 Describing household scenarios, demand response strategy, and protocol

In this first scenario, a residential household with a deterministic set of actions is considered. These actions are described in natural language in Table 14.3 and the load profiles of four selected appliances are shown in Figure 14.9.

The control strategy that manage energy usage of the controllable appliances in the household is based on a *soft threshold* constraint in total power. If the total power consumption (controllable and non-controllable appliances) is below a threshold

*Table 14.3 Timed natural language description of a consumers' activities*

| Time (hh:mm) | Activity description |
|---|---|
| 18:01 | Consumer opens the front door and turns ON light. |
| 18:05 | Consumer starts the stove and cooks food. |
| 18:06 | Consumer switches ON the oven. |
| 18:30 | Consumer plugs the electric car in its charging station. |
| 18:35 | Consumer finishes cooking and switches OFF the stove. |
| 18:40 | Consumer switches OFF the oven. |
| 18:45 | Consumer puts the laundry in the washing machine and turns it ON. |
| 18:50 | Consumer eats the food while watching television. |
| 19:30 | Consumer finishes eating and switches OFF the television. |
| 20:10 | The washing machine finishes and consumer puts cloths into the tumble dryer. |
| 20:30 | The car is fully charged. |
| 22:05 | Consumer switches ON the tumble dryer. |
| 23:05 | The consumer switches OFF the lights and goes to sleep. |
| 23:30 | The tumble dryer has finished its operation. |

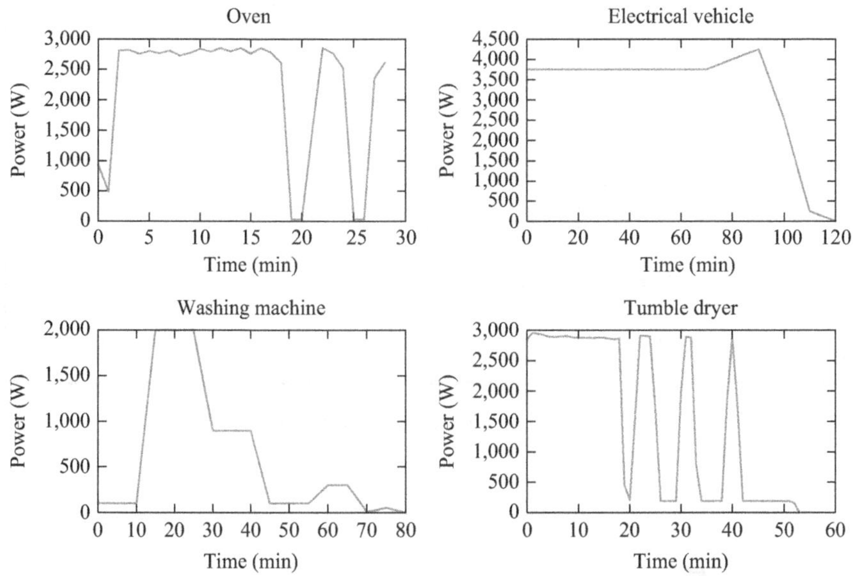


*Figure 14.9 Load of four different appliances used in the simulations. All loads are based on real data except the electrical vehicle. The washing machine profile is extracted from Reference 41, the oven and tumble dryer profiles from Reference 42, and the electrical vehicle from Reference 43*

(3,000 W) the demand response system allows appliances to run in the immediate future.

The demand response protocol to be tested is SEP2. This protocol is detailed in the specification document [6]. In the evaluating process, two different parameters from SEP2 have been considered: time between pooling events and randomized starting time. The time between pooling events is the time that a client waits before sending a new petition to start running (with a maximum value of 5 min). The randomized starting time is a randomly chosen time (with a maximum of 60 min) added/subtracted to the scheduled starting time of an appliance. This randomized time is meant to avoid a rebound peak due to a massive reconnection of several clients after a demand response event.

#### 14.6.1.2 Platform-independent and executable descriptions

Figure 14.10 shows the timed scenario of the consumers' interactions with its appliances. Worth mentioning that, the power consumption function of *SmartAppliance* classifier (see Figure 14.7) is extracted from real appliances consumption profiles. Appliances can be divided in four categories: controllable (e.g., electrical vehicle), non-controllable (e.g., television), shiftable (e.g., washing machine), and non-shiftable (e.g., lights).

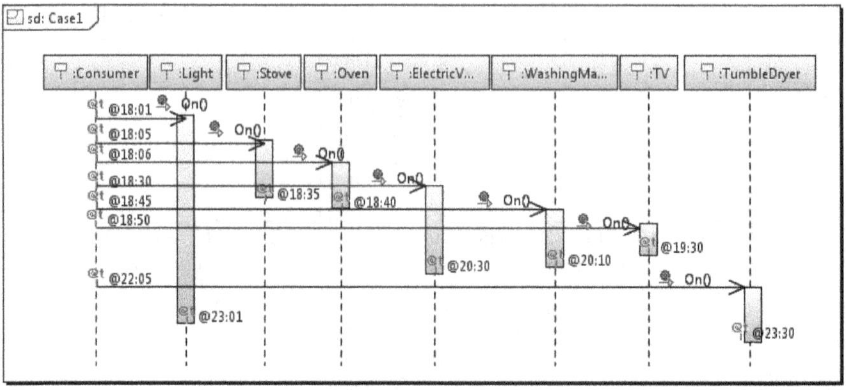


*Figure 14.10 The consumer's scenario description*

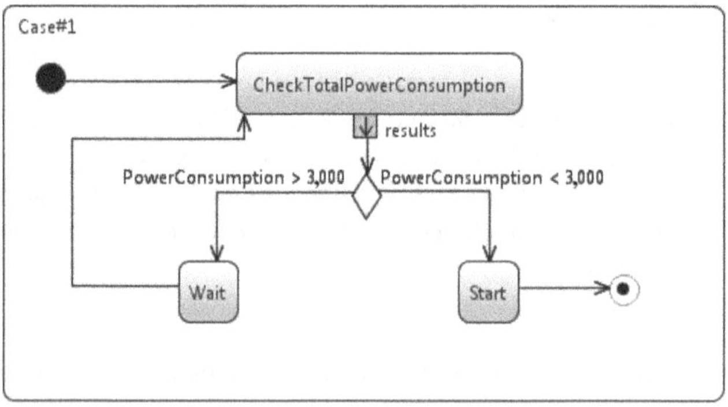


*Figure 14.11 Activity diagram of the demand response strategy*

The UML activity diagram has been used to model the demand response strategy as shown in Figure 14.11. This demand response strategy can only manage controllable appliances thus leading in some circumstances to exceed the threshold.

The UML sequence diagram has been used to model the SEP2 communication protocol between the appliances and a demand response server as shown in Figure 14.12. During the high-level synthesis of the protocol, a subset of the SEP2 specification has been implemented considering the Demand Response and Load Control (DRLC) function set with HTTP as the carrying request–response protocol and XML as a data format. SimPy[10] Python library has been used as a simulation environment to model the case study by creating both HTTP servers and clients having socket communication.

[10]http://pypi.python.org/pypi/simpy

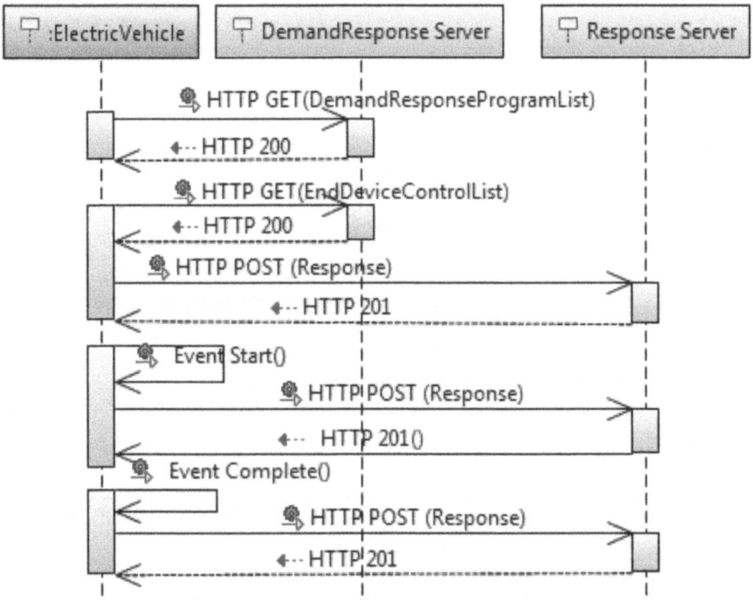


*Figure 14.12 Communication between the electric vehicle and demand response server with SEP2 [6]*

#### 14.6.1.3 Evaluating demand response strategy and protocol

An iterative process is conducted, considering different values of the tuning parameters: randomized starting time (0, 20, and 40 min) and time between pooling events (1 and 5 min). The process of deciding the protocol parameters has been done following a somewhat naive approach. The previously introduced performance metrics have been used to evaluate the control strategy with different protocol parameters. The used metrics are: RMS of the difference between the soft threshold (3,000 W) and the power, standard deviation (std. dev.) on the power, the power peak, the PAR, and the consumption overflow. The consumption overflow is the area above the horizontal-dashed line in Figure 14.13(a) and is defined as the percentage of the total energy used (kWh) above the soft threshold. Two metrics are considered from the protocol side: total number of HTTP GET messages sent by all controllable appliances in the household and the total waiting time of all appliances. This last metric is the difference between actual starting time and the time when the appliance requests to start.

Table 14.4 contains the results for all considered combinations of the tuning parameters as well as the baseline (without demand response control strategy). Table 14.4 shows that the values of RMS and standard deviation of the baseline are reduced in comparison to the demand response cases. Additionally, it can be observed that the larger the randomized start is, the smaller the values of RMS and standard deviation are. A small value on standard deviation of the power can be understood as a signal of having a more even distribution in power. The simulation results also illustrate that the power peak, the PAR, and the consumption overflow are decreased

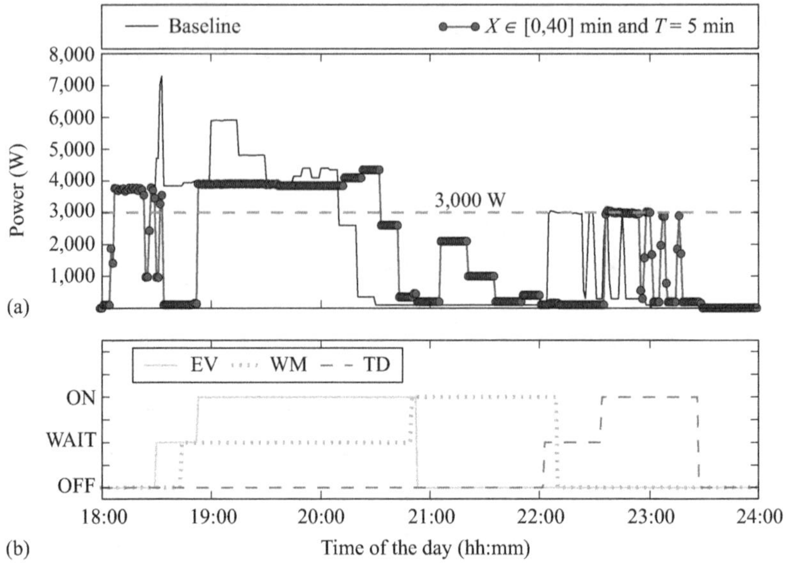


*Figure 14.13 (a) Comparison of electricity consumption with and without demand response strategy. (b) Appliances status for $X \in [0, 40]$ and $T = 5$. The soft threshold in (a) is exceeded due to uncontrollable loads*

*Table 14.4 Results of the evaluation of the demand response strategy and protocol for single household scenario*

| **Metric** | **Baseline** | $X = 0$ | | $X \in [0,20]$ | | $X \in [0,40]$ | |
|---|---|---|---|---|---|---|---|
| | | $T = 1$ | $T = 5$ | $T = 1$ | $T = 5$ | $T = 1$ | $T = 5$ |
| RMS (Ref-Pow) (W) | 2,332 | 2,059 | 2,059 | 2,053 | 2,052 | 2,049 | 2,049 |
| Std. dev. (W) | 2,030 | 1,709 | 1,709 | 1,702 | 1,701 | 1,697 | 1,697 |
| Peak (W) | 7,295 | 4,350 | 4,350 | 4,350 | 4,350 | 4,350 | 4,350 |
| PAR (–) | 3.94 | 2.35 | 2.35 | 2.35 | 2.35 | 2.35 | 2.35 |
| Consumption overflow (%) | 25.16 | 16.56 | 16.56 | 16.56 | 16.56 | 16.52 | 16.56 |
| HTTP GET (–) | 0 | 98 | 27 | 128 | 32 | 176 | 41 |
| Waiting time (min) | 0 | 93 | 97 | 125 | 132 | 173 | 177 |

$X$: uniform random variable for the starting delay in minutes when permitted to start.
$T$: waiting time in minutes between a client's request for a possible permission to start.

when comparing the baseline operation and the demand response cases. For the specified scenario, there are no differences in these metrics between different values on the tuning parameters.

Table 14.4 also displays values on the total number of HTTP GET messages and waiting time. It is observed that for low values of time between pooling events

($T = 1$ min) the number of GET messages is larger but the waiting time of the appliances is lower. The reasoning behind this is that the appliance asks every minute to start, it will overall send more messages but it would ensure to start as soon as possible. This high message rate introduces overflow in the network that in some cases may not be desired. This is a drawback on client-initiated protocols like SEP2.

For illustrative purposes, the power profile of the household and the appliances status are displayed in Figure 14.13. Figure 14.13(a) shows power for two different cases: baselines (solid line) and demand response with randomized starting time of maximum 40 min and time between pooling events of 5 min (circled line). The soft power threshold of 3,000 W is also displayed (horizontal-dashed line). In the baseline, there are several peaks in power due to simultaneous usage of appliance with large loads (i.e., stoves, oven, electric vehicle, and laundry). It can be observed that these peaks are shifted in time by adding the soft threshold control strategy in a demand response scenario. The soft threshold is exceeded due to the presence of loads that the energy service provider (demand response server) cannot control (i.e., light, stoves, television, and oven). It is important to remark that the energy used (kWh) in both presented cases is the same, the only difference is the time where this energy is used.

Figure 14.13(b) illustrates the status of three controllable loads in the household: electrical vehicle (EV), washing machine (WM), and tumble dryer (TD). The status OFF indicates that an appliance is not being used, when the appliance wants to start running is placed into WAIT until the energy service provider (demand response sever) allows it to run and it is labeled with ON status. This plot illustrates the inconvenience that the consumers' experience for providing demand response. For a successful deployment of a new demand response strategy, it is important to consider the trade-off between grid needs (e.g., peak reduction) and consumer needs (e.g., low waiting time). This trade-off can be evaluated by simulations using the presented methodology.

## *14.6.2 Case 2: load aggregation*

### 14.6.2.1 Describing scenario, demand response strategy, and protocol

In this case study, an energy service provider can manage a portfolio of 10,000 of washing machines in Denmark. The load profile of each washing machine is considered the same as shown in Figure 14.9. The starting time of each of the 10,000 appliances is decided in a stochastic manner following the probability distribution. This probability distribution is extracted from a survey on washing machine usage in Denmark [44].

The Energy Service Provider aims at deploying a peak clipping control strategy, i.e., reduce the power during peak hours. To that extend, the service provider limits the amount of clients (washing machines) that can start their operation between 18:00 and 20:00. The typical peak experienced in residential consumers in Denmark is at 18:00, coinciding with the dinner time. Outside the defined time period, there is no limit on the amount of washing machines that can start.

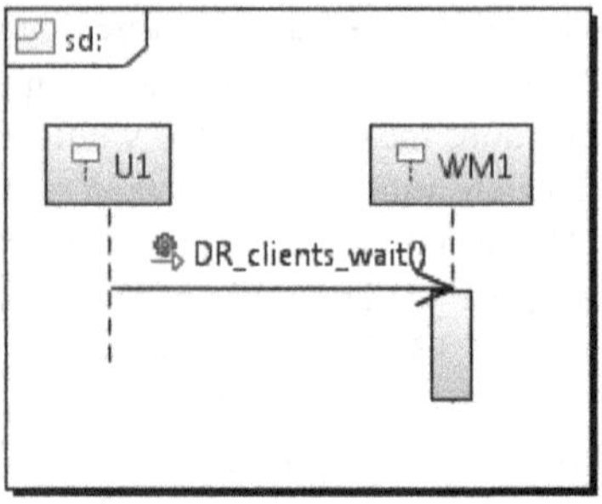


*Figure 14.14 The consumer's scenario description*

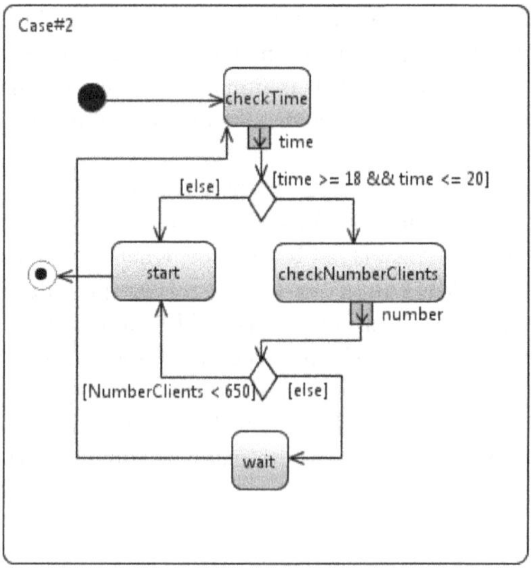


*Figure 14.15 Activity diagram of the load aggregation control strategy*

SEP2 is again chosen as a protocol to be analyzed. This time there is just one parameter to be tuned: the randomized starting time. The time between pooling events is fixed to 1 min.

#### 14.6.2.2 Platform-independent and executable descriptions

The timed scenario of one washing machine is capture by the UML sequence diagram shown in Figure 14.14. The randomized starting time of an appliance is modeled by using **@random(x)** as a placeholder for the function that generates random numbers.

The control strategy is modeled in the activity diagram in Figure 14.15. Outside the peak period (18:00–20:00) every appliance, which wants to start, is granted

*Table 14.5 Results of the evaluation of the demand response strategy and protocol for load aggregation with fixed time between pooling events (1 min)*

| Metric | Baseline | $X = 0$ | $X \in [0,20]$ | $X \in [0,40]$ | $X \in [0,60]$ |
|---|---|---|---|---|---|
| Standard deviation (kW) | 125 | 138 | 172 | 179 | 180 |
| Peak (kW) | 509 | 739 | 926 | 886 | 810 |
| PAR (–) | 1.49 | 2.17 | 2.65 | 2.48 | 2.22 |
| HTTP GET (–) | 5,388 | 15,896 | 53,788 | 93,387 | 135,611 |
| Waiting time (min) | 0 | 10,683 | 51,074 | 92,667 | 137,719 |

$X$: uniform random variable for the starting delay in minutes when permitted to start.

permission. During the peak period, the SEP2 demand response server checks the amount of clients that have been granted permission to start and those that are already running. If the total amount of clients is higher than a certain threshold (650 clients) new clients are placed in the waiting list. When the number of clients is below the threshold, the clients in the waiting list are allowed to start until the threshold is reached again. It should be noted that once a washing machine has started its operation the demand response server should not stop it (i.e., clothes could get damaged). However, if an appliance has been granted to start but has not yet start its operation the demand response server can ask to cancel the event as described in Reference 6. Since SEP2 is a client-initiated protocol, this is done by periodically (1 min) asking for an update on the status of the demand response event.

#### 14.6.2.3 Evaluating demand response strategy and protocol

A similar evaluation procedure than in the first case has been followed. This time four values of the randomized start (tuning parameter) have been considered: 0 (no randomized start), 20, 40, and 60 min. The randomized start delays the scheduled starting time of the appliances by randomly selecting a delay time between 0 min and the specified maximum (i.e., 20, 40, or 60 min). The time between pooling event status is fixed to 1 min. The performance metrics considered in the evaluation process are: standard deviation of the power, power peak, the PAR, the total number of HTTP GET messages sent, and the aggregated waiting time of appliances to start.

Table 14.5 summarizes the results of the evaluation process. The metrics displayed in Table 14.5 are calculated in the time window from 17:00 to 24:00. Table 14.5 shows that the larger the randomized starting time is, the larger the standard deviation of the power is. This indicates that by deploying the control strategy the power presents a higher variation proportional to randomized starting time. The power peak is also affected by the randomized starting time. To better illustrate this effect the reader is referred to Figure 14.16. The top part of this figure shows the total power over time for three cases: baseline (solid line), demand response without randomized starting time (circled line), and demand response with randomized starting time of maximum 20 min (stared line). It can be observed how the power usage in the demand response period delimited by dashed lines (18:00–20:00) is in general lower than the

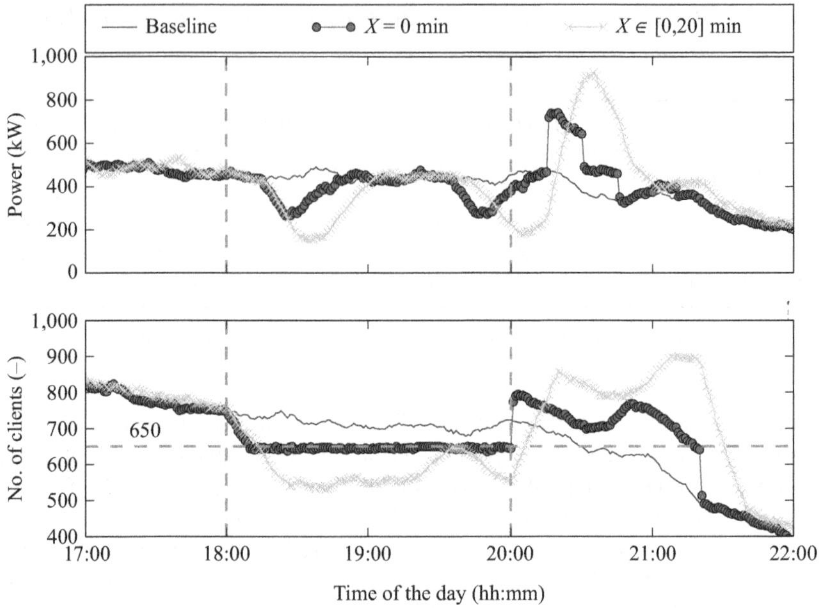


*Figure 14.16 Total power consumption and number of clients over time for three cases: baseline (solid line), demand response without randomized starting time $X = 0$ (circled line), and demand response with randomized starting time $X \in [0, 20]$ min (stared line). The dashed lines show the event time limits (18:00 and 20:00) and the clients threshold (650)*

baseline. The power oscillations in this period are due to the power consumption profile of a washing machine (see Figure 14.9). Limiting the number of appliances that can start running during a time window generates a rebound peak after the event is completed: all appliances that have been in waiting list want to start. To avoid the rebound peak issue, SEP2 uses two mechanisms: the randomized starting time and randomized duration of an event. From the simulation on this specific case study, it can be observed that the minor rebound peak is achieved by disabling the randomized starting time. This can be seen in both Figure 14.16 and in Table 14.5 by a lower PAR and peak value.

The defined control strategy can be observed in the bottom plot of Figure 14.16. From 18:00 to 20:00, a constraint of a maximum of 650 appliances (dashed line) is set. In the first minutes of the event, the number of appliances decreases until the threshold is met: running appliances cannot be switched off by the demand response server. For the demand response case without randomized start, the appliances running are always equal to the threshold: when an appliance finishes the demand response server allows one appliance from the waiting list to start right afterward. In the 20 min case (stared line), the number appliances running is lower than the threshold because

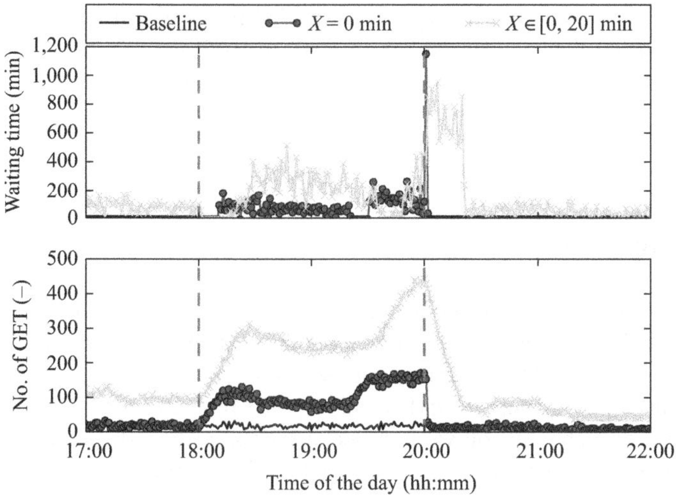


*Figure 14.17 Total waiting time and total number of HTTP GET messages over time for three cases: baseline (solid line), demand response without randomized starting time* $X = 0$ *(circled line), and demand response with randomized starting time* $X \in [0, 20]$ *min (stared line). The dashed lines show the event time limits (18:00 and 20:00)*

there are appliances that have been allowed to start but that have not done that yet due to the randomized starting time.

Looking at the protocol metrics in Table 14.5, it can be observed that the higher randomized starting time is, the more HTTP GET messages are sent and the longer the waiting time is. This tendency is also illustrated in Figure 14.17. The waiting time for the demand response without randomized starting time (circled line) is lower than the one with randomized starting limit of 20 min (stared line). For the 20 min case, there is a peak on waiting time just after event is finished. This is caused by having more than 200 appliances starting in a very short time window.

The reason for a longer waiting time for a larger randomized start is straightforward but the larger number of HTTP GET messages is not. The number of HTTP GET messages is influenced by the randomized starting time for the following reason. A client sends a HTTP GET message when it has not been allowed to start. Additionally, a client also sends messages when it has been allowed to start but it has not done so due to the randomized starting time. This is done because client periodically contacts the demand response server to know if it has changed the status of an event (e.g., stop running an appliance before it has actually started). Therefore, a larger randomized starting time leads to a larger number of HTTP GET messages.

This case study illustrates a possible application domain of the model-driven evaluation methodology: assessment of a control strategy by simulations before deployment. A decision that could be taken for this specific scenario would be to disable randomize starting time feature of SEP2. In the presented case, the randomized starting time causes the opposite effect that it was designed for. However, this may not apply to other scenarios. It is therefore very important that a holistic assessment like the one here presented is carried out before deployment of any new strategy.

## 14.7 Conclusion

Balancing the smart grid is a continuous process that requires to equate the energy production with the consumption of electricity in the grid. Demand response can support the stability of the grid by incorporating the customer side. In order to communicate with the customers, standardized demand response protocols are used. However, the performance of such protocols varies based on the application, demand response strategy, and their tuning parameters. Therefore, this chapter presents a methodology for evaluating the performance of demand response protocols combined with a demand response strategy for the smart grid.

The methodology shows how to formalize, model, and simulate a household scenario with demand response strategies. It reuses existing specifications of demand response protocols, and strategies for its evaluation. The methodology rests on a model-driven approach for evaluating demand response that combines strategy, scenario, and protocol models. Moreover, the methodology offers benefits by speeding up the design process and decreasing error risks of implementing protocols, scenarios, and strategies models by automatically synthesizing them into executable code. Using a simulation environment, the protocol is evaluated by observing a set of performance metrics. The results are used to optimize the protocol behavior by tuning its parameters. Furthermore, the methodology is validated through two comprehensive case studies using the demand response and load control function set of the SEP2 protocol. The case studies showed that it is possible to apply the methodology on the SEP2 protocol for a household scenario description and a specified demand response strategy for evaluating its performance.

## Acknowledgments

The research leading to these results has received funding from the European Union Seventh Framework program (FP7/2007-2013) under grant agreement no. 619560 (SEMIAH), no. 317761 (SmartHG), and the Danish Energy Agency project no. 12019 (VPP4SGR).

## Bibliography

[1] H. Farhangi, "The path of the smart grid," *Power and Energy Magazine, IEEE*, vol. 8, no. 1, pp. 18–28, Jan 2010.

[2] S. Rotger-Griful and R. H. Jacobsen, "Control of smart grid residential buildings with demand response," in *Chaos Modeling and Control Systems Design*, ser. Studies in Computational Intelligence, A. T. Azar and S. Vaidyanathan, Eds. Springer, Berlin, vol. 581, pp. 133–161, 2015 [Online]. Available from http://link.springer.com/10.1007/978-3-319-13132-0. Accessed on May 2016.

[3] R. Jacobsen and S. Mikkelsen, "Infrastructure for intelligent automation services in the smart grid," *Wireless Personal Communications*, vol. 76, no. 2, pp. 125–147, 2014 [Online]. Available from http://dx.doi.org/10.1007/s11277-014-1682-6. Accessed on May 2016.

[4] Association of Home Appliance Manufactures, "Assessment of communication standards for smart appliances: The home appliance industry's technical evaluation of communication protocols," Association of Home Appliance Manufactures, Tech. Rep., 2010 [Online]. Available from http://www.aham.org/ht/a/GetDocumentAction/i/50696. Accessed on May 2016.

[5] OpenADR Alliance, "OpenADR 2.0b profile specification," OpenADR Alliance, Tech. Rep., 2013 [Online]. Available from http://www.openadr.org/specification. Accessed on May 2016.

[6] ZigBee Alliance, Inc. and HomePlug Powerline Alliance, Inc., "SEP 2 application protocol standard," ZigBee Alliance, Tech. Rep. Document 13-0200-00, 2013.

[7] IEEE standard Association "IEEE Adoption of Smart Energy Profile 2.0 Application Protocol Standard," pp. 1–348, Nov 2013 [Online]. Available from https://standards.ieee.org/findstds/standard/2030.5-2013.html

[8] E. Ebeid, S. Rotger-Griful, S. Mikkelsen, and R. Jacobsen, "A methodology to evaluate demand response communication protocols for the smart grid," in *2015 IEEE International Conference on Communication Workshop (ICCW)*, London, UK, June 2015, pp. 2012–2017.

[9] S. Schutte, S. Scherfke, and M. Troschel, "Mosaik: A framework for modular simulation of active components in smart grids," in *2011 IEEE First International Workshop on Smart Grid Modeling and Simulation (SGMS)*, Brussels, Belgium, October 2011, pp. 55–60.

[10] J. Dede, K. Kuladinithi, A. Förster, O. Nannen, and S. Lehnhoff, "OMNeT++ and Mosaik: Enabling simulation of smart grid communications," *CoRR*, vol. abs/1509.03067, 2015 [Online]. Available from http://arxiv.org/abs/1509.03067. Accessed on May 2016.

[11] F. Andren, M. Stifter, and T. Strasser, "Towards a semantic driven framework for smart grid applications: Model-driven development using CIM, IEC 61850 and IEC 61499," *Informatik-Spektrum*, vol. 36, no. 1, pp. 58–68 2013 [Online]. Available from http://dx.doi.org/10.1007/s00287-012-0663-y. Accessed on May 2016.

[12] A. Nieße, M. Tröschel, and M. Sonnenschein, "Designing dependable and sustainable smart grids – How to apply algorithm engineering to distributed control in power systems," *Environmental Modelling & Software*, vol. 56, pp. 37–51, Jun 2014 [Online]. Available from http://dx.doi.org/10.1016/j.envsoft.2013.12.003. Accessed on May 2016.

[13] Object Management Group (OMG), "UML, Superstructure V2.5," Object Management Group (OMG), Tech. Rep., Sep 2013 [Online]. Available from http://www.uml.org. Accessed on May 2016.

[14] E. Ebeid, F. Fummi, and D. Quaglia, "Model-driven design of network aspects of distributed embedded systems," *IEEE Transactions on Computer-Aided Design of Integrated Circuits and Systems*, Vol. 34, no. 4, pp. 603–614, Apr 2015 [Online]. Available from http://ieeexplore.ieee.org/xpl/articleDetails.jsp?reload=true&arnumber=7017501

[15] M. de Miguel, T. Lambolais, M. Hannouz, S. Betgé-Brezetz, and S. Piekarec, "UML extensions for the specification and evaluation of latency constraints in architectural models," in *Proceedings of the Second International Workshop on Software and Performance*, ser. WOSP'00. New York, NY: ACM, 2000, pp. 83–88 [Online]. Available from http://doi.acm.org/10.1145/350391.350411. Accessed on May 2016.

[16] A. Hennig, D. Revill, and M. Ponitsch, "From UML to performance measures – Simulative performance predictions of IT-systems using the JBoss application server with OMNET++," *Al-Dabass*, 2003.

[17] K. Mets, J. Ojea, and C. Develder, "Combining power and communication network simulation for cost-effective smart grid analysis," *Communications Surveys Tutorials, IEEE*, vol. 16, no. 3, pp. 1771–1796, Mar 2014.

[18] Object Management Group, "A UML profile for MARTE (version 1.1)," OMG document number: formal/2011-06-02, Tech. Rep., Jun 2011 [Online]. Available from http://www.omg.org/spec/MARTE/1.1/. Accessed on May 2016.

[19] G. Ghatikar and E. Koch, "Deploying systems interoperability and customer choice within smart grid," Berkeley Lab, Tech. Rep. LBNL-6016E, Nov 2012.

[20] W. Xiang, T. Kunz, and M. St-Hilaire, "Flexible residential smart grid simulation framework," in *2013 IEEE International Conference on Smart Energy Grid Engineering (SEGE)*, Oshawa, ON, Aug 2013, pp. 1–7 [Online]. Available from http://ieeexplore.ieee.org/xpl/articleDetails.jsp?arnumber=6707902. Accessed on May 2016.

[21] M. Pipattanasomporn, H. Feroze, and S. Rahman, "Multi-agent systems in a distributed smart grid: Design and implementation," in *2009 IEEE/PES Power Systems Conference and Exposition*, Seattle, WA, Mar 2009, pp. 1–8 [Online]. Available from http://ieeexplore.ieee.org/xpl/articleDetails.jsp?arnumber=4840087. Accessed on May 2016.

[22] CEN-CENELEC-ETSI Smart Grid Coordination Group, "Smart grid reference architecture contents," CEN-CENELEC-ETSI Smart Grid Coordination Group, Tech. Rep. November, 2012 [Online]. Available from http://ec.europa.eu/energy/sites/ener/files/documents/xpert_group1_reference_architecture.pdf. Accessed on May 2016.

[23] S. J. Olivieri, G. P. Henze, C. D. Corbin, and M. J. Brandemuehl, "Evaluation of commercial building demand response potential using optimal short-term curtailment of heating, ventilation, and air-conditioning loads," *Journal of Building Performance Simulation*, vol. 7, no. 2, pp. 100–118,

2014 [Online]. Available from http://www.tandfonline.com/doi/abs/10.1080/19401493.2013.783118. Accessed on May 2016.
[24] M. D. Knudsen and S. Rotger-Griful, "Combined price and event-based demand response using two-stage model predictive control," in *IEEE International Conference on Smart Grid Communications*, Miami, Florida, USA, Nov 2015.
[25] Michael P. Lee, O. Aslam, B. Foster, D. Kathan, and C. Young, "Assessment of demand response and advanced metering," Federal Energy Regulatory Commission (FERC), Tech. Rep. 5, 2014 [Online]. Available from https://www.ferc.gov/legal/staff-reports/2014/demand-response.pdf. Accessed on May 2016.
[26] V. Giordano, A. Meletiou, C. F. Covrig, *et al.*, "Smart grid projects in Europe: Lessons learned and current developments 2012 update," Join Research Centre – Institute for Energy and Transport, Petten, Tech. Rep. EUR 25815 EN, 2013.
[27] C. Greer, D. A. Wollman, D. E. Prochaska, *et al.*, "NIST framework and roadmap for smart grid interoperability standards, release 3.0," National Institute of Standards and Technology (NIST), Tech. Rep., Oct 2014 [Online]. Available from http://nvlpubs.nist.gov/nistpubs/SpecialPublications/NIST.SP.1108r3.pdf. Accessed on May 2016.
[28] ZigBee Alliance, "ZigBee IP specification," ZigBee Alliance, Tech. Rep., 2013 [Online]. Available from http://www.zigbee.org/zigbee-for-developers/network-specifications/zigbeeip/. Accessed on May 2016.
[29] S. Cheshire and M. Krochmal, "DNS-based service discovery," Internet Requests for Comments, RFC Editor, RFC 6763, Feb 2013 [Online]. Available from http://www.rfc-editor.org/rfc/rfc6763.txt. Accessed on May 2016.
[30] D. Cooper, S. Santesson, S. Farrell, S. Boeyen, R. Housley, and W. Polk, "Internet X.509 public key infrastructure certificate and certificate revocation list (CRL) profile," Internet Requests for Comments, RFC Editor, RFC 5280, May 2008 [Online]. Available from http://www.rfc-editor.org/rfc/rfc5280.txt. Accessed on May 2016.
[31] ANSI/ASHRAE Standard 135-2012, "BACnet: A data communication protocol for building automation and control networks," ANSI/ASHRAE, Tech. Rep., 2012.
[32] Y.-C. Li and S. H. Hong, "BACnet-EnOcean smart grid gateway and its application to demand response in buildings," *Energy and Buildings*, vol. 78, pp. 183–191, Aug 2014 [Online]. Available from http://linkinghub.elsevier.com/retrieve/pii/S0378778814003326. Accessed on May 2016.
[33] A. Claassen, S. Rohjans, and S. Lehnhoff Member, "Application of the OPC UA for the smart grid," in *2011 Second IEEE PES International Conference and Exhibition on Innovative Smart Grid Technologies*. Piscataway, NJ: IEEE, Dec 2011, pp. 1–8 [Online]. Available from http://ieeexplore.ieee.org/xpl/articleDetails.jsp?arnumber=6162627. Accessed on May 2016.
[34] J.-p. Chou, L.-c. Chen, Y. J. Zhang, and L.-h. Pan, "OPC unified architecture for industrial demand response," *International Journal of Security and its*

*Applications*, vol. 6, pp. 275–280, 2012 [Online]. Available from http://www.sersc.org/journals/IJSIA/vol6_no2_2012/37.pdf. Accessed on May 2016.

[35] G. Thanos, M. Minou, T. Ganu, *et al.*, "Evaluating demand response programs by means of key performance indicators," *2013 Fifth International Conference on Communication Systems and Networks (COMSNETS)*, Bangalore, India, pp. 1–6, Jan 2013 [Online]. Available from http://ieeexplore.ieee.org/xpl/articleDetails.jsp?arnumber=6465597. Accessed on May 2016.

[36] A.-H. Mohsenian-Rad, V. Wong, J. Jatskevich, and R. Schober, "Optimal and autonomous incentive-based energy consumption scheduling algorithm for smart grid," in *Innovative Smart Grid Technologies (ISGT), 2010*, Gaithersburg, Maryland, USA, Jan 2010, pp. 1–6.

[37] W. Tan, J. Liu, T. Chen, and H. J. Marquez, "Comparison of some well-known PID tuning formulas," *Computers & Chemical Engineering*, vol. 30, no. 9, pp. 1416–1423, Jul 2006 [Online]. Available from http://linkinghub.elsevier.com/retrieve/pii/S0098135406000718. Accessed on May 2016.

[38] Association of Edison Illuminating Companies, "DR measurement & verification," Association of Edison Illuminating Companies, Tech. Rep., 2009.

[39] W. Shi, N. Li, X. Xie, C.-C. Chu, and R. Gadh, "Optimal residential demand response in distribution networks," *IEEE Journal on Selected Areas in Communications*, vol. 32, no. 7, pp. 1441–1450, Jul 2014.

[40] E. Ebeid, F. Fummi, and D. Quaglia, "HDL code generation from UML/MARTE sequence diagrams for verification and synthesis," *Springer Design Automation for Embedded Systems (DEAS)*, 2015.

[41] M. E. Kobber, "Undersøgelse af Demand Response strategier implementeret med Smart Energy Profile 2," Master's thesis, Aarhus University, Denmark, 2014.

[42] M. Pipattanasomporn, M. Kuzlu, S. Rahman, and Y. Teklu, "Load profiles of selected major household appliances and their demand response opportunities," *IEEE Transactions on Smart Grid*, vol. 5, no. 2, pp. 742–750, Mar 2014 [Online]. Available from http://ieeexplore.ieee.org/lpdocs/epic03/wrapper.htm?arnumber=6576261. Accessed on May 2016.

[43] F. Marra, G. Y. Yang, C. Traholt, E. Larsen, C. Rasmussen, and S. You, "Demand profile study of battery electric vehicle under different charging options," in *Power and Energy Society General Meeting, 2012 IEEE*, San Diego, CA, USA, Jul 2012, pp. 1–7.

[44] R. Stamminger, G. Broil, C. Pakula, *et al.*, "Synergy potential of smart appliances," Smart-A, Tech. Rep. D2.3 Smart-A Project, 2008 [Online]. Available from http://www.smart-a.org/WP2_D_2_3_Synergy_Potential_of_Smart_Appliances.pdf. Accessed on May 2016.

# *Chapter 15*
# Energy-efficient smart grid communications

*Shengrong Bu*[1] *and F. Richard Yu*[2]

Information and communication technologies (ICTs) are playing an important role in the modernization of the power grid. Communication networks for the smart grid should be energy-efficient (EE), so that these extra associated networks themselves will not significantly increase energy consumption of the total grid. More importantly, communication devices should not have to be recharged frequently and ideally would keep working even during extreme situations such as blackouts. A significant amount of research efforts have been put into reducing energy consumption of networks powered by the smart grid. However, the impact of green ICTs on the smart grid communications and applications remains to be explored. In this chapter, we surveyed EE smart grid communication networks, which are divided into three domains: smart grid home area networks (SG-HANs), smart grid neighborhood area networks (SG-NANs), and smart grid wide area networks (SG-WANs). To the best of our knowledge, this is the first survey of EE smart communications networks. Due to the importance of SG-NANs, this chapter focuses on SG-NANs, where data aggregation units (DAUs) communicate with home gateways (HGWs). Moreover, a multicell orthogonal frequency-division multiple access (OFDMA) cellular network is proposed for an SG-NAN. In order to improve energy-efficiency of the SG-NAN, a distributed resource allocation scheme is proposed, which also takes HGW fairness and priority into consideration. Average data rates are considered, since they are more appropriate from the HGWs' perspective. The EE resource allocation with fairness optimization problem is transformed from a fractional to equivalent subtractive form, which is subsequently modeled as a non-cooperative game. Interference pricing functions are used to drive the Nash equilibrium (NE) to Pareto optimal. An EE resource allocation iterative algorithm is designed for the resource allocation optimization problem. Simulation results show the effectiveness of the proposed scheme on energy-efficiency and HGW fairness.

[1]School of Engineering, University of Glasgow, Glasgow, Scotland
[2]Department of Systems and Computer Engineering, Carleton University, Ottawa, ON, Canada

## 15.1 Introduction

The power grid infrastructure is experiencing a significant shift from the traditional electricity grid to the smart grid. The electricity demand of consumers has sharply increased in recent years. Moreover, there is increasing interest in integrating renewable resources into the power grid, in order to decrease carbon emissions. In addition, demand side management (DSM), such as dynamic pricing and demand response (DR) programs will be used to improve the reliability of the grid by dynamically changing or shifting electricity consumption. These new requirements and aging of existing grids make the modernization of the grid infrastructure a necessity. The smart grid can optimize electricity generation, transmission, and distribution; reduce peaks in power usage; and sense and prevent power blackouts.

ICTs are playing an important role in the modernization of the power grid. Different data communications and network infrastructures are being developed for the smart grid: home energy management systems (HEMSs), advanced metering infrastructure (AMI), wide-area measurement systems (WAMSs), and sensor and actuator networks (SANETs) [1]. A HEMS can be used to monitor and control home electrical appliances. AMI, a key component of the smart grid, supports data communication architecture between smart meters installed at home to a meter data management system (MDMS) adopted by a utility company [2]. A WAMS monitors transmission system conditions by using phasor measurements units to measure current, voltage, and frequency of the systems over large areas, and therefore detects and further prevents grid instabilities for the entire power grid [3]. SANETs can be used to monitor and control the behaviors of different systems in the smart grid so that outage or disturbance can be prevented [4].

Communication networks in the smart grid are generally divided into three domains based on the network size and applications: SG-HANs, SG-NANs, and SG-WANs [5]. An SG-HAN is a single residential unit, which connects with intelligent devices, such as a smart meter, a HGW, load control devices, plug-in electric vehicles as well as sensors and actuators [5]. The HGW can manage communications within the SG-HAN and between SG-HANs. In the SG-NAN, the metering and service information are collected from multiple HAN gateways and transmitted to the DAUs, which can act as the SG-NAN gateways to transfer data to the MDMS [6]. Real-time price information is sent by the service providers to all of the smart meters. An SG-WAN provides a backhaul communication infrastructure for distribution systems in the smart grid, e.g., MDMS and AMI [7]. Wireless technologies have been widely used for smart grid communications, due to their low installation cost, flexible deployment, environmental adaptiveness, and growing performance [8].

Communication networks for the smart grid should be energy-efficient. The smart grid is originally proposed to save energy, but the extra associated communication networks in the power grid will consume a significant amount of energy. For example, the number of deployed smart meters will increase in an exponential rate, which will in turn contribute large greenhouse gas emissions [9]. Communication devices in the smart grid might not be easily recharged. Communication devices are expected to keep working even during extreme situations such as blackouts, since it is critical for service providers to know the status of power lines and to be able to send out control messages. Therefore, green wireless communication is essential for the smart grid.

Currently, a significant amount of academic and industry efforts have been put into reducing the energy consumption of core and access networks, due to the associated significant increase of the electricity bills of network operators [10,11]. Since smart grid traffic is machine generated and therefore previous research on EE network design based on human-generated traffic cannot directly applicable to the smart grid domains. Also, much work has been proposed to improve energy-efficiency of networks powered by the smart grid. However, the direction of how green ICTs impact on the smart grid communications and applications need more exploration. Moreover, energy-efficiency of smart grid communication and its impact on spectrum efficiency, quality of service (QoS), and security is worthy of research.

The rest of this chapter is organized as follows. Current existing work on EE smart grid communications are surveyed in Section 15.2. In this chapter, we focus on an SG-NAN, since it is an important bridge connecting an SG-HAN and an SG-WAN. To improve energy-efficiency of an SG-NAN, a distributed EE resource allocation scheme is proposed. The system model is described in Section 15.3. The EE resource allocation optimization problem is transformed to its subtractive form in Section 15.4. Then, the problem in the subtractive form is formulated as a non-cooperative EE resource allocation game in Section 15.5, and analyzed in Section 15.6. An iterative algorithm is designed for the EE resource allocation optimization problem in Section 15.7. Simulation results are discussed in Section 15.8. Finally, Section 15.9 presents our conclusions.

## 15.2 Energy-efficient wireless smart grid communications

A typical SG application in an SG-HAN is home automation. Potential candidates for wireless SG-HAN communications include Zigbee, WiFi and small cell networks [5]. Various approaches have been proposed to improve energy-efficiency of Zigbee, WiFi, and cellular networks. Energy-efficiency is an essential property of Zigbee, which employs duty-cycle scheduling to increase total network lifetime [12]. Power consumption of a node with WiFi will be further reduced with the continuous development of system-on-chip technology [13]. The deployment of heterogeneous networks based on small cells is an important technique for increasing energy-efficiency of cellular networks [14]. However, very limited research has been done in EE SG-HANs, especially taking smart grid concepts (e.g., dynamic pricing and DSM) into consideration. Erol-Kantarci and Mouftah utilized a Zigbee-based wireless sensor network (WSN) for DSM in an SG-HAN [15]. A WiFi Direct technique was used to improve power saving and to enhance reliable communications for the smart grid [16]. More EE SG-HAN approaches need to be proposed and also comparison of various wireless technologies for SG-HAN in terms of energy-efficiency need to be investigated.

Several SG applications operate in SG-NANs, including smart metering, distribution automation, and power outage management and DR [17]. Strong candidates for SG-NAN are 3G, 4G, and WiMAX standards [5]. For urban areas, the IEEE802.11

family has also the potential use for SG-NANs. General energy-efficiency of these communication systems has been widely researched [18,19]. However, the research findings for these systems cannot be directly applied to SG-NANs, since SG-NANs have different requirements compared to these general communication systems in term of demanding data rate and low latency. EE SG-NAN research is still in its infant stage. Ye *et al.* proposed two algorithms to minimize the total cost of an EE self-sustaining wireless NAN [6]. In this work, solar panels and batteries were used to power network equipment, since renewable energy can further improve energy-efficiency and enhance adaptiveness and flexible deployment of NAN nodes. In order to achieve both energy efficiency and data rate reliability requirements, Ye *et al.* proposed a Stackelberg game-theoretical power control scheme for the uplink transmission from NAN gateways to the concentrator connecting the MDMS [2]. One potential interesting research topic is to investigate how various EE methods impact the performance of SG-NANs. For example, power saving mode (PSM) can be adopted by IEEE802.11b and IEEE802.11s to save energy, by allowing wireless nodes to sleep when they are not receiving or transmitting [20]. It is worth to research how the additional delay caused by the PSM would impact performance of SG-NANs for various applications.

SG-WAN applications include wide-area monitoring, control and protection [21]. Among various wireless networks, cellular networks and WiMAX are strong candidates for SG-WAN communications. Since technologies in smart grid communications are not separated from other communications, many cutting-edge technologies originally developed for 5G networks, such as relaying techniques, cooperative communications, and coordinated multi-point (CoMP) are considered as promising techniques for smart grid communications. OFDMA and multiple-input and multiple-output (MIMO) are two common techniques for cellular networks and WiMAX [5]. For most of these listed technologies, limited research has been done in their energy-efficiency aspects. The optimal placement of remote radio units (RRUs) in terms of energy-efficiency was investigated in Reference 22 and energy-efficiency of CoMP and relaying under an average outage constraints was studied in Reference 23. Adaptively changing the number of active antennas was proposed to provide energy-efficiency for MIMO [24]. However, potential energy savings due to these 5G network technologies have not been well explored, and especially not in the smart grid applications. Moreover, how the proposed EE approaches, such as putting Base-Station (BS)s to sleep during low traffic volume, or adaptively changing the number of antennas, would impact the QoS requirements of smart grid applications remain unknown.

In wireless networks, spectral efficiency, QoS, and security are generally conflicting requirements with energy-efficiency [25,26]. These key trade-offs need to be considered when designing EE solutions for wireless networks. Little consideration has been given to achieving joint objectives in smart grid communications networks. Xiang *et al.* [27] proposed a trust-based EE geographical routing protocol named Dynamic Trust Elective Geo Routing (DTEGR) to tackle security issues in wireless mesh smart grid communication networks, so as to achieve energy-efficiency by avoiding packet loss and packet latency from various threats. Alohali and Vassialkis [28] presented a secure and EE multicast routing protocol for smart grids. Cognitive radio (CR) was originally proposed to improve spectrum efficiency. An EE and reliable CR-based Multiple Access Control (MAC) protocol design was proposed for CR-equipped sensor networks in the SG [29]. Zheng *et al.* [30] used a clustering-based grid network topology to analyze packet loss performance of communication networks and discussed the optimal network design considering both QoS requirement of depend response and energy consumption of the communications. However, study on these key trade-offs remains to be an open problem. For example, future research needs to explore the trade-off between EE techniques and the QoS of different types of smart grid applications.

## 15.3 System model

We consider an SG-NAN OFDMA cellular network with $M$ DAUs serving $K$ HGWs, which are randomly located across the network, illustrated in Figure 15.1. The SG-NAN can collect data from the smart meters and also relay the control information to the smart meters. Let DAU set $\mathcal{M} = \{1, \ldots, M\}$ and total HGW set $\mathcal{K} = \{1, \ldots, K\}$. The number of HGWs in cell $m$ is denoted as $K_m$ and the set is denoted as $\mathcal{K}_m$. We assume that all the cells in the network share the same frequency band. For each DAU, the total bandwidth $B$ is equally divided into $N$ subchannels (namely, subchannel set $\mathcal{N} = \{1, \ldots, N\}$), each with a bandwidth of $W$. For an arbitrary DAU $m$, the transmit power of subchannel $n$ is denoted as $p_{mn}$. The transmit power of DAU $m$ can be denoted vector $\mathbf{p}_m = (p_{m1}, \ldots, p_{mN})$. The transmit power matrix of these $M$ DAUs is denoted $\mathbf{P} = [\mathbf{p}_1, \ldots, \mathbf{p}_M]'$. We assume that at each time slot in each cell, each subchannel is exclusively assigned to at most one HGW to avoid interference within DAUs. Let $a_{mnk}$ indicate whether subchannel $n$ is assigned to HGW $k$ in DAU $m$ or not. Therefore, the subchannel assignment indicator matrix $\mathbf{A}_m$ ($\mathbf{A}_m = [a_{mnk}]_{N \times K_m}$) for DAU $m$ should meet the following requirements:

$$\sum_{k \in \mathcal{K}_m} a_{mnk} \leq 1, \quad \forall n \in \mathcal{N}, \tag{15.1}$$

$$a_{mnk} \in \{0, 1\}, \quad \forall n \in \mathcal{N}, \ \forall k \in \mathcal{K}_m. \tag{15.2}$$

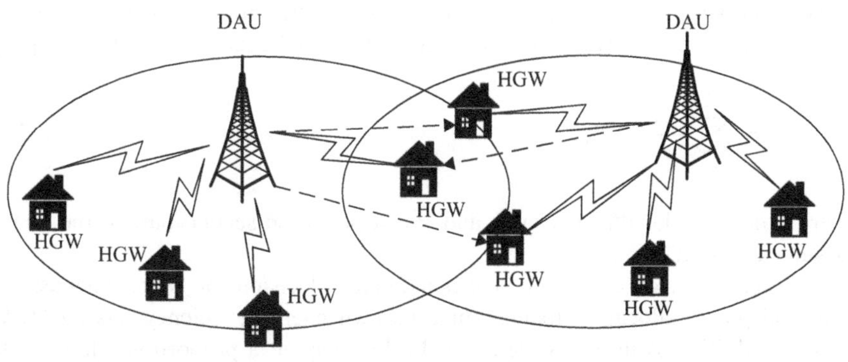


*Figure 15.1 An example of an OFDMA SG-NAN network*

The subchannel assignment matrix **A** denotes the subchannel assignment for all the DAUs.

The maximum achievable data rate of HGW $k$ on subchannel $n$ in DAU $m$ is

$$R_{mnk} = W \log\left(1 + \frac{p_{mn} h_{mmnk}}{\sigma^2 + \sum_{j \neq m} p_{jn} h_{jmnk}}\right), \tag{15.3}$$

where $h_{mmnk}$ denotes the channel gain from DAU $m$ to HGW $k$ on subchannel $n$ in DAU $m$ and $h_{jmnk}$ denotes the channel gain from DAU $j$ to HGW $k$ on subchannel $n$ in DAU $m$. The aggregate data rate for HGW $k$ in DAU $m$ and the overall data rate of DAU $m$ are

$$R_{mk} = \sum_{n=1}^{N} a_{mnk} R_{mnk}, \tag{15.4}$$

$$R_m = \sum_{k=1}^{K_m} R_{mk} = \sum_{k=1}^{K_m} \sum_{n=1}^{N} a_{mnk} R_{mnk}. \tag{15.5}$$

In practice, the total transmit power of an arbitrary DAU $m$ is non-negative and is less than $P_m^{max}$, namely,

$$p_m = \sum_{n=1}^{N} p_{mn} \leq P_m^{max}, \quad \forall m \in \mathcal{M}. \tag{15.6}$$

The transmit power of each subchannel in the DAUs is also constrained to some value. For example, the constraint of subchannel $n$ in DAU $m$ is defined as $p_{mn} \leq P_{m,n}^{max}$, where $P_{m,n}^{max}$ is the power mask on subchannel $n$ in DAU $m$.

For downlink transmissions, the overall power consumed by each DAU can be divided into two parts [31]: the first part is the power conversion (PC) power, which is equal to the total transmit power over the power amplifier (PA) efficiency. The second part is independent of the transmit power, and includes the circuit power, signal processing power, the power consumption of the cooling system, and so on. Therefore, the overall power consumption at DAU $m$ can be calculated as follows [31]:

$$p_m^{tot} = \sum_{n=1}^{N} \frac{p_{mn}}{\eta} + P_{mc}, \tag{15.7}$$

where $\eta$ denotes the PA efficiency and $P_{mc}$ denotes the second part of the power consumed by DAU $m$.

At each time slot, each DAU makes individual decisions on subchannel assignment and power allocation to maximize its own energy-efficiency, taking HGW priority and fairness into consideration. In this chapter, a proportional fairness to the HGWs is deployed in all cells, since it is one of the most representative measures for fairness [32]. Therefore, the optimization objective of DAUs at an arbitrary time slot $t$ is to solve the following optimization problem with constraints:

$$\max_{\mathbf{P}[t],\mathbf{A}[t]} \left\{ \frac{\sum_{m=1}^{M} \sum_{k\in\mathcal{K}_m} \alpha_{mk} \log (\bar{R}_{mk}[t])}{\sum_{m=1}^{M} \left( \sum_{n=1}^{N} \frac{p_{mn}[t]}{\eta} + P_{mc} \right)} \right\}, \tag{15.8}$$

s.t.

$$C1: \; p_m[t] \le P_m^{max}, \quad \forall m, \tag{15.9}$$

$$C2: \; 0 \le p_{mn}[t] \le P_{m,n}^{max}, \quad \forall m, n, \tag{15.10}$$

where $\alpha_{mk}$ is a weight corresponding to the QoS target for HGW $k$ in cell $m$ and $R_{mk}[t]$ is the instantaneous data rate assigned to HGW $k$ in DAU $m$ at time slot $t$. $\bar{R}_{mk}[t]$ is the long-term throughput for HGW $k$ up to time slot $t$ and $\bar{R}_{mk}[t] = \frac{1}{t}\sum_{\tau=1}^{t} R_{mk}[\tau] = \bar{R}_{mk}[t-1] + \epsilon_t[R_{mk}[t] - \bar{R}_{mk}[t-1]]$ (by letting $\epsilon_t = 1/t$). To simplify the presentation, parameter $[t]$ is omitted when equations refer to a single time slot.

## 15.4 Problem transformation

In the above problem, a brute-force method can be used to obtain its global optimal solution. However, this method might be computationally infeasible for large networks. Therefore, we introduce the following transformation in order to obtain an efficient iterative resource allocation algorithm.

Based on (15.8), the maximal generalized energy-efficiency $\beta^*$ of the network can be defined as follows:

$$\beta^* = \frac{f_n(\mathbf{P}^*, \mathbf{A}^*)}{f_d(\mathbf{P}^*)}, \tag{15.11}$$

where $\mathbf{P}^*$ is the optimal power allocation policy and $\mathbf{A}^*$ is the optimal HGW scheduling policy, $f_n(\mathbf{P}, \mathbf{A}) = \sum_{m=1}^{M}\sum_{k\in\mathcal{K}_m} \alpha_{mk} \log(\bar{R}_{mk}[t]), f_d(\mathbf{P}) = \sum_{m=1}^{M}\left(\sum_{n=1}^{N} \frac{p_{mn}}{\eta} + P_{mc}\right)$.

**Theorem 15.1.** *The necessary and sufficient condition to achieve the maximal energy-efficiency $\beta^*$ is as follows [33]:*

$$\max_{\mathbf{P},\mathbf{A}} \left\{ f_n(\mathbf{P}, \mathbf{A}) - \beta^* f_d(\mathbf{P}) \right\} = f_n(\mathbf{P}^*, \mathbf{A}^*) - \beta^* f_d(\mathbf{P}^*) = 0, \tag{15.12}$$

*for $f_n(\mathbf{P}, \mathbf{A}) \ge 0$ and $f_d(\mathbf{P}) > 0$.*

*Proof.* Theorem 15.1 is proved in Appendix A. [33]. □

Theorem 15.1 shows that, for any EE resource allocation optimization problem, if the objective function is in a fractional form, there always exists an objective function in the subtractive form that can lead to the same resource allocation policies as that in the fractional form.

## 15.5 Non-cooperative game formulation

For a multicell scenario, the sum of the instantaneous data rates for an arbitrary HGW at each time slot depends on the subchannel assignment in the cell it belongs to and power allocation in all cells. When the transmit power of the DAU increases, this sum data rate increases, but it causes higher co-channel interference in the neighboring cells. This conflict among the DAUs can be resolved by using non-cooperative game theory [34]. In this section, we first describe the utility function of each DAU and then formulate the problem as a non-cooperative EE resource allocation game.

### *15.5.1 Utility function of each DAU in the multicell OFDMA cellular network*

Based on the objective function in the subtractive form, the utility function of an arbitrary DAU $m$ at time slot $t$ can be defined as follows:

$$u_m(\mathbf{A}_m, \mathbf{p}_m, \mathbf{p}_{-m}) = \sum_{k \in \mathcal{K}_m} \alpha_{mk} \log(\bar{R}_{mk}[t]) - \beta \left( \sum_{n=1}^{N} \frac{p_{mn}}{\eta} + P_{mc} \right), \qquad (15.13)$$

s.t. $C1$, $C2$,

where $\mathbf{p}_{-m}$ denotes the transmit power matrix for all DAUs except DAU $m$. A standard gradient-based algorithm [35] can be used to maximize the sum of the weighted rates where the weights are marginal utilities, which leads to the optimal solution for the above optimization problem.

Because each DAU behaves selfishly, the resulting NE, one of the most commonly used solution concepts in non-cooperative game, may be far from the Pareto optimum [34]. Therefore, in this chapter, interference pricing is used as a penalty to encourage selfish DAUs to work in a cooperative manner, and to drive the NE to the Pareto optimum [36]. Furthermore, linear interference pricing functions are used due to their simple implementations. Therefore, for an arbitrary DAU $m$, the utility function with interference pricing is transformed as follows:

$$\widetilde{u}_m(\mathbf{A}_m, \mathbf{p}_m, \mathbf{p}_{-m}) = \sum_{k \in \mathcal{K}_m} \frac{\alpha_{mk} R_{mk}[t]}{\bar{R}_{mk}[t-1]} - \beta \left( \sum_{n=1}^{N} \frac{p_{mn}}{\eta} + P_{mc} \right) - \sum_{n=1}^{N} \mu_{mn} p_{mn}, \qquad (15.14)$$

s.t. $C1$, $C2$,

where $\mu_{mn}$ is the linear pricing function factor for subchannel $n$ of DAU $m$.

### *15.5.2 Game formulation within each time slot*

The EE resource allocation problem with proportional fairness can be modeled as a non-cooperative resource allocation game (NRAG). These $M$ DAUs are considered as selfish, rational players, and each of them tries to maximize its individual utility without considering the impact on the other players. Therefore, the proposed non-cooperative game can be described as follows:

$$\mathcal{G} = \langle \mathcal{M}, \{\mathbb{P}_m \times \mathbb{A}_m\}, \{\widetilde{u}_m\} \rangle,$$

where $\{\mathbb{P}_m \times \mathbb{A}_m\}$ is the strategy space of an arbitrary DAU $m$, defined by $\mathbb{P}_m = \{\mathbf{p}_m | \sum_{n=1}^{N} p_{mn} \leq P_m^{max}, 0 \leq p_{mn} \leq P_{m,n}^{max}\}$ and $\mathbb{A}_m = \{\mathbf{A}_m | a_{mnk} = 0 \text{ or } 1, \ \forall n, k, \ \text{and} \ \sum_{k \in \mathcal{K}_m} a_{mnk} = 1\}$, and $\widetilde{u}_m$ denotes the payoff function of DAU $m$. The proposed NRAG can be expressed as follows:

$$NRAG: \max_{\mathbf{p}_m \in \mathbb{P}_m, \mathbf{A}_m \in \mathbb{A}_m} \widetilde{u}_m(\mathbf{A}_m, \mathbf{p}_m, \mathbf{p}_{-m}), \quad \forall m \in \mathcal{M}.$$

An arbitrary DAU $m$ tries to determine its transmit power vector $\mathbf{p}_m$ and assignment matrix $\mathbf{A}_m$ (i.e. $\mathbf{p}_m \in \mathbb{P}_m$ and $\mathbf{A}_m \in \mathbb{A}_m$) to maximize its own utility.

One of the most commonly used solution concepts in non-cooperative game is called an NE, which is an equilibrium where every player plays the best-response strategy when taking others' decisions into account.

**Definition 15.1.** *In the NPAG, a network transmit power matrix* $\mathbf{P}^*$ *can be called an NE, if for every DAU* $m \in \mathcal{M}$ *and all* $\mathbf{p}_m \in \mathbb{P}_m$*, we have*

$$\widetilde{u}_m(\mathbf{p}_m^*, \mathbf{p}_{-m}^*, \mathbf{A}_m^*(\mathbf{P}^*)) \geq \widetilde{u}_m(\mathbf{p}_m, \mathbf{p}_{-m}^*, \mathbf{A}_m^*(\mathbf{p}_m, \mathbf{p}_{-m}^*)).$$

This definition indicates that no DAU can improve its payoff by a unilateral deviation from the NE, given that the other DAUs adopt the NE.

## 15.6 Analysis of the proposed EE resource allocation game with fairness

We first analyze the proposed EE resource allocation game with proportional fairness. Then, we investigate the form of the interference pricing functions, which can drive the NE of the proposed game to Pareto optimum. Finally, we prove the existence of the NE in the proposed non-cooperative game, and also propose a parallel iterative algorithm to obtain the NE in the real systems.

### *15.6.1 Subchannel assignment algorithm*

We first consider for a given network power matrix $\mathbf{P}$, how each DAU assigns its subchannels to its HGWs to maximize its own utility defined in (15.14). Once the transmit powers are given, the second item and the third item of the utility are fixed. Therefore, DAU $m$ only needs to make scheduling decisions to maximize the first item of its utility. Because only one HGW can be active in a given subchannel at each time slot, the HGW scheduling algorithm should assign HGW $k$ in each $n$ as follows:

$$k^* = \arg\max_k \left\{ \frac{\alpha_{mk}}{\bar{R}_{mk}[t-1]} R_{mnk}[t] \right\}. \tag{15.15}$$

In other words, the scheduler simply chooses $k$ in each subchannel such that the weighted instantaneous rate is maximized. Therefore, for subchannel $n$,

$$a_{mnk}^* = \begin{cases} 1, & \text{if } k = \arg\max_k \left\{ \frac{\alpha_{mk}}{\bar{R}_{mk}[t-1]} R_{mnk}[t] \right\}, \\ 0, & \text{otherwise.} \end{cases} \tag{15.16}$$

This algorithm leads to a significant reduction in complexity by using only $NK_m$ searches to allocate all the subchannels in each DAU.

### 15.6.2 Non-cooperative EE power allocation game

Since the optimal subchannel assignment matrix $\mathbf{A}^*$ can be determined once the network power matrix $\mathbf{P}$ is known, the NRAG reduces to a non-cooperative power allocation game:

$$\mathcal{G}' = \langle \mathcal{M}, \{\mathbb{P}_m\}, \{\widetilde{u}_m\} \rangle. \tag{15.17}$$

The corresponding objective of an arbitrary DAU $m$ can be written as follows:

$$\max_{\{p_{mn}\}\in\mathbb{P}_m} \sum_{n=1}^{N} \varpi_{mk^*}[t] R_{mnk^*}[t] - \beta\left(\sum_{n=1}^{N} \frac{p_{mn}}{\eta} + P_{mc}\right) - \sum_{n=1}^{N} \mu_{mn} p_{mn}, \tag{15.18}$$

$$\text{s.t. } C1, C2,$$

where

$$\varpi_{mk^*}[t] = \frac{\alpha_{mk^*}}{\bar{R}_{mk^*}[t-1]}. \tag{15.19}$$

For a given subchannel assignment matrix, the utility function of an arbitrary DAU $m$ is a convex function of $\mathbf{p}_m$, and the strategy set is also convex. Therefore, the problem is a convex optimization problem, and the Karush–Kuhn–Tucker (KKT) condition [37] can be applied to obtain its solution. The Lagrangian associated with this problem can be described as follows:

$$\mathcal{L}(\mathbf{p}_m, \lambda_m) = \sum_{n=1}^{N} \varpi_{mk^*}[t] R_{mnk^*}[t] - \beta\left(\sum_{n=1}^{N} \frac{p_{mn}}{\eta} + P_{mc}\right) - \sum_{n=1}^{N} \mu_{mn} p_{mn} - \lambda_m\left(\sum_{n=1}^{N} p_{mn} - P_m^{max}\right), \tag{15.20}$$

where $\lambda_m$ is the Lagrangian multiplier for the maximum power constraint. Then, the KKT condition can be written as follows [37,38]:

$$\frac{\partial \mathcal{L}(\mathbf{p}_m, \lambda_m)}{\partial p_{mn}} = \frac{\varpi_{mk^*}[t] h_{mmk^*n}}{\sigma^2 + \sum_{j\neq m} p_{jn} h_{jmk^*n} + p_{mn} h_{mmk^*n}} - \frac{\beta}{\eta} - \mu_{mn} - \lambda_m = 0, \tag{15.21}$$

$$\lambda_m \geq 0, \tag{15.22}$$

$$\sum_{n=1}^{N} p_{mn} - P_m^{max} \leq 0, \tag{15.23}$$

$$0 \leq p_{mn} \leq P_{m,n}^{max}, \quad \forall n, \tag{15.24}$$

$$\lambda_m\left(\sum_{n=1}^{N} p_{mn} - P_m^{max}\right) = 0. \tag{15.25}$$

Therefore,

$$p^*_{mn} = \left[ \frac{\varpi_{mk^*}[t]}{\frac{\beta_m}{\eta} + \mu_{mn} + \lambda_m} - \frac{\sigma^2 + \sum_{j \neq m} p_{jn} h_{jmk^*n}}{h_{mmk^*n}} \right]_0^{P^{max}_{m,n}}, \tag{15.26}$$

$$\text{s.t.}(15.22), (15.23), (15.25),$$

where $[\cdot]_0^{P^{max}_{m,n}} = \min(P^{max}_{m,n}, \max(\cdot, 0))$. This is a utility-based water filling with water level determined by the interference pricing factor and the maximal total transmit power constraint.

### 15.6.3 Properties of the interference pricing function factors

We will present the characteristics the interference pricing function factors should have in order to ensure the NE of the NRAG is Pareto optimal. The NE defines the best-response strategy of each DAU as stated below.

**Proposition 15.2.** *In the NPAG, for a given subchannel assignment matrix, the best-response function $\mathcal{B}_m$ for an arbitrary DAU m can be defined as follows:*

$$\mathbf{p}_m = \mathcal{B}_m(\mathbf{p}_{-m}) = [\mathcal{B}_m(\mathbf{p}_{-m})(1), \ldots, \mathcal{B}_m(\mathbf{p}_{-m}(N)], \tag{15.27}$$

*with*

$$\mathcal{B}_m(\mathbf{p}_{-m})(n) = p^*_{mn} \text{ defined in (15.26).} \tag{15.28}$$

**Proposition 15.3.** *If there exists an NE for game $\mathcal{G}'$ and if this NE is Pareto optimal, then the linear interference pricing function factor $\mu_{mn}$ ($\forall m, \forall n$) should be*

$$\mu_{mn} = \sum_{j \neq m}^{M} \frac{\varpi_{jk^*}[t] p_{jn} h_{jjk^*n} h_{mjk^*n}}{\sigma^2 + \sum_{m \neq j} p_{mn} h_{mjk^*n} + p_{jn} h_{jjk^*n}}. \tag{15.29}$$

*Proof.* See Appendix C. □

When $\mu_{mn}$ is chosen according to Proposition 15.3, the solutions for the optimization problem maximizing the total utilities of all DAUs and the optimization problem maximizing the individual utility of each DAU defined in (15.18) are always the same, because they have the same KKT conditions.

### 15.6.4 Existence of the NE in the proposed game

**Theorem 15.4.** *In the proposed NPAG, there exists an NE.*

*Proof.* A NE exists in game $\mathcal{G}' =< \mathcal{M}, \{\mathbb{P}_m\}, \{\widetilde{u}_m\} >$ if, $\forall m \in \mathcal{M}$, the following three conditions are met [39]:

1. $\mathbb{P}_m$ is compact and convex.
2. $\widetilde{u}_m(\mathbf{p}_m, \mathbf{p}_{-m})$ is continuous in $\mathbf{p}_{-m}$.
3. $\widetilde{u}_m(\mathbf{p}_m, \mathbf{p}_{-m})$ is continuous and quasi-concave in $\mathbf{p}_m$.

It is clear that the first condition is satisfied and the utility function $\tilde{u}_m(\mathbf{p}_m, \mathbf{p}_{-m})$ is a continuous function of $\mathbf{p}_{-m}$ and $\mathbf{p}_m$. Let $\widehat{R}_{mn}(p_{mn}, \mathbf{p}_{-m}) \equiv \varpi_{mk^*}[t]R_{mnk^*}[t]$. Each $\widehat{R}_{mn}$ is strictly monotonically increasing function of $p_{mn}$, and therefore $\widehat{R}_{mn}$ is a quasi-concave function of $p_{mn}$ [40]. Since the utility function $\tilde{u}_m(\mathbf{p}_m, \mathbf{p}_{-m})$ can be written as a sum of the quasi-concave functions in the corresponding $p_{mn}$, the utility function is quasi-concave in $\mathbf{p}_m$ [41]. □

### *15.6.5 Proposed parallel iterative algorithm*

A parallel iterative algorithm (i.e., Algorithm 1) is proposed to achieve the NE in real systems, since the sequential updating strategy algorithms may suffer from slow convergence when the number of DAUs in the network is large [42]. At each iteration, each DAU updates its own subchannel assignment and power allocation strategies simultaneously by using the interference pricing and transmit power generated by the others in the previous iteration. The condition $||\mathbf{p}_m^l - \mathbf{p}_m^{l-1}||/||\mathbf{p}_m^{l-1}|| \leq \varepsilon$ is the stopping criteria. Parameter $\varepsilon$ is preset to a small value, such as 1%. If the condition is not met after $L_{max}$ iterations, the algorithm terminates.

---

**Algorithm 1**. The proposed distributed algorithm with parallel iteration to achieve the NE

---

1. Initialization: $p_{mn} = P_{mn}^{max}, \forall m, n$; iteration count $l = 0$
2. **Repeat** iterations:
   (a) $l = l + 1$
   (b) **for** $m = 1$ to $M$ DAUs do
   (c) **for** $n = 1$ to $N$ subchannels do
   (d) Obtain $a_{mnk}^*$
   (e) Calculate the interference pricing factor $\mu_{mn}$
   (f) **end**
   (g) $\mathbf{p}_m^{(l)} = \mathcal{B}_m\left(\mathbf{p}_1^{(l-1)}, \ldots, \mathbf{p}_{m-1}^{(l-1)}, \mathbf{p}_{m+1}^{(l-1)}, \ldots, \mathbf{p}_M^{(l-1)}\right)$
   (h) **end**
   (i) **until** $l \geq L_{max}$ or $||\mathbf{p}_m^{(l)} - \mathbf{p}_m^{(l-1)}||/||\mathbf{p}_m^{(l-1)}|| \leq \varepsilon$
3. **End** iteration

---

## 15.7 EE resource allocation iterative algorithm

The proposed non-cooperative game is used to obtain the optimal resource allocation policies in the subtractive form for a given $\beta$. Based on the above sections, functions $\hat{f}_n(\mathbf{A})$ and $\hat{f}_d(\mathbf{P})$ can be defined as follows:

$$\hat{f}_n(\mathbf{P}, \mathbf{A}) = \sum_{m=1}^{M} \sum_{k \in \mathcal{K}_m} \frac{\alpha_{mk} R_{mk}[t]}{\bar{R}_{mk}[t-1]}, \tag{15.30}$$

$$\hat{f}_d(\mathbf{P}) = \sum_{m=1}^{M} \left( \sum_{n=1}^{N} \frac{p_{mn}}{\eta} + P_{mc} \right). \tag{15.31}$$

The equivalent objective function $\hat{F}(\beta)$ can be defined as follows:

$$\hat{F}(\beta) = \max_{\mathbf{P},\mathbf{A}} \left\{ \hat{f}_n(\mathbf{P}, \mathbf{A}) - \beta \hat{f}_d(\mathbf{P}) \right\}. \tag{15.32}$$

**Proposition 15.5.** *$\hat{F}(\beta)$ is a strictly monotonic decreasing function of $\beta$.*

*Proof.* Please refer to Appendix B. [33]. □

An iterative algorithm (namely, Algorithm 2) is proposed to solve the above EE resource allocation problem. This algorithm is based on the Dinkelbach method [33], which is used to generate a sequence of parameters converging to the optimal energy-efficiency $\beta^*$.

---

**Algorithm 2**. Iterative EE resource allocation algorithm

---

1. Initialization: Set energy-efficiency parameter $\beta = 0$
   iteration count $q = 0$
2. **Repeat** iterations:
   (a) $q = q + 1$
   (b) For a given $\beta$, obtain the optimal
   resource allocation policies $\{\mathbf{P}^*, \mathbf{A}^*\}$ by using Algorithm 1
   (c) **if** $|\hat{f}_n(\mathbf{P}^*, \mathbf{A}^*) - \beta \hat{f}_d(\mathbf{P}^*)| \leq \delta$, **then**
   (d) Convergence = true
   (e) **return** $\beta^* = \frac{\hat{f}_n(\mathbf{P}^*, \mathbf{A}^*)}{\hat{f}_d(\mathbf{P}^*)}$
   (f) **else**
   (g) Set $\beta = \frac{\hat{f}_n(\mathbf{P}^*, \mathbf{A}^*)}{\hat{f}_d(\mathbf{P}^*)}$
   (h) Convergence = false
   (i) **end if**
   (j) **until** $q \geq Q_{max}$ or Convergence = true
3. End iteration

---

The condition (Convergence = true) is the stopping criteria for this proposed iterative algorithm. Parameter $\delta$ is set to a small value, such as 0.0001. If the condition is not met after $Q_{max}$ iterations, the algorithm terminates.

## 15.8 Simulation results and discussions

The system parameters are listed in Table 15.1, unchanged in all simulations unless stated. We show results for two subchannels of each DAU. While the exact shape of the figures would change with the parameters, the insight remains the same.

The feasibility of the linear interference pricing functions is studied. We assume that two DAUs exist in the network. The following results are obtained in the first time slot (i.e., $t = 1$) when $\beta$ is set to 0.55. Figure 15.2 illustrates that each

*Table 15.1 System parameters*

| | |
|---|---|
| Total bandwidth | 10 MHz [43] |
| Number of subchannels | 50 [43] |
| Subchannel bandwidth | 0.2 MHz [43] |
| Thermal noise power | −141 dBW/MHz [44] |
| Maximum transmit power $P_m^{max}$ | 46 dBm [45] |
| $P_{mc}$ | 119 W [45] |
| $\eta$ | 1/2.4 [45] |

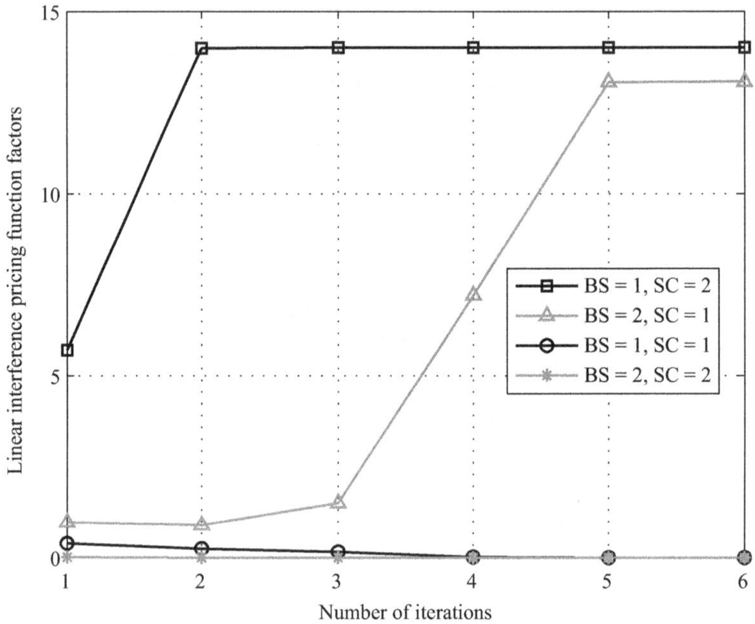


*Figure 15.2 Linear interference pricing factors received by subchannels 1 and 2 of the DAUs in a high interference scenario. $\beta = 0.55$, $\alpha_{mk} = 1$, $\forall m, k$*

subchannel of each DAU might have a different linear interference pricing function factor. If transmit power of one subchannel in the DAU has a greater impact on the utility of the other DAU, the other DAU will provide a higher interference pricing factor, and the first DAU's transmit power is more likely to be reduced to maximize the total energy-efficiency of the network.

The feasibility of Algorithm 1 to obtain the maximal generalized energy-efficiency of the network $\beta^*$ is also studied in a higher interference scenario with different total transmit power constraints. Figure 15.3 illustrates the results for

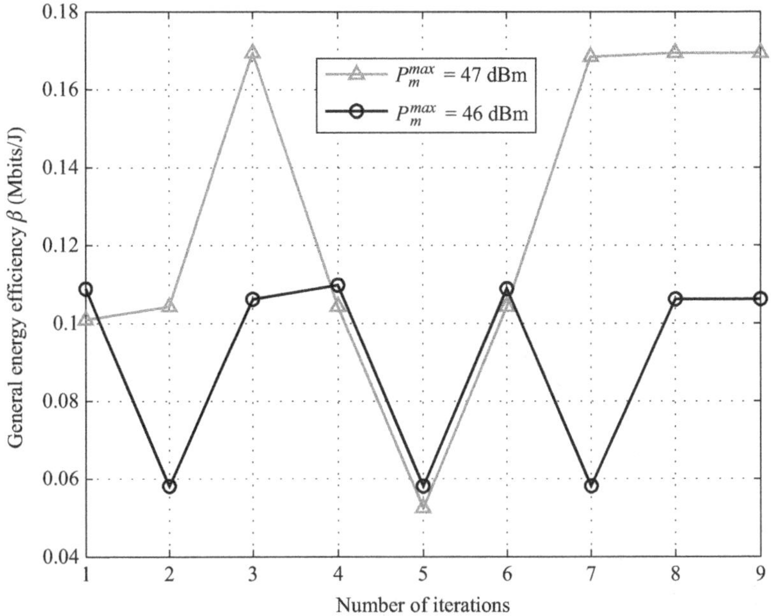


*Figure 15.3 Convergence of the generalized energy-efficiency in a high interference scenario.* $\alpha_{mk} = 1, \ \forall m, k$

the first time slot (i.e., $t = 1$). We assume that all HGWs have the same priority. Figure 15.3 shows that after a small number of iterations, energy-efficiency parameter $\beta$ converges to its maximal value $\beta^*$. Figure 15.3 also evaluates the impact of the total transmit power constraint $P_m^{max}$ on generalized energy-efficiency of the network, and shows that in this setting, when $P_m^{max} = 47$ dBm, the energy-efficiency $\beta^*$ has higher value than that when $P_m^{max} = 46$ dBm. The reason is that in this high interference scenario, only one of the two DAUs transmits on each subchannel, and $P_{mc}$ is relatively big compared to the total transmit power constraint. However, increasing $P_m^{max}$ does not always improve the energy-efficiency of the network.

The energy-efficiency of the network and the corresponding throughput performance are compared with various numbers of HGWs in each cell among three different schemes: the spectrum-efficiency scheme, the proposed energy-efficiency scheme without fairness and the proposed EE scheme with fairness. In the spectrum-efficiency scheme, the HGWs which can obtain the total highest throughput are scheduled. Therefore, spectral-efficiency scheme generates the highest throughput among these three schemes, which is normally not EE, is shown in Figure 15.4. For the EE scheme without considering fairness, the objective is to maximize the energy-efficiency of the network at each time slot. Therefore, this scheme has the best energy-efficiency performance as illustrated in Figure 15.5. The EE scheme

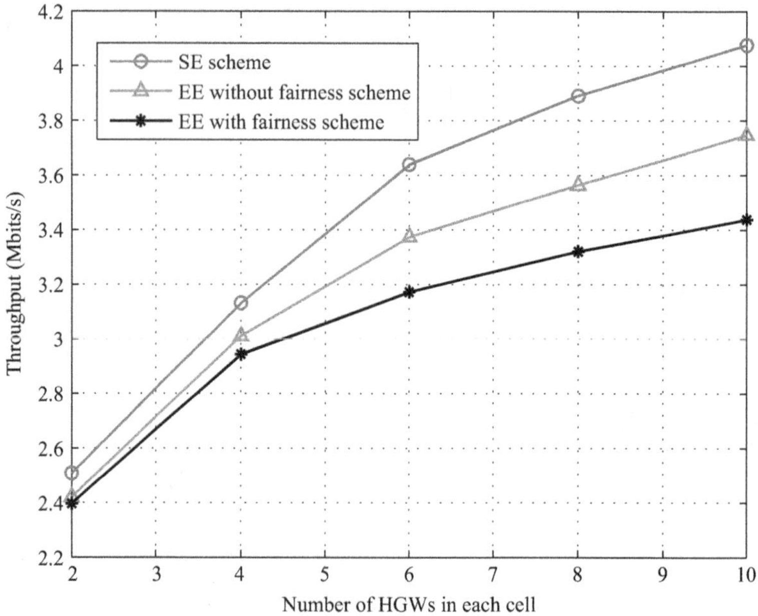


*Figure 15.4 Throughput comparison among three schemes with various numbers of HGWs in each cell*

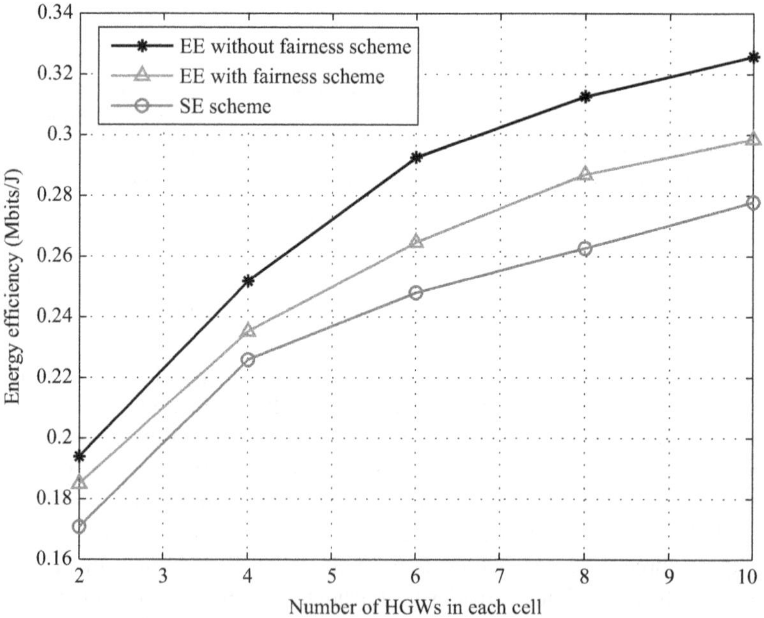


*Figure 15.5 Energy-efficiency comparison among three schemes with various numbers of HGWs in each cell*

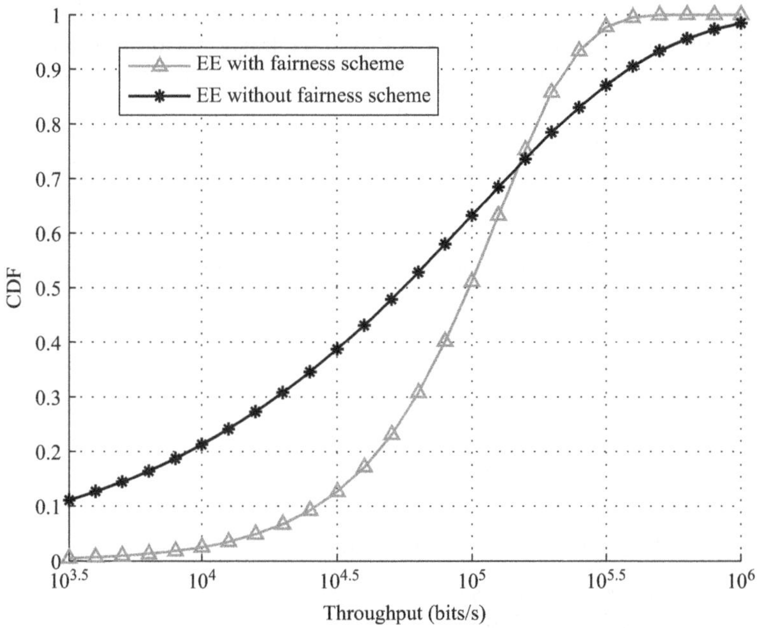


*Figure 15.6 CDF comparison between the EE schemes with/without fairness*

with fairness takes fairness to the HGWs into consideration, and so the throughput and energy-efficiency are not as high as the second scheme.

The cumulative distribution functions (CDFs) of the total throughput for EE schemes with/without fairness are shown in Figure 15.6. The scheme with fairness maximizes the low-throughput range. Figure 15.6 shows that the EE scheme with fairness does not favor one HGW at the expense of others.

## 15.9 Conclusions

In this chapter, we first surveyed EE wireless smart grid communications, which are divided into SG-HANs, SG-NANs, and SG-WANs. Due to the importance of the SG-NAN, we focused on EE SG-NANs. A strong candidate, a multicell OFDMA cellular network, was proposed for an SG-NAN, where DAUs communicate with HGWs. In the SG-NAN, an EE resource allocation approach was studied to improve its energy-efficiency, also taking the priority and fairness of HGWs into consideration. Since the average rates are more important for HGW, the HGW's average data rates were considered. At each time slot, each DAU makes HGW scheduling and power allocation decisions for each subchannel to maximize its generalized energy-efficiency. For any EE resource allocation optimization problem, if the objective function is in a fractional form, there always exists an objective function in the subtractive form that can lead to the same resource allocation policies as that in the fractional form. Non-cooperative game theory was used to model the distributed EE resource allocation of the SG-NAN in the subtractive form. Interference pricing functions were proposed to drive the NE of the non-cooperative game to the Pareto optimal. The existence of the proposed game equilibrium has been proved. An EE resource allocation iterative algorithm was proposed to obtain the optimal resource allocation policies for the SG-NAN. Simulation results show that the proposed scheme improves the energy-efficiency of the network compared to that of the spectral-efficient design, and improve the fairness to the HGWs.

## Appendix

### A. Proof of Theorem 15.1

(a) Necessary condition

Let $\beta^*$ be the solution of (15.11). Therefore, we have

$$\beta^* = \frac{f_n(\mathbf{P}^*, \mathbf{A}^*)}{f_d(\mathbf{P}^*)} \geq \frac{f_n(\mathbf{P}, \mathbf{A})}{f_d(\mathbf{P})},$$

namely, $f_n(\mathbf{P}^*, \mathbf{A}^*) - \beta^* f_d(\mathbf{P}^*) = 0$ and $f_n(\mathbf{P}, \mathbf{A}) - \beta^* f_d(\mathbf{P}) \leq 0$. Therefore, $\max\{f_n(\mathbf{P}, \mathbf{A}) - \beta^* f_d(\mathbf{P})\} = 0$.

(b) Sufficient condition

Let $f_n(\mathbf{P}^*, \mathbf{A}^*) - \beta^* f_d(\mathbf{P}^*) = 0$. Based on (15.12), we can obtain

$$f_n(\mathbf{P}, \mathbf{A}) - \beta^* f_d(\mathbf{P}) \leq f_n(\mathbf{P}^*, \mathbf{A}^*) - \beta^* f_d(\mathbf{P}^*) = 0, \tag{A.1}$$

namely,

$$\beta^* = \frac{f_n(\mathbf{P}^*, \mathbf{A}^*)}{f_d(\mathbf{P}^*)} \geq \frac{f_n(\mathbf{P}, \mathbf{A})}{f_d(\mathbf{P})}.$$

### B. Proof of Proposition 15.5

Let $\{\mathbf{P}', \mathbf{A}'\}$ and $\{\mathbf{P}'', \mathbf{A}''\}$ be two distinct optimal resource allocation policies for $\hat{F}(\beta')$ and $\hat{F}(\beta'')$, respectively. Without loss generality, we assume that $\beta' > \beta''$.

$$\begin{aligned}\hat{F}(\beta'') &= \max_{\mathbf{P},\mathbf{A}}\{\hat{f}_n(\mathbf{P}, \mathbf{A}) - \beta''\hat{f}_d(\mathbf{P})\}\\ &= \hat{f}_n(\mathbf{P}'', \mathbf{A}'') - \beta''\hat{f}_d(\mathbf{P}'')\\ &> \hat{f}_n(\mathbf{P}', \mathbf{A}') - \beta''\hat{f}_d(\mathbf{P}')\\ &> \hat{f}_n(\mathbf{P}', \mathbf{A}') - \beta'\hat{f}_d(\mathbf{P}')\\ &= \hat{F}(\beta')\end{aligned} \tag{B.1}$$

Therefore, $\hat{F}(\beta)$ is a strictly monotonic decreasing function of $\beta$.

## C. Proof of Proposition 15.3

To maximize the total utilities of all DAUs, the following optimization problem needs to be solved:

$$\max_{\{\mathbf{P},\mathbf{A}\}} \sum_{m=1}^{M} \left( \sum_{n=1}^{N} \varpi_{mk^*}[t] R_{mnk^*}[t] - \beta \left( \sum_{n=1}^{N} \frac{p_{mn}}{\eta} + P_{mc} \right) \right), \tag{C.1}$$

s.t. $C1, C2$.

The Lagrangian function can be written as follows:

$$\mathcal{L}(\mathbf{P},\mathbf{A},\lambda_m) = \sum_{m=1}^{M} \left( \sum_{n=1}^{N} \varpi_{mk^*}[t] R_{mnk^*}[t] - \beta \left( \sum_{n=1}^{N} \frac{p_{mn}}{\eta} + P_{mc} \right) \right) - \sum_{m=1}^{M} \lambda_m \left( \sum_{n=1}^{N} p_{mn} - P_m^{max} \right). \tag{C.2}$$

The KKT conditions for the optimization problem in (C.1) can be written as follows [37,38]:

$$\frac{\partial \mathcal{L}(\mathbf{P},\mathbf{A},\lambda_m)}{\partial p_{mn}} = \varpi_{mk^*}[t] \frac{\partial R_{mnk^*}[t]}{\partial p_{mn}} + \sum_{j \neq m}^{M} \varpi_{jk^*}[t] \frac{\partial R_{jnk^*}[t]}{\partial p_{mn}} - \frac{\beta}{\eta} - \lambda_m = 0, \tag{C.3}$$

$$\lambda_m \geq 0, \quad \forall m, \tag{C.4}$$

$$\sum_{n=1}^{N} p_{mn} - P_m^{max} \leq 0, \quad \forall m, \tag{C.5}$$

$$0 \leq p_{mn} \leq P_{m,n}^{max}, \quad \forall m, \ \forall n, \tag{C.6}$$

$$\lambda_m \left( \sum_{n=1}^{N} p_{mn} - P_m^{max} \right) = 0, \quad \forall m. \tag{C.7}$$

By comparing KKT conditions in (15.21) and (C.3), to obtain the same solution, we must have

$$\mu_{mn} = -\sum_{j \neq m}^{M} \varpi_{jk^*}[t] \frac{\partial R_{jnk^*}[t]}{\partial p_{mn}}, \tag{C.8}$$

namely,

$$\mu_{mn} = \sum_{j \neq m}^{M} \frac{\varpi_{jk^*}[t] p_{jn} h_{jjk^*n} h_{mjk^*n}}{\sigma^2 + \sum_{m \neq j} p_{mn} h_{mjk^*n} + p_{jn} h_{jjk^*n}}. \tag{C.9}$$

## Bibliography

[1] D. Niyato and P. Wang, "Cooperative transmission for meter data collection in smart grid," *IEEE Communications Magazine*, vol. 50, pp. 90–97, Apr. 2012.

[2] F. Ye, Y. Qian, R. Q. Hu, and S. Das, "Reliable energy-efficient uplink transmission for neighborhood area networks in smart grid," *IEEE Transactions on Smart Grid*, vol. 6, pp. 2179–2188, Sep. 2015.
[3] B. Singh, N. K. Sharma, A. N. Tiwari, K. S. Verma, and S. N. Singh, "Applications of phasor measurement units (PMUs) in electric power system networks incorporated with facts controllers," *International Journal of Engineering, Science and Technology*, vol. 3, no. 3, pp. 64–82, 2011.
[4] E. Hossain, Z. Han, and H. V. Poor, *Smart Grid Communications and Networking*. Cambridge University Press, Cambridge, 2012.
[5] M. Erol-Kantarci and H. T. Mouftah, "Energy-efficient information and communication infrastructures in the smart grid: a survey on interactions on open issues," *IEEE Communication Survey and Tutorials*, vol. 17, no. 1, pp. 179–197, First Quarter 2015.
[6] F. Ye, Y. Qian, and R. Q. Hu, "Energy efficient self-sustaining wireless neighbourhood area network design for smart grid," *IEEE Transactions on Smart Grid*, vol. 6, pp. 220–229, Jan. 2015.
[7] N. Kayastha, D. Niyato, E. Hossain, and Z. Han, "Smart grid sensor data collection, communication, and networking: a tutorial," *Wireless Communications and Mobile Computing*, vol. 14, pp. 1055–1087, Aug. 2014.
[8] P. Parikh, M. G. Kanabar, and T. S. Sidhu, "Opportunities and challenges of wireless communication technologies for smart grid applications," in *Proceedings of the IEEE Power and Energy Society General Meeting* (Minneapolis, MN, USA), Jul. 2010.
[9] F. Saghezchi, A. Radwan, J. Rodriguez, and T. Dagiuklas, "Coalition formation game toward green mobile terminals in heterogeneous wireless networks," *IEEE Wireless Communications*, vol. 20, pp. 85–91, Oct. 2013.
[10] S. Bu, F. R. Yu, Y. Cai, and P. Liu, "When the smart grid meets energy-efficient communications: green wireless cellular networks powered by the smart grid," *IEEE Transactions on Wireless Communications*, vol. 11, pp. 3014–3024, Aug. 2012.
[11] Z. Zhang, Y. Bejerano, and S. Antonakopoulos, "Energy-efficient IP core network configuration under general traffic demands," *IEEE/ACM Transactions on Networking*, vol. 24, no. 2, pp. 745–758, April 2016.
[12] G. Anastasi, M. Conti, M. D. Francesco, and V. Neri, "Reliability and energy efficiency in multi-hop IEEE 802.15.4/Zigbee wireless sensor networks," in *Proceedings of the IEEE Symposium on Computers and Communications* (Riccione, Italy), Jun. 2010.
[13] L. Li, X. Hu, C. Ke, and K. He, "The applications of WiFi-based wireless sensor network in internet of things and smart grid," in *Proceedings of the IEEE ICIEA* (Beijing, China), Jun. 2011.
[14] S. Bu and F. R. Yu, "Green cognitive mobile networks with small cells for multimedia communications in the smart grid environment," *IEEE Transactions on Vehicular Technology*, vol. 63, pp. 2115–2126, Jun. 2014.
[15] M. Erol-Kantarci and H. T. Mouftah, "Wireless sensor networks for cost-efficient residential energy management in the smart grid," *IEEE Transactions on Smart Grid*, vol. 2, pp. 314–325, Jun. 2011.

[16] Z. Li, Q. Liang, and X. Cheng, "Emerging WiFi Direct technique in home area networks for smart grid: power consumption and outage performance," *Ad Hoc Networks*, vol. 22, pp. 61–68, Nov. 2014.

[17] M. Kuzlu, M. Pipattanasomporn, and S. Rahman, "Communication network requirements for major smart grid applications in HAN, NAN and WAN," *Computer Networks*, vol. 67, pp. 74–88, Jul. 2014.

[18] D. Feng, C. Jiang, G. Lim, L. J. Cimini, G. Feng, and G. Y. Li, "A survey of energy-efficient wireless communications," *IEEE Communications Surveys & Tutorials*, vol. 15, pp. 167–178, First Quarter 2013.

[19] R. Y. Kim and S. Mohanty, "Advanced power management techniques in next-generation wireless networks," *IEEE Communications Magazine*, vol. 48, pp. 94–102, May 2010.

[20] M. N. Alam, R. Jantti, J. Kneckt, and J. Nieminen, "Performance analysis of the IEEE 802.11s PSM," *Journal of Computer Networks and Communications*, pp. 712–716, vol. 2012, 2012.

[21] S. Borlase, *Smart Grids: Infrastructure, Technology and Solutions*. CRC Press, Boca Raton, FL, 2012.

[22] C. Zhang, T. Zhang, Z. Zeng, L. Cuthbert, and L. Xiao, "Optimal locations of remote radio units in CoMP systems for energy efficiency," in *Proceedings of the IEEE VTC-Fall* (Ottawa, Canada), Sep. 2010.

[23] D. Cao, S. Zhou, C. Zhang, and Z. Niu, "Energy saving performance comparison of coordinated multi-point transmission and wireless relaying," in *Proceedings of the IEEE GLOBECOM* (Miami, FL, USA), Dec. 2010.

[24] H. Kim, C. B. Chae, G. de Veciana, and R. W. Heath, "A cross-layer approach to energy efficiency for adaptive MIMO systems exploiting spare capacity," *IEEE Transactions on Wireless Communications*, vol. 8, pp. 4264–4275, Aug. 2009.

[25] Y. Zhang, R. Yu, M. Nekovee, Y. Liu, S. Xie, and S. Guessing, "Cognitive machine-to-machine communications: Visions and potentials for the smart grid," *IEEE Network*, vol. 26, pp. 6–13, May–Jun. 2012.

[26] Y. Chen, S. Zhang, S. Xu, and G. Li, "Fundamental trade-offs on green wireless networks," *IEEE Communications Magazine*, vol. 49, pp. 30–37, Jun. 2011.

[27] M. Xiang, Q. Bei, and W. Liu, "Self-adjustable trust-based energy efficient routing for smart grid systems," in *Proceedings of the IEEE/WIC/ACM International Conferences on Web Intelligence and Intelligent Agent Technology* (Macau, China), Dec. 2012.

[28] B. A. Alohali and V. G. Vassialkis, "Secure and energy-efficient multicast routing in smart grids," in *Proceedings of the IEEE Tenth International Conference on Intelligent Sensors, Sensor Networks and Information Processing* (Singapore), Apr. 2015.

[29] A. Aijaz, S. Ping, M. Akhavan, and H. Aghvami, "CRB-MAC: a receiver-based MAC protocol for cognitive radio equipped smart grid sensor networks," *IEEE Sensors Journal*, vol. 14, pp. 4325–4333, Dec. 2014.

[30] L. Zheng, S. Parkinson, D. Wang, L. Cai, and C. Crawford, "Energy efficient communication networks design for demand response in smart grid," in

*Proceedings of the International Conference on Wireless Communications and Signal Processing* (Nanjing, China), Nov. 2011.

[31] C. Xiong, G. Y. Li, S. Zhang, Y. Chen, and S. Xu, "Energy-efficient resource allocation in OFDMA networks," *IEEE Transactions on Communications*, vol. 60, pp. 3767–3778, Dec. 2012.

[32] H. Shi, R. Prasad, E. Onur, and I. Niemegeers, "Fairness in wireless networks: issues, measures and challenges," *IEEE Communications Surveys & Tutorials*, vol. 16, pp. 5–24, May 2013.

[33] W. Dinkelbach, "On nonlinear fractional programming," *Management Science*, vol. 13, pp. 492–498, Mar. 1967.

[34] M. J. Osborne, *An Introduction to Game Theory*. Oxford University Press, Oxford, Aug. 2003.

[35] K. Son, S. Chong, and G. de Veciana, "Dynamic association for load balancing and interference avoidance in multi-cell networks," *IEEE Transactions on Wireless Communications*, vol. 8, pp. 3566–3576, Jul. 2009.

[36] F. Wang, M. Krunz, and S. Cui, "Price-based spectrum management in cognitive radio networks," *IEEE Journal of Selected Topics in Signal Processing*, vol. 2, pp. 74–87, Feb. 2008.

[37] S. Boyd and L. Vandenberghe, *Convex Optimization*. Cambridge University Press, Cambridge, 2004.

[38] P. Tsiaflakis, M. Diehl, and M. Moonen, "Distributed spectrum management algorithms for multiuser DSL networks," *IEEE Transactions on Signal Processing*, vol. 56, pp. 4825–4843, Oct. 2008.

[39] D. Fudenberg and J. Tirole, *Game Theory*. MIT Press, Cambridge, MA, 1991.

[40] G. Miao, N. Himayat, and G. Li, "Energy-efficient link adaptation in frequency-selective channels," *IEEE Transactions on Communications*, vol. 58, pp. 545–554, Feb. 2010.

[41] H. Kwon and B. G. Lee, "Distributed resource allocation through noncooperative game approach in multi-cell OFDMA systems," in *Proceedings of the IEEE ICC* (Istanbul, Turkey), Jun. 2006.

[42] G. Scutari, D. P. Palomar, and S. Barbarossa, "Asynchronous iterative water-filling for Gaussian frequency-selective interference channels," *IEEE Transactions on Information Theory*, vol. 54, pp. 2868–2878, Jul. 2008.

[43] H. Holtkamp, G. Auer, S. Bazzi, and H. Haas, "Minimizing base station power consumption," *IEEE Journal on Selected Areas in Communications*, vol. 32, pp. 297–306, Feb. 2014.

[44] G. Miao, N. Himayat, G. Li, and S. Talwar, "Distributed interference-aware energy-efficient power optimization," *IEEE Transactions on Wireless Communications*, vol. 10, pp. 1323–1333, Apr. 2011.

[45] H. Holtkamp, G. Auer, and H. Haas, "On minimizing base station power consumption," in *Proceedings of the IEEE Vehicular Technology Conference Fall* (San Francisco, CA, USA), Sep. 2011.

# Index